T3-BWO-377

STP 1033

Environmental Aspects of Stabilization and Solidification of Hazardous and Radioactive Wastes

Pierre Côté and Michael Gilliam, editors

ASTM
1916 Race Street
Philadelphia, PA 19103

Library of Congress Cataloging-in-Publication Data

Environmental aspects of stabilization and solidification of hazardous and radioactive wastes/Pierre Côté and Michael Gilliam, editors.
(STP; 1033)
"ASTM publication code number (PCN)"—T.p. verso.
Papers presented at the Fourth International Hazardous Waste Symposium on Environmental Aspects of Stabilization/Solidification of Hazardous and Radioactive Wastes held 3–6 May, 1987 in Atlanta, Georgia.
Includes bibliographical references.
ISBN (invalid) 0-8031-1261-0
1. Hazardous wastes—Environmental aspects—Congresses.
2. Radioactive wastes—Environmental aspects—Congresses. 3. Sewage sludge digestion—Environmental aspects—Congresses. I. Côté, Pierre. II. Gilliam, T. M. III. International Hazardous Waste Symposium on Environmental Aspects of Stabilization/Solidification of Hazardous and Radioactive Wastes (4th: 1987: Atlanta, Ga.)
IV. Series: ASTM special technical publication; 1033.
TD1060.E58 1989
628.4'2--dc20 89-17761
CIP

Copyright © by AMERICAN SOCIETY FOR TESTING AND MATERIALS 1989

NOTE

The Society is not responsible, as a body,
for the statements and opinions
advanced in this publication

Peer Review Policy

Each paper published in this volume was evaluated by three peer reviewers. The authors addressed all of the reviewers' comments to the satisfaction of both the technical editor(s) and the ASTM Committee on Publications.

The quality of the papers in this publication reflects not only the obvious efforts of the authors and the technical editor(s), but also the work of these peer reviewers. The ASTM Committee on Publications acknowledges with appreciation their dedication and contribution of time and effort on behalf of ASTM.

Printed in Baltimore, MD
November 1989

Foreword

The papers in this publication, *Environmental Aspects of Stabilization and Solidification of Hazardous and Radioactive Wastes,* were presented at the Fourth International Hazardous Waste Symposium on Environmental Aspects of Stabilization/Solidification of Hazardous and Radioactive Wastes held 3–6 May 1987 in Atlanta, Georgia. The symposium by ASTM Committee D-34 on Waste Disposal, Environmental Canada, Oak Ridge National Laboratory, Alberta Environmental Centre, and U.S. Department of Energy. The Cooperating organizations included U.S. Environmental Protection Agency and Imperial College of Science and Technologies. Michael Gilliam, Oak Ridge National Laboratory, and Pierre Côté, Environmental Canada, presided as symposium chairmen and are editors of this publication.

Contents

LABORATORY EVALUATION

Overview

Stabilization and solidification have been used for two to three decades as final treatment steps prior to land disposal of radioactive and chemically hazardous wastes. These technologies are also playing an increasingly important role for on-site or in-situ treatment and remediation of waste lagoons or contaminated soils. *Stabilization* refers to those aspects of the technology which result in rendering a waste less toxic through fixation of the contaminants that contain and/or by providing a stable chemical environment. *Solidification* is related to those operations which improve the physical and handling characteristics of the waste.

This Special Technical Publication contains 33 peer-reviewed papers out of 62 that were presented at the 4th International Hazardous Waste Symposium on Environmental Aspects of Stabilization/Solidification of Hazardous and Radioactive Wastes, held in Atlanta, Georgia, 3–6 May 1987. The symposium was sponsored by ASTM Committee D-34 on Waste Disposal, with Environment Canada and the Alberta Environmental Centre representing the chemically hazardous waste management field, and the U.S. Department of Energy and the Oak Ridge National Laboratory (ORNL) representing the radioactive waste management field. Although the two scientific communities, working on chemically hazardous or low-level radioactive waste, are faced with similar problems and basically work with the same technology, the symposium represented the first forum for technology exchange. This first encounter was an occasion to understand the gap that separate two groups that use a different vocabulary and are subject to markedly different regulations. The later fact has resulted in the creation of a new category of wastes in the United States, "mixed wastes" that fall under both the Nuclear Regulatory Commission radioactive wastes and the Environmental Protection Agency hazardous wastes regulations, which are not always compatible. Following the symposium, the contacts have intensified between the two communities, which should certainly result in a better and wider use of the technology.

Land disposal of waste should be avoided when possible. Waste reduction, recycle, and reuse (for nonradioactive wastes) are obviously superior alternatives. There are, however, instances where the physical nature of the waste, and the type and concentration of contaminants that it contains do not technically or economically allow avoidance of the land disposal option. In these instances, the waste should be treated to the maximum extent possible, to reduce its volume. Solidification of large quantities of contaminated water, although possible, should be avoided. When properly managed, land disposal after stabilization and solidification will ensure long-term containment of the waste constituents.

A wide spectrum of wastes can be stabilized and solidified prior to land disposal. These include radioactive and/or chemically hazardous wastes such as incinerator ash and other combustion ashes, sludges, and filter cakes from physical and chemical wastewater treatment operations, contaminated soils, foundry sands, spent catalysts, tank bottoms, mine tailings, and dredged sediments. These wastes may contains inorganic contaminants (for example, lead, mercury, cadmium) and a wide variety of organic contaminants. Although, stabilization, solidification, and land disposal are normally not considered a prime alternative for organic wastes, organic contaminants are often associated in small quantities with inorganic matrices (for example, PCB contaminated soils). There is a fundamental difference between radioactive and chemically toxic contaminants. While the hazards associated with some of the former decrease with time as a result of decay and eventually disappear, chemically hazardous contaminants remain so for ever. This fact stresses the need for effective long term containment.

The stabilization/solidification technology involves mixing a waste with additives to chemically stabilize it and physically contain it, producing a waste form suitable for shallow land burial. Stabilization and solidification processes are often classified based on the principal additives used to obtain a solid matrix. Various systems based on inorganic and organic additives are listed below:

Inorganic-based Systems	*Organic-based Systems*
portland cement	bitumen
soluble silicates-cement	urea formaldehyde
pozzolan-lime	polybutadiene
pozzolan-cement	polyester
clay-cement	epoxy
gypsum	polyethylene

Technologies based on vitrification and disposal in deep geological formations have mostly been considered for high-level radioactive waste and are beyond the scope of this publication.

Inorganic-based systems all use some kind of hydraulic cement that allows the waste form to develop sufficient strength to be self-supporting in a landfill. Portland cement is most frequently used. Pozzolanic cements used in stabilization and solidification are often themselves waste materials such as power plant fly ash, cement kiln dust, and ground blast furnace slag. Cement-based processes are well suited to the treatment of aqueous wastes since cement needs water for hardening. A cement-based waste form is a porous matrix whose hydraulic conductivity is a function of the pore structure and the amount of water originally present in the waste. The leaching of a contaminant thus depends on whether it remains in solution in the pore system or is immobilized through chemical reaction. Cement-based processes create an alkaline environment suitable to the containment of several toxic metals. The additives, which are often themselves waste materials are inexpensive and can be blended with the waste using simple equipment. Cement-based processes have widely been used commercially for chemically hazardous wastes and low-level radioactive wastes.

Organic-based processes consist in encapsulating waste elements in an impervious matrix that is placed in a container to maintain dimensional stability in the landfill. Organic-based additives are normally hydrophobic. When used with aqueous waste streams, water is either evaporated in thermosetting processes (for example, bitumen) or encapsulated within the polymer matrix in catalyst-based processes. Organic-based processes normally show very low leaching of contaminants since the matrices are impervious. However, there is usually no reaction between the waste constituents and the polymer. Since hazardous constituents are not destroyed or insolubilized, the long-term stability of the waste form depends entirely on its physical integrity in the disposal environment. Organic-based processes are energy-intensive, requiring high cost additives and sophisiticated blending equipment. Their use has been practically limited to special types of high-hazard, low-volume radioactive wastes.

Most of the attention of the scientific community with regard to the stabilization and solidification technology has focused on understanding and predicting long-term containment. The most likely exposure route is water, and leaching tests play a central role in evaluating waste forms. A well designed land disposal operation often contains multiple barriers between the waste and the environment. Liners, either low permeability soils or polymeric membranes, are used for macrocontainment. Wastes may also be placed in engineered containers such as concrete tanks or jacketed drums. A waste form, either cement or organic based, is therefore the last element in a line of defense to prevent release of contaminants to the environment. Long-term containment studies can be broadly categorised in two groups, leaching studies and durabilities studies. Leaching is the study of the transfer of contaminants from the waste form to water. Durability refers to the ability of the waste form to resist wear and remain intact for a long period of time.

The content of this Special Technical Publication is well balanced. Fourteen papers deal with radioactive wastes, 16 with chemically hazardous wastes, and 3 with mixed wastes. The papers were grouped in four chapters to deal with: (1) Processes, (2) Regulatory Aspects and Testing Methods, (3) Laboratory Evaluation, and (4) Large-scale Evaluation or Demonstration.

Processes

There are numerous research and development efforts being spent in formulating stabilization and solidification processes. Endless variations of basic processes based on cement or pozzolan have been developed to address specific containment problems. Most of that information is however proprietary and was discussed more during coffee breaks than within formal presentations at the symposium! Of the seven papers included in this section, two are cement-based (an under-representation), four are organic-based (an over-representation) and one is about drum jacketing. This latter paper is borderline between the topic of this publication, micro-containment, and macrocontainment through engineered barriers.

Regulatory Aspects and Testing Methods

Regulations pertaining to the land disposal of stabilized and solidified wastes are still being developed. The radioactive waste community is ahead in this area as reflected by two papers describing the rules in the United States and in Italy. The lack of regulations in the chemically hazardous waste field has often been blamed for the relatively small role that the technology has played overall in the management of hazardous waste.

Testing methods for stabilized and solidified wastes are needed at different levels of control. At the *research level,* a wide variety of testing methods may be used to gain an understanding of the morphological, chemical, and engineering properties of waste forms. Several papers in this section describe the development or illustrate the utilization of testing methods used to provide fundamental information. The next level is the *regulatory level* where standard testing methods are used in conjunction with set performance criteria to determine the suitability of a technology applied to a specific waste stream in a given disposal scenario. Several testing methods that would meet the conditions to be used as regulatory tools are discussed in this section. The following level is the *treatability study level* where simple and rapid testing methods are needed to select an optimal formulation that would meet regulatory levels (without having to run the regulatory test on every formulation). The last level is the very important *Quality Assurance/Quality Control level* where tests are needed to verify that the performance obtained in the laboratory is reproduced under large-scale or field conditions. The two last levels are addressed in two papers but would certainly deserve more attention in the future.

Laboratory Evaluation

The seven papers included in this section present process development or application work done at bench scale, which involves the selection of specific additives to optimize containment or the investigation of the mechanisms of containment or leaching.

Large-scale Evaluation or Demonstration

It is important to remember that laboratory leaching tests only partially evaluate a waste containment system. Their results cannot be directly interpreted in terms of field leachate concentration. As a result, large-scale evaluation or demonstration studies are an important part in the determination of the environmental suitability of waste forms. The eight papers in this section were all contributed by researchers from the radioactive waste field, a clear indication of the leading role that they are playing in putting stabilization and solidification to use.

Pierre Côté
Environment Canada
Burlington, Ontario, Canada;
symposium chairman and editor.

Michael Gilliam
Oak Ridge National Laboratory
Oak Ridge, Tennessee;
symposium chairman and editor.

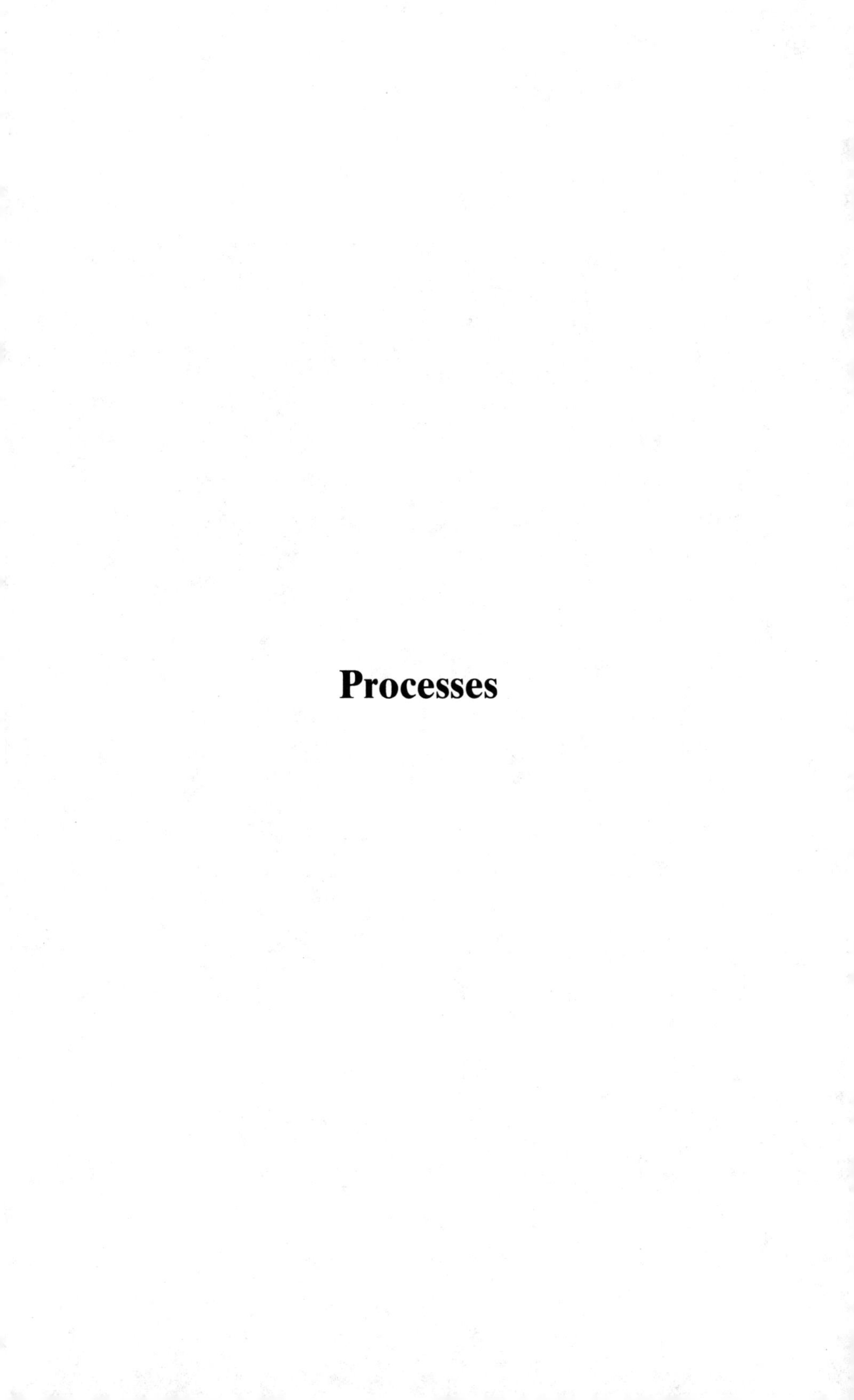

Processes

Doug Ezell[1] *and Piero Suppa*[2]

The Soliroc Process in North America: A Stabilization/Solidification Technology for the Treatment of Metal-Bearing Wastes with Reference to Extraction Procedure Toxicity Testing

REFERENCE: Ezell, D. and Suppa, P., **"The Soliroc Process in North America: A Stabilization/Solidification Technology for the Treatment of Metal-Bearing Wastes with Reference to Extraction Procedure Toxicity Testing,"** *Environmental Aspects of Stabilization and Solidification of Hazardous and Radioactive Wastes, ASTM STP 1033,* P. L. Côté and T. M. Gilliam, Eds., American Society for Testing and Materials, Philadelphia, 1989, pp. 7–14.

ABSTRACT: The Soliroc Process is a proven European stabilization/solidification technology which is used in the treatment of liquid wastes containing heavy metals. Tricil has carried out laboratory-scale studies of the Soliroc Process at three of its North American waste treatment facilities and also has contracted an independent consultant to carry out similar studies.

Waste samples at each facility were treated, using the Soliroc technology, and then subjected to the Extraction Procedure Toxicity Test (EP Tox). In all the lab tests, identical waste samples were also treated using a conventional lime precipitation process and compared with the Soliroc results. The results indicated superior performance of the Soliroc process to lime precipitation with respect to the EP Tox test. In addition, chelating agents were found not to adversely affect the EP Tox results of the Soliroc process.

KEY WORDS: stabilization, silica reagent, extraction procedure toxicity test

The Soliroc Process is a proven European stabilization/solidification technology which is used in the treatment of hazardous liquid wastes. The technology is especially effective in the treatment of heavy metals in the wastes. Although the Soliroc Process is not designed to treat organics, the presence of these materials and chelating agents do not affect its performance in treating the heavy metals wastes.

There are four commercial plants operating in Europe, and numerous studies on the Soliroc process have been carried out throughout the world. However, information on the application of Soliroc in North America has been limited.

This paper presents laboratory-scale results of the Soliroc Process on North American liquid wastes. Comparison between Soliroc and lime precipitation, a conventional stabilization/solidification technology, is also investigated.

[1] Tricil Environmental Services, 1640 Antioch Pike, Antioch, TN 37013.
[2] Senior process engineer, 59 Queensway West, Messessaugn, Ontario 15B 2V2, Canada.

Background

The Soliroc Process has been in commercial operation in Europe for over ten years. The newest plant is located in Modena, Italy, and went into full-scale operation in 1985. The wastes treated in the European plants are similar to North American wastes and come mainly from the metal-finishing industry. The resulting slurries from many of these facilities are dewatered and disposed of in a sanitary landfill site. However, the criteria for disposal (which in some European countries are nonexistent, while in others the Dynamic Leach Test [ANSI/ANS 16.1, 1986] is used) are not as stringent as the Environmental Protection Agency (EPA) Extraction Procedure Toxicity test (EP Tox), or current U.S. delisting standards.

Tricil Limited is developing the commercial application of the Soliroc Technology for the treatment of North American wastes. As part of this development laboratory-scale studies were carried out at three existing facilities which use lime precipitation. These plants are located in Michigan, Ohio, and Tennessee. Typical wastes were treated at each plant utilizing both lime precipitation and the Soliroc process. The resulting residues were subjected to the EP Tox Test and checked for compliance with current and proposed delisting requirements. A consultant laboratory also was contracted to compare lime precipitation with Soliroc and to evaluate the effect of chelating agents on each process.

Soliroc Process Description

Many existing waste treatment processes involve the addition of a base, usually caustic or lime, to precipitate metal ions as hydroxides. However, the stability of the metallic hydroxides is pH sensitive and will resolubilize with small changes in pH.

The Soliroc Technology involves a three-stage treatment process in which a chemical reaction takes place between metals in an acidic medium and a silicious reagent. The process flow chart is shown in Fig. 1.

Initially, silica reagents are added to an acidic liquid waste to produce a monosilic acid solution. The amount of silica reagent added is a function of the silica content of the reagent

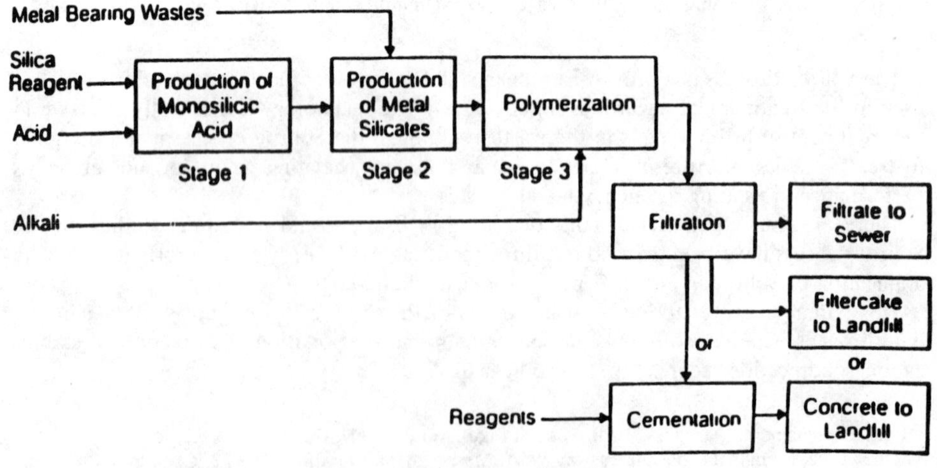

FIG. 1—*Soliroc Process flow chart.*

TABLE 1—*Analysis results of raw waste submitted to consultant laboratory.*

Element	Metal-Bearing Waste, mg/L	Acid Waste (Pickle Liquor), mg/L
Cadmium	950	0.05
Chromium	2640	800
Nickel	5	29
Lead	1	1.5
Iron	480	37 200
Copper	3	6
Zinc	1130	76

and the metal concentration in the liquid waste (that is, the ratio of silicon oxide to metal). During this stage, the pH is maintained at 1.5. Next, metal-bearing liquid wastes which have undergone chrome reduction and cyanide treatment are added to the monosilic acid to initiate the production of inert metal silicates.

The pH is then raised to 11 by the addition of alkali. The rapid rise in pH causes the metal silicates to precipitate and the monosilicic acid to polymerize to form an insoluble bulk mass. Following the polymerization step, the treated material can be filtered or a setting agent such as portland cement may be added to enhance hardening before landfilling.

It should be noted that both the acid used to dissolve the silica reagent and the alkali used for polymerization can themselves be waste materials.

Experimental Procedures

The laboratory studies consisted of two phases. The first phase involved the use of Battelle, a consultant laboratory. The lab was provided with a blend of metal wastes, acid wastes, a silica reagent, hydrated lime, and step-by-step procedures for the Soliroc Process and the lime precipitation process (see Appendix for step-by-step procedures).

Battelle was asked to stabilize the wastes provided using both processes and to test the resulting treated materials using the EP Tox [*1*]. This test was selected to determine if wastes treated with Soliroc could pass this leachate procedure and, therefore, be likely candidates for disposal at a sanitary landfill site. The consultant was also instructed to vary the endpoint

TABLE 2—*Consultant results comparison of Soliroc to lime precipitation.*

Run No.	Final Process pH	EP Toxicity Extraction Data, mg/L			Filtrate Quality, mg/L		
		Cd	Cr	Zn	Cd	Cr	Zn
1. Lime	9	5.40	<0.1	0.49	3.84	<0.1	0.06
2. Lime	11	0.04	<0.1	0.01	0.05	<0.1	0.10
3. Soliroc	9	0.02	0.11	0.51	<0.01	0.11	0.07
4. Soliroc	11	<0.01	<0.01	<0.05	<0.01	<0.1	0.10

TABLE 3—*Consultant results effect of chelating agents.*[a]

Run No.	Chelating Agent Added	EP Toxicity Extraction Data, mg/L			Filtrate Quality, mg/L		
		Cd	Cr	Zn	Cd	Cr	Zn
5. Lime	1000 mg/L ammonia	0.3	0.12	<0.05	0.02	<0.1	0.05
6. Lime	1000 mg/L EDTA	2.70	<0.10	<0.05	1.09	<0.1	0.14
7. Soliroc	1000 mg/L ammonia	<0.01	<.1	0.06	<0.01	<0.1	0.10
8. Soliroc	1000 mg/L EDTA	<0.01	<.1	0.89	0.01	<0.1	0.08

[a] Note that the final process pH for all runs in this table was 11.

pH of the Soliroc Process to determine the effect of this parameter on EP Tox results and to add chelating agents such as ethylenediaminetetraacetate (EDTA) and ammonia to the raw waste streams to see if they affected either process. The raw data of the waste appear in Table 1. The comparison of Soliroc and lime precipitation is given in Table 2 and the effects of the chelating agents on each process in Table 3.

In the second phase of the laboratory-scale studies, waste samples were blended at each Tricil operating plant. The blended samples consisted of a combination of wastes of moderate to high metal concentrations. The raw data of the blended wastes are presented in Table 4. The tests carried out at each plant focused on comparison of Soliroc to lime precipitation.

At each location, samples of the blended wastes were treated using both processes, the resultant residues were filtered, and the EP Tox test performed on the filter cake (Table 5). At the Tennessee location, further tests were carried out to evaluate the effect of the final reaction pH to the Siliroc Process. The purpose was to determine if the final reaction pH had an effect on the results of the EP Tox test. The results are presented in Table 6.

The effect of varying the ratio of silicon oxide to metal ion was explored at the Ohio plant. The results are illustrated in Table 7.

The benchmarks chosen to evaluate process performance were the proposed EP Tox-

TABLE 4—*Analysis results of plant raw waste samples in mg/L.*

Sample	Cd	Cr	Cu	Fe	Ni	Pb	Zn
MICHIGAN PLANT							
Metal-bearing waste	...	18 650	241	1 023	1 837	912	15
Waste acid (pickle liquor)	...	155	3	72 300	31	3	334
OHIO PLANT							
Metal-bearing waste	158	5 600	832	2 480	376	20	4 700
Waste acid (pickle liquor)	...	92	...	29 000	16	...	...
TENNESSEE PLANT							
Metal-bearing waste	91	3 042	921	51 028	482	3	257
Waste acid (pickle liquor)	...	170	...	145 000	...	...	...

TABLE 5—*Comparison of Soliroc to lime precipitation.*

Run No.	Final Process pH	EP Toxicity Extraction Data, mg/L Cd	Cr	Ni	Pb
		OHIO WASTE			
1. Lime	9.8	6.40	0.7	21.7	0.06
2. Lime	12.5	5.20	0.9	8.4	0.01
3. Soliroc	9.5	0.15	0.2	1.5	0.03
4. Soliroc	11.4	0.04	0.2	1.8	0.01
		MICHIGAN WASTE			
1. Lime	10.8	0.04	11.3	62.2	0.17
2. Lime	12.5	0.03	0.2	3.9	0.07
3. Soliroc	10.9	0.02	0.4	14.3	0.06
4. Soliroc	12.2	0.02	0.2	0.1	0.15
		TENNESSEE WASTE			
1. Lime	9	0.40	1.4	19.0	0.29
2. Lime	11	0.80	0.5	16.1	0.28
3. Soliroc	8.4	0.10	0.08	10.8	0.36
4. Soliroc	11	0.12	0.41	3.1	0.50

icity delisting limits for one of the Tricil facilities. These were 0.06 mg/L for cadmium, 0.3 mg/L for chromium, 2.2 mg/L for nickel, and 0.3 mg/L for lead.

Discussion

The results of the work carried out by Battelle show the superior performance of the Soliroc Process when compared with lime precipitation, especially in the treatment of cadmium. The efficiency of removing heavy metals is demonstrated in the filtrate quality data given in Tables 2 and 3. The Soliroc leachate results for all the testing program were all below EP Toxicity limits.

Table 3 illustrates Soliroc's capability in overcoming any sequestering effects due to che-

TABLE 6—*Effect of pH on Soliroc.*

Tennessee Waste Run No.	Final Process pH	EP Toxicity Extraction Data, mg/L Cd	Cr	Ni	Pb
5	8.3	0.17	0.08	13.5	0.33
6	10	0.14	0.20	6.5	0.34
7	11	0.27	0.13	5.8	0.50
8	12	0.08	0.05	0.5	0.63

TABLE 7—*Effect of silica-to-metal ratio on Soliroc.*

Ohio Waste Run No.	Final Process pH	SiO_2 to Metal Ratio	EP Toxicity Extraction Data, mg/L			
			Cd	Cr	Ni	Pb
5	11.1	0.5	5.60	3.9	9.2	0.16
6	11.1	1.5	0.67	3.3	5.1	0.04
7	11.0	2.5	0.29	0.7	4.0	0.07
8	10.0	3.5	0.12	0.4	2.5	0.04
4	11.4	5.0	0.04	0.2	1.8	0.01

lating agents such as ammonia and EDTA. The results confirm that the presence of chelating agents do not effect the chemical reactions of the Soliroc Process. However, the adverse effect of chelating agents on the leachability of metals from lime precipitation products is clearly shown.

The comparison tests carried out at Tricil's facilities confirm Soliroc relative superiority. At comparable pH levels, the leachability of the lime precipitation products was greater than that of the Soliroc Process products. The only exception was with respect to lead at both the Michigan and Tennessee studies, although this discrepancy was minor.

In all of the samples listed, the leachate results for Soliroc are below EPA EP Tox limits. Only on occasion did the Soliroc leachate results fail the benchmark "delisting limits." However, as Table 7 shows, with the increase in slag addition, the more stringent limits were achieved.

Conversely, lime precipitation on many occasions failed the EP Tox limits even when alkalinity content was increased (see Ohio Run 2 in Table 5 at pH 12.5). The leachate from the lime process never passed the delisting limits for nickel.

Table 6 illustrates the effect of pH on the Soliroc Process. It can be seen that as the pH was increased the performance of the process was generally superior, except for lead.

Table 7 shows dramatically the effect of increasing the dosage of the silica reagent. The most stringent limits were passed at silica to metal ratios greater than 3.5. The tradeoff for achieving such chemically inert material is the cost of reagent and the subsequent increase in sludge production per waste litre treated. The volume increase can be anywhere in the range of 15 to 35%.

Tricil evaluated these and other economic issues such as sources of silica reagent and characteristics of optimum agents. This information will be presented after Tricil carries out its next phase of development.

The laboratory studies and the consultant's work clearly show the potential for Soliroc to treat a segment of the North American waste market and meet North American environmental regulations. The next phase involves carrying out pilot-scale studies. Two pilot plants have been built at the Tennessee facility and are capable of treating batches from as low as 76 L (20 gal) to as high as 38 000 L (10 000 gal).

The laboratory studies have confirmed Soliroc as a fixation technology for the treatment of metal-bearing wastes, in terms of meeting the "characteristic" requirement for delisting. The Demonstration plant will now generate the required economic data in the North American commercial application of the Soliroc Process.

Conclusions

1. The Soliroc Process has shown superior performance to lime precipitation in the treatment of the wastes used for the laboratory studies (the wastes being representative of actual market quality).

2. Chelating agents have not adversely affected the Soliroc Process in treating these wastes.

3. The Soliroc Process products can pass existing leachate limits and, at increased doses of silica reagent, the treated residue meets the most stringent delisting limits imposed on Tricil's operating facilities.

APPENDIX

Soliroc Test Procedure

1. Place 100 mL of Cleveland waste in a 1-L beaker. Add 110 mL of pickle liquor to the Cleveland waste. Mix the two liquids together thoroughly.
2. Add 35 g of slag to the Cleveland and pickle liquor solution and mix. When the pH of the solution is between 1.5 and 2, begin to add the lime slurry.
3. Add enough lime slurry to raise the pH to 5.
4. Add 25 mL of Cleveland cyanide waste to the Cleveland, pickle liquor, and slag solution.
5. Add the rest of the lime slurry to raise the pH of the solution to the specified level.
6. Determine the weight of the suspension.
7. Filter the suspension as described in the filtration procedures of the EP Toxicity Test, with the exception of using a Whatman No. 4 Filter as opposed to the Millipore Membrane Filter.
8. Weigh and record the weight of the filter cake, making sure the cake does not dry out in the process.
9. Analyze the filtrate for the following metals: cadmium, chromium, zinc.
10. Split the filter cake into two equal portions.
11. Perform the EP Toxicity Test in duplicate, using one-half of the split equal portions of the filter cake. Analyze for the following metals: cadmium, chromium, zinc.

Lime Precipitation Procedure

1. Place 100 mL of blended Cleveland waste in a 1-L beaker. Add 110 mL of pickle liquor to the Cleveland waste and mix thoroughly for five minutes.
2. Add 25% lime solution to raise the pH of the slurry to the specified pH value. No Cleveland cyanide waste is to be added.
3. Determine the weight of the slurry.
4. Filter the slurry as described in the filtration procedure of the EP Toxicity Test, with the exception of using a Whatman No. 4 Filter as opposed to the Millipore Membrane Filter.
5. Weigh and record the weight of the filter cake.
6. Analyze the filtrate for the following metals: cadmium, chromium, zinc.
7. Carry out the EP Toxicity Test in duplicate on the filter cake and analyze for the

following metals: cadmium, chromium, zinc. No further solidifying or curing is to be done on these tests.

Reference

[1] "Extraction Procedure Toxicity Test," *U.S. Code of Federal Regulations for the Protection of the Environment*, Vol. 40, part 261, Appendix 2, Washington, DC, p. 365.

Terry L. Sams[1] *and T. Michael Gilliam*[2]

Systematic Approach for the Design of Pumpable, Cement-Based Grouts for Immobilization of Hazardous Wastes

REFERENCE: Sams, T. L. and Gilliam, T. M., **"Systematic Approach for the Design of Pumpable, Cement-Based Grouts for Immobilization of Hazardous Wastes,"** *Environmental Aspects of Stabilization and Solidification of Hazardous and Radioactive Wastes, ASTM STP 1033,* P. L. Côté and T. M. Gilliam, Eds., American Society for Testing and Materials, Philadelphia, 1989, pp. 15–20.

ABSTRACT: Cement-based grouts have proven to be an economical and environmentally acceptable means of waste disposal. Costs can be reduced if the grout is pumped to the disposal site. This paper presents a systematic approach to guide the development of pumpable grouts.

KEY WORDS: waste disposal, immobilization, cement-based grouts, grout formulation development, hazardous, radioactive, and mixed wastes

Researchers at the Oak Ridge National Laboratory (ORNL) have developed versatile and inexpensive processes to solidify large quantities of radioactive (low and intermediate level) liquids, sludges, and fine solids (<0.6 mm diameter) in cement-based grouts. Using engineered integrated systems composed of off-the-shelf equipment, these batch or continuous processes are compatible with a wide range of disposal methods such as above-ground intermediate storage, shallow-land burial, deep geological disposal, and bulk *in situ* solidification. The resulting grout monoliths can be tailored to be less permeable than the surrounding host soil and resistant to (*a*) leaching by ground water, (*b*) deterioration on exposure to radiation or thermal cycling, and (*c*) overburden subsidence. Disposal costs can be greatly reduced when these monoliths are formed on a large scale (for example, in vaults or earthen trenches).

Research at ORNL has shown that cement-based grouts can be tailored to tolerate wide fluctuations in waste feed compositions and still maintain properties that are compatible with standard processing equipment and regulatory performance criteria. The 20-year operational history of the ORNL grouting program has demonstrated this resilience and reliability with a spectrum of waste feeds that bracket the developmental history of the nuclear fuel cycle. These wastes include ion-exchange media, evaporator bottoms, filter media, waste slags, incinerator ashes, waste calcines, shredded metallic pieces, basaltic glass frit, and contaminated pump oils. Research efforts have been recently expanded to address the solidification of hazardous chemical and mixed (that is, radioactive and chemically hazardous) wastes [*1–3*].

[1] Project engineer, Chemical Technology Division, Oak Ridge National Laboratory, Oak Ridge, TN 37831.

[2] Manager, Engineered Waste Disposal Technologies, Chemical Technology Division, Oak Ridge National Laboratory, Oak Ridge, TN 37831-7273.

Research personnel at ORNL develop grout formulas that are compatible with both the waste chemistry and the process and environmental (for example, regulatory) constraints [*4–6*]. In many cases, it is cost-effective to formulate the grout to be sufficiently fluid so that it can be pumped to the disposal area where it hardens *in situ* to form large grout monoliths, such as in the ORNL hydrofracture process [*7*] and the Westinghouse Hanford Operations Transportable Grout Facility [*8*]. In these two applications, the disposal sites are situated some distance (300 and 1000 m, respectively) from the grout facility, making the use of a fluid (that is, pumpable) grout imperative. The development of pumpable grouts has long been considered an art; however, a systematic approach to developing these grouts has been applied at ORNL and is presented in this paper.

Basis for Rheological Limits

A typical grout facility consists of four major modules: (1) a dry solids blending and feed facility that prepares and delivers the solids blend (such as cement and fly ash); (2) a liquid/slurry waste feed system; (3) a mixing facility where the dry solids blend and the liquid waste are brought into contact with each other; and (4) a grout distribution system for pumping the fluid grout to the disposal area, where it hardens *in situ*. Of these four modules, the grout distribution system incurs the greatest limitations on the development of pumpable grouts because the non-Newtonian rheological behavior of the fluid grout must be tailored to be compatible with this system [*4,8*].

Typically, three parameters—density, flow rate, and viscosity—define the rheological behavior of a fluid. However, defining the rheological behavior of a cement-based grout is more difficult because of its non-Newtonian behavior. For non-Newtonian grouts, shear stress is dependent on shear rate as first described by the Ostwald-de Waele model [*9*]. Field experience has shown that the relationship between shear stress and shear rate is adequately represented by the power-law model [*10*]

$$S_s = K'(S_r)^{\eta'} \tag{1}$$

where

S_s = shear stress, Pa,
K' = fluid consistency index, $\text{Pa} \cdot \text{s}^{\eta'}$,
S_r = shear rate, s^{-1}, and
η' = flow behavior index ($0 < \eta' < 1.0$), dimensionless.

From Eq 1, the Reynolds number for flow through a cylindrical pipe can be calculated by

$$Re = \frac{8000\rho U^{2-\eta'}}{K'(8/d_i)^{\eta'}} \tag{2}$$

where

Re = Reynolds number, dimensionless,
U = fluid velocity, m/s,
ρ = fluid density, g/cm^3, and
d_i = inside pipe diameter, *m*.

Although grouts can be pumped in the laminar flow regime, the turbulent flow regime is preferred because radial components of velocity aid in keeping these particulate slurries well mixed. These radial flow components promote mixing at the pipe wall and, hence, minimize caking at the pipe wall. This, in turn, minimizes the operational process flushing requirements.

In general, turbulent flow begins at a Reynolds number greater than or equal to 2100 [10]. If one assumes that critical velocity (U_c) is achieved at a Reynolds number of 2100, then the minimum pumping velocity necessary to achieve turbulent flow in the grout distribution system can be calculated by

$$U_c = \left[\frac{2100K'(8/d_i)^{\eta'}}{8000\rho}\right]^{\frac{1}{2-\eta'}} \tag{3}$$

For a given grout formula, K', η', and ρ can be calculated from laboratory measurements using standard techniques [5] to determine the minimum desired pumping velocity for onset of turbulent flow as described by Eq 3.

The pressure drop resulting from pumping the fluid grout through the distribution pipe can be calculated by

$$\Delta P_f = \frac{1000\rho U^2 f}{d_i} \tag{4}$$

where

ΔP_f = frictional pressure drop through a straight pipe, Pa/m, and
f = Fanning Friction Factor, dimensionless (f is a function of Reynolds number).

The maximum pumping velocity can be determined from Eq 4 by comparing the calculated pressure drop with the discharge pressure rating for a given pump. Thus, for an established grout formula, Eqs 3 and 4 can be used to determine the maximum and minimum pumping required for the grout distribution system.

However, the design criteria for the process facility are commonly selected independently of the grout's properties, for example, when facility design proceeds before the optimum grout formula has been determined or when use of an existing facility is desired. In such an instance, the grout formula must be tailored so that it will be compatible with an existing plant design. Then, Eqs 3 and 4 establish rheological criteria for grout formulation development. As such, it is necessary to determine limits on K', η', and ρ because these characteristic parameters define the flow behavior of the grout and thereby determine the compatibility of the grout with the design of the grout distribution system.

In the past, an enormous number of experiments has been required to develop grouts compatible with the distribution system design due to the lack of a systematic approach to formulation development. However, experience at ORNL has led to the development of a systematic approach which greatly reduces the experimentation required to develop a grout formula compatible with an established design.

For a given grout distribution system design, three pertinent parameters have been established:

(1) inside diameter of the distribution pipe (d_i),
(2) maximum permissible pressure drop across the distribution pipe (ΔP_m), and
(3) maximum distribution pump throughput capacity (U_m).

These values in combination with Eq 4 can be used to establish limits on fluid grout density

$$\rho_m = \frac{\Delta P_m\, d_i}{1000\, U_m^{\,2} f} \tag{5}$$

where ρ_m is the maximum fluid grout density. Although the selection of an appropriate fanning friction factor (f) is somewhat arbitrary, experience at ORNL has shown that, 0.008, the value for smooth pipe at a Reynolds number of 2100 [*11*] is generally appropriate [*4,8*]. Thus, Eq 5 reduces to

$$\rho_m = \text{constant} = C \tag{6}$$

The density of a fluid grout can be approximated by

$$\rho_g = \frac{W_g}{(W_s/\rho_s + W_w/\rho_w)} = \frac{W_s + W_w}{(W_s/\rho_s) + (W_w/\rho_w)} \tag{7}$$

where

ρ_g = density of fluid grout,
W_g = total mass of grout,
W_s = mass of dry solids blend,
W_w = mass of waste,
ρ_s = true density of dry solids blend, and
ρ_w = density of waste.

Substituting Eq 6 into Eq 7 and rearranging yields

$$\frac{W_s}{W_w} = \frac{\rho_s\,(\rho_w - C)}{\rho_w\,(C - \rho_s)} \tag{8}$$

Thus Eq 8 establishes the upper limit of dry solids blend per unit of waste for formulation development studies. This ratio expressed in units of kilograms dry solids blend per liter of waste is typically referred to as the mix ratio (M). The minimum mix ratio is determined by product performance criteria such as minimum required compressive strength[3] and rate of set.[4] The determination of this minimum is still somewhat of an art, and one generally relies on the experience of the researcher. However, in the absence of other guidance, civil engineering applications of grouting technology have shown that the minimum mix ratio must result in a fluid-grout water-to-cement ratio (weight/weight) of at least 0.38 [*12*].

At this point, both the minimum and maximum mix ratios for the dry solids blend under study have been established. However, guidance on limits for K' and η' are still required. This guidance is supplied by another parameter of an established distribution system: the nominal throughput. Assuming turbulent flow is desired at or below this throughput and substituting the throughput for critical velocity in Eq 3 results in a three parameter equation involving K' and η', and ρ. However, Eq 3 is relatively insensitive to density [*13*]. Consequently, the waste density can be substituted into Eq 3, and it then becomes a two-parameter equation which can guide formulation development efforts.

For example, if a distribution system is characterized by 0.05 m inside diameter of the grout distribution pipe, a nominal throughput of 3.15 L/s, and a waste density of 1 g/cm^3,

[3] See the American Society for Testing and Materials (ASTM) Test Method for Compressive Strength of Hydraulic Cement Mortars (Using 2-in. or 50-mm Cube Specimens) (C 109-80).

[4] See ASTM Test Method for Time of Setting of Concrete Mixtures by Penetration Resistance (C 403-80).

then Eq 3 becomes

$$1.61 \geq \left[\frac{2100K'(160)^{\eta'}}{8000}\right]^{\frac{1}{2-\eta'}} \tag{9}$$

Equation 9 is shown graphically in Fig. 1. The combinations of K' and η' values which satisfies Eq 9 lies on or below the straight line (Fig. 1) and bounds the area representing an acceptable window of grout flow characteristics.

This window, coupled with limits on mix ratio, becomes the basis for developing grout formulas with acceptable flow behavior. This systematic approach greatly reduces the amount of experimentation required to develop a grout with acceptable flow characteristics. In addition, measurements required by this approach are performed on freshly prepared grouts. Thus, this phase of grout formulation requires no waiting on the grout cure period.

Summary

Versatile and inexpensive processes to solidify large quantities of radioactive liquids, sludges, and fine solids in cement-based grouts have been developed and are used routinely for the ultimate disposal of radioactive wastes. Using engineered integrated systems composed of off-the-shelf equipment, these batch or continuous processes are compatible with a wide range of disposal methods such as above-ground intermediate storage, shallow-land burial, deep geological disposal, and bulk *in situ* solidification. The resulting grout monoliths have been shown to be less permeable than the surrounding host soil and resistant to: (*a*) leaching by ground water, (*b*) deterioration on exposure to radiation or thermal cycling, and (*c*) overburden subsidence. Recent studies have shown that similar product performance can be obtained upon application of these processes to the disposal of chemically hazardous or mixed wastes.

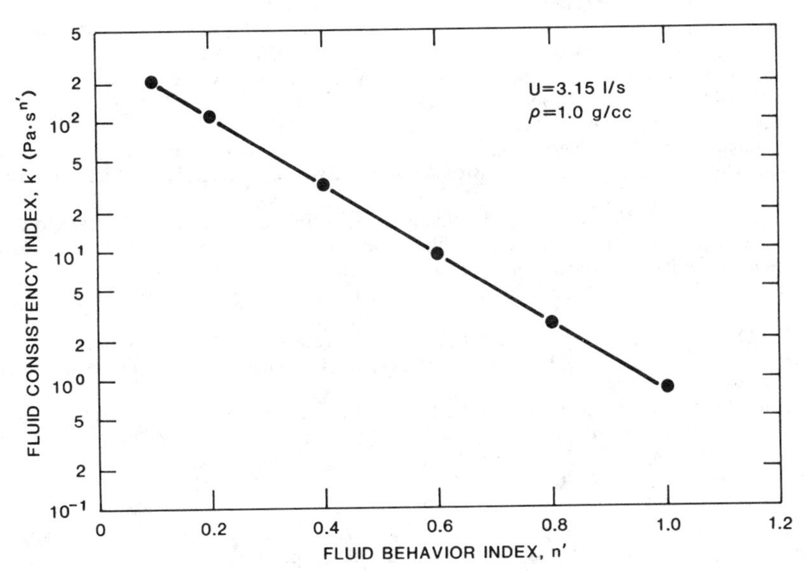

FIG. 1—*Combination of* K' *and* η' *which results in turbulent flow at reference design conditions.*

Disposal costs can be further reduced if the cement-based grout is formulated to be sufficiently fluid so that it can be pumped to the disposal area where it hardens *in situ* to form large grout monoliths. The development of pumpable grouts has long been considered an art; however, a systematic approach to developing these grouts has been applied at ORNL and presented in this paper. The approach greatly reduces the experimentation required for grout formulation development. As such, both the cost and time for formulation development are reduced. In addition, the approach provides a tool for application of quality control/assurance to the development process. Thus, the approach is a powerful tool which can be used to rapidly extend the application of grout technology to the ultimate disposal of chemically hazardous waste.

Acknowledgment

This research was sponsored by the Office of Defense Waste and Transportation Management, U.S. Department of Energy, under Contract No. DE-AC05-84OR21400 with Martin Marietta Energy Systems, Inc.

References

[*1*] Gilliam, T. M. and Loflin, J. A., "Leachability Studies of Hydrofracture Grouts," ORNL/TM-9879, Martin Marietta Energy Systems, Inc., Oak Ridge National Laboratory, Oak Ridge, TN, 1986.

[*2*] Mrochek, J. E., Gilliam, T. M., McDaniel, E. W., Land, J. F., and Godsey, T. T., "Grout Formulation and Leaching for the Immobilization of Sludges from the F-1 Tank at Y-12," ORNL/TM-9843, Martin Marietta Energy Systems, Inc., Oak Ridge National Laboratory, Oak Ridge, TN, in press.

[*3*] Gilliam, T. M., Dole, L. R., and McDaniel, E. W., "Waste Immobilization in Cement-Based Grouts," *Hazardous and Industrial Solid Waste Testing and Disposal: Sixth Volume, ASTM STP 933,* American Society for Testing and Materials, Philadelphia, 1986.

[*4*] "Immobilization of Hanford Facility Decontamination Waste in a Cement-Based Grout," Gilliam, T. M., McDaniel, E. W., Sams, T. L., Tallent, O. K., and Dole, L. R. in *Proceedings,* Spectrum '86, American Nuclear Society, 14–18 Sept. 1986, in press.

[*5*] Gilliam, T. M., Sams, T. L., and Pitt, W. W., "Testing Protocols for Evaluating Monolithic Waste Forms Containing Mixed Waste," *Ceramic Advances,* Vol. 20, 1986, p. 220.

[*6*] Mattus, A. J., Gilliam, T. M., and Dole, L. R., "A Review of EPA, DOE, and NRC Regulations Regarding Establishment of Solid Waste Performance Criteria," ORNL/TM-9322, Martin Marietta Energy Systems, Inc., Oak Ridge National Laboratory, Oak Ridge, TN.

[*7*] de Laguna, W. et al., "Engineering Development of Hydraulic Fracturing as a Method for Permanent Disposal of Radioactive Wastes," ORNL-4259, Oak Ridge National Laboratory, Oak Ridge, TN, Aug. 1968.

[*8*] Sams, T. L., McDaniel, E. W., and Gilliam, T. M., "Immobilization of Neutralized Cladding-Removal Waste in a Cement-Based Grout," *Proceedings of the Second International Conference on Radioactive Waste Management,* Canadian Nuclear Society, 7–11 Sept. 1986.

[*9*] Bird, R. B., Stewart, W. E., and Lightfoot, E. N., *Transport Phenomena,* Wiley, New York, 1960, pp. 10–15.

[*10*] Smith, D. K., "Cementing," *Society of Petroleum Engineers of AIME,* New York, 1976.

[*11*] Cheremisinoff, N. P., "Fluid Flow, Pumps, Pipes, and Channels," Ann Arbor Science Publishers, Ann Arbor, MI, 1982.

[*12*] Gilliam, T. M., McDaniel, E. W., Dole, L. R., Friedman, H. A., Loflin, J. A., Mattus, A. J., Morgan, I. L., Tallent, O. K., and West, G. A., "Summary Report on the Development of a Cement-Based Formula to Immobilize Hanford Facility Waste," ORNL/TM-10141, Martin Marietta Energy Systems, Inc., Oak Ridge National Laboratory, Oak Ridge, TN, 1987.

[*13*] Gilliam, T. M., McDaniel, E. W., Dole, L. R., and West, G. A., "Preliminary Data on Rheological Limits for Grouts in the Transportable Grout Facility," ORNL/TM-9117, Martin Marietta Energy Systems, Inc., Oak Ridge National Laboratory, Oak Ridge, TN, 1987.

Deborah P. Swindlehurst,[1] *R. D. Doyle,*[1] *and A. J. Mattus*[2]

The Use of Bitumen in the Stabilization of Mixed Wastes

REFERENCE: Swindlehurst, D. P., Doyle, R. D., and Mattus, A. J., **"The Use of Bitumen in the Stabilization of Mixed Wastes,"** *Environmental Aspects of Stabilization and Solidification of Hazardous and Radioactive Wastes, ASTM STP 1033,* P. L. Côté and T. M. Gilliam, Eds., American Society for Testing and Materials, Philadelphia, 1989, pp. 21–27.

ABSTRACT: Mixed wastes pose a problem to generators since there are no burial sites or treatment facilities currently accepting this type of waste. One potential disposal method is treating the waste to render it nonhazardous and disposing of it in accordance with radioactive waste requirements. A possible means of accomplishing this transformation is solidifying the waste in asphalt (bitumen).

Associated Technologies, Incorporated, in cooperation with Oak Ridge National Laboratory (ORNL), solidified in asphalt a surrogate sodium nitrate-based waste, spiked with extraction procedure (EP) toxic metals and nonradioactive cesium and strontium. This paper reports the characteristics of the spiked ORNL solution that was solidified as well as the properties of the solidified end product. The waste samples generated underwent EP toxicity testing as well as American Nuclear Society (ANS) 16.1 leach testing for 90 days, and the results of these tests are presented. Also, a discussion of the criteria for classifying a waste as hazardous is included in order to demonstrate that the waste, once solidified in asphalt, can no longer be considered hazardous.

KEY WORDS: hazardous, radioactive, extraction procedure (EP) toxicity, compressibility, leachability

Oak Ridge National Laboratory (ORNL) currently has eight 190 000-L (50 000 gal) tanks full of waste that is both radioactive and hazardous. The waste is produced by the regeneration of ion exchange columns, yielding a nitric acid-based solution. This solution is heated to remove any reclaimable nitric acid, and the remaining slurry is adjusted to a pH of 12 and stored in the Melton Valley storage tanks. Oak Ridge has initiated a plan to characterize the waste and determine the best methodology to dispose of this waste due to the fact that these tanks are filling rapidly and a more permanent means of disposal is required.

One of the treatment methods that Oak Ridge is investigating is the solidification of this waste in asphalt. Asphalt solidification may be used to render the waste nonhazardous, allowing it to be disposed of in accordance with radioactive waste requirements.

Tests toward this end were conducted using a thin-film evaporator-based asphalt solidification process owned and operated by Associated Technologies, Incorporated (ATI), of Charlotte, North Carolina. This paper discusses the results of the testing performed to support the assertion that the waste produced will no longer be hazardous and hence will be suitable for disposal as radioactive waste.

[1] Process engineer and manager, Business Development, respectively, Associated Technologies, Inc., 212 S. Tryon St., Charlotte, NC 28281.

[2] Oak Ridge National Laboratory, P.O. Box P, Oak Ridge, TN 37831.

TABLE 1—*Melton Valley storage tank surrogate waste.*

Species	Concentration, ppm
Arsenic	43
Barium	47
Cadmium	48
Calcium	13 000
Cesium	46.3
Chloride	2 200
Chromium	49
Lead	50
Magnesium	920
Mercury	31.6
Nitrate	212 000
Selenium	24
Silver	49
Sodium	81 000
Strontium	47
Sulfate	200

Sample Preparation

The actual Oak Ridge waste contains both hazardous and radioactive constituents. In order for equipment contamination (due to radioactivity) to be avoided, a surrogate was prepared according to a formulation developed at Oak Ridge (Table 1). The surrogate waste was a 27 weight percent solution of primarily sodium nitrate with eight heavy metals plus nonradioactive cesium and strontium.

The asphalt used in sample preparation is an oxidized asphalt classified by the American Society of Testing and Materials as ASTM D312 Type I roofing asphalt. Two waste form loadings of 40.9 and 43.3 weight percent solids were achieved during operation of the system. The waste was solidified into 0.10-m (4-in.) diameter molds to be used for testing and analysis. The waste specimens were cut after cooling to a 0.10 m (4 in.) height. All solidified waste forms were returned to Oak Ridge for analysis and evaluation.

Process Description

The TVR III System (Fig. 1) which was used in the sample preparation is an over-the-road, self-contained, trailer-mounted system which is centered around a thin-film evaporator. It also contains subsystems for the waste feed tank, distillate collection tank, heating fluid, and the heating, ventilating and air conditioning (HVAC) system. The asphalt tank is separate from the trailer and is electrically heated. The system is a full-scale unit capable of solidifying up to 1900 L (500 gal) of waste in an 8-h shift.

The waste is transferred from the client via temporary connections to TVR's waste batch tank, where it is pretreated for optimum solidified waste properties. The waste is then recirculated through the tank while a smaller side stream is metered from the recirculation line into the evaporator at a rate determined by the solids concentration in the waste. Simultaneously, hot asphalt is pumped into the evaporator at a controlled rate to yield the target solids content in the solidified product. The evaporator is a 1.7-m (5.5-ft) tall, 0.4-m (16-in.) diameter cylinder that is jacketed with heating fluid. The waste and asphalt are supplied to the internal surface of the cylinder where they are mixed by the rotor blades, which pass within 2 mm (0.078 in.) of the evaporative surface. As the waste and the asphalt enter at the top of the evaporator through distributors, they are swept into a thin film which

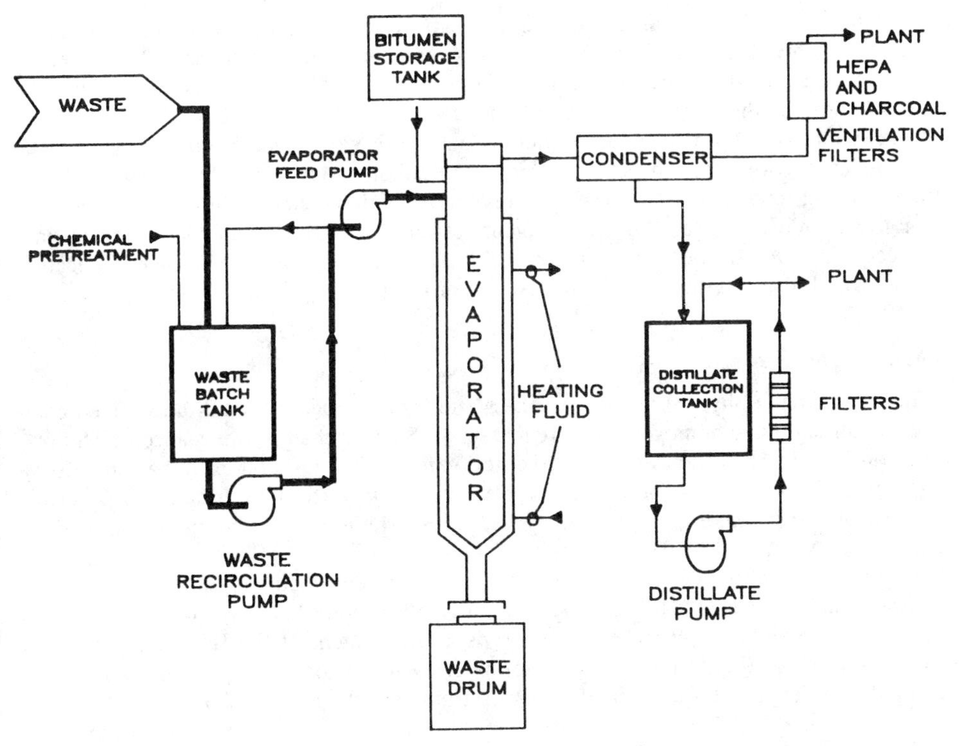

FIG. 1—*Basic flow diagram, TVR III System.*

allows rapid evaporation of the waste water, leaving the waste solids embedded in asphalt. The asphalt and solids mixture exists through a cone-shaped nozzle at the bottom of the evaporator. The evaporated water (distillate) is pulled up through the evaporator by a fan and through a heat exchanger, where the moisture is condensed. The condensate flows to a collection tank for treatment, if necessary, before being returned to the client. The air which was pulled up through the evaporator with the distillate is filtered, monitored for activity, and released to the environment or to the client's HVAC system.

Testing

Leachability

In order for the solidified product to be established as nonhazardous, it must meet established limits in the performance of an EP (extraction procedure) toxicity test [*1*]. EP toxicity tests were conducted on each of the two waste-loadings, using deionized water as the extracting solvent and maintaining a pH of 5 ± 0.2 with acetic acid, as specified in the procedure.

Both waste loadings were also tested according to the American Nuclear Society's leaching test, Measurement of the Leachability of Solidified Low-Level Radioactive Wastes by a Short-Term Procedure (ANS 16.1-1986). This procedure, performed as a 90-day test, is used in the determination of waste form stability as required by the Nuclear Regulatory Commission's (NRC) Branch Technical Position (BTP) [*2*]. The BTP specifies the stability requirements which a solidified radwaste product must meet in order to be considered acceptable at a shallow land radwaste burial facility. The test prescribes sample dimensions, leachate volume, leachate changeout frequency, and evaluation mathematical parameters. The test measures the effective diffusivity of radionuclides from the waste form, and the "leach index" is then defined as an arithmetic average of a negative log function of the effective diffusivity for a given time period. (Consult ANS 16.1-1986 for further detail.) The leach index provides a "yardstick" by which waste solidification agents can be measured for their leaching tendencies. The BTP requires that all radionuclides exhibit leach indices of 6 or greater in order for a solidification agent's leachability to be acceptable. In this testing, the leach index for all waste constituents was measured so as to provide a general indication of the overall leaching tendency of the asphalt product.

Physical Properties

In order for the physical integrity of the waste forms to be evaluated, unconfined compressive strength and homogeneity were measured. Since asphalt is a substance which flows under applied force, a widely recognized criterion for asphalt is that it exhibit a compressive strength of 3.45×10^5 Pa (50 lb/in.^{2}) or greater at 10% vertical deformation [3]. Testing was performed in accordance with the ASTM Test Method for Compressive Strength of Bituminous Mixtures (D 1074-83).

Homogeneity in the waste form is also of interest in determining its stability. Settling of the waste salts could lead to higher concentrations in the product bottom, potentially resulting in higher leachabilities. Waste homogeneity is a requirement of the BTP. Samples taken from the top and the bottom of waste forms of each salt loading were analyzed for salt content in order to determine the degree of settling that the asphalt allowed.

Results

Leachability

The results of the EP toxicity test for the eight hazardous metals in the waste are presented in Table 2. As can be seen from the table, the concentration in the leachate of each of the metals is far less than the allowable concentration; thereby, meeting one criterion for a nonhazardous waste.

A far more stringent requirement on waste form leachability is the ANS 16.1 leach test, the results of which are presented in Table 3. Three waste forms at each of the two waste loadings (40.9 and 43.3 weight percent solids in product) were tested in order to provide

TABLE 2—*EP toxicity results for heavy metals, ppm.*

	Salt Loading, %			
Species	40.9	43.3	0.0 (Pure Asphalt)	Limit
Arsenic	<0.005	0.010	<0.005	5
Barium	0.120	0.140	<0.0010	100
Cadmium	0.010	0.022	<0.0030	1
Chromium	<0.010	<0.010	<0.010	5
Lead	0.010	0.022	0.005	5
Mercury	<0.0002	<0.0002	<0.0002	0.2
Selenium	0.009	0.008	<0.005	1
Silver	<0.0060	<0.0060	<0.0060	5

TABLE 3—*Average leach indices*[a] *(LI) for all surrogate constituents after 90-day ANS 16.1 leach test.*

Species	Waste Loading, 40.9%		Waste Loading, 43.3%	
	Avg LI	Standard Deviation[b]	Avg LI	Standard Deviation[b]
Arsenic	>10.7	0.8	>10.8	0.8
Barium	8.4	0.1	8.9	0.3
Cadmium	>10.5	0.5	>10.7	0.5
Calcium	8.3	0.1	8.7	0.3
Cesium	>7.8	0.2	>7.8	0.3
Chloride	>5.4	1.1	>5.8	0.9
Chromium	10.1	0.8	>10.2	0.8
Lead	9.5	1.0	9.6	0.9
Magnesium	9.3	0.2	9.5	0.3
Mercury	>13.6	0.8	>13.6	0.8
Nitrate[c]	8.3	0.3	9.0	0.5
Selenium	>10.3	0.4	>10.6	0.6
Silver	>10.6	0.8	>10.6	0.8
Sodium	8.2	0.2	8.7	0.4
Strontium	8.3	0.1	8.7	0.9

[a] Average of three replicate leach indices.
[b] Average of the three standard deviations of the incremental leach indices.
[c] Analyses available for only two replicates.

an indication of consistency in the data. Some of the tabulated leach indices are reported as "greater than" values, which correspond to a leachate analysis which is less than the limit of detection. Leach indices for all of the waste constituents were evaluated using chemical analysis techniques. As the data show from the table, all of the leach indices, except for chloride, were above 6.0, and reasonably consistent indices were achieved for all three replicates. No appreciable differences can be seen due to the difference in waste loadings, but since the weight percentages are close to one another, this is not unexpected.

Physical Properties

The results of the unconfined compressive strength tests are tabulated in Table 4. The compressive strengths of the two waste loadings exceeded the 3.45×10^5 Pa (50 psi) requirement at 10% vertical deflection. Higher compressive strengths would normally be expected at higher waste loadings since the addition of waste salts has been seen previously to strengthen the solidified product [*4*]. However, since the two weight percentages are relatively close to one another, this overall trend is not readily observable.

A sample from the top and the bottom of each waste loading was analyzed for percentage waste solids to determine product homogeneity, the results of which are tabulated in Table 5. The higher salt loading, which increases the strength of the product in most cases, also

TABLE 4—*Unconfined compressive strength at 10% vertical deflection, psi.*

Salt Loading	Compressive Strength, Pa
40.9	$1.93 (10^6)$ (280 psi)
43.3	$1.83 (10^6)$ (265 psi)
0.0 (pure asphalt)	$3.24 (10^5)$ (47 psi)

TABLE 5—*Salt loading homogeneity.*

Specimen Waste Loading Weight % Solids	Top Weight % Solids	Bottom Weight % Solids
40.9	38.6	43.1
43.3	42.1	44.4

produced a product less likely to settle. The 40.9 weight percent solids product concentration varied by 5.6% from the average concentration, whereas the 43.3 weight percent solids product varied 2.8% from the average. Previous testing using 0.21-m^3 (55-gal) drums showed that there was no appreciable settling and that the solids remained evenly distributed within the asphalt matrix [*4*].

Hazardous Waste Characteristics

According to Vol. 40 Part 261 of the *Code of the Federal Regulations* [*1*], a hazardous waste is a waste which exhibits the characteristics of ignitability, corrosivity, reactivity, EP toxicity, or has been assigned an EPA Hazardous Waste Number. A solid waste generated from the treatment of a hazardous waste, classified according to Subpart C, can be classified as nonhazardous if it does not exhibit the characteristics of Subpart C, namely ignitability, corrosivity, reactivity, and EP toxicity [40 CFR 261.3 b(3)(d)(1)]. The Oak Ridge waste is considered hazardous because it contains EP toxic heavy metals—a hazardous waste characteristic according to Subpart C. If the asphalt product generated from the solidification of this waste is no longer EP toxic and is not corrosive, ignitable, or reactive due to the addition of asphalt to the waste, then the waste is rendered nonhazardous and can be disposed of as low-level radioactive waste. The concern, then, is whether or not the asphalt solidification process creates a solidified product which is hazardous under the characteristics of Subpart C.

Ignitability is defined, for substances other than a liquid, as capable of causing fire through friction, moisture absorption, or spontaneous chemical changes under standard temperature and pressure. Asphalt is not capable of causing fire under any of these circumstances and therefore would not be considered ignitable.

Corrosivity is not well-defined for solids under Subpart C, so it is not readily refutable when discussing asphalt. However, asphalt is generally not considered a corrosive material as evidenced by the use of asphalt as a corrosion protective coating on underground tanks.

Reactivity, under Subpart C, is defined at length in terms of instability; violent reactivity with water to form explosive mixtures or toxic fumes; capable of detonation at standard temperature and pressure; etc. The fact that asphalt is used in the paving of roads and roofing of buildings verifies the assertion that it cannot be considered a reactive substance.

As proven in the testing at the Oak Ridge National Laboratory, the stabilization of EP Toxic compounds in asphalt results in a solid waste which does not exhibit the characteristic of EP Toxicity. This rounds out the requirements needed to render a waste nonhazardous, allowing its treatment as solely a radioactive waste.

Radioactive Waste Requirements

In order for a Class B or C radioactive waste to be considered stable and suitable for shallow land burial, according to the BTP, it must meet specified criteria in the areas of leachability, thermal cycling, biodegradation, compressive strength, radiation stability, and immersion in water. Testing in all of these areas has been completed for nonradioactive waste surrogates solidified over a range of waste loadings by the TVR III system. A topical report on the results of this testing is currently under review by the NRC.

Summary

It is evident, from the testing performed in conjunction with Oak Ridge, that the solidification of certain mixed wastes in asphalt can result in a waste no longer considered hazardous. This can permit the disposal of the waste in accordance with regulations for radioactive wastes alone. Primarily, it permits the final disposal of the waste rather than requiring its indefinite storage.

References

[*1*] "Identification and Listing of Hazardous Waste," *Code of Federal Regulations,* 40 CFR 261, Environmental Protection Agency, 1 July 1987.

[*2*] "Branch Technical Position on Waste Form," Rev. 0, Nuclear Regulatory Commission, Washington, DC, May 1983.

[*3*] "An Evaluation of the Stability Tests Recommended in the Branch Technical Position on Waste Forms and Container Materials," BNL-NUREG 51784, Brookhaven National Laboratory, Upton, NY, Jan. 1985.

[*4*] "Bitumen as a Radwaste Solidification Agent," Topical Report Supplement No. 2 to No. ATI-VR-001-P-A, Associated Technologies, Incorporated, Charlotte, NC, 1985.

A. J. Mattus,[1] *M. M. Kaczmarsky,*[2] *and C. K. Cofer*[2]

Leaching and Comprehensive Regulatory Performance Testing of an Extruded Bitumen Containing a Surrogate, Sodium Nitrate-Based, Low-Level Waste

REFERENCE: Mattus, A. J., Kaczmarsky, M. M., and Cofer, C. K., **"Leaching and Comprehensive Regulatory Performance Testing of an Extruded Bitumen Containing a Surrogate, Sodium Nitrate-Based, Low-Level Waste,"** *Environmental Aspects of Stabilization and Solidification of Hazardous and Radioactive Wastes, ASTM STP 1033,* P. L. Côté and T. M. Gilliam, Eds., American Society for Testing and Materials, Philadelphia, 1989, pp. 28–39.

ABSTRACT: Performance test results obtained from laboratory testing of an extruded bitumen containing a surrogate, sodium nitrate-based waste are presented. A relatively viscous form of oxidized bitumen (ASTM D 312, Type III) has been tested and has been shown to meet all of the current regulatory performance criteria. Molded specimens were obtained using a 53-mm Werner & Pfleiderer extruder, operated by personnel of WasteChem Corporation of Paramus, New Jersey. A surrogate, low-level, mixed, liquid waste, formulated to represent an actual on-site waste at the Oak Ridge National Laboratory, was used. The surrogate waste contained ~30 weight percent sodium nitrate, in addition to eight heavy metals, cold cesium, and strontium. Waste form specimens contained three levels of waste loading: 40, 50, and 60 weight percent salt. Results include thermal testing, extraction procedure toxicity tests, and 90-day American Nuclear Society 16.1 leach tests, as well as compressive strength tests.

KEY WORDS: bitumen, extruder, leach index, EP toxicity

At the Oak Ridge National Laboratory (ORNL), ion-exchange processes used primarily to remove cesium-137 (^{137}Cs) and strontium-90 (^{90}Sr), as well as trace amounts of cobalt-60 (^{60}Co) and rare earths, from building process wastes result in the production of a mixed, low-level, nitric acid-based waste. The acid is heated in evaporators to recover as much acid as possible for reuse in the ion-exchange circuit. Following acid recovery, the resulting slurry is made alkaline at approximately pH 12 using sodium hydroxide.

The resulting alkaline sodium nitrate solution is then pumped to storage tanks on-site at ORNL, where it is stored in tanks which are referred to as the Melton Valley storage tanks. The radioactive solution contains activity in the range of 9.8 E+7 to 9.8 E+8 Bq/L and ~30 weight percent sodium nitrate.

Concern over the rate at which eight, 190 000-L storage tanks were filling prompted an investigation into those fixation media which were immediately available for on-site use and offered a good volumetric-reduction efficiency. Bitumen was then chosen as one of the possible fixation media which could potentially immobilize the first 378 000 L of waste.

The decision-making process used in choosing a fixation technology or vendor involves laboratory performance testing of prospective vendors' small-scale waste form specimens.

[1] Martin Marietta Energy System, Inc., P.O. Box 2008, Oak Ridge, TN 37831-6046.
[2] WasteChem Corporation, One Kalisa Way, Paramus, NJ 07652.

Performance testing includes compliance with test methods centered primarily around the Nuclear Regulatory Commission's (NRC's) Branch Technical Position (BTP) Paper [1], which recommends methods which are acceptable for demonstrating compliance with the 10 CFR Part 61 [2] waste-form stability criteria. In addition, tests required by the Environmental Protection Agency (EPA) [3] have been performed on the bitumen-based waste forms.

Background

During June 1986, waste-form specimens were prepared at the Werner & Pfleiderer site in northern New Jersey, by WasteChem Corporation personnel, using a 53-mm, twin-screw extruder. The waste forms were prepared in the presence of ORNL personnel to corroborate that conditions were as close as possible to those which might be encountered in the field.

Because contamination of equipment and transportation of radioactive materials presented a major problem, a nonradioactive, surrogate waste was prepared to represent the Melton Valley waste solution. The resulting chemical analysis of the as-prepared surrogate waste used in the demonstration of the extruder technology is presented in Table 1.

Extruder Technology

The extruder/evaporator is a one-step volume-reduction and solidification system which utilizes bitumen as the fixation medium. The extruder process equipment is very simple and consists of a direct-current, geared motor which drives two nitrided steel screws that mesh together very closely. The screws are slightly offset and move in the same direction. The surfaces of the screws are designed such that they are self-wiping. The screws impart mechanical energy into the viscous bitumen, while mixing and transporting the product forward toward an orifice at the end of the machine. The screws actually effect a mixing and kneading action on the bitumen product as it moves along.

The outer steel casing of the extruder is heated in separate sections along its entire length, thereby facilitating the addition of heat to the bitumen and liquid waste. The heat lowers

TABLE 1—*Surrogate waste composition representing the Melton Valley nitrate-based waste.*

Species	Concentration, mg/L
arsenic	47
barium	42
cadmium	50
calcium	13 000
cesium	48.9
chloride	2 200
chromium	46
lead	47
magnesium	990
mercury	38.4
nitrate	238 000
selenium	43
silver	48
sodium	77 000
strontium	47
sulfate	560
Density at 25°C	1.25 g/cc

the viscosity of the American Society for Testing and Materials (ASTM), Type III, air-blown bitumen used in the demonstration and, at the same time, evaporates the water from the waste. Typically, <1 weight percent water remains in the bitumen exiting the extruder.

The electrically heated sections of the casing were maintained at controlled temperatures to evaporate the water at the desired rate. The steam produced entered domes on top of the machine, which lead to water-cooled condensers. A maximum temperature of 177°C was obtained in the hottest section, while the product exited the extruder at 138°C.

Methods

Waste Form Preparation and Tests

Cylindrical waste-form specimens were cast into the proper size and geometry to facilitate regulatory testing using aluminum molds with Neoprene bottoms. Aluminum sheets (10 mils thick) were cut to form tubes, and aluminum-backed tape was used to hold the seams together. Disks molded from Neoprene were tightly fitted into the bottom of the tubes, and a thin layer of silicone-based cement was placed on the edge of the disks to ensure a complete seal.

The extruded bitumen flowed from the extruder directly into the aluminum molds. Target loadings were 40, 50, and 60 weight percent salt, while, in addition to these targets, specimens containing 26 weight percent salt were also obtained during equipment line-out.

Compressive Strength

Because the BTP suggests that waste forms exhibit a minimum unconfined compressive strengths of 50 psi (3.45 E+5 Pa), compression testing was performed. The recognized test for bituminous mixtures is the ASTM Test Method for Compressive Strength of Bituminous Mixtures (D 1074-83). This test was performed on right circular cylinders of the bitumen product molded in soft-drink cans, with a height-to-diameter ratio of 1.

The vertical rate of deformation (crosshead speed) for these tests was set at 0.127 cm/min/cm of specimen height. The Nuclear Regulatory Commission (NRC) has recommended that because bitumen flows, rather than fractures, the unconfined compressive strength should be evaluated at the point where 10% deformation in specimen height occurs.

Waste Loading

Waste-form loading was determined by dissolving the waste forms in dry xylene and aniline. Then the salts were washed repeatedly with solvent until all the organic material was dissolved. The salts were then allowed to come to constant weight in the open air, where they absorbed their normal water of hydration. Correction was made from the water content of the salts in all pertinent calculations.

The waste loading is, however, reported in terms of the total weight percent salt in equilibrium with air. Specimens were removed from the top and bottom of the waste forms and analyzed for salt content to determine if the waste forms were homogeneous in regards to salt loading.

Volume-Reduction Efficiency

We have found that the volume-reduction efficiency (VRE) correlates directly with the percentage of waste loading. We have defined the VRE in this study as the ratio of the initial liquid waste volume treated to the final volume of the waste form. As a point of reference, using grout as a fixation medium generally results in a VRE of ~0.7, which translates to an ~43% increase in volume. Comparatively, a thermal process such as bitumenization always results in a VRE greater than unity (a volume reduction) for wastes containing volatile solvents.

The reduction efficiency has been determined on the basis of the nitrate content of the solid waste forms. From the analytical concentration of the feed solution and the salt content of the waste forms, the VRE was calculated from these data.

Thermal Testing

Using a Cleveland open-cup flash-point test apparatus in accordance with the ASTM Test Method for Flash and Fire Points by Cleveland Open Cup (D 92-72), the flash point and ignition points of the waste form were determined. The apparatus was composed of an electrically heated brass cup with a propane flame above for ignition of the molten waste form. A chromel-alumel thermocouple was placed inside the cup, near the bottom, to monitor the temperature of the molten waste form. The temperature change was constantly recorded on a strip-chart recorder during the tests.

Leaching

Two kinds of regulatory leach tests were performed in this performance evaluation: the EPA's extraction procedure (EP) toxicity test, and the American Nuclear Society (ANS)-American National Standards Institution (ANSI) Measurement of the Leachability of Solidified Low-Level Radioactive Wastes by a Short-Term Procedure (16.1) recommended by the NRC.

EP toxicity tests were performed using acetic acid, according to the regulatory procedure, over 24 h. In this test, high-density polyethylene vessels were used and were connected to an automatic acetic acid delivery system composed of a burette, set-point controller, and electric solenoid valve.

The leachates from the tests were analyzed for all waste species, not only the eight regulatory metals of current concern to the EPA. Chemical analysis of the leachates from both leach tests were performed using EPA-approved methods in an EPA certified laboratory. In addition, the 90-day version of the ANSI/ANS 16.1 leach test was performed on all waste forms. The leachant used in these tests was distilled water having an electrical conductivity of <5 μho/cm at 25°C and a total organic content of <3 ppm.

Calculated leach indices from these leach tests are reported for all species after 90 days, even though this test was established primarily for radionuclides. In this way, the indices serve as a "figure of merit" which may be used for modeling the expected fractional release rate following disposal.

The ANSI/ANS 16.1 leach tests were performed in ½-L Teflon vessels prepared in accordance with the Materials Characterization Center (MCC) Static Leach Test Method-Draft (MRB-0326) cleaning procedure. The waste forms were suspended in stainless-steel wire baskets in the center of the leachant. These tests are continuing past the regulatory 90 days until essentially all the nitrate has leached from the waste forms. During the extended testing, the effects of the observed physical changes occurring to the waste forms during prolonged contact with water will continue to be investigated.

Results

Compressive Strength

One of the first tests performed was the unconfined compressive strength test. The results of the tests are shown in Table 2, with the pure, salt-free bitumen sample serving as the blank.

The regulatory minimum compressive strength is 345 kPa (50 psi). As shown, all waste forms easily exceeded this minimum, including the blank. The greater resistance offered by increasing amounts of salt is evident from the upward trend in the compressive strength with loading. The waste form containing ~60% salt was the maximum loading achievable during the demonstration; above 60% salt, the bitumen exiting the extruder became clumpy and less fluid.

Waste Loading

Salt loading is important in two ways: a high percent loading is desirable to the point that the leachability of the waste form is negatively impacted, and homogeneity along the length of the waste form is desirable because salt crystals will then be evenly coated with bitumen.

Because the density of the bulk salt, sodium nitrate, is 2.3 g/cm^3 and the bitumen density is <1 at the operating temperature, one can understand why salts might settle to the bottom of the waste form while the bitumen is cooling, especially considering that the thermal conductivity of bitumen is very low. A larger amount of salt at the bottom of the waste form may enhance leachability because the film of bitumen surrounding each salt crystal would have to be thinner and therefore present less of a barrier toward leaching. Table 3 presents the results of the salt content analyses 2 cm below the top and 2 cm above the bottom of the waste forms.

As shown in Table 3, the waste forms were very homogeneous and close to the targeted waste loading. The harder, more viscous bitumen used in immobilizing this waste likely contributed much to the degree of homogeneity obtained, together with the mixing and kneading provided by the extruder process equipment.

Volume-Reduction Efficiency

Analyzing the feed solution and comparing these data with the amount of nitrate present per unit volume of waste form, the VRE has been calculated. The results of these calculations are presented in Table 4. As shown in Table 4, the VRE is high, as one might expect of a thermal process when a relatively dilute aqueous waste stream is involved. Such a reduction can potentially offer a desirable cost savings for transportation or interim storage.

TABLE 2—*Table of unconfined compressive strength data at 10% deformation.*

Salt Loading, weight %	Compressive Strength, kPa[a]
0.0 (blank)	1720
41.6	2480
50.6	3520
60.2	4290

[a] 50 psi = 345 kPa.

TABLE 3—*Waste form salt loading distribution.*

Target Loading, weight %	Loading Achieved, weight %	
	Top	Bottom
40	41.5	41.6
50	49.7	50.6
60	59.2	60.2

Thermal Test Results: Flash and Ignition Points

Due to concerns surrounding the use of bitumen in the presence of the oxidant sodium nitrate, thermal stability tests were of interest. Results of the open-cup flash and ignition tests are presented in Table 5. The data reveal an apparent trend in the flash points and ignition points that is not consistent with expectations. For samples containing larger amounts of salt, it was expected that both the flash point and ignition point would increase with higher salt loading. Instead, the flash point appears to drop while the ignition point is raised as loading is increased.

These tests are subject to problems since gassing of nitrogen oxides raises the level of the molten bitumen in the cup, thereby resulting in premature flash points. Material was constantly removed to maintain a constant level during the tests.

In addition, as the temperatures of the bitumen increases during the test, its density steadily decreases becoming less than 1 g/cm^3 compared with the density of the sodium nitrate salt. As a result, the salt must move to the bottom of the cup with time and in this way influences the results since the material becomes more nonhomogeneous with time.

During the course of the flash and ignition point tests, at the higher temperatures, on one occasion the bitumen was observed to burn rapidly due to reaction with the nitrates. Because of this, a chromel-alumel thermocouple was placed inside, near the bottom of the cup to monitor the temperature on a strip chart recorder. The temperature profile results of such a controlled burn is presented in Fig. 1.

As the strip chart profile of the burn shows in Fig. 1, the spontaneous burn occurred at approximately 320°C, as indicated by the rapid movement of the pen off the chart paper. Although this burn occurred around this temperature, the highest temperature obtained along the length of the evaporator-extruder during the preparation of the waste forms during the demonstration was 177°C.

Of the controlled burns that occurred, note that the combustion temperature was always just above the melting point of sodium nitrate (307°C). Under the conditions of this test, the sodium nitrate was pooled on the bottom of the brass cup, with the molten bitumen on top. A molten sample, which was allowed to cool to room temperature, revealed that the salts formed a distinct solid phase on the bottom of the cup.

TABLE 4—*Calculated volumetric reduction efficiency ratio based on waste form nitrate loading.*

Salt Loading, weight %	Efficiency Ratio, unitless
41.6	1.51
50.6	1.68
60.2	2.43

TABLE 5—*Open-cup, flash, and ignition-point test data as function of salt loading.*

Salt Loading, weight %	Flash point, °C	Ignition Point, °C
blank	310	343
41.6	296	304
50.6	293	327
60.2	288	332

Data resulting from these tests are complicated by a number of uncontrollable variables. The influence of the hot brass cup on initiating the combustion, as well as the settling of salts in a medium having poor heat transfer properties, cannot be easily assessed.

In addition to these tests, thermal gravimetric (TG) and differential thermal analyses (DTA) were performed on samples of the waste forms at all levels of salt loading. Data revealed that all curves looked nearly identical, and, therefore, results appeared to be insensitive to the level of nitrate salt loading. The overlaid curves of the DTA and TG are presented in Fig. 2 for a waste form specimen containing 60 weight percent salt and with a heating rate of 10°C/min under a nitrogen atmosphere.

Figure 2 shows that two exotherms occur—one starts at ~300°C and the other, larger exotherm at 415°C. Interestingly, the first exotherm occurred at approximately the melting point of sodium nitrate. As the TG curve shows, the total sample weight loss was ~46% over the duration of both exotherms.

Similar thermal stability studies performed by researchers in the People's Republic of China [*4*] corroborate our observations. For upon heating a 40% sodium nitrate-bitumen sample, two exotherms were observed—the first exotherm occurred at 375°C, while the

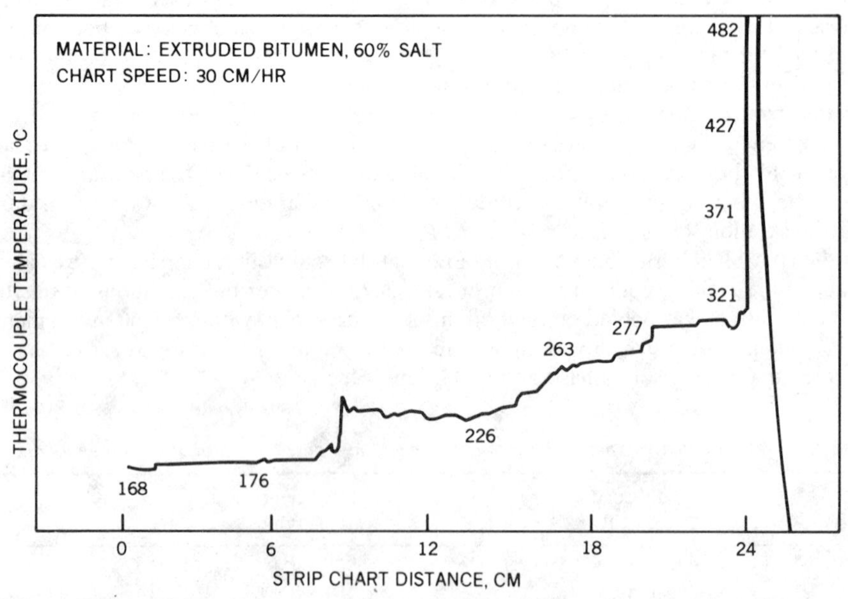

FIG. 1—*Temperature profile of a controlled burn of extruded bitumen containing 60 weight percent nitrate-based salt.*

second occurred near 448°C. Their combustion temperature was said to occur between 410 and 470°C. This work also reports the initial formation of ammonia, the reduction product, at 410°C, along with other gases such as N_2, NO_x, CH_4, and CO_2.

Leaching

EP Toxicity—Because the Melton Valley waste may be considered a mixed waste, interest in the leachability of the eight EPA metals is of interest. Planned treatment schemes incorporating "ultrafiltration" may cause the alkaline waste not to be classified in this way prior to waste fixation. For this reason, regulatory EP toxicity tests were performed on all waste forms, including a waste form obtained during the fixation demonstration which only contained 26.1% salt. Tests performed with this waste form permitted us to establish the lower limit for waste loading and to determine the resulting nitrate present in the EP toxicity test leachate as a function of salt loading.

Although the concentration of nitrate is not of concern in this regulatory test, we have tried to meet drinking water standards [5] in the leachate. The exact application of this standard for nitrate is still being debated among regulators, as well as the associated point of compliance in a waste disposal area. Without guidance, we have used the regulatory limit of 10-ppm nitrogen or 44.4-ppm nitrate as our upper limit. The results of our EP toxicity tests are presented in Table 6.

As the data in Table 6 show, all waste forms easily passed the EP toxicity test. The data show that for most of the metals, the concentration in the leachates was at or near the analytical detection limit.

The drinking water standard for nitrate, as applied, was exceeded for all waste forms, except for that waste specimen containing 26.1% salt. The sensitivity of the leachability of the very mobile nitrate anion to waste form loading is probably a result of differing bitumen wall thicknesses surrounding salt crystals at various levels of loading.

In an effort to interpolate that level of salt loading likely to pass the nitrate drinking water

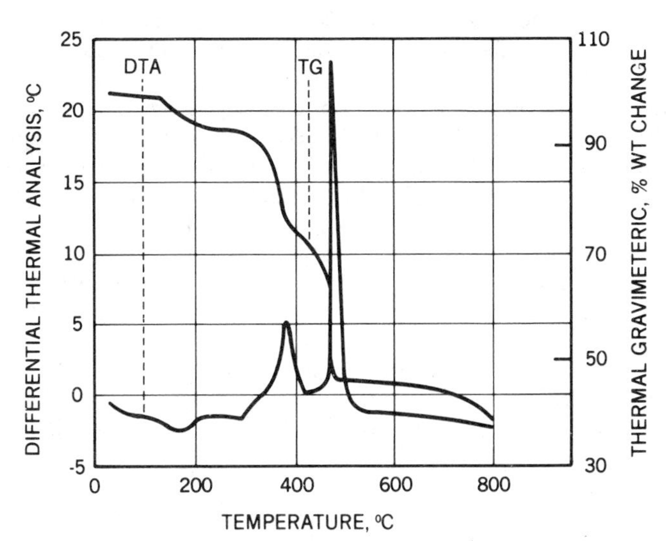

FIG. 2—*Thermal gravimetric and differential thermal analysis scans of extruded bitumen containing 60 weight percent nitrate salts.*

TABLE 6—*Comparison of EP toxicity test results as function of salt loading.*

Loading, weight %	Concentration, ppm				
Species	26.1	41.6	50.6	60.2	Limit
arsenic	0.008	<0.005	<0.005	<0.005	5
barium	0.017	0.010	0.024	0.049	100
cadmium	<0.0030	0.0032	0.0081	0.024	1
chromium	<0.010	<0.010	<0.010	<0.010	5
lead	0.010	0.033	0.010	0.016	5
mercury	<0.0002	0.0003	0.0003	<0.0002	0.2
selenium	<0.0050	0.027	<0.005	0.005	1
silver	<0.0060	<0.0060	<0.0060	<0.0060	5
nitrate[a]	12.9	54	160	320	44.3

[a] Nitrate is not of regulatory concern in the EP toxicity test.

standard, the data in Table 6 were correlated. Plotting the logarithm of the nitrate concentration in the leachate versus the waste form salt loading, a nearly straight line is obtained.

Interpolation reveals that a waste form loading as high as 38% theoretically will pass the drinking water standards. This observation, coupled with the fact that the exact application of the nitrate concentration limit by regulators is currently uncertain, is promising.

ANSI/ANS 16.1 Test—The NRC requires that a radionuclide possess a leach index of at least 6 when leached in accordance with the ANS 16.1 leach procedure. Because the leach index is the negative logarithm of the effective diffusion coefficient, it can be used in modeling the expected leaching behavior of any waste species if certain assumptions are made (ANSI/ANS 16.1). The leach index then becomes a "figure of merit" used to ascertain the ability of a given waste form or treatment scheme to impede the leachability of a waste component.

Because the leach indices are based on a logarithmic relationship, each unit change in the leach index is a change in the effective diffusion coefficient of a factor of 10. The higher the leach index, the better the resistance of the waste form toward leaching of that waste species.

The resultant leach indices obtained upon leaching for 90 days in distilled water are presented in Table 7. From the data, one sees that both cesium and strontium exceeded the NRC's minimum leach index of 6. Applying this limit to all remaining species, even though they are not radionuclides, one sees that they have also favorably exceeded this leach index by a large margin.

The greater-than symbols preceding some of the indices are present as a result of analytical detection limit constraints. As a consequence of nonradiological species, the less-than concentrations becomes greater-than symbols when reporting the indices due to detection limit constraints.

With the exception of those indices in Table 7 which are preceded by a greater than symbol, a general trend toward lower leach indices with increased salt loading is apparent. This trend is probably the result of thinner walls of bitumen between salt crystals when salt loading is increased.

Leaching Mechanism and Surface Changes—The exact mechanism involved in the leaching of bitumen-based waste forms is unknown. The authors believe, however, that activated diffusion of water into the bitumen precedes the entry and dissolution of salt inside.

Figure 3 depicts water passing through the bitumenous material into an area containing

TABLE 7—*Tabulation of average leach indices after 90 days of leaching in distilled water.*

Loading, weight %	41.6		50.6		60.2	
	Average[a]	S.D.[b]	Average[a]	S.D.[b]	Average[a]	S.D.[b]
arsenic	>10.9	0.8	>11.1	0.8	>11.3	0.8
barium	10.7	0.5	10.2	0.4	9.8	0.3
cadmium	>10.7	0.3	10.6	0.3	10.4	0.5
calcium	10.5	0.4	9.8	0.3	9.4	0.3
cesium	> 8.0	0.4	> 8.4	0.7	> 8.6	0.8
chloride	> 9.7	0.7	> 9.7	0.5	> 9.6	0.4
chromium	>10.3	0.8	>10.5	0.8	>10.7	0.8
lead	9.7	1.1	10.0	0.9	10.2	0.9
magnesium	10.7	0.5	10.5	0.5	10.4	0.5
mercury	>13.7	0.8	>13.9	0.8	>14.1	0.8
nitrate	10.6	0.5	9.8	0.2	9.3	0.3
selenium	>10.9	0.8	>11.1	0.7	>11.2	0.7
silver	>10.8	0.8	>11.0	0.8	>11.2	0.8
sodium	10.4	0.4	9.5	0.2	9.2	0.3
strontium	10.7	0.5	10.0	0.3	9.4	0.3

[a] Arithmetic average of three replicates.
[b] Root-mean-square of the standard deviations of the replicates.

salt crystals. When the water dissolves the salt, the resulting saturated solution expands into a cavity, forming through a tearing action, interconnecting channels to the outer surface. This schematic demonstrates what is believed to be occurring during the leaching process.

After 90 days of leaching in distilled water, the waste forms were observed to evaluate surface changes. Waste forms containing 60% salt after 90 days had become swollen with the outer surface grooved and spongy. When the outer surface was pared away, a hard surface was observed underneath. Similar results were obtained at Chalk River Nuclear Laboratories in Canada [*6*] with sodium nitrate; however, little swelling was observed in their work. The swelling and outer surface effect are attributable to osmotic pressure effects.

Bitumen-based waste forms containing water-soluble salts are known to swell as a result of osmotic forces which are in effect when the waste form is immersed in water [*7*]. These forces can be related to those involved when two solutions of different concentration are separated by a semipermeable membrane.

The vapor pressure of pure water at 25°C is 23.8 mm of mercury, while the vapor pressure of water above a saturated solution of sodium nitrate is only 17.8 mm of mercury at this temperature [*8*]. This difference in pressure can result in an osmotic pressure maximum as high as 390 atm when calculated using a modified Van't Hoff relationship [*9*]. Equation 1 was used to calculate the osmotic pressure maximum:

$$\pi = \frac{RT}{\overline{V}} \ln (P^{\circ}/P) \tag{1}$$

where

π = osmotic pressure, atm,
R = gas constant, atm/deg·mol,
$\overline{V}$ = partial molal volume of solvent, cm^3/mol,
P° = vapor pressure of pure water at temperature, mmHg,
P = vapor pressure of solution at temperature, mmHg, and
T = temperature, K.

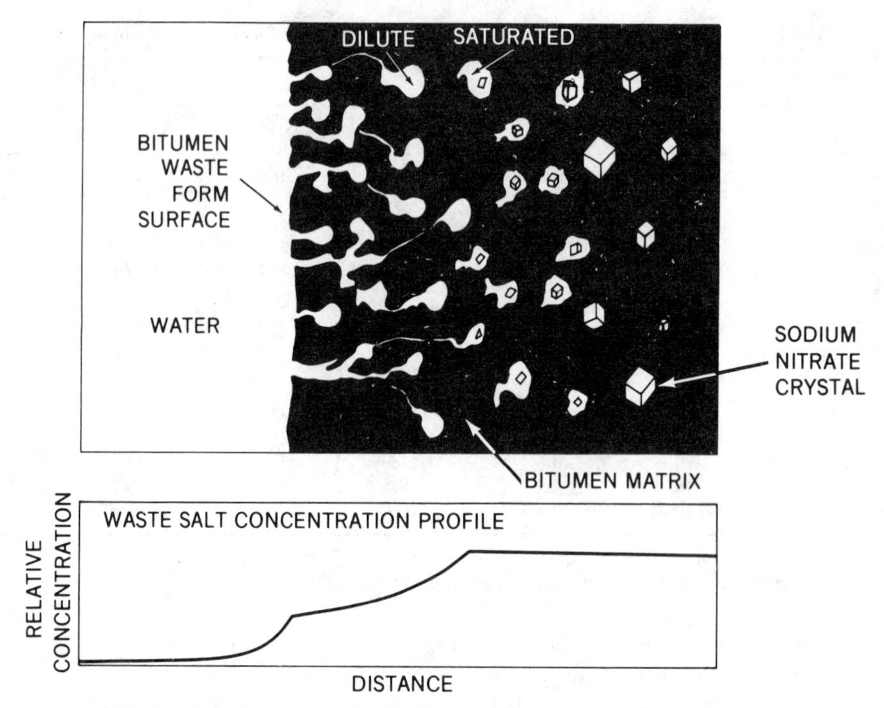

FIG. 3—*Staged sequence of sodium nitrate leaching from a bitumen matrix.*

The calculation of the osmotic pressure, based on solvent vapor pressure depression, is an accurate means of calculating this parameter in concentrated solutions. The use of water vapor pressure-reduction measurements should be considered to gage whether osmotic forces will be a concern, especially when highly soluble salts are present.

The twin screw extruder imparts a high shear on the salt which reduces the particle size, while providing excellent mixing with bitumen. These two actions permit efficient coating of the salt particles. Lower salt loading or use of softer asphalt types (Type I or II) or both with the extruder may reduce the swelling and improve leach resistance.

Work performed at Chalk River has also revealed the potential for swelling when anhydrous salts such as sodium carbonate are present, for this salt will take up a large amount of water of crystallization. Bitumen containing 60% sodium carbonate was observed to swell to twice its original size in 24 h when soaked in water.

Such salts as sodium carbonate or sodium sulfate should not be used at high waste loadings without first considering methathesis to another salt not containing as much water of hydration. Effects such as osmosis and rehydration of anhydrous salts must not be overlooked when using bitumen.

Conclusions

Regulatory performance testing of extruded bitumen has shown that the relatively viscous form of oxidized bitumen used has been able to meet all required performance tests. The extruder technology used to prepare the test specimens was shown to combine both superior physical and thermal processing capabilities in one unit, thereby permitting the mixing of high-viscosity materials in an extremely homogeneous manner.

Even at relatively high levels of waste loading, the waste forms were capable of offering superior resistance to the leachability of those current EPA metals of current concern while showing promising results for very problematic nitrate.

Thermal analysis of the waste forms has revealed that the material is capable of combustion at temperatures which are almost twice as high as the highest process temperature and greater than the melting point of sodium nitrate. The magnitude of the flash-point and ignition-point data were also found to generally agree with data reported elsewhere.

The high VRE observed with this type of thermal waste fixation process undoubtedly offers potential transportation and storage cost savings. Long-term leach testing is continuing in order to observe the effects of the physical changes occurring to the waste forms, as described, over a period of one year.

Acknowledgment

This research was sponsored by the Office of Defense Waste and Transportation Management, U.S. Department of Energy, under Contract No. DE-AC05-84OR21400 with Martin Marietta Energy Systems, Inc.

References

[1] "Low-Level Waste Licensing Branch Technical Position on Waste Form," U.S. Nuclear Regulatory Commission, May 1983.

[2] "Licensing Requirements for Land Disposal of Radioactive Waste," 10 CFR 61, *Code of Federal Regulations*, U.S. Government Printing Office, Washington, DC, 1987.

[3] "Test Methods for Evaluating Solid Waste—Physical/Chemical Methods," SW-846, 2nd ed., U.S. Environmental Protection Agency, 1982.

[4] Xu, Y.-C., Wang, S.-J., and Luo, S.-G., "Study of the Thermal Stability of Sodium Nitrate Bitumenized Waste System," *Heh Hua Hsueh Yu Fang She Hua Hsueh*, Vol. 2, No. 1, Peking, People's Republic of China, 1980, pp. 28–38.

[5] "National Interim Primary Drinking Water Regulations," 40 CFR 141, *Code of Federal Regulations*, U.S. Environmental Protection Agency, 1983.

[6] Buckley, L. P., "Waste Packages and Engineered Barriers for the Chalk River Nuclear Laboratories Disposal Program," *Proceedings of the International Seminar on Radioactive Waste Products—Suitability for Final Disposal*, 1985, pp. 184–201.

[7] Broderson, K., "The Influence of Water Uptake on the Long-Term Stability of Conditioned Waste," *Radioactive Waste Management—A Series of Monographs and Tracts, Testing, Evaluation and Shallow Land Burial of Low and Medium Radioactive Waste Forms*, Vol. 13, Harwood Academic Publishers, New York, 1986.

[8] *International Critical Tables*, 1st ed., Vol. 3, National Research Council, McGraw-Hill, New York, 1928.

[9] Farrington, D. and Alberty, R. A., *Physical Chemistry*, 3rd ed., Wiley, New York, 1956, pp. 223–224.

Samuel L. Unger,[1] Rodney W. Telles,[1] and Hyman R. Lubowitz[1]

Surface Encapsulation Process for Stabilizing Intractable Contaminants

REFERENCE: Unger, S. L., Telles, R. W., and Lubowitz, H. R., **"Surface Encapsulation Process for Stabilizing Intractable Contaminants,"** *Environmental Aspects of Stabilization and Solidification of Hazardous and Radioactive Wastes, ASTM STP 1033,* P. L. Côté and T. M. Gilliam, Eds., American Society for Testing and Materials, Philadelphia, 1989, pp. 40–52.

ABSTRACT: This paper reports advanced technology for stabilization/solidification of pollutants for their final disposal in the earth. Pollutants are secured in modules fabricated with organic resins. The modules exhibit high-performance pollutant stability versus stresses of handling, transportation, and final disposal. They are set forth for managing pollutants because present stabilization/solidification products do not satisfactorily address such stresses.

Organic resins are materials of choice for module fabrication because of their excellent chemical stability. But cost-effective, high-performance pollutant stabilization/solidification by resin modules requires judicious resin selection and module design. Polybutadiene resin was identified as an excellent material for binding pollutant particles and forming aggomerates. Agglomerates holding more than 90% by weight pollutants were readily prepared. They were reinforced by fusing polyethylene resin pellets onto their surfaces, thereby encapsulating aggomerates by seamless resin jackets.

Testing of laboratory-scale modules holding toxic wastes demonstrated their outstanding stability under harsh physical, chemical, and mechanical stresses. Based on successful stabilization/solidification of many types of pollutants on a laboratory-scale, the process was scaled up under Department of Energy sponsorship. The work included design and construction of a prototype apparatus for fabricating commercial-size modules. A moldule is cylindrical in shape, has a capacity of 182 L, and is 61 cm high and 61 cm in diameter. In securing sodium sulfate, a pollutant formed in nuclear plants, the module weighs 325 kg, holds 87% by weight salt in a thermoset polybutadiene matrix, and is reinforced by seamless, 0.95-cm-thick, high-density polyethylene jackets.

KEY WORDS: toxic wastes, low-level radioactive wastes, sludges, stabilization, solidification, encapsulation

Waste-reduction processes, such as incineration and precipitation, do not totally eliminate toxic material and thus yield pollutants. Because stabilizing and solidifying the pollutants for final disposal in the earth, in many cases, is the only viable option for their management, there is a vital need for relevant advanced, cost-effective technology.

Stabilization/solidification (S/S) circumvents dispersion of pollutants into the ecology by preventing their dissolution by leachates and entrainment by water action. The first consideration is effective stabilization by treating pollutants to resist the solubilizing action of regulatory-specified leachates. But leachates vary in composition, and thus pollutants effectively stabilized for laboratory-formulated leachates may not be similarly stabilized for leachates that exist in the ecology. Furthermore, it may not be possible to treat many pollutants to prevent their dissolution by leachates.

[1] Chief engineer, chemical analyst, and general manager, respectively, Environmental Protection Polymers, Inc., 13414 Prairie Ave., Hawthorne, CA 90250.

The second consideration is solidification to prevent entrainment of pollutants by the physical action of water. Because pollutant migration in water is facilitated by peptizing, pollutant migration potential is greater than that estimated due to physical force of water. This enhanced entrainment is further augmented by physical and mechanical stresses such as freeze-thaw, wet-dry, impact, and compression. These stresses reduce the size of pollutant particles and sensitize particle surfaces, chemically causing particle dispersion in water.

In our work, considerations regarding dissolution and entrainment of pollutants are addressed by housing pollutants in monolithic structures called modules. To secure pollutants under all conditions, modules must exhibit stability to mechanical, physical, and chemical stresses of handling, transportation, and final disposal. Because current products do not provide high-performance pollutant S/S, the purpose of the work is to realize this goal.

The Surface Encapsulation Process was invented and developed for fabricating high-performance modules, whose advanced stability is due to their unique composite structure fashioned from selected organic resins. To form modules, pollutant particles are agglomerated by thermosetting resin and encapsulated by thermoplastic resin. Polybutadiene is employed for aggomeration because it tolerates high concentrations of many different types of pollutants and yields tough aggomerates. They are then encapsulated by polyethylene. This resin is employed for agglomerate encapsulation because it is mechanically tough, resistant to chemical corrosion, and impermeable to leachates. Polyethylene is applied onto the agglomerate surfaces by thermal fusion of commercial pellets. The module thus consists of a core holding pollutants in chemically cross-linked polybutadiene, and a thick, seamless, polyethylene jacket that encapsulates the core.

Several performance advantages stem from the composite configuration of modules: Landfill overburden is carried by the core rather than the jacket, thus modules provide greater load bearing capability for overburden than containers. Product deformation and creep due to compressive loads are mitigated by the chemically cross-linked nature of the core resin. Resin encapsulation of the core markedly advances product mechanical properties, particularly impact strength. Chemical stability is achieved by the chemical nature of polybutadiene and polyethylene, and thus product corrosion is addressed, notwithstanding the multiplicity of chemical species in the environment that may affect the modules. Water penetration of modules is mitigated by the hydrophobic nature of the resins due to their aliphatic composition, and modules thereby effectively stabilize hydrophilic pollutants. The above advantages are realized in the context of preferred product performance and cost parameters:

(*a*) high-performance modules of toxic wastes,
(*b*) minimal volume increase for modules relative to bulk volume of pollutants,
(*c*) simple processing operations,
(*d*) processability of many different waste types,
(*e*) excellent product reproducibility, and
(*f*) employment of low cost, commercial resins.

Laboratory-Scale Investigations

Resin modules with unique composite structure were fabricated to stabilize toxic wastes. They are cylindrical 7.6-cm (3-in.) diameter by 10.2-cm (4-in.) high monoliths. Seamless resin jackets encapsulate cores that contain pollutants up to 90% by weight. Cores were formed by blending pollutant particles with polybutadiene resin and then agglomerating the mixture by thermosetting reaction. Encapsulation was carried out by fusing polyethylene resin onto the agglomerate surfaces to form a 0.6-cm (¼-in.) thick seamless jacket. An important aspect of this operation concerns the heat distortion temperature of the core material. It is greater than the polyethylene resin fusion temperature due to the chemically cross-linked polybutadiene matrix. Thus, polyethylene encapsulation can be carried out without causing module distortion. The resin jackets mechanically reinforce the agglomerates and advance performance of waste modules versus handling and environmental stresses.

Sludge Management

Toxic waste sludge treatment involves two steps. In the first step, the contaminants are chemically stabilized and the sludge solidified. Chemical stabilization of contaminants addresses requirements for delisting toxic wastes. Although modules exhibit excellent stabilization of contaminants under harsh leaching stresses, delisting may require grinding modules and therewith showing contaminant stability in the leachates. Therefore, the sludges are chemically treated to stabilize pollutants versus leachates specified in the regulations. Precipitation of heavy metals as sulfides and polysilicates has been successful generally for meeting Extraction Procedure (EP) toxicity and Toxicity Characteristics Leaching Procedure (TCLP) requirements. Absorbents are employed to stabilize organic liquid pollutants.

Solidification converts the sludge into a solid state that is free of mobile liquid. In the second step, the solidified sludges are particulated and the particles are agglomerated by polybutadiene binder and encapsulated by high density polyethylene. Figures 1 through 4 show products in fabricating laboratory size modules securing heavy metal sludges. Table 1 lists the wide variety of wastes managed on a laboratory scale.

Low-cost dehydrating agents such as lime, kiln dust, or portland cement are employed for sludge solidification. This operation does not increase the volume of the sludges significantly because the quantity of agents needed to dewater the sludges is not appreciable. The limited amount of dehydrating agent employed does not give rise to load bearing solids; the cured mixtures are very friable and they particulate readily. The pollutant particles are then agglomerated and encapsulated.

FIG. 1—*Aqueous sludge in container holding mineral and small amounts of organic contaminants.*

FIG. 2—*Particulated sludge solidified by lime.*

FIG. 3—*Sludge particles agglomerated by polybutadiene.*

FIG. 4—*Agglomerate reinforced by polyethylene encapsulation.*

TABLE 1—*Examples of toxic waste sludges treated by surface encapsulation.*

Source	Major Contaminants
Salt cake from wastewater evaporators	radionuclides[a]
Ion-exchange resins for wastewater demineralization	radionuclides[a]
Spent activated carbon	radionuclides,[a] heavy metals
Ion-exchange glasses for wastewater demineralization	radionuclides[a]
Pesticides	monosodium methanearsonate
Electroplating sludge	Cu, Cr, Zn
Nickel-cadmium battery sludge	Ni, Cd
Pigment production sludge	Cr, Fe, Cn^-
Chlorine production brine sludge	Hg, Na, Cl^-
SO_x scrubber sludge, double alkali process, western coal	Cu, Na, $SO_4^=/SO_3^=$
SO_x scrubber sludge, limestone process, eastern coal	Cu, $SO_4^=/SO_3^=$, heavy metals
SO_x scrubber sludge, lime process, eastern coal	Ca, $SO_4^=/SO_3^=$, heavy metals
SO_x scrubber sludge, double alkali process, eastern coal	Na, Ca, $SO_4^=/SO_3^=$, heavy metals
SO_x scrubber sludge, limestone process, western coal	Ca, $SO_4^=/So_3^=$, heavy metals
Calcium fluoride sludge	Ca, F^-
Mixed waste sludge	heavy metals, radionuclides[a]

[a] Simulants.

Binder Properties

The polybutadiene binder resins thermoset in the presence of many different contaminants to yield a chemical structure analogous to chemically cross-linked polyethylene. The binder shows excellent chemical stability and withstands mechanical stresses without distortion due to creep. The low surface energy of liquid polybutadiene facilitates good wet-out of pollutant particles, thus yielding agglomerates was high waste loadings.

The ability of polybutadiene to react in the presence of pollutants and to withstand pH and compositional variations of input material is due to the resin's water incompatibility (hydrophobicity) and high functionality. Since the resin does not absorb polar and mineral impurities, these reaction-interfering materials do not materialize at the reaction sites to impede curing of resin/waste mixtures. Furthermore, the high functionality of the resin gives rise to three-dimensional resin networks at relatively low extents of reaction, thereby readily yielding dimensionally stable agglomerates.

The nonpolar nature of the resin's functional groups provides additional processing benefits. In contrast to polar functional groups of resins such as epoxy and urea-formaldehyde, whose reactivities are sensitive over a broad temperature range including atmospheric temperatures, the nonpolar groups are stable up to a threshold temperature that is moderately above atmospheric temperatures thereon they are triggered into reactivity by free radical initiators. This feature permits blending of resin and particles to be carried out at atmospheric temperatures without being impeded by increasing viscosity and premature gelling of the mixture. By alleviating these operations from time-sensitive constraints, one gains the options of storing, adjusting, and inspecting the mixture prior to formation of dimensionally stable agglomerates. In addition, mixtures can be made by pouring resin onto a bed of pollutants, thus eliminating mechanical mixing.

Upon the initiation of the agglomerating reaction of mixtures of pollutants and polybutadiene, the temperature of the waste/resin does not significantly increase during resin thermosetting because the cure rate of mixtures depends upon the heat transfer rate. The mixtures cure incrementally with the surrounding material, providing a heat sink for heat generated at the reaction site. In contrast, fixatives such as thermosetting polyesters, epoxides, and cements begin to react upon addition of the reactants; consequently, the mixtures cure *en toto* rather than incrementally. The simultaneous release of chemical energy causes a significant increase in the temperature of the mixtures that may lead to process-control problems. Polybutadiene mixtures, on the other hand, exhibit reaction exotherm temperatures that do not significantly exceed the reaction initiation temperatures.

Module Properties

Microscopic examination of modules showed 0.6-cm (¼-in.) thick high-density polyethylene (HDPE) jackets to be free of flaws. The modules exhibited dimensional stability under compressive loads of about 12.4 MPa (1800 psi) [*1*]. Appreciable distortion of modules by severe compressive loads did not produce flaws; modules remained watertight and contaminants secure. Figures 5 to 8 show module behavior under uniaxial compression, partially compressed, and then compressed to 20% of its original height; the module remained intact. The compressed modules recovered to 90% of their original height upon release of the compressive load. Figure 9 shows a module greatly compressed laterally, a severe mode of compression. In this mode of compression the module remained intact. The unconstrained modules regained over 80% of the dimensions of their original state [*1*].

Uniaxial compression strengths and toughness of resin modules holding high content of pollutants exceeded that of pozzolanics holding low pollutant content. The modules' cores sustained the applied mechanical load and resisted creep due to the chemical cross-linked

FIG. 5—*Onset of vertical compression of module.*

structure of the resin. Dropped onto a steel plate, the modules remained intact on impact. Modules also resisted penetration by a pointed steel probe because the resin jackets are tough and reinforced by impregnation onto the surfaces of the cores. It was noted that modules exhibited high damage tolerance in withstanding severe impact and puncture stresses. In every situation, no spillage of pollutants occurred due to their being locked in the core matrix resin [*1*].

Modules were dimensionally stable in leachate immersion tests, notwithstanding the waste type managed or its concentration. During immersion of ion-exchange resin (IER) laden modules, no dimensional change occurred. Figure 10 shows module stability during a 45-

FIG. 6—*Partial compression of module.*

FIG. 7—*Module compressed to 20% of original height with no jacket rupture.*

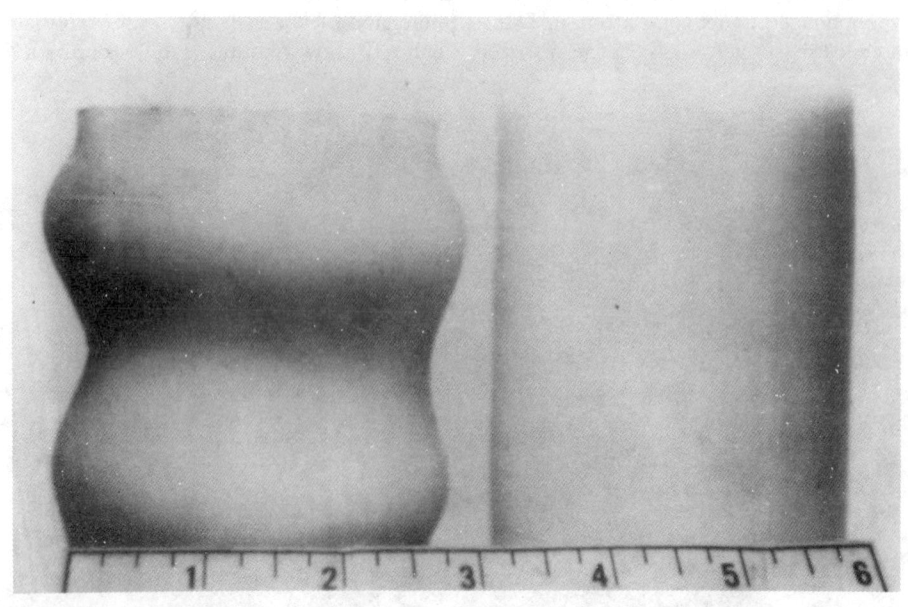

FIG. 8—*Comparison of module after release of compressive load* (left) *with unperturbed module* (right).

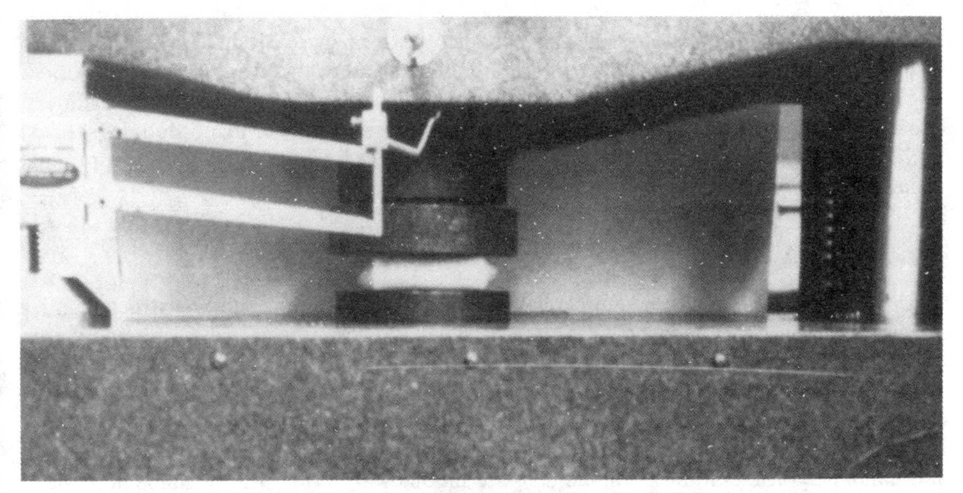

FIG. 9—*Module greatly compressed laterally with no jacket rupture.*

day water immersion test. These modules held 70% by weight IER. In comparative tests, IER-laden modules fabricated with asphalt and cement underwent appreciable dimensional change, the former swelling and the latter cracking and spalling. These modules contained only 20% and 8% IER, respectively [*2,3*].

In Environmental Protection Agency (EPA) -sponsored work, static leaching of modules holding concentrated contaminants was carried out in leachates that were harsher than those currently specified by EPA or the Nuclear Regulatory Commission (NRC). The leaching solutions included strong mineral acids and bases, simulated seawater, an organic solvent, and organic acid. The studies showed excellent retention by modules for highly mobile heavy metals [*1*].

The Army Corps of Engineers conducted column leaching tests for a two-year period on surface encapsulation modules and comparative ones for other solidified wastes from various laboratories; the modules exhibited excellent performance [*4*]. Testing of modules was also

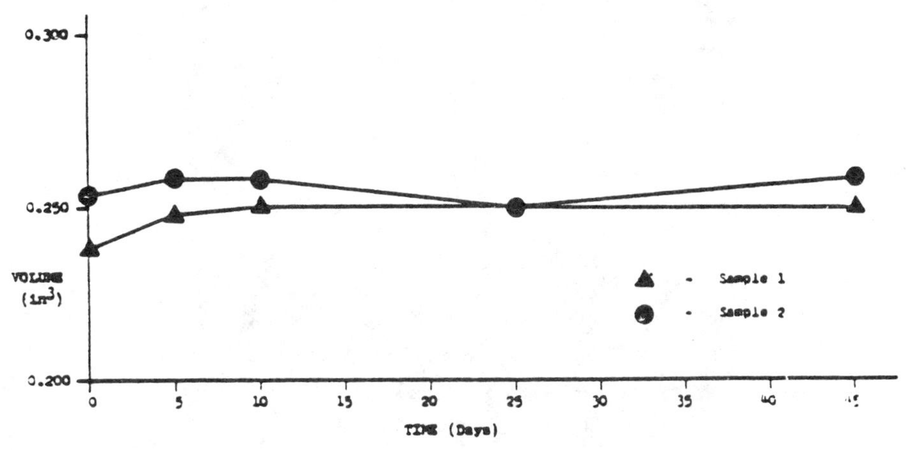

FIG. 10—*Dimensional stability of modules immersed in water. Sample volume versus duration of immersion in water (stabilized ion-exchange resin powders).*

carried out to assess their physical and chemical performance. Modules were subject to wet/dry and freeze/thaw cycling. They were frequently water-soaked and then rapidly dried. They were cycled from −10°C salt/ice bath to 100°C boiling water, with cycles lasting 15 min at a rate of 16 cycles/day—modules were stored in a freezer overnight. Modules showed no visible signs of damage, and their compressive strengths were within 95% of their original values.

Scale-Up Development

An apparatus for fabricating commercial-size modules was designed and constructed under Department of Energy sponsorship. Large modules give rise to cost-effective employment of resins and processing operations. The first pollutant managed in scale-up development was sodium sulfate salt, a low-level radioactive contaminant that is a byproduct of wastewater evaporation in nucler plants.

Commercial-scale module size and configuration were determined by product processing and handling considerations. Commercial-size modules are cylindrical in shape and 182 L (48 gal) in volume. The module shape approximates that of a 55-gal drum, thus making

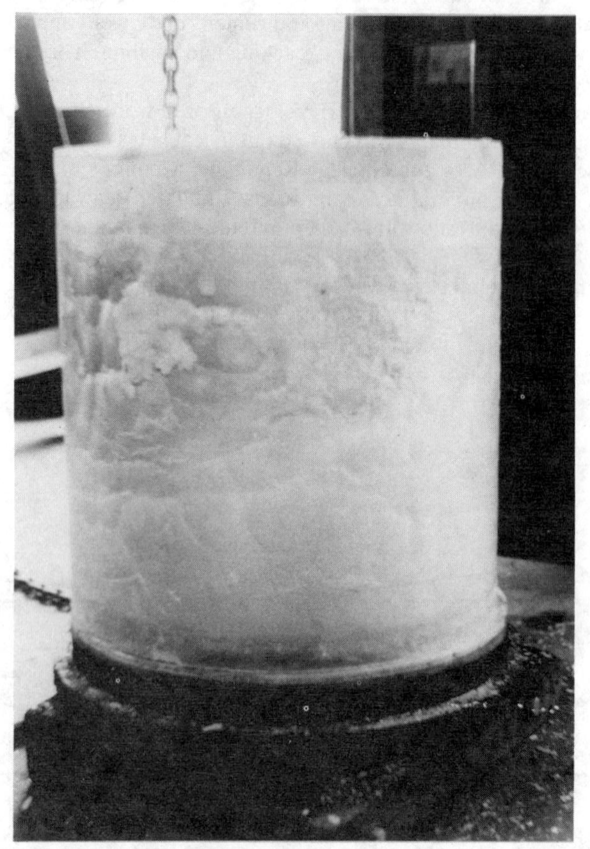

FIG. 11—*Toxic waste module fabrication (module 324 kg, 61 by 61 cm): Na_2SO_4 agglomerated by polybutadiene resin.*

FIG. 12—*Toxic waste module fabrication (module 324 kg, 61 by 61 cm): agglomerate encapsulated by 9.5-mm-thick, black high-density polyethylene.*

modules manageable by conventional drum-handling equipment. Figure 11 shows a module core consisting of agglomerated sodium sulfate. Figure 12 shows the complete module, 61 cm (2 ft) in diameter by 61 cm (2 ft) in height. Additional module specifications are

(*a*) weight, 325 kg (715 lb),
(*b*) content, 87% by weight salt, and
(c) exterior, 0.95-cm-thick (3/8 in.) HDPE, with 2% carbon black.

Toxic Waste Processing

Figure 13 provides a schematic for managing sludges. Pollutants are stabilized and sludges dewatered by dehydrating agents. The particles are loaded into the agglomerating mold by an auger, with polybutadiene binder introduced by a metering pump. With confinement and moderate temperature, agglomerates of particulated wastes are fabricated in approximately two hours. The agglomerates are then encapsulated by molding polyethylene resin onto their surfaces. This is carried out in a matched die mold providing interstitial space for introducing resin powder or beads.

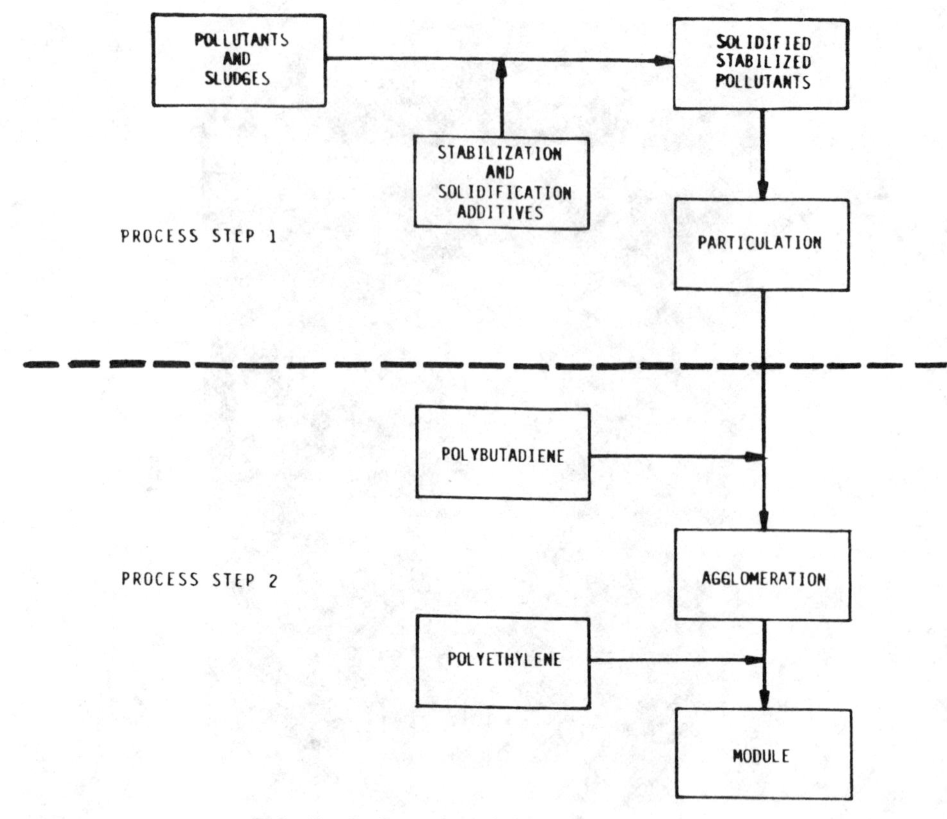

FIG. 13—*Surface encapsulation process schematic.*

Apparatus

Figure 14 shows the apparatus and operator stand. The only utility requirements are electricity (440 V, three-phase) and tap water for cooling the molds. The apparatus features

(*a*) molds mounted on a single frame,
(*b*) an indexing table for product manipulation, and
(*c*) hydraulics for mold actuation.

The molds for agglomeration and encapsulation have electrical band heaters and drilled-in channels for water cooling. The agglomerating mold is a cylindrical steel shell 60 cm (23.5 in.) in diameter. A platen fits inside the mold body to confine the waste/binder mixture. A linear positioning table transports the fabricated agglomerate to the encapsulating mold. The jacketing mold is split vertically to facilitate product demolding.

The apparatus is remotely operable by a microprocessor which controls the following functions: actuating and heating the molds, maintaining mold temperature, regulating cooling water flow, indexing the agglomerate to the encapsulating mold, and indexing the module to the unloading station.

FIG. 14—*Surface encapsulation process apparatus.*

Material Cost Estimate

The Surface Encapsulation Process employs low-cost, mass-produced resins. For pollutant agglomeration, the polybutadiene employed is the type used to coat steel automobile frames to make them corrosion-proof. The polyethylene employed for module encapsulation is the type marketed to the rotomolding and blow-molding industries for producing holding and transportation containers for fertilizers and harsh chemicals. Resin cost of $80.00 is estimated in order to fabricate a 182-L (48 gal) module. Since crude polybutadiene rather than the refined polybutadiene is employable for module fabrication, there is a potential to significantly reduce resin costs.

Material costs for current low-level radioactive waste (LLRW) fixation processes and containerization were compared to those for surface encapsulation. Table 2 presents material costs for fixation processes employing cement, asphalt, polyester, polybutadiene/polyethylene (surface encapsulation), and high-integrity containers (HIC). Costs for materials and HIC are based on vendor quotations. The values given for ratios of materials to wastes represent those reported in the technical literature for managing spent ion-exchange resins.

The cost advantage for module fabrication stems from the minimal amount of materials employed. This advantage manifests in managing many other toxic wastes as well as spent ion-exchange resins. In addition to cost advantages for materials, surface encapsulation yields modules that incur lower transportation and final disposal costs due to their management of concentrated pollutants; their high performance in service also will provide additional cost advantages.

TABLE 2—*Comparative material costs for LLRW management.*

Materials	Fixative Material Costs, \$/lb[a]	Materials/ Ion-Exchange Resins Weight Ratio	Management Cost for Ion-Exchange Resins, \$/lb[a] waste
Cementitious sodium silicate	0.06	10:1	0.60
Asphalt	0.20	3:2	0.30
Polyester	0.75	2:3	0.50
Surface encapsulation	1.25	2:10	0.24
High-integrity containers	225 (per 55-gal drum)	420 lb/drum[a]	0.60

[a] 1 lb = 0.45 kg.
[b] 1 gal = 3.8 L.

Conclusions

The Surface Encapsulation Process provides a unique option for high-performance, cost-effective stabilization/solidification of toxic wastes. In its present state of development, laboratory investigations and scale-up have been successfully concluded. The process is now postured for pilot plant studies. In these studies, processing parameters will be defined for commercial application of the process.

References

[1] Telles, R. W. et al., "Contaminant Fixation Process," *Proceedings*, 40th Annual Purdue Industrial Waste Conference, May 1985, pp. 685–691.
[2] Matsura, H. et al., "Improvement of Asphalt Waste Products in Leachability," JAERI-M8664, Japan Atomic Research Institute, Tokai, 1980.
[3] Columbo, P. and Neilson, R. M., Jr., "Waste Form Development Program Annual Progress Report, Oct 1981–Sept 1982," BNL-51614, Brookhaven National Laboratory, Upton, NY, Sept 1982.
[4] Malone, P. G., Mercer, R. B., and Thompson, D. W., "The Effectiveness of Fixation Techniques in Preventing Loss of Contaminants from Electroplating Wastes," First Annual Conference on Advanced Pollution Control for Metal Finishing Industry, American Electroplaters' Society and U.S. Environmental Protection Agency, Jan 1978.

J. H. Heiser, III,[1] *Eena-Mai Franz,*[1] *and Peter Colombo*[1]

A Process for Solidifying Sodium Nitrate Waste in Polyethylene

REFERENCE: Heiser, J. H., III, Franz, E. M., and Colombo, P., **"A Process for Solidifying Sodium Nitrate Waste in Polyethylene,"** *Environmental Aspects of Stabilization and Solidification of Hazardous and Radioactive Wastes, ASTM STP 1033,* P. L. Côté and T. M. Gilliam, Eds., American Society for Testing and Materials, Philadelphia, 1989, pp. 53–62.

ABSTRACT: A laboratory-scale process has been developed for the solidification of nitrate salt wastes in polyethylene. The process uses a commercially available single-screw extruder which continuously discharges prescribed polyethylene-waste mixtures from the hoppers to the output die, where it is extruded into a container while still in the molten form. The molten mixture (~110 to 120°C) conforms to the shape of the container and solidifies upon cooling. Proportional feeders maintain waste-to-binder ratio and homogeneity of the waste form. Present studies use dry wastes, although wet solid wastes can be processed using vented extruders of the type used for the bitumen solidification process.

Tests were performed to determine leachability and mechanical stability. Emphasis is placed upon leaching of nitrates from the waste forms. Leach tests were performed according to ANS 16.1 as well as the EPA extraction procedure (EP). For polyethylene waste forms containing 30 to 70 weight percent sodium nitrate, ANS 16.1 leach indices range from 11 to 7.8, respectively. Compressive yield strengths range from 18 to 5 MPa (after 90 days' water immersion). The results of the EP indicate that the nitrate release levels for waste forms containing as much as 70 weight % sodium nitrate are not defined as characteristically hazardous waste.

Differential scanning calorimetry (DSC) was used to confirm the compatability of polyethylene and simulated salt wastes at elevated temperatures. Components of the polyethylene/sodium nitrate system, alone and in combination, were tested by DSC at temperatures to 400°C. At these temperatures, no chemical interactions were observed between polyethylene and nitrate waste compositions.

KEY WORDS: polyethylene, radioactivity, low-level waste, salt waste, hazardous waste, waste forms, extrusion process, leaching

Large quantities of nitrate salt waste are produced at various Department of Energy (DOE) facilities [*1*] as a result of nuclear fuel processing and reprocessing. Recent regulations regarding the amount of nitrates allowed into the ground have imposed severe restrictions on the types of solidification agents that can be used. In this study, polyethylene has been identified as an improved solidification agent for immobilization of nitrate wastes.

The selection of polyethylene was based on such considerations as materials properties, compatability with the waste, solidification efficiency, ease of processibility, availability of materials, and economic feasibility. Earlier studies at Brookhaven National Laboratory (BNL) have demonstrated the applicability of polyethylene for the solidification of pressurized water reactor (PWR) and boiling water reactor (BWR) evaporator concentrates using the extruder process [*2–4*].

[1] Chemistry associate, radiochemist, and chemist, respectively, Nuclear Waste Research Group, Fuel Cycle Analysis Division, Brookhaven National Laboratory, Associated Universities, Inc., Upton, NY 11973.

TABLE 1—*Nitrate waste composition* [9].

SRP Nitrate Waste		Rocky Flats Plant Nitrate Waste	
Component	Weight Percent	Component	Weight Percent
$NaNO_3$	48.9	$NaNO_3$	53.9
$NaNO_2$	12.2	KNO_3	33.9
NaOH	13.3	NaCl	3.2
Na_2CO_3	5.3	KCl	2.2
$NaAl(OH)_4$	11.3	Na_2SO_4	2.8
Na_2SO_4	6.0	K_2SO_4	1.8
NaF	0.2	Na_3PO_4	0.9
NaCl	0.4	K_3PO_4	0.6
Na_2SiO_3	0.1	NaF	0.4
Na_2CrO_4	0.2	KF	0.3
$Na_2C_2O_4$	1.0		
Na_3PO_4	0.4		
$NaB(C_6H_5)_4$	0.2		
other salts	0.6		

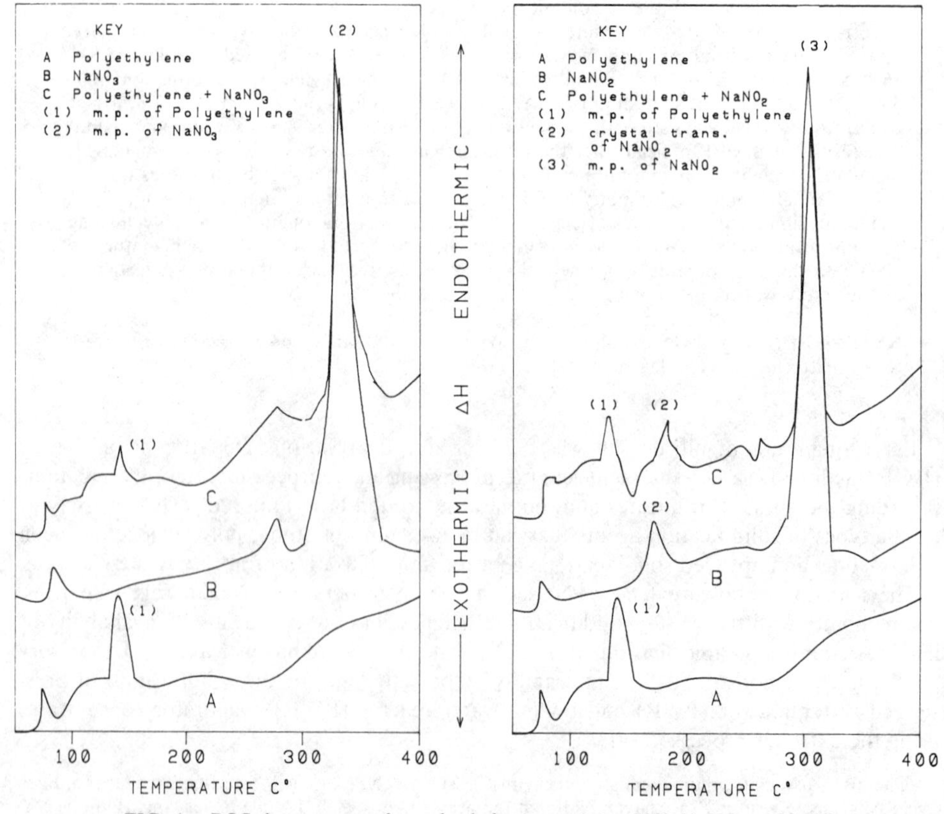

FIG. 1—*DSC thermograms for polyethylene containing 50% $NaNO_3$ or $NaNO_2$.*

The use of polyethylene as an encapsulation material for low-level waste (LLW) was originally suggested by researchers at Oak Ridge National Laboratory (ORNL) [5]. Polyethylene has also been used for solidification of low- and intermediate-level wastes in the Netherlands (unpublished) and Argentina [6,7]. Currently, researchers in Japan [8] have been investigating various methods for the encapsulation of LLW in polyethylene.

Materials Compatibility

Differential scanning calorimetry (DSC) was used to determine the thermal behavior of nitrate wastes and waste components in the presence of polyethylene. Waste stream components comprising greater than 5% of the simulated waste compositions shown in Table 1 were characterized by DSC. In addition, actual Savannah River Plant (SRP) nitrate waste was used. The waste solution obtained from SRP was vacuum dried and combined with polyethylene for thermal studies. Testing was performed at temperatures to 400°C, which is above the decompositional temperature of the major nitrate waste components, such as sodium nitrate ($NaNO_3$) and sodium nitrite($NaNO_2$). Several of the thermograms obtained using the DSC are shown in Figs. 1 and 2. In all cases, the only observable peaks were characteristic endotherms corresponding to melting points or crystal transitions of the ma-

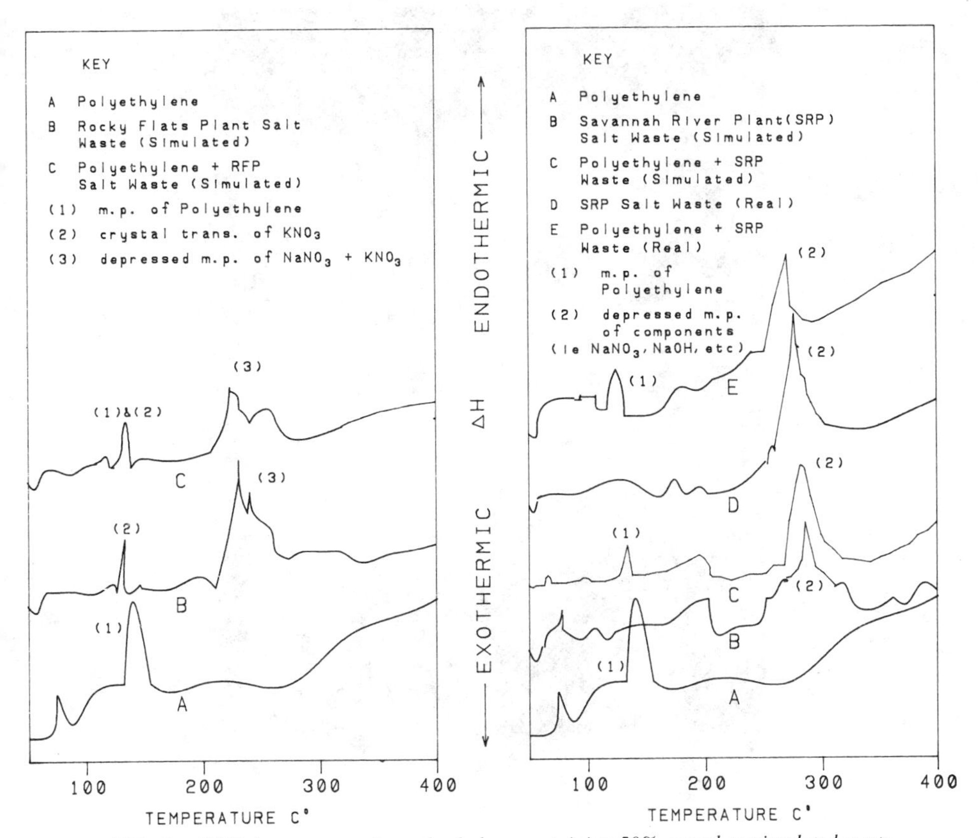

FIG. 2—*DSC thermograms for polyethylene containing 50% actual or simulated waste.*

terials. At temperatures up to 400°C, the absence of exothermic peaks is an indication of the thermal stability of the polyethylene/nitrate waste mixtures.

Solidification Process

An extrusion process was developed which simultaneously heats and blends the polyethylene and waste materials to produce a solid homogeneous waste form. Laboratory-scale specimens were produced using a commercially available 3.2-cm single-screw extruder. A photograph of the extruder and hoppers is shown in Fig. 3. A simplified schematic of the polyethylene extrusion process is given in Fig. 4.

The extrusion process for solidification of radioactive waste in polyethylene involves the heating, mixing, and extrusion of materials in one basic operation. The process is broken down into the following steps:

1. The polyethylene binder and dry waste materials are transferred from either a single hopper or individual hoppers in which they are stored to the extruder feed throat. Metering of waste-to-binder ratios is accomplished at this step.
2. The mixture is conveyed through a heated cylinder by the motion of the rotating screw. The initial portion of the cylinder is controlled at a temperature below the polyethylene

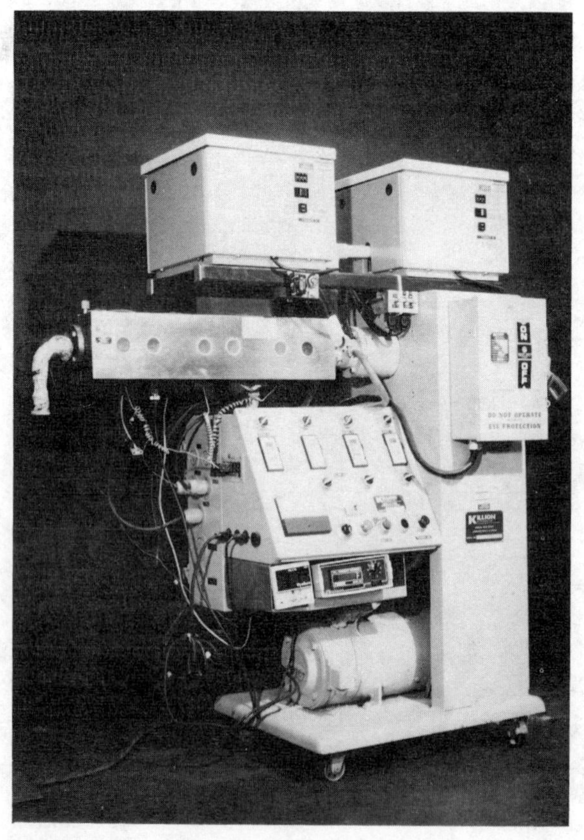

FIG. 3—*Photograph of laboratory-scale single-screw extruder with dual hopper/feeders.*

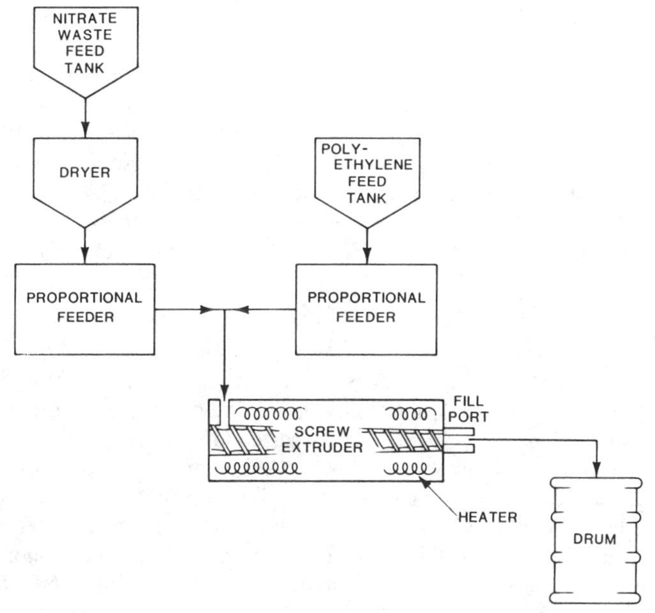

FIG. 4—*Schematic of extrusion process for incorporating nitrate waste into polyethylene.*

melting point of 120°C. This gradually preheats the materials, but at the same time assures proper transport of the mixture.

3. As the waste-binder mixtures moves forward past the initial preheating zone, it is masticated under pressure due to the compressive effects of a gradual reduction in the channel area between the screw and cylinder. Screw rotation also assists in the mixing of the materials to a homogeneous state.

4. The gradual transfer of thermal energy by the combined effects of the barrel heaters and frictional heat melts to the polyethylene. The frictional heat input is difficult to control and must be compensated for by the regulation of the resistance band heaters. In some cases it is necessary to remove excessive heat by the use of external blowers.

5. The melted thermoplastic-waste mixture is forced through an output die into a mold and is allowed to cool and solidify.

Based on experience, a processing temperature of ~120°C was established. At this temperature, good processibility is obtained and problems associated with the decomposition-volatilization of wastes and waste form components are minimized.

The extruder employed in this study was not vented, requiring predrying of the aqueous nitrate waste prior to mixing with polyethylene. Commercially available vented extruders of the type used in bitumen processes would allow incorporation of wet nitrate waste as feed material.

Product Evaluation

Stability of a waste form is an important factor in controlling the release of $NaNO_3$ waste into the environment. A series of waste-form evaluation tests were suggested in the Nuclear

Regulatory Commission's (NRC) Branch Technical Position Paper on Waste Form [*10*] in support of 10 CFR 61. Two of these tests, the 90-day immersion in water and the American Nuclear Society (ANS) 16.1 Leach Test [*11*], were used in this study to evaluate the polyethylene/$NaNO_3$ waste forms. The method outlined in the American Society for Testing and Materials (ASTM) Test Method for Compressive Properties of Rigid Plastics (D 695) was used to measure the compressive yield strength after immersion testing. Since some of the states have adopted the Environmental Protection Agency (EPA) drinking water standards as a guideline for the amount of nitrates allowed to be released into the environment, the EPA extraction procedure [*12*] was performed to determine compliance.

ANS 16.1 Leach Test

Leach testing was performed for 90 days in accordance with the procedures in ANS 16.1 Standard, "Measurement of the Leachability of Solidified Low-Level Radioactive Wastes" [*11*]. The test was designed to provide a standardized laboratory method for characterizing the leaching behavior of low-level waste forms. Although the test procedures do not necessarily simulate waste form leaching under actual burial conditions, it allows a comparison of the relative leachability of various waste/binder combinations. This test addresses specifically the leaching of radionuclides, but it is also applicable to nonradioactive chemical species and is used in these studies to determine the release rate of $NaNO_3$ from the waste form.

Cylindrical specimens, measuring 4.5 to 4.9 cm in diameter and ~9 cm in length and containing 30, 50, 60, and 70 weight percent sodium nitrate, were leached in a volume of deionized distilled water such that the leachant volume to specimen surface area was 10 ± 0.2 (cm). The leachant was changed at prescribed time intervals and the leachate was analyzed for sodium using atomic absorption spectrophotometry (AAS). Since the only source of sodium in the samples is $NaNO_3$, the amount of sodium released from the specimens corresponds directly to the amount of $NaNO_3$ released. At the end of 90 days, the leach data were calculated as cumulative fraction leached (CFL), leach rate, and leach index. The CFL data are presented as a function of time in Fig. 5 and clearly show the dependence of leaching on increased waste loadings. Arithmetic averages of CFL, leach rate, and leach index are given in Table 2. The indices were calculated according to ANS 16.1 and ranged from 11.1 to 7.8 for 30 to 70 weight percent $NaNO_3$ loaded specimens, respectively. The leach index is a dimensionless figure-of-merit, which quantifies the relative leachability for a given waste type/solidification agent combination. Since the index is inversely proportional to the effective diffusivity, higher index values represent better retention and reduced leachability of a waste species.

EPA Extraction Procedure

The EPA extraction procedure (EP) for hazardous wastes was designed to characterize leaching performance of hazardous waste packages under landfill conditions [*12*]. In order for the attenuation and dilution expected during migration of the leachate to the groundwater to be taken into account, the amount of contaminant released during the test is to be combined with a generic attenuation/dilution factor of 100 [*13*].

Prior to leaching, specimens (3.3 cm diameter by 7.1 cm length) containing 30, 50, 60, and 70 weight percent $NaNO_3$ were subjected to the EPA structural integrity procedure (SIP) as specified in the EP. The SIP involves dropping a 0.33-kg weight 15 times onto the

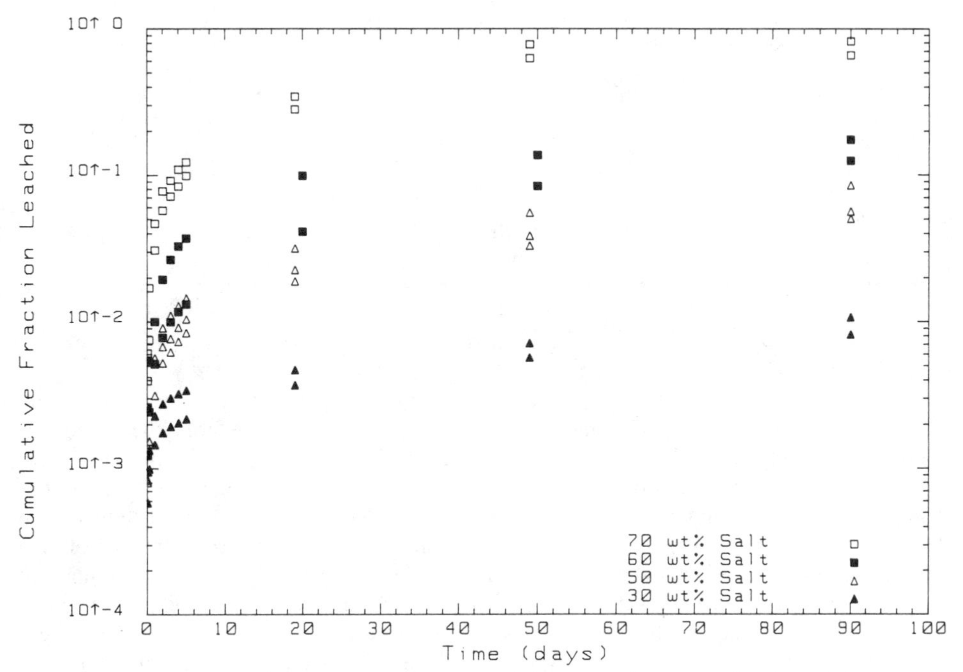

FIG. 5—*CFL of sodium nitrate from polyethylene waste forms as a function of time.*

face of the specimen to be leached. This procedure is designed to ascertain the specimens' mechanical ability to remain a monolith. Details of the procedure and a schematic of the SIP apparatus are found in 10 CFR 40 Part 261, Appendix II. None of the samples subjected to the SIP showed any signs of damage.

The specimens were leached in demineralized water, equal to 16 times the specimen weight. The leachate was made acidic (pH = 5.0 ± 0.2) using 0.5 N acetic acid and monitored as prescribed to maintain a constant pH of 5. After the initial pH correction, further acid additions were not required. The leachate was agitated by tumbling at 30 rpm (±2 rpm). At the end of the 24-h leaching period, the leachate was analyzed for sodium using AAS. The calculated nitrate release levels given in Table 3 are below the drinking water standard of 44 ppm of nitrate [*14*] when the attenuation/dilution factor of 100 is applied. This indicates the specimen would not be classified as a characteristically hazardous waste under the Resource Conservation and Recovery Act (RCRA).

TABLE 2—*ANS 16.1 Leach test data for sodium nitrate in polyethylene waste forms.*

Weight Percent $NaNO_3$	Cumulative Fraction Leached	Leach Rate (s^{-1})	Leach Index
30	0.9	8.4×10^{-10}	11.1
50	6.3	6.0×10^{-9}	9.7
60	15.0	1.1×10^{-8}	9.0
70	73.4	1.5×10^{-7}	7.8

TABLE 3—*Sodium nitrate releases from polyethylene waste forms using EPA extraction procedure.*

Waste Loading, weight percent	Nitrate in Leachate, ppm	Passes Drinking Water Standard with Attenuation/Dilution Factor of 100
30	11	yes
50	40	yes
60	64	yes
70	492	yes

Mechanical Integrity

Water immersion tests were performed on cylindrical specimens (~4.6 cm diameter and 9 cm length) using deionized distilled water. After 90 days of immersion, the specimens were destructively tested to determine compressive yield strength. All measurements were made according to ASTM D 695. The results, which are given in Table 4, are compared to compressive yield strengths of similar specimens which had not been exposed to water. In the case of LLW disposal, the NRC recommends a minimum compressive strength of 0.35 MPa (50 psi) to ensure that a waste form remains stable under the compressive loads inherent in a disposal environment. As can be seen from the data in Table 4, the compressive yield strength does not change when waste loadings are increased from 30 to 60 weight percent or when the samples are immersed in water for 90 days. Only the 70 weight percent loaded samples exhibited reduced compressive yield strength at ~5 MPa.

Dimensional changes due to swelling were also investigated after 90 days' immersion. The average dimensional change in length was +0.2%; the average change in diameter was −0.1%. These are negligible differences and show no swelling problems for the polyethylene/sodium nitrate samples investigated. Figure 6 shows a 70 weight percent specimen before and after 90 days' immersion in water.

Conclusions

Polyethylene is an effective solidification agent for nitrate salt wastes. High loading efficiencies are achievable using simple processing techniques. Leaching indices calculated for polyethylene/sodium nitrate waste forms ranged from 11.1 to 7.8 for 30 to 70 weight percent waste loadings, respectively. All specimens subjected to the EPA extration procedure were classified as a "characteristically hazardous waste" according to RCRA.

TABLE 4—*Compressive yield strength of polyethylene/sodium nitrate waste forms.*

Waste Loading, weight percent	Compressive Strength, MPa	
	No Treatment	After 90 Days' Immersion
30	16.3	17.6
50	13.2	13.2
60	15.4	15.9
70	7.03	4.96

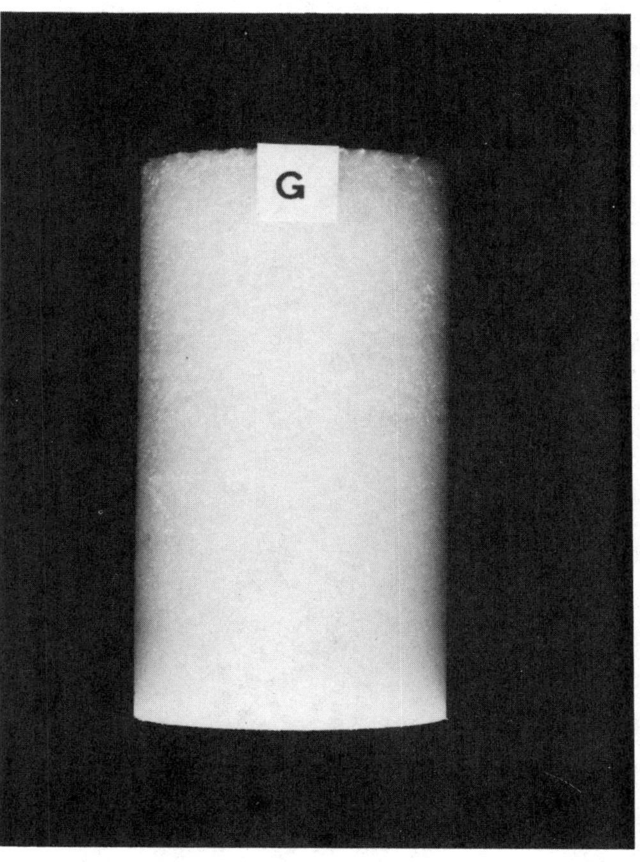

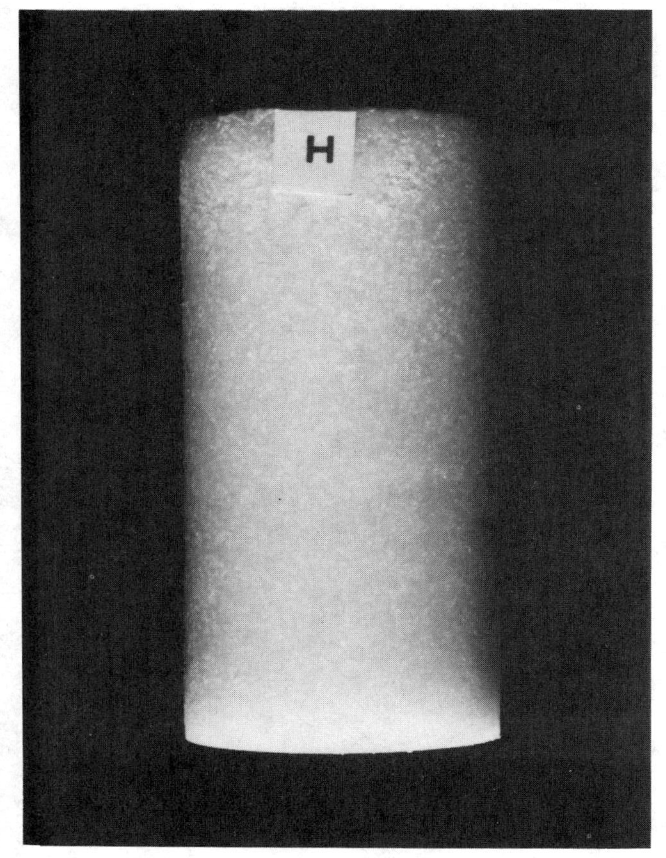

FIG. 6—*Polyethylene waste forms containing 70 weight percent sodium nitrate, untreated* (left) *and after 90 days' immersion in water* (right).

References

[1] Franz, E. M. and Colombo, P., "Identification, Characterization and Selection of DOE Low-Level Radioactive Problem Wastes," BNL-38496, Letter Report, Brookhaven National Laboratory, Upton, NY, July 1986.

[2] Kalb, P. and Colombo, P., "Polyethylene Solidification of Low-Level Wastes," BNL-51867, Topical Report, Brookhaven National Laboratory, Upton, NY, Oct. 1984.

[3] Franz, E. M. and Colombo, P., "Waste Form Evaluation Program," BNL-51954, Final Report, Brookhaven National Laboratory, Upton, NY, Sept. 1985.

[4] Kalb, P. D. and Colombo, P., "An Economic Analysis of Volume/Reduction Polyethylene Solidification of LLW," BNL-51866, Brookhaven National Laboratory, Upton, NY, Jan. 1984.

[5] Fitzgerald, C. L., Godbee, H. W., Blanco, R. E., and Davis, W., Jr., "The Feasibility of Incorporating Radioactive Wastes in Asphalt or Polyethylene," *Nuclear Applications and Technology,* Vol. 9, Dec. 1970.

[6] Colombo, P. and Neilson, R. M., Jr., "Properties of Radioactive Wastes and Waste Containers," BNL-NUREG-50692, Quarterly Progress Report, U.S. Nuclear Regulatory Commission, Washington, DC, July/Sept. 1976.

[7] Colombo, P. and Neilson, R. M., Jr., "Properties of Radioactive Wastes and Waste Containers," BNL-NUREG-50957, First Topical Report, Brookhaven National Laboratory, Upton, NY, June 1977.

[8] Moriyama, N. et al., "Incorporation of an Evaporator Concentrate in Polyethylene for a BWR," *Nuclear and Chemical Waste Management,* Vol. 3, No. 1, 1982, pp. 23–28.

[9] Johnson, A. J. and Arnold, P. M., *Waste Generation Reduction—Nitrates. Comprehensive Report of Denitrification Technologies,* RFP-3899, Rockwell International, Pittsburgh, PA, 15 March 1986.

[10] "Technical Branch Position on Waste Forms," Final Waste Classification and Waste Form Technical Position Papers, U.S. Nuclear Regulatory Commission, Washington, DC, May 1983.

[11] *Measurement of the Leachability of Solidified Low-Level Wastes,* ANS Standards Committee, Working Group 16.1, American Nuclear Society, June 1984.

[12] 40 CFR Part 261, *Code of Federal Regulations,* App. 11, Rev., 1 July 1984.

[13] *Federal Register,* Vol. 51, No. 114, Friday, 13 June 1986, Proposed Rules, p. 21649 (as referenced from U.S. EPA, "Guidelines for Performing Regulatory Impact Analysis," Washington, DC, Dec. 1983).

[14] "National Interim Primary Drinking Water Regulations," 40 CFR Part 141, *Code of Federal Regulations,* July 1984.

Homer Lowenberg[1] *and Mark D. Shaw*[2]

Development of a Composite Polyethylene—Fiberglass-Reinforced-Plastic High-Integrity Container for Disposal of Low-Level Radioactive Waste

REFERENCE: Lowenberg, H. and Shaw, M. D., **"Development of a Composite Polyethylene—Fiberglass-Reinforced-Plastic High-Integrity Container for Disposal of Low-Level Radioactive Waste,"** *Environmental Aspects of Stabilization and Solidification of Hazardous and Radioactive Wastes, ASTM STP 1033,* P. L. Côté and T. M. Gilliam, Eds., American Society for Testing and Materials, Philadelphia, 1989, pp. 63–73.

ABSTRACT: Bondico Nuclear (Los Angeles, California) has initiated a program to develop a high-integrity container (HIC) for handling, transportation, and disposal of low-level radioactive wastes. The HIC, made of a composite material, consists of an inner layer of polyethylene bonded to an outer casing of fiberglass-reinforced plastic.

Preliminary handmade prototype units containing about 0.22 m^3, called HIC-7, have been fabricated and exposed to some of the most demanding U.S. Nuclear Regulatory Commission (NRC) and state tests. The HICs withstood over twice the external pressure from maximum burial conditions and twice the Type A package internal pressure requirements. In addition, free drops on compacted soil and an unyielding surface showed no deleterious effects.

The composite material has been tested for mechanical properties, such as tensile, compressive, and bond shear strengths, creep, thermal expansion, and hardness. In addition, specimens have been exposed to environments including thermal cycling, gamma and ultra violet radiation, biodegradation from fungi and bacteria, and internal and external chemical corrosion followed by mechanical testing. Prototype production units have been tested to the full range of NRC and state requirements. A topical report showing how this HIC meets all NRC and state requirements has been submitted to the NRC.

KEY WORDS: high-integrity container (HIC), low-level radioactive waste (LLW), composite material, polyethylene (PE), fiberglass-reinforced plastic (FRP), U.S. Nuclear Regulatory Commission (NRC), Type A package requirements, mechanical properties, environmental exposures, topical report

The Nuclear Regulatory Commission (NRC) Regulation 10 CFR Part 61 covers low-level waste (LLW) licensing requirements for land disposal. This part of the NRC regulations provides criteria that require stability and integrity for acceptance of some waste forms. Such properties can be achieved either by immobilization in an inert media or by enclosure in a structure or container.

Such a container, called a high-integrity container (HIC), can provide a convenient and economical means for handling, transportation, and disposal of LLWs because it can elim-

[1] President, Lowenberg Associates, Inc., 10901 Rosemont Drive, Rockville, MD 20852.
[2] 5469 Running Creek Lane, Jacksonville, FL 32223.

inate the need for solidification into a monolith. Since the nuclear industry is interested in HICs, NRC has provided guidance on acceptable methods of evaluating their capabilities in a Branch Technical Position (BTP). Further, NRC has provided a means for generic review of HIC capabilities to meet NRC and state requirements through a topical report review and approval process.

Bondico Nuclear (originally of Jacksonville, Florida) has been manufacturing special composite containers for hazardous chemical wastes. They are fabricated from a composite material made of an inner layer of polyethylene (PE) bonded to an outer casing of fiberglass-reinforced plastic (FRP). Their use has been approved by the Environmental Protection Agency (EPA) for containing leaking or damaged packages of hazardous materials, such as PCBs, dioxins, and corrosives. In addition they have been qualified under Department of Transportation (DOT) regulations as 7A Type A packaging and are being used as specialty containers by the nuclear industry. The loading of one of these units is shown in Fig. 1.

FIG. 1—*Hazardous waste container loading.*

HIC Development and Preliminary Testing

In response to requests from several nuclear waste generators, Bondico Nuclear embarked on a program to develop and qualify various sizes of HICs. The first portion of this program involved fabrication of handmade prototype units. The size HIC selected for this work, called HIC-7, is similar to a 200-L (55 gal) drum and has a capacity over 0.22 m^3 (7.8 ft^3).

An elevation sketch of this HIC is shown in Fig. 2. Figure 3 depicts a handmade prototype HIC-7 showing the full open lid and the equipment for making the closure weld. The all PE seal of the lid to the body is made by electrically heating the PE flanges of the two HIC

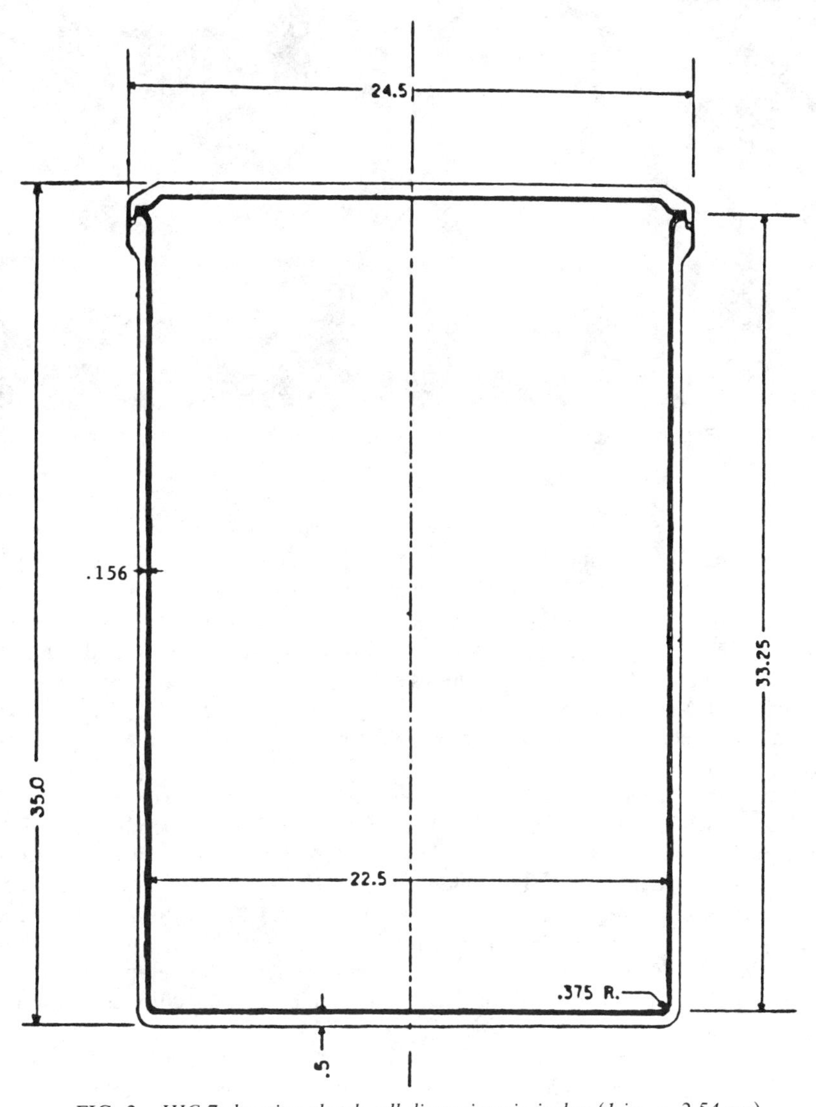

FIG. 2—*HIC-7 elevation sketch, all dimensions in inches (1 in. = 2.54 cm).*

FIG. 3—*HIC-7:* (left) *elevation showing closure-weld equipment;* (right) *full open lid and connection for welded closure.*

parts. This bonds the PE inner liners of the parts by welding the flanges together. A family of HICs from 0.28 to 5.7 m^3 (10 to 200 ft^3) is planned for the future.

Handmade prototypes of HIC-7s have been fabricated, loaded, sealed, and tested preliminarily to determine their ability to meet the most stringent NRC and state requirements of free drop and pressure. The NRC requires a Type A package drop test on an unyielding surface from a 1.2-m (4 ft) height (Figs. 4 to 6, on concrete). The States of South Carolina and Washington require free drop tests of HICs from a height of 7.6 m (25 ft) on compacted soil (Figs. 7 to 10).

The maximum external pressure requirements result from burial conditions of HICs at the Richland, Washington, commercial LLW burial site. This burial depth of 16.8 m (55 ft) is equivalent to an external pressure of 3.23 kg/cm^2 (46 psig). To establish the HIC's outstanding physical strength, we tested handmade prototypes to an external hydrostatic pressure of 7.0 to 8.4 kg/cm^2 (100 to 120 psig) without failure. This pressure is from 2 to 2½ times the equivalent maximum burial pressure of the deepest U.S. commercial burial conditions and clearly demonstrates the HIC's excellent strength. Photographs of the external hydrostatic pressure test setup and the HIC-7 and welded seal condition after the test are shown in Figs. 11 to 13.

In addition, regulations require a Type A package to withstand an absolute external pressure of 0.25 kg/cm^2 (3.5 psi). This is equivalent to an internal gage pressure of 0.79 kg/cm^2 (11.2 psi). Accordingly, the handmade HICs have been tested to internal gage pressures of about 1.76 kg/cm^2 (25 psig) with no sign of failure or effect.

FIG. 4—*Bondico Nuclear HIC, 1.2-m drop test on concrete: moment of drop.*

FIG. 5—*Bondico Nuclear HIC, 1.2-m drop test on concrete: moment of impact.*

FIG. 6—*Bondico Nuclear HIC, 1.2-m drop test on concrete: inspection showing no effects.*

FIG. 7—*Bondico Nuclear HIC, 7.6-m drop test on sand: lifting for drop test.*

FIG. 8—*Bondico Nuclear HIC, 7.6-m drop test on sand: moment of impact.*

FIG. 9—*Bondico Nuclear HIC, 7.6-m drop test on sand: after impact.*

FIG. 10—*Bondico Nuclear HIC, 7.6-m drop test on sand: inspection showing no effects.*

Development of Production HIC's

Encouraged by these outstanding preliminary test results using handmade prototypes, Bondico Nuclear initiated an extensive material testing program with a large national testing laboratory in late 1986. Test specimens have been machined from composite plate material made by production processes.

These specimens have been tested for physical properties such as tension, compression, bond shear, creep, thermal expansion, and hardness. Preliminary results indicate average tensile and compressive strengths in the range of 2722 kg/cm^2 (38 700 psi). In addition, tests of the composite material's performance after exposure to the following environments have been carried out:

(*a*) thermal cycling,
(*b*) gamma and ultraviolet radiation,
(*c*) biodegradation from fungi and bacteria, and
(*d*) chemical corrosion (internal and external).

Following procurement and installation of production equipment, production HICs have been fabricated. Full-scale production prototype units have been loaded with simulated solid wastes, sealed and tested to demonstrate performance under a full range of package conditions as required by NRC and the states as follows:

(*a*) compression,
(*b*) external and internal pressure,
(*c*) penetration,

FIG. 11—*Bondico Nuclear HIC, external hydrostatic test: before test.*

(*d*) free drops to NRC and state requirements,
(*e*) passive venting,
(*f*) lift accident deceleration, and
(*g*) transportation vibration.

Topical Report Preparation and Submittal

A topical report has been prepared to document the results of the previously mentioned production material and prototype test programs, as well as

(*a*) engineering design,
(*b*) fabrication methods,
(*c*) handling and loading procedures,

FIG. 12—*Bondico Nuclear HIC, external hydrostatic test: after test, section through intact walls and lid.*

FIG. 13—*Bondico Nuclear HIC, external hydrostatic test: after test inspection—no effect on intact closure weld.*

(*d*) use requirements, and
(*e*) quality assurance program.

We believe that this work, as summarized in the topical report, will convincingly demonstrate the HIC's ability to meet all NRC and state requirements for the disposal of LLW including:

(*a*) integrity of the welded PE closure seal,
(*b*) outstanding HIC corrosion and radiation resistance,
(*c*) high HIC strength-to-weight ratio,
(*d*) excellent volumetric efficiency of 85 to 90%, and
(*e*) more than adequate physical and chemical properties.

Regulatory Aspects and Testing Methods

Thomas Jungling[1] *and John Greeves*[1]

Nuclear Regulatory Commission Regulations and Experience with Solidification/Stabilization Technology

REFERENCE: Jungling, T. and Greeves, J., **"Nuclear Regulatory Commission Regulations and Experience with Solidification/Stabilization Technology,"** *Environmental Aspects of Stabilization and Solidification of Hazardous and Radioactive Wastes, ASTM STP 1033,* P. L. Côté and T. M. Gilliam, Eds., American Society for Testing and Materials, Philadelphia, 1989, pp. 77–82.

ABSTRACT: Two important areas in the U.S. Nuclear Regulatory Commission regulation for low-level waste management, 10 CFR Part 61, involve the requirements for waste classification and waste form. The waste classification system establishes three categories of wastes acceptable for near-surface disposal. These categories are determined by the concentrations of nuclides important for disposal. Class A wastes have low concentrations and need only meet minimum waste-form requirements. Class B and C wastes have higher concentrations and are required to have stability to minimize disposal trench subsidence effects. This paper discusses the approaches recommended to demonstrate that such wastes meet the stability criteria, including acceptable tests and test criteria which could be used by waste generators to demonstrate waste stability. Discussion is also included on the solidification technologies and high-integrity containers that have been developed to meet the requirement.

KEY WORDS: radioactive waste, solidification, stabilization

In 1985, about 51 000 m^3 (1 800 000 ft^3) of commercial low-level radioactive waste (LLW) was shipped to disposal sites in the United States by nuclear reactor operations, hospitals, research laboratories, universities, industry, and the Federal Government. Typical materials comprising LLW include ion-exchange resins from reactors, solidified and absorbed liquids, failed equipment, compacted trash, contaminated protective clothing, animal carcasses, scintillation fluids, and decontamination wastes.

To deal with these wastes, in December 1982 the Nuclear Regulatory Commission (NRC) published the rule, "10 CFR Part 61—Licensing Requirements for Land Disposal of Radioactive Waste." This rule established performance objectives for the land disposal of radioactive waste, minimum technical requirements for a near-surface disposal facility, and the licensing procedures NRC will follow in licensing new disposal capacity for low-level radioactive waste.

The performance objectives established in 10 CFR Part 61 are

(*a*) Protection of the general population from releases of radioactivity,
(*b*) Protection of individuals from inadvertent intrusion, and
(*c*) Protection of individuals during operations.

[1] Division of Low-Level Waste Management and Decommissioning, U.S. Nuclear Regulatory Commission, Washington, DC 20555.

Stability of the site after closure is accomplished by requiring that the disposal facility shall be sited, designed, utilized, operated, and closed to achieve long-term stability of the site and to eliminate, to the extent practicable, the need for ongoing, active maintenance following closure, so that only surveillance, monitoring, or minor custodial care are required (10 CFR Part 61.44).

A principal requirement for disposal of radioactive wastes in the United States is that the waste provide structural stability to the disposal trench in order to minimize the infiltration of water. For most wastes, structural stability is currently provided by either processing to produce a solidified waste form or disposal within a container, designed to provide long-term structural stability.

Two of the most important technical criteria in the rule involve the classification of wastes and the waste-form requirements. These requirements are implemented for NRC licensees through the manifest requirements in 10 CFR Part 20.311.

Regulations

The waste classification system establishes three categories for wastes acceptable for disposal at a near-surface burial facility. This classification system is based on the concentrations of radionuclides important for disposal. In this system, wastes having greater radiologic hazards are required to be disposed of with greater protection.

The waste classification system in 10 CFR Section 61.55 consists of three classes, A, B, and C. Class A wastes contain the lower concentrations of the nuclides important for disposal. These wastes can be disposed of with only the minimum waste form requirements. Among these are the limitation for free liquids and restrictions on pyrophoric, explosive, and hazardous characteristics of the waste. Because of the lower concentrations, waste instability will not produce significant hazards to public health and safety. Risks associated with intrusion into Class A wastes are considered acceptable after an interval of 100 years.

Class B wastes have higher activities than do the Class A waste materials. Because of this, Class B wastes must be segregated from Class A wastes and must be structurally stable to minimize waste degradation and the resultant subsidence of the waste trenches. Trench subsidence, as a result of little or no waste-form control, has resulted in a large infiltration of water into trenches at the now closed Sheffield, Illinois, West Valley, New York, and the Maxey Flats, Kentucky, disposal sites. In order to reduce the need for major remedial actions in future sites and potentially increased ground-water pathway impacts, 10 CFR Part 61 requires segregation and stability for the higher concentration wastes. Waste-form stability can be obtained by processing (such as, solidification), use of a container or structure to provide stability (such as, high-integrity container), or by the waste itself (for example, a large activated component). The free liquid requirement for solidified wastes is limited to 0.5% by volume. To the extent practicable, stable wastes should maintain gross physical properties and identity for 300 years.

Class C wastes have higher concentrations than do the Class B wastes. In addition to stability, Class C wastes must be disposed of with an intruder barrier to minimize the possibility of inadvertent intruders contacting the wastes following the loss of institutional control of the disposal site. Burial of at least 5 m from the surface provides an acceptable intruder barrier for Class C wastes. Engineered barriers used against inadvertent intrusion should have an effective life of at least 500 years.

Additional Guidance

The waste classification and waste-form requirements in 10 CFR Part 61 are relatively general. The NRC refrained from listing detailed prescriptive requirements to allow some flexibility because of the multitude of different waste forms and types of waste generators that ship to commercial disposal sites. In order to provide more specific guidance, the NRC prepared Technical Positions (TP's) on waste classification and waste form in 1983. The TPs were modified for publication as Regulatory Guides (RGs), which have been available for public comment since early 1987.

The objective of the waste-form TP and RG is to provide guidance on acceptable methods for demonstrating stability for Class B and C wastes, although it should be noted that these methods are not the only methods for demonstrating compliance. The documents provide guidance on developing and qualifying process-control programs for waste solidification processes, designing high-integrity containers, packaging filter cartridges, and loading organic ion-exchange resins.

For solidification of Class B and C wastes, the documents (TP and RG) indicate that each waste stream/solidification media combination should be prepared and subjected to the recommended tests. The recommended tests address the following properties: compressive strength, leach resistance, compressive strength following immersion in water, resistance to biodegradation and radiation, thermal cycling stability, as well as inspection to verify that the 0.5% free liquid requirement for solidified products is met. Acceptable test methods and test results for demonstrating stability are given in Table 1. The compression strength of 0.41 MPa (60 psi) is based on a conservative mechanical loading a waste form would undergo in a disposal trench. The radiation accumulated dose of 10^8 rad is based on the upper limit of current, routinely generated wastes. The biodegradation tests are short-term tests that provide go/no-go results. If microbial attack is observed, additional testing may need to be performed to confirm that waste forms are not subject to substantial biodegradation. The leachability index (LIX) of 6 is calculated using the method described in American Nuclear Society (ANS) 16.1. The immersion test can be performed in conjunction with the leaching test. The thermal degradation will be important for wastes held in storage prior to disposal. During storage, the stability of the waste form should not be affected by changes in temperature. The free-liquid test limit is the only prescriptive waste form requirement in 10 CFR Part 61 and is intended to allow trace quantities of liquids which might result from condensation during handling and transport.

Testing may be performed on simulated, nonradioactive wastes. If laboratory-size specimens are used, these specimens should be correlated with full-scale waste forms to demonstrate that the actual waste products will have properties similar to those of the specimens tested.

An acceptable approach for implementing the TP/RG tests for solidified Class B and C

TABLE 1—*Solidified product guidance.*

Test	Method	Result
1. Compression strength	ASTM C 39 or adequate backfill (bitumen)	[a]0.4 MPa (60 psi)
2. Radiation stability		0.4 MPa after 10^8 rad
3. Biodegradation	ASTM G 21 and G 22	no growth
4. Leachability	ANS 16.1	LIX of 6
5. Immersion		0.4 MPa after 90 days
6. Thermal cycling	ASTM B 553	0.4 MPa after 30 cycles from −40 to 60°C
7. Free-liquid	ANS 55.1	0.5%
8. Full-scale tests		homogeneous and correlates to lab size test results

wastes would be through the qualification of a process control program (PCP). The full battery of tests need only be performed at the time of PCP qualification. The PCP, however, should be reverified periodically to ensure that the system is operating as designed. The reverification could be accomplished with a short series of tests. The use of generic test data for PCP qualification would be acceptable.

As an option to solidification, the rule (10 CFR 61.56) allows waste generators the flexibility of using high-integrity containers (HICs) to provide stability. The TP provides guidance on the design of HICs. This guidance is summarized in Table 2. The HIC should have as a design goal a lifetime of 300 years and also should be able to meet the Department of Transportation (DOT) Type A package qualification. An important aspect in the design and use of HICs is a quality-assurance (QA) program. The QA program should give special consideration to fabrication, testing, and use of the containers. For certain containers, materials-specific wastes may need to be restricted from contact with the containers. The QA program should carefully address this area. Finally, a PCP should be developed to ensure that the 1% free-liquid requirements will be met.

Changes in Regulatory Guide

Since the Technical Position on Waste Form was issued in May of 1983, it has been recognized that additional guidance or modifications to the existing guidance was necessary. These needs have been considered in the preparation of the Regulatory Guide on Waste Form Stability. One area that has been clarified is the definition of major waste streams needing qualification. Certain wastes are specifically identified as separate waste streams, including bead resin, powdered resins, boric acid and sodium sulfate evaporator bottoms, and individual decontamination solutions. Guidance is also included regarding compositional variations within each waste formulation which should be considered for requalification. In the performance of the full-scale correlation tests, additional guidance is provided regarding the use of the most conservative waste stream for individual solidification media and the use of compression and immersion tests to provide the comparison. Additions with respect to the leach test include the use of seawater and demineralized water to bracket the range of ionic strengths to be encountered, and the recommendation that cobalt, cesium, and strontium be used as tracers.

The major change contained in the Regulatory Guide on Waste Form Stability addresses the subject of viscoelastic material (bitumen) stability. Because viscoelastic creep can contribute to trench instability, additional measures must be taken to ensure that potential voids around bitumen solidified wastes are filled to minimize the creep of the waste forms.

TABLE 2—*High-integrity container design guidance.*

1. Design for 300-year lifetime objective
2. Design to withstand corrosive and chemical environment
3. Design to withstand loads of burial
4. Materials designed to withstand 10^8 rad
5. Materials resistant to biodegradation
6. Design to DOT[a] Type A package qualification
7. Conduct prototype testing
8. Have quality-assurance program
9. Use process control program to demonstrate

[a] Department of Transportation.

Alternative Disposal Methods

As alternative disposal concepts are identified, less reliance may be placed on waste forms and more placed on the engineered structures in providing structural stability. Since 10 CFR Section 61.56 allows stability to be provided by a structure, in addition to processing, a container, or the inherent stability of the waste, no change in the regulation is anticipated as a result of alternative disposal methods. Additional guidance regarding alternative disposal methods was prepared in response to the Low-Level Waste Policy Amendment Act of 1985. The guidance, in the form of a Standard Review Plan, was released at the end of 1987. No changes to the 10 CFR Part 61 stability requirements were needed for engineered disposal concepts.

Standardization

The NRC has adopted a policy of standardization, where appropriate, in order to conserve the increasingly limited resources. Although NRC desires to give waste generators the flexibility to select solidification media and HIC designs, standardization through the use of topical report qualification test data can help conserve resources of waste generators, vendors, and regional inspectors.

Technology Status

Solidification Processes

The NRC has received ten topical reports (TRs) on solidification processes: six on cement or related media, two on polymeric materials, and two on bitumen solidification media. Most of the cement products utilize conventional cement technology augmented by the use of proprietary additives and methods. All of the cement TRs remain under review. A generic question remaining to be resolved concerns the effect of cure period on cement waste forms.

One of the polymeric solidification media (General Electric Company) has been approved while the other (Dow Chemical Company) is very near completion. The General Electric process encapsulates the wastes after removal of the water by azeotropic distillation. The Dow process encapsulates the waste and water in a vinyl ester stryene matrix. One of the remaining questions about the Dow process is the resolution of indications of biological attack observed in NRC-funded programs.

The two bitumen solidification processes will utilize blown (oxidized bitumen). Both of these topical reports, from Waste Chem Corporation and Associated Technologies, Inc., are in mid-review while the vendors respond to NRC questions. One major concern with bitumen, as mentioned earlier, has been its viscoelastic properties which have been resolved by the implementation of controlled backfill procedures at the disposal sites.

High-integrity containers are designed and constructed to meet the structural stability requirements of Part 61. The NRC has received seven HIC TR's using four material types: polyethylene (PE), fiberglass-reinforced plastic (FRP), polymer-impregnated concrete, and a duplex stainless steel. Two additional HICs are expected to be submitted within the year which are composites of the above materials—one is stainless steel with a polyethylene liner and the second a fiberglass reinforced plastic bonded to an inner polyethylene liner.

Four vendors have proposed use of PE: Chem-Nuclear Systems, Inc. (CNSI), NUS Process Services Corporation (now LN Technologies Corporation), TFC Nuclear, and Westinghouse Hittman Nuclear (WHN). The designs utilize high-density, cross-linked polyethylene and vary in size from 209 L (55 gal) to cylinders 1.8 m (6 ft) in diameter by 1.8 m (6 ft) high—8.5 m^3 (300 ft^3). All of the PE HICs are made by a rotational molding process. They are proposed for use in disposal of bead and powered ion-exchange resins, filter sludge, solid wastes, fibrous filter media, and solidified waste forms.

Because of generic questions involving the structural stability of PE HICs, the NRC contracted with Brookhaven National Laboratory (BNL) to recommend stress-strain criteria and to develop a finite-element computer code incorporating long-term creep and buckling criteria to evaluate the response of polymeric HICs to the burial loads expected at the burial sites. Although further work is being proposed to extend the BNL Code, initial indications are that potential concerns may exist with the long-term creep of PE HICs. NRC staff will be asking the vendors to address and resolve these issues.

The Chichibu Cement Company has designed a steel fiber polymer, impregnated concrete (SFPIC) HIC. The SFPIC material has significantly better strength and toughness properties relative to plain concrete. The polymer also greatly decreases the permeability of the composite. This HIC is currently designed in 209 and 418-L (55 and 110-gal) sizes. The TR was approved on 25 June 1986 for the disposal of ion-exchange media, filter media, solid wastes, filter cartridges, and solidified resins and sludges.

Nuclear Packaging has submitted two TRs for HICs made from a duplex stainless steel, Ferralium alloy 255. (Ferralium is a registered trademark of Bonar Langley Alloys, Ltd.) The first TR covered the design of a 1.4 m^3 (50 ft^3) HIC designated as the FL-50/EA-50 HIC. The second TR describes larger HICs to 5.9 m^3 (210 ft^3) designated as the Enviralloy Family. All of the Ferralium HICs have excellent corrosion resistance to potential chemical environments including liquids with pH as low as 3. The FL-50/EA-50 HIC was approved on 7 November 1985. Subsequent minor modifications to the structural criteria were made and the final TR was submitted in late 1986. Responses to the remaining questions on the Enviralloy Family came in early 1987.

CNSI also submitted a TR on a fiberglass-reinforced plastic (FRP) HIC. Comments were sent to CNSI in October 1985, but the TR was withdrawn in May 1986.

Summary

Solidification technology has been employed successfully for many years to stabilize a variety of radioactive wastes. The regulations of 10 CFR Part 61 and the guidance of the Technical Position on Waste form helped to clarify the criteria regarding solidified waste forms and introduce the alternative of a high-integrity container. In the past four years, since the issuance of these documents, additional information from laboratory data, topical report reviews, and field information has indicated the need to update some of the guidance which will be provided in the Regulatory Guide on Low-Level Waste Form Stability. There are no foreseen changes to the regulations of 10 CFR Part 61 regarding the solidification requirements.

Giuseppe A. Ricci[1] *and A. Donato*[1]

Characterization and Control of Solidified Radioactive Wastes According to the Italian Rules: Organization of a Characterization Facility and First Results

REFERENCE: Ricci, G. A. and Donato, A., **"Characterization and Control of Solidified Radioactive Wastes According to the Italian Rules: Organization of a Characterization Facility and First Results,"** *Environmental Aspects of Stabilization and Solidification of Hazardous and Radioactive Wastes, ASTM STP 1033,* P. L. Côté and T. M. Gilliam, Eds., American Society for Testing and Materials, Philadelphia, 1989, pp. 83–92.

ABSTRACT: According to Technical Guide No. 26, officially published in 1987 by the Italian Regulatory Body (ENEA-DISP), the radioactive wastes are now classified in Italy in accordance with the identities and the concentrations of radionuclides, no distinction being made from the point of view of the physical state.

Three "radwaste" categories have been established: The first refers to wastes that decay below exemption limits for free discharge, as defined by current Italian laws, within a few months or within a few years, maximum. The second category of radwastes, which is the most important in terms of overall production, refers to those wastes whose radioactivity decays to radioactive levels comparable with the exemption limits set up for natural radioactive solids within a few centuries, maximum. The radwastes falling into this second category, which are generated mainly by nuclear power plants, must be solidified by means of suitable solidification agents such as cement, bitumen, or polymers. Moreover, a test program must be performed to assure the compliance of the waste forms with a set of specified waste form characteristics.

The third category of radwastes refers to the high-activity liquids arising from the first cycle of the fuel reprocessing facilities and to the alpha-bearing wastes.

The tests to be applied for the characteristics control of real radioactive solidified wastes of the second category are briefly described. They require, in some cases, that a noncommercial apparatus be operated in radiation-controlled areas. A glove-box chain, also suitable for the characterization of solidified alpha wastes, has been designed and built.

The first characterization test campaign has been carried out on radwastes produced by boiling water reactor power plants solidified in cement and the results are presented.

KEY WORDS: conditioned radwastes characterization, quality assurance, boiling water reactor radwastes, ISO long-term leaching test, cesium leachability in cement, strontium leachability in cement, cobalt leachability in cement

In Italy, the regulatory problems concerning radioactive waste management began to be solved with the issue of clear rules from the authorities. The Italian body responsible for licensing and regulating nuclear facilities and materials on behalf of the Minister of Industry, is the Comitato Nazionale per la Ricerca e lo Sviluppo dell'Energia Nucleare e delle Energie Alternative Direzione Sicurezza A Nucleare e Protezione Sanitaria (ENEA/DISP). Its

[1] Chemist researcher and research manager, respectively, ENEA, Nuclear Fuel Cycle Division, Rome, Italy.

responsibilities include protecting public health and safety, protecting the environment, and protecting and safeguarding materials and plants in the interest of national security.

This regulatory body initiated in 1983 a large program aimed at the issuance of official documents, such as Technical Guides or Technical Positions, on the different aspects of the National Waste Management Policy.

Technical Guide No. 26 (TG 26) is the first document of the aforementioned program. A draft was issued in April 1985 [*1*] and specifies both the fundamental criteria about the waste's conditioning and the requirements with which the conditioned "radwastes" must comply in order to be considered acceptable for disposal.

In this context, a program for the characterization of real cemented radioactive wastes, according also to the quality-assurance (QA) requirements, has been planned by the Radwaste Treatment and Conditioning Laboratory (RIFIU) of the ENEA, Nuclear Fuel Cycle Division.

Technical Guide No. 26, ENEA/DISP

TG 26, from the point of view of a general management strategy and in particular of the different disposal methods, classifies radioactive wastes in accordance with the identity and concentration of radionuclides, making no distinction between the point of view of physical states and that of origin.

Three radwaste categories have been identified, as shown in Fig. 1, where the radwaste producers also are identified. First-Category Wastes are wastes that decay below the exemption limits (10 nCi/g) for free discharge, as defined by current Italian laws, within a few months or within a few years, maximum. First-Category Wastes are mainly generated by biomedical and research activities.

Second-Category Wastes are those that take 10 to 1000 years to decay to radioactivity levels of about 10 nCi/gr. This radwaste category is mainly generated by

1. Nuclear power plants' operation,
2. Fuel cycle plants' operation,
3. Nuclear facilities' decommissioning, and
4. Medical and industrial uses.

Third-Category Wastes decay to radioactivity levels comparable with the natural radioactivity background within thousands or millions of years.

TG 26 largely deals with the Second-Category Wastes because of their high production rate and their greater significance from the radiological and environmental point of view. In particular, about the waste form characteristics, TG 26 requires the control of the following characteristics:

1. Compressive strength.
2. Thermal cycling.
3. Radiation resistance.
4. Fire resistance.
5. Leaching rate.
6. Free liquids.
7. Biodegradation resistance.
8. Immersion resistance.

With reference to the waste form qualification, TG 26 establishes that the testing may be performed on radioactive or nonradioactive specimens and, if small simulated laboratory

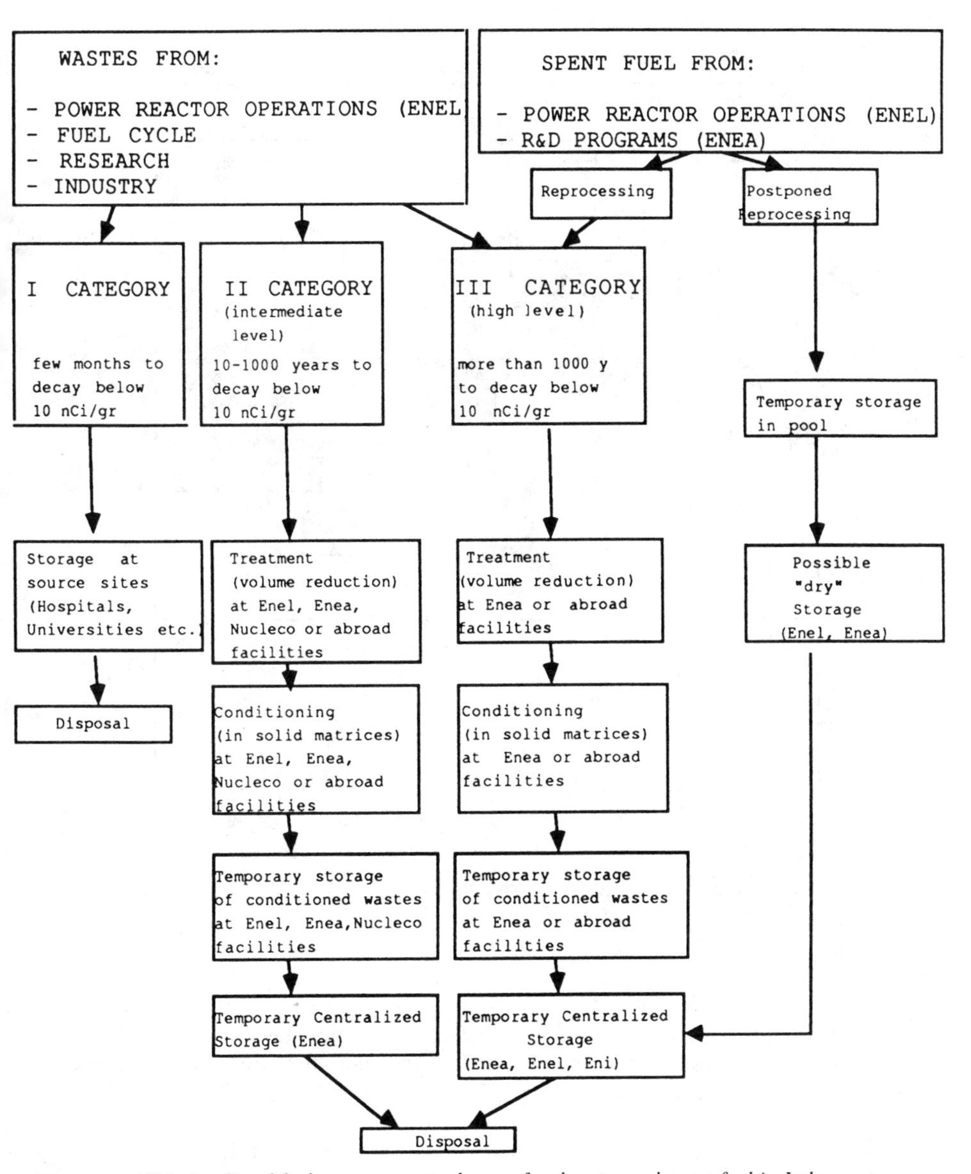

FIG. 1—*Simplified management scheme of radwastes and spent fuel in Italy.*

size specimens are used, their characteristics must be correlated with those of the actual size products.

ENEA-RIFIU Laboratory Quality-Assurance Program on Radioactive Waste Forms Characterization

In the context of the above-mentioned program on the radwaste form characterization, performed both on simulated and on actual specimens, RIFIU takes on the role shown in Fig. 2. In order to operate in compliance with the QA for actual radioactive specimens,

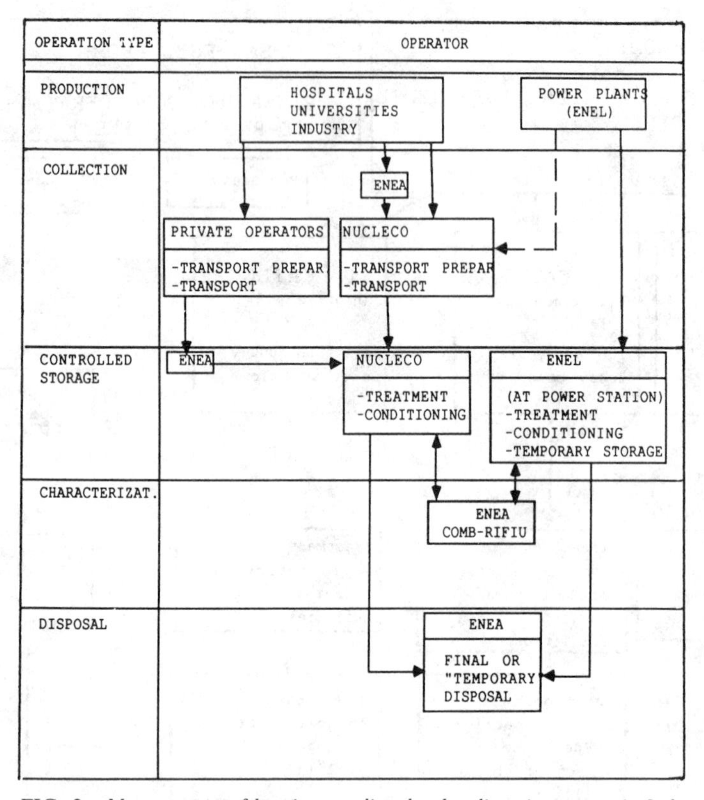

FIG. 2—*Management of low-intermediate-level radioactive wastes in Italy.*

RIFIU both drafted a series of Internal Technical Procedures about the test's execution and supplied the needed equipments for calibration and control.

Technical Procedures

The solidified 2ht category wastes characterization is carried out by RIFIU according to the test requirements established in TG 26. In particular, the test procedures and the minimum requirements are the following:

1. Compressive strength: This test is performed according to the Unificazione Italiana (UNI) Destructive Test on Concrete for Compressive Strength (6132-72). The minimum compressive strength required is 50 kg/cm^2 ($\approx$5000 kN/m^2).
2. Thermal cycles: Thirty thermal cycles, each lasting 24 h, from −40 to +60°C (−40 to +140°F) at 90% relative humidity (RU) are required. At the end of the cycles, the compressive strength must be at least 50 kg/cm^2.
3. Radiation resistance: The radiation resistance test exposes the specimens, wrapped in tin foil to minimize water loss, to the required 10^8 rads (10^6 Gy) radiation dose by means of a cobalt apparatus irradiator. At the end of the test, the compressive strength must be higher than 50 kg/cm^2.
4. Fire resistance: The American Society for Testing and Materials (ASTM) Test for Rate of Burning and/or Extent and Time of Burning of Self-Supporting Plastics in a Horizontal

Position (D 635-81) is required. The conditioned waste must be incombustible or, at least, self-extinguishing.

5. Leaching rate: The measurement of this parameter must be performed according to long-term method. So the International Standards Organization (ISO) 6961/82 Standard Method for Long-Term Testing of Solidified Radioactive Waste Forms was chosen, operating at the test temperature of 40°C (104°F). The methodology used for the sample immersion in the leachant is shown in Fig. 3.

6. Free liquids: This test is performed according to American National Standards Institute/American Nuclear Society (ANSI/ANS) Method for Solid Radioactive Waste Processing System for Liquid Cooled Reactor Plants Appendix 2: Free Liquid Tests (55-1). Conditioned waste must be exempt from free liquids.

7. Biodegradation resistance: ASTM Recommended Practice for Determining Resistance of Synthetic Polymeric Materials to Fungi (G 21-70) and Recommended Practice for Determining Resistance of Plastics to Bacteria (G 22-76) procedures have been adopted. The final compressive strength after the test must be at least 50 kg/cm^2.

Finally, the immersion resistance test is carried out leaving the samples for 90 days dipped in fresh water. The immersion container is a tank made of stainless steel, in which every specimen is completely surrounded by a 5-cm water layer. The final compressive strength after the test must be at least 50 kg/cm^2.

Apparatus Used to Perform Tests on Actual Radioactive Waste Forms

For the above-mentioned tests to be performed on actual radioactive specimens, a glove-box chain was equipped with the required apparatus: environmental chamber, compressive strength testing machine, and immersion tanks. In this glove-box chain, shown in Fig. 4, the specimens, after the required immersion period, are sent to the compression testing machine to determine the final compressive strength. The same procedure is used for the samples coming from the thermal cycles, which are produced using a patented environmental chamber.

First Experimental Results

Type and Origin of the Conditioned Radwastes to be Characterized

At present, three nuclear power plants are in operation in Italy, while a fourth stopped activity some time ago. A new 1000 MW BWR plant is under construction at Montalto di

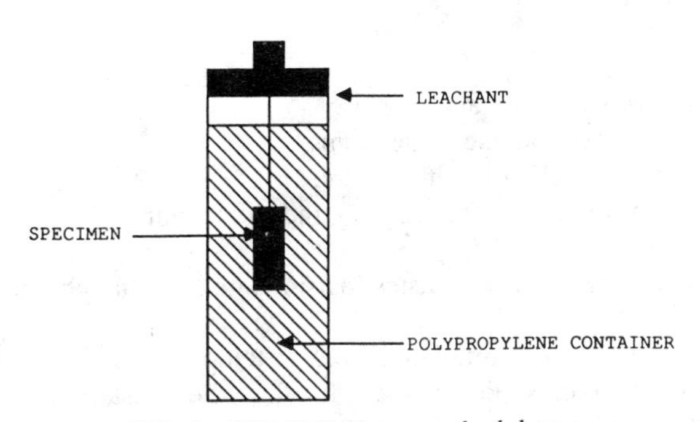

FIG. 3—*ISO 6961-82 test methodology.*

FIG. 4—*Glove-box chain for characterization of solidified radwastes.*

Castro, and its operation is foreseen to begin on 1990, if the conclusion of the political debate presently in Italy will permit it.

The liquid and the wet radioactive wastes produced at Montalto di Castro will be solidified by means of a continuous cementation plant provided by the Italian firm Castagnetti SpA, which is also responsible for the good exploitation results.

In order to guarantee the attainment of a safe and satisfactory management of the radwastes which will be produced there, a collaboration has been established between ENEA and Castagnetti SpA [*1*], focusing on the formulation and characterization of the cement products.

The radioactive wastes that will be cemented at the Montalto di Castro boiling water reactor (BWR) plant are

(*a*) evaporation concentrates,
(*b*) exhausted Powdex ion-exchange resins,
(*c*) mixed-bed exhausted ion-exchange resins,
(*d*) sludges from the traveling belt filter, and
(*e*) concentrates from the decontamination and the laundry.

The chemical composition of these wastes, as simulated for the characterization work, is given in Table 1.

The evaporation concentrates must be stored at temperatures higher than 40°C (104°F) in order to avoid sulphate crystallization. As for the other waste types, the wastes coming from the traveling belt filter are nearly solid, due to the relatively low water content after

TABLE 1—*Chemical composition of Montalto di Castro radioactive wastes.*

Waste Form	Composition	Percentage (w/w)
evaporation concentrates	Na_2SO_4	25
	NaCl	3.5
carbonates, phosphates, silicates		0.5
powdex ion-exchangers	dry powdex	30
	Fe_2O_3 (crud)	6
	H_2O	64
bead resins	Montedison Kastell C383 SM (dry)	24
bead resins	Montedison A503 SM (dry)	27
	H_2O	49
traveling belt	diatomaceous earths	41.6
filter sludges	PAO Termokimik	4.2
	PCH Termokimik	4.2
	H_2O	50
laundry concentrates	Solka Floc	16
	Rolfon 230	4
	H_2O	80

filtration. The laundry concentrates are in form of a very dense sludge composed of water-embedded detergents and cotton wool.

In general, for the simulated waste preparation the same chemicals used at the Italian power plants have been employed.

The overall production per year of the radioactive wastes is estimated to be

Evaporation concentrates:	740 m^3
Powdex ion exchangers:	17.1 tonnes (t)
Bead resins:	66.4 t
Filter sludges:	13.8 t
Laundry concentrates:	1.4 t

In order to overcome the problems due to the transfer of some waste type, and to optimize the overall reduction factors, the waste solidification by the cementation plant will be operated on mixtures of the different waste types, excluding the laundry wastes, according to the following scheme:

1. Cement + evaporation concentrates.
2. Cement + evaporation concentrates + Powdex resins.
3. Cement + evaporation concentrates + bead resins.
4. Cement + evaporation concentrates + filter sludges.
5. Cement + laundry concentrates.

The cement that will be used for the solidification is Portland 425 or Pozzolanic 325, both being easily available and of reliable supply.

Product Formulation and Characterization

The radwaste cement solidification described previously requires a preliminary product formulation and a careful product characterization whether to comply with TG 26 requirements or to optimize the volume reduction factors. The cement mixture selection has been

TABLE 2—*Waste/cement mixtures selected by mechanical strength.*

Characteristics	Evaporation Concentrates	Powdex	Bead Ion-Exchangers	Filter Sludges	Laundry Concentrates
waste/C	0.8	0.28	0.2	0.3	0.5
evaporation concentrates/C	...	0.8	0.55	0.8	...
antifoam, mL/100 g waste	...	...	...	...	1.0
compressive strength after 28 days, kN/m^2	17 000	12 000	15 000	15 500	13 000

done, taking into account the experimental results obtained in the characterization work, using first the compressive strength at 28 days as the screening parameter. Thus, the mixtures listed in the Table 2 have been selected. The selected mixtures have been tested accordingly to TG 26 following the above-mentioned procedures. Table 3 gives an overview of the test results.

The characterization of the solidified evaporation concentrates and filter sludge waste type seems to give the best results mainly as far as the compressive strength, immersion

TABLE 3—*Overview of main characteristics of Montalto di Castro cementation products.*

Characteristics	Evaporation Concentrates	Powdex	Bead Ion-Exchangers	Filter Sludges	Laundry Concentrates
waste/C	0.8	0.28	0.2	0.3	0.5
evaporation concentrates/C	...	0.8	0.55	0.8	...
antifoam (mL/100 g waste)	...	...	...	...	1.0
compressive strength, kN/m^2,					
after 28 days,	17 000	12 000	15 000	15 500	13 000
at 90 days, seawater immersion	22 000	8 500	7 000	18 000	16 000
at 90 days, table water immersion	22 000	10 000	9 000	21 000	14 500
after 30 freeze thawing cycles	12 500	8 500	distr.	17 500	13 500
volume change, %					
at 28 days	+0.2	+1.94	+0.8	+0.6	0
at 90 days, sea water immersion	+1.0	+3.6	+6.2	+0.9	+0.7
at 90 days, table water immersion	+0.5	+3.6	+1.8	+1.4	+0.3
after 30 freeze thawing cycles	+0.5	−1.1	...	+0.7	+0.6
free-standing water	no	no	no	no	no
leaching rate after 214 days, kg/(m^2 · s)					
cesium	3.1E-8	2.2E-8	2.3E-8	1.5E-8	
strontium	1.1E-9	6.0E-10	1.8E-9	4.0E-10	
cobalt	1.0E-10	4.0E-10	5.0E-10	1.0E-10	

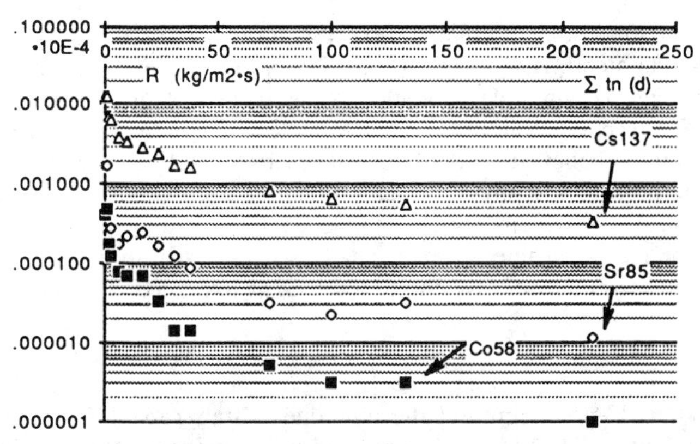

FIG. 5—^{137}*Cs,* ^{58}*Co,* ^{85}*Sr leaching factors of evaporation concentrate incorporated in CPZ 325* (*W/C* = *0.8*).

resistance, and thermal cycles are concerned. On the contrary, in the case of the solidified Powdex and particularly of the solidified bead ion-exchange resins, some problems arise regarding the thermal cycles. In fact, because of the different expansion coefficients between resins and cement and because of the lower thermal conductivity of cement containing ion-exchange resins, cracks are produced in the samples.

Regarding the results obtained from the long-term leaching test, as described before, the procedure adopted to perform the test is the ISO 6961-82 procedure at 40°C. The leaching has been performed placing the samples, as shown in Fig. 3, in accordance with ISO procedures.

The results shown in Table 3 and in Fig. 5, which refer to evaporation concentrates waste type only, were obtained using cylindrical samples of 5 cm in both diameter and height, traced with ^{137}Cs, ^{85}Sr, and ^{58}Co isotopes, and leached in Montalto di Castro tap water. The leachant volume was 250 cc.

The obtained results show a very high leaching rate for ^{137}Cs, whose cumulative leached fractions, in the case of the evaporation concentrate wastes, is more than 20% after about 200 days (Fig. 6). This difference was expected because, whereas the cobalt and strontium

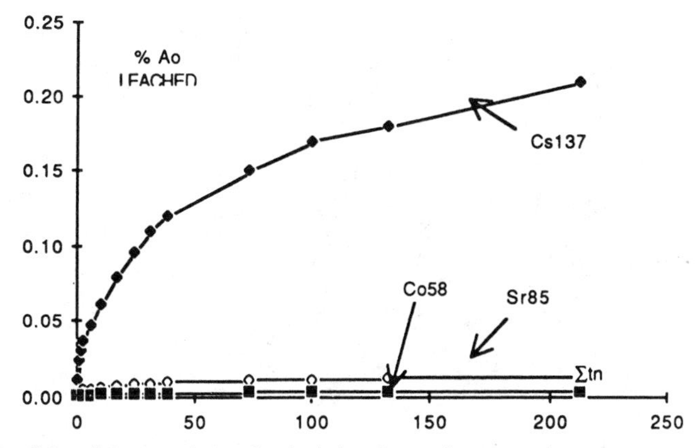

FIG. 6—^{137}*Cs,* ^{58}*Co,* ^{85}*Sr cumulative leached fractions of evaporation concentrates in CPZ 325* (*W/C* = *0.8*).

in the cementitious environment are in the form of insoluble idroxides, the cesium chemical form, as an alkaline metal, is of a very soluble ion in the cement pore liquid solution.

Although the high cesium release could appear to be not negligible, it is difficult to correlate the ISO leaching conditions to a possible accident scenario in a real working situation. The authors believe that the ISO test results should be considered very conservative, taking into account the physico-chemical parameters involved in the test: 40°C temperature, high surface-to-volume ratio, sample completely dipped in the leachant, and frequent leachant renewal (nine leachant changes in the first six weeks). Moreover, no correlation is possible at all in Italy, due to the lack of reference disposal sites.

Conclusion

The facility at ENEA is at present the only one in Italy capable of performing all the tests required by ENEA-DISP TG 26 on simulated and actual radioactive conditioned wastes.

Relevant information from the first series of results presented here is that liquid wastes produced at the BWR Montalto power plant, such as evaporation concentrates, filter sludges, and laundry concentrates, may be conditioned in cement also at relatively high ratio wet-waste/cement (W/C), surpassing the minimum requirements imposed by TG 26. The same results, on the other hand, confirm the difficulty for the direct conditioning of bead ion-exchange resins and Powdex ion-exchange resins in cement: the amount of exhausted ion-exchangers in fact should be kept quite low, in the range of 0.15 to 0.25 wet-waste/cement ratio.

References

[1] *Sicurezza e Protezione,* Vol. 3, No. 7, ENEA-DISP, Jan./April 1985.
[2] Association Contract, ENEA-Castagnetti SpA, No. 1558, 18 Sept. 1984.

Philip E. Rushbrook,[1] *Grant Baldwin,*[1] *and Christopher B. Dent*[1]

A Quality-Assurance Procedure for Use at Treatment Plants to Predict the Long-Term Suitability of Cement-Based Solidified Hazardous Wastes Deposited in Landfill Sites

REFERENCE: Rushbrook, P. E., Baldwin, G., and Dent, C. G., **"A Quality-Assurance Procedure for Use at Treatment Plants to Predict the Long-Term Suitability of Cement-Based Solidified Hazardous Wastes Deposited in Landfill Sites,"** *Environmental Aspects of Stabilization and Solidification of Hazardous and Radioactive Wastes, ASTM STP 1033,* P. L. Côté and T. M. Gilliam, Eds., American Society for Testing and Materials, Philadelphia, 1989, pp. 93–113.

ABSTRACT: This paper is the first detailed presentation of the work conducted during a three-year research program at Harwell Laboratory and sponsored by the U.K. Department of the Environment. The purpose of the work was to investigate the physical and leaching characteristics of the solidified products derived from mixing hazardous wastes with cement and pulverized fuel ash (PFA). The proportions of cement, PFA, and hazardous waste were varied over a wide range. Where possible, relationships between the physical characteristics of the solidified waste and its chemical composition were elucidated. The work has culminated in the development of a new concept to predict, at the time of mixing, whether the future physical and leaching behavior of a batch of solidified hazardous waste will satisfactorily meet regulatory requirements.

It has long been appreciated by regulatory authorities and operators in the United Kingdom that using engineering "standards," such as setting rate and compressive strength determinations, to define the acceptability of a newly prepared waste mix presents many disadvantages. Some of the components commonly found in hazardous wastes are known to retard the setting of cements such that the long-term physical characteristics of a solidified waste mix are not apparent until many days after preparation. The "quality-assurance" method described in this paper is considered to be a significant advance in enabling both operators to assess the acceptability of their waste mixes, and regulators to verify that any waste mix is likely to conform to the relevant regulatory requirements.

The technique presented is based upon establishing those proportions of waste, cement, and PFA which will produce a solidified product within the locally relevant "acceptable" limits (as laid down in site licenses or local regulations). Findings are based on the results of many trial mixes which were prepared using actual hazardous wastes commonly treated by solidification in the United Kingdom. For each mix, five parameters were measured: rate of setting, hydraulic conductivity, compressive strength, leachate composition, and supernatant retention. The results were used to prepare a "quality-control chart" based upon reasonable standards which could be expected in the United Kingdom for a solidified product. Naturally, different standards may be set by other regulatory agencies outside the United Kingdom. The method of derivation of a quality-control chart and the situations where an existing chart should be revised are discussed. On each control chart there is a "region" defining acceptable proportions

[1] Section leader-Waste Operation, higher scientific officer, and section leader-Waste Research Unit, respectively, Environmental Safety Centre, Harwell Laboratory, DIDCOT, Oxfordshire, OX11 ORA, United Kingdom.

of waste, cement, and PFA. This region of acceptability is likely to be dependent upon the chemical nature of the bulk waste streams at a treatment plant site. It will be necessary to prepare a separate quality-control chart for each individual plant; this chart will have to be redrawn if the plant is modified or if the composition of a bulk waste constituent, the cement, or PFA varies by more than a predetermined percentage.

KEY WORDS: hazardous wastes, acceptance testing, solidification, stabilization, cone penetrometer, permeability, sample preparation, laboratory tests, leaching tests

Given the current widespread production of hazardous wastes and the likelihood that much of its production will continue in the foreseeable future, it is perhaps surprising to realize that all waste-disposal methods currently are in a state of intensive development. This is apparent even for the "established" disposal methods such as landfill co-disposal and high-temperature incineration, all of which are being subjected to detailed technical evaluation and investigation. Such events would have been impossible to predict a few years ago.

Solidification has been used to a lesser or greater extent in most industrialized nations to immobilize a variety of hazardous wastes. In the United Kingdom, the only processes used commercially involve mixing pretreated wastes with ordinary portland cement (OPC) or both OPC and pulverized fuel ash (PFA). Other similar processes reported to be in use in North America include PFA with lime and soluble silicates with cement [*1*].

Solidification is viewed not as providing indefinite containment, but rather as one which retards the rate of release of hazardous chemical species and prevents them from entering the environment at unacceptable concentrations. The cement-based solidification process is purported to be suitable for a wide range of inorganic wastes. When properly prepared, the solidified product should allow only a very gradual release of toxic components into the environment [*2*]. Furthermore, the rate of release should not lead to a measurable adverse effect on the environment.

In addition to cement- and silicate-based processes, various other materials have been used, although to date many appear to have found only specialized applications, for example, vitrification of liquid radioactive wastes and organic polymer preparations to clear up oil spills.

Much of the information on the physical behavior of solidified waste products and the mechanisms by which they retain waste compounds is unpublished or cloaked behind commercial confidentiality. This has caused difficulties for regulators trying to evaluate the environmental suitability of a solidification technology and its ability to retain a particular waste. Furthermore, an intrinsic difficulty, reported in both the United Kingdom [*3*] and the United States [*4*], is that the long-term physical characteristics of solidified wastes do not become established until several days after preparation of the product. These characteristics include the rate of formation of the cement-silicate matrix, long-term compressive strength, leaching behavior, and hydraulic conductivity.

The U.K. Department of the Environment (DoE) was aware of the problems and weaknesses in the underlying scientific knowledge of solidification technology encountered by waste disposal authorities when seeking to license and regulate treatment plants. Historically, this led to operations being licensed to produce solidified waste products which had to conform to minimum physical characteristics (that is, typically setting rate and compressive strength) which were only apparent after 28 days.[2] In reality, a waste mix which did not meet the minimal criteria might have already been landfilled for some time and might be no longer recognizable or retrievable.

[2] This period of time seems to have been taken from the testing of concrete in the construction industry, where it is deemed that concretes have obtained 95% of their final strength after 28 days. (See British Standards Institution Methods of Testing Concrete [BS 4550].)

Over the past four years, the DoE has sponsored research to achieve a better understanding of the behavior of cement-based solidified wastes. The Harwell Laboratory has conducted research to investigate the major physical properties resulting from various compositions of cement, wastes, and other ingredients. The purpose of this work was twofold:

(*a*) produce an improved means to predict the likely long-term field characteristics and
(*b*) identify those parameters which could be measured most easily by waste-disposal regulators.

The range of proportions of constituents and the methods of waste pretreatment used in producing the solidified waste samples was designed to resemble the conditions likely to be found at plants operated commercially.

A brief overview of the results obtained has been presented elsewhere [5] together with an introduction to the concept of "quality-control charts" for use at solidification treatment plants. This paper provides the first detailed presentation of the Harwell research program and its findings. Running in parallel with this work has been a study conducted at Imperial College (University of London) on the microstructure of solidified waste products. This was designed to elucidate some of the physico-chemical mechanisms by which selected heavy metals are retained within the solidified cement matrix [6]. Work of this nature, which is continuing, is important since the measurement and prediction of "external," physical properties of solidified wastes as reported here would be considerably enhanced by an improved knowledge of the "internal" mechanisms by which the waste components are retained in the cement matrix.

Experimental Procedures

The research program was designed to address several areas of interest:

1. How do the results from waste/cement mixes designed to simulate well-managed operations compare with those designed to simulate poorly managed operations (including inadequate pretreatment)?
2. How do the physical and leaching properties of solidified waste vary with different proportions of waste, cement, and other components?
3. How representative are the results from laboratory samples when compared with those from products with similar waste/cement proportions prepared at full-scale treatment plants?
4. How will the solidified waste product deteriorate over several years by being subjected to continued leaching and exposure to extremes of temperature?
5. How can it be properly ensured at the time of preparation that the solidified waste ultimately will meet acceptable physical and leaching criteria set by regulatory agencies? Furthermore, in the past, these criteria have often been set for a product which has been allowed to cure for, say, 28 days, and not for the quality of the original mix immediately after preparation.

This paper discusses the work undertaken to date which has been designed to improve our scientific understanding in each of the above areas. Progress in some areas has been marked, namely, the comparison between pretreatment practices, comparison between the physical and leaching properties of different waste/cement compositions, and the establishment of an approach to better ensure at the time of preparation, production of "acceptable" solidified products. However, the research program is still under way, and the results of current work will be reported in due course.

In the laboratory program, five physical and leaching parameters were investigated for each of the waste/cement proportions used:

(*a*) cone penetrometry to measure setting rate (BS 4550),[3]
(*b*) applied mechanical pressure to measure compressive strength (BS 4550),
(*c*) laboratory techniques to measure hydraulic conductivity [*7*],
(*d*) leaching tests and subsequent analysis of heavy metal concentrations in leachates [*8*], and
(*e*) liquid retention and its analysis.

The tests listed were selected because they have either been conducted on solidified wastes during previous studies (for examples, Refs *2* and *4*) or are commonly used in testing similar materials such as concretes, bitumens, and clays. Wherever possible, recognized test procedures were used to measure each parameter. Furthermore, it was considered that these tests could also be performed satisfactorily in any modestly equipped laboratory at a treatment plant, and collectively, the results should provide a reasonable working assessment of the likely field characteristics of a solidified waste product.

The research program was conducted in two stages. Stage 1 involved the preparation of 20, 0.75-kg samples, each containing varying proportions of cement, liquid and solid wastes, and PFA. The intention at this stage was to prepare a solidified waste product which resembled those that might be expected from a *poorly managed* treatment operation. This was achieved by performing only the minimum of waste pretreatment and mixing. The physical and leaching characteristics observed in each solidified waste product, therefore, were regarded as a reasonable worst case situation for the waste streams used and the method of solidification. In addition to the basic 20 samples, variants were produced for some waste/cement compositions. These addressed related areas of interest, such as changes in the rate of setting due to the substitution of PFA by spent foundry sand, and a comparison between samples cured in the laboratory and replicates exposed to the vagaries of the prevailing weather conditions outdoors.

Stage 2 of the laboratory program provided a contrast to Stage 1. The intention from the outset was to produce solidified wastes which were representative of what was considered to be a *well-managed and carefully controlled* operation. Consequently, another 20, 0.75-kg mixes were prepared with similar, though not identical, compositions to those in Stage 1. In addition, for three of the waste/cement compositions (Samples F11 [7% cement], F14 [9%], and F15 [12%]) (Table 1), "bulk," 5-kg samples were also prepared and their physical characteristics were compared with the "standard" 0.75-kg samples used throughout the rest of the program. The purpose was to assess the effect of variations in sample size on the reproducibility of experimental results. In general it was found that the standard samples set at about twice the rate of the bulk samples over the first seven days, although ultimately, after 28 days the setting rates had converged (Fig. 1). There also appeared to be little difference between the performance of standard and bulk samples for the low, intermediate, and higher cement proportions.

Solidified Mix Components

A fundamental intention of this research was to prepare solidified wastes which resembled as closely as possible the characteristics of those produced by commercial operators. Consequently, a mixture of five hazardous wastes was chosen and incorporated into each sample,

[3] In addition to BS 4550, see Penetration of Bituminous Materials (BS 4691) and Methods of Test for Soils for Civil Engineering Purposes (BS 1377).

TABLE 1—*Percentage composition of each sample waste mix.*[a]

			Liquid Waste				Solid Waste		
Stage of Study	Sample No.	Cement	Oxidized Solid Cyanide Waste (in Liquid Form after Oxidation)	Oxidized Liquid Cyanide Waste	Neutralized Acid/Alkali	Total Liquid Waste	PFA	Metal Slags	Total Solid Waste
Stage 1	A1	16	22	23	32	77	—	7	7
	A2	4	22	23	32	77	12	7	19
	A3	25	26	26	17	69	—	6	6
	A4	11	17	17	17	51	33	5	38
	A5	33	19	20	22	62	—	6	6
	A6	16	19	20	22	62	16	6	22
	B1	11	34	21	20	75	—	14	14
	B2	5	24	16	15	53	32	10	42
	B4	14	36	21	20	72	—	14	14
	B6	17	32	19	19	70	—	13	13
	C1	27	17	17	10	44	27	2	29
	C2	4	26	29	16	71	21	4	25
	C3	7	9	9	5	23	69	1	70
	D1	2	36	28	9	71	12	15	27
	D2	4	24	19	7	50	36	10	46
	D3	5	24	19	7	52	33	10	43
	E1	3	12	24	31	67	30	0	30
	E2	6	13	27	37	77	17	0	17
	E3	11	6	13	17	36	53	0	53
	E4	8	9	20	25	54	38	0	38
Stage 2	F1	2	23	21	21	65	17	16	33
	F2	2	18	16	16	50	24	24	48
	F3	2	14	13	13	40	29	29	58
	F4	2	10	10	10	30	34	34	68
	F5	4	19	18	18	55	21	20	41
	F6	4	15	15	15	45	26	25	51
	F7	4	13	11	11	35	31	30	61
	F8	4	9	8	8	25	36	35	71
	F9	7	20	20	20	60	17	16	33
	F10	7	18	16	16	50	22	21	43
	F11	7	14	13	13	40	27	26	53
	F12	7	10	10	10	30	32	31	63
	F13	9	19	18	18	55	18	18	36
	F14	9	15	15	15	45	23	23	46
	F15	12	18	16	16	50	19	19	38
	F16	12	14	13	13	40	24	24	48
	F17	14	23	21	21	65	11	10	21
	F18	14	19	18	18	55	16	15	31
	F19	14	15	15	15	45	21	20	41
	F20	17	18	16	16	50	17	16	33

[a] All percentages are rounded to the nearest whole number.

rather than using laboratory reagents as simulants. Although we recognize that this introduced a potentially greater heterogeneity into the solidified waste compositions, it was considered that their use provided an opportunity to assess in a more representative manner the likely performance of solidified wastes deposited in a landfill site. The individual wastes selected were considered to be representative of large volume arisings processed by cement-based solidification operators in the United Kingdom, that is,

(*a*) liquid and solid cyanide wastes,
(*b*) caustic waste solution,
(*c*) mixed inorganic acid solution (nitric and chromic), and
(*d*) metal bearing slags (from lead and arsenic smelters).

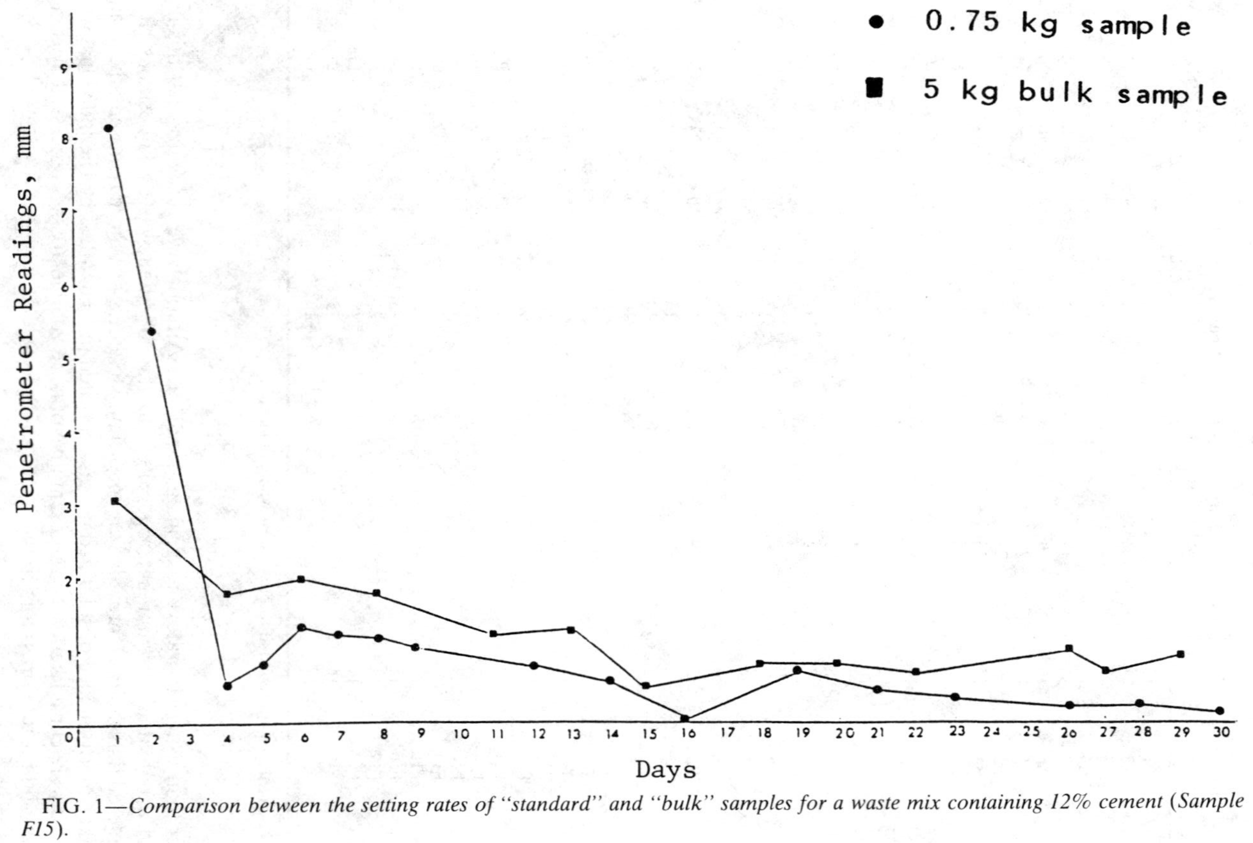

FIG. 1—*Comparison between the setting rates of "standard" and "bulk" samples for a waste mix containing 12% cement (Sample F15).*

All mixes were prepared using OPC,[4] the cement widely (if not exclusively) used by commercial operators in the United Kingdom, and possibly also in the United States [*9*]. Nonstandard formulations or those with specific properties (for example, resistance to sulphate) were not considered.

In addition to hazardous wastes and cement, PFA was also added. Several solidification process patents have claimed that the inclusion of PFA enhances the setting processes [*10–12*]. The benefits presumably arise from the pozzolanic characteristics of PFA which improve the setting rate of cements [*9*]. However, the quality of PFA varies depending on the characteristics of the coal-fired furnace producing it and the origin of the coal. Separate experiments were conducted in a side study to investigate the reproducibility of setting rates with cements mixed with varying proportions of PFA. PFA from three sources were mixed with cement and water in a ratio of 3:1:1 (cement:water:PFA) and their setting rates measured over an eight-day period (Fig. 2). A wide variability in characteristics was found. Consequently, PFA from one source, Drakelow power station, was used predominantly throughout the research program.

Mixing Procedure

Broadly, the research program in both stages followed the same procedure:

1. Pretreatment: —oxidation to cyanate of cyanide-containing solid and liquid wastes using sodium hypochlorite
   —neutralization of acid waste with alkali waste solutions
   —size reduction of other solid wastes by grinding
2. Sample production:—combine wastes and mix
   —add PFA and cement, and mix
   —dispense prepared waste mixes into molds
3. Analysis —store product at between 15 and 20°C, away from direct sunlight and drafts
   —perform physical and leaching tests on the prepared waste mixes

Pretreatment and Preparation of Solidified Wastes

Cyanide-containing wastes were treated with sodium hypochlorite and sodium hydroxide. These originally contained 4.9% by weight of cyanide in the solid waste and 2.8% in the liquid waste. In the "poorly managed" scenario used in Stage 1, a significant concentration of cyanide (approximately 20% total cyanide) remained. This was as a result of permitting only a short reaction time (for example, less than two days), the establishment of only a moderate alkaline pH in the reaction tank and the infrequent mixing of reactants. In contrast, Stage 2 reactants were continuously mixed and maintained at pH 10.5 for eight days. Correspondingly, only a small residual concentration of cyanide (less than 1% total cyanide) remained.

The liquid caustic waste was adjusted to pH 7 by the controlled addition of the mixed acid waste.

[4] See BS12—1978 ordinary and rapid-hardening portland cement, BS146—specification for portland-blastfurnace cement, BS915—1972 specification for high alumina cement, BS1370—1979 specification for low heat portland cement, BS4027—1980 specification for sulphate-resisting portland cement, BS4246—specification for low heat portland-blastfurnace cement, and BS4248—1974 specification of supersulphated cement.

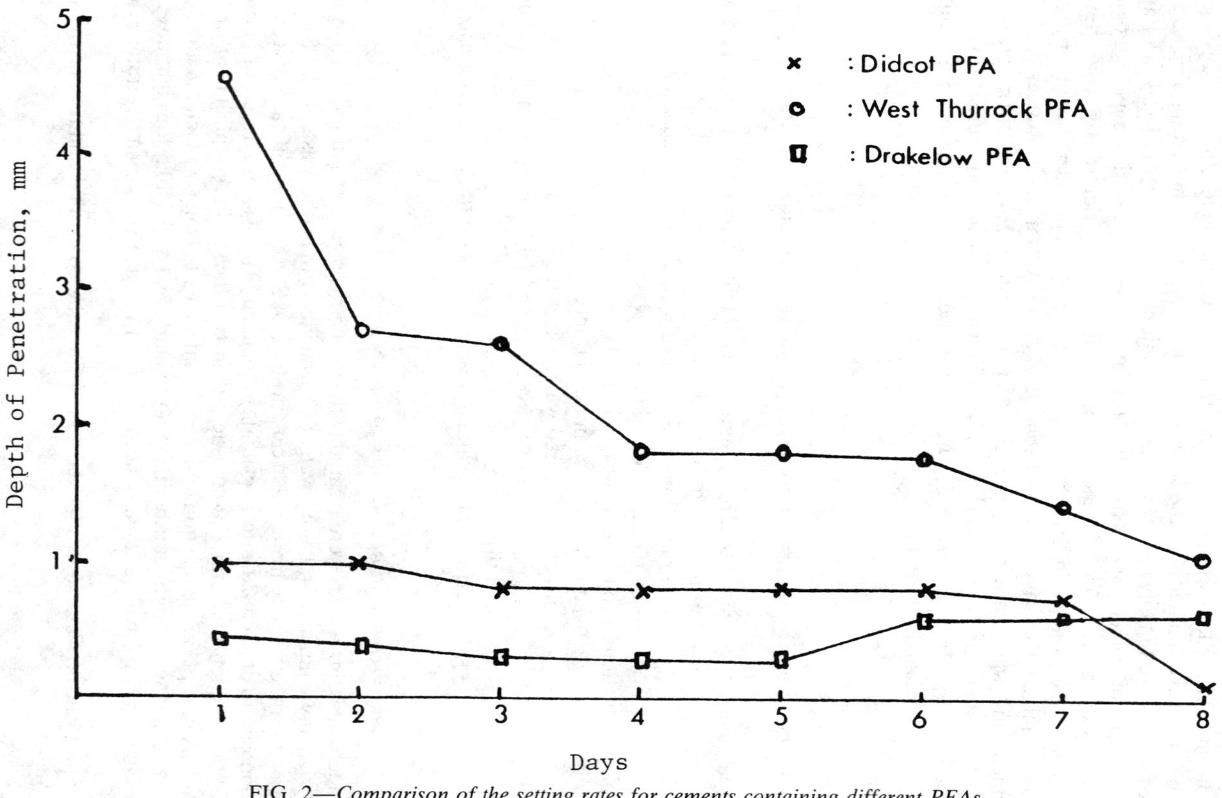

FIG. 2—*Comparison of the setting rates for cements containing different PFAs.*

The preparation of samples for solidification was as follows: Individual pretreated waste streams were first combined and mixed. PFA was then added and mixed. Last, cement was added, and again the sample was thoroughly mixed. Details of the compositions of the solidified waste samples are given in Table 1. Approximately 1.5 kg of each waste mix was prepared. A 0.75-kg specimen was poured into a tray and left undisturbed in a cool room maintained at between 15 and 20°C. This specimen was used for determining the rate of setting and the rate of supernatant production, and later provided material for compressive strength tests and leaching experiments. A further 400-g specimen was placed into a cylindrical mold and allowed to set for later measurement of hydraulic conductivity. The relative ease or difficulty in applying each test method is discussed in an earlier publication [5].

Results

Setting Rate

During the first ten days, the rates of setting were comparatively rapid, but declined quickly thereafter. This is a typical response found in cements. By Day 28, the penetrometry measurements were approximated (as per BS 4550) to 95% of their final strength. The improvement in setting characteristics found beyond this time was negligible. An example for one waste mix is presented in Fig. 3. The most significant influence on the rate of setting was the proportion of liquid waste in the mix. Five days after preparation, Stage 2 mixes with over 40% liquid waste generally achieved no better than 60% of their final, Day 28 strength. Those mixes with less than 40% liquid waste set faster and produced between 72% to 98% of their final strength. Figures 4 and 5 show the results of a seven-day cone penetrometer test in Stages 1 and 2. Depending on the "local" criteria of acceptability set by a regulatory authority (for example, less than 1.5 mm penetration is acceptable), those mixes which "pass" or "fail" the criteria can be read off from these figures. Predominantly, mixes with higher cement (more than 7% by weight) and lower liquid content (less than 55% by weight) were the most successful in achieving penetration values below 1.5 mm. Mixes with less than 40% liquid waste gave penetration values of below 0.5 mm. A similar comparison was also made between samples after 28 days.

Mixes containing very low cement contents (4%) and high liquid waste loadings (50%) either did not cure or cured only after a considerable period of time.

A brief comparison between the setting rates of identical specimens, one left to set in the laboratory and the other exposed to winter weather conditions (between −4 and 13°C) exhibited marked differences. The setting rates of weathered specimens exhibited a more variable and slower trend than unweathered laboratory specimens. No clear evidence was found to indicate that any of the meteorological factors which were continually monitored, such as maximum and minimum air temperatures and rainfall, were dominant in controlling the setting rates. However, given the greater variability of properties exhibited by weathered samples, it was considered prudent to use unweathered samples which had been cured in a controlled environment for the measurement of the performance of a waste mix.

In general, PFA improved the rate of setting of waste mixes compared with the performance of analogous samples containing spent foundry sand. This behavior was most significant in those mixes containing a low proportion of cement (that is, less than 4%). For example, in Fig. 6 comparisons are given for two waste mixes (Sample C1, 27% cement, 44% liquid waste, and Sample C2, 4% cement, 71% liquid waste). The PFA and foundry sand samples containing a high proportion of Cement C1 both set rapidly at a similar rate with penetrations of less than 1 mm by Day 3. Conversely, the samples containing a low proportion of Cement C2 showed a wide disparity between using PFA and foundry sand. The specimen containing

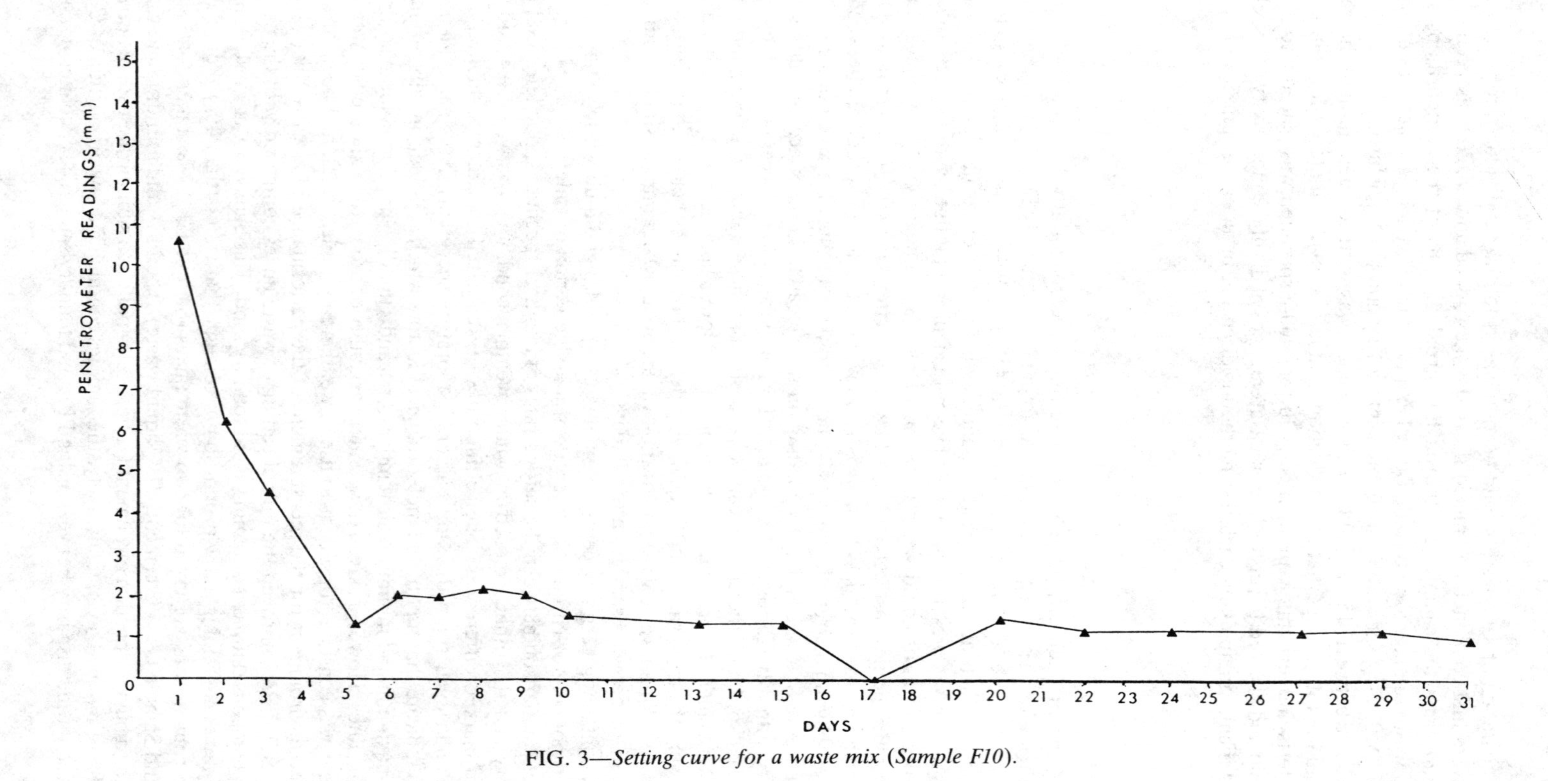

FIG. 3—*Setting curve for a waste mix (Sample F10).*

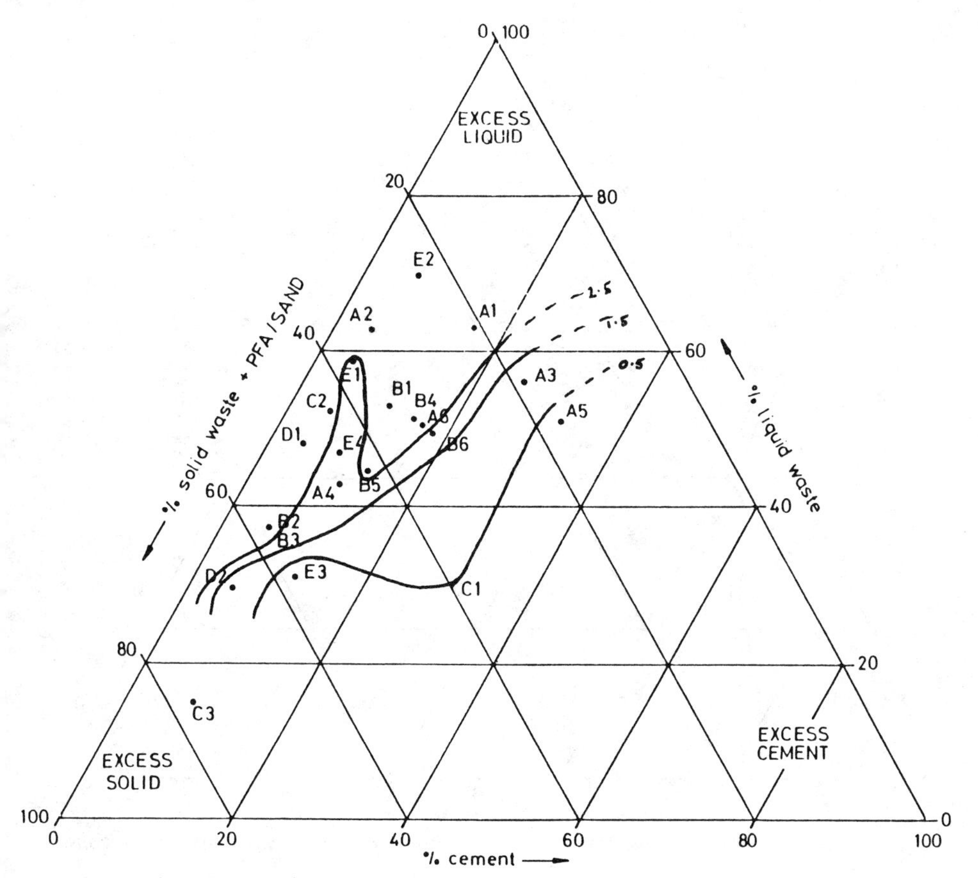

FIG. 4—*Day 7 penetrometer measurements plotted against the Stage 1 waste mix compositions.*

foundry sand cured slowly and by Day 7, 12 mm penetration was measured, which improved to only 3 mm penetration by Day 28. By contrast, the PFA specimen had a penetration of 8 mm at Day 7 and less than 1 mm by Day 28. It was concluded that the incorporation of foundry sand had little effect on the rate of setting of waste mixes containing high proportions of cement, but significantly delayed the setting performance of waste mixes containing low proportions of cement.

Compressive Strength

All solidified wastes when subjected to compression failed in a brittle/compliant manner and exhibited lower compressive strengths than traditional formulations of concrete. Although insufficient material was available to prepare British Standard (BS) size samples (that is, 1000 cm^3), the compressive strength values for mixes containing a high proportion of liquid waste (greater than 55%) were in the region of below 1×10^6 Nm^{-2}. Higher cement contents (greater than 10%) or lower liquid contents or both gave better compressive strength values, between 3 and 5×10^6 Nm^{-2}, that is, approximately the strength of a sugar cube. Other work at Harwell has achieved compressive strengths up to 17.5×10^6 Nm^{-2} for some solidified waste (20% cement, 30% liquid waste).

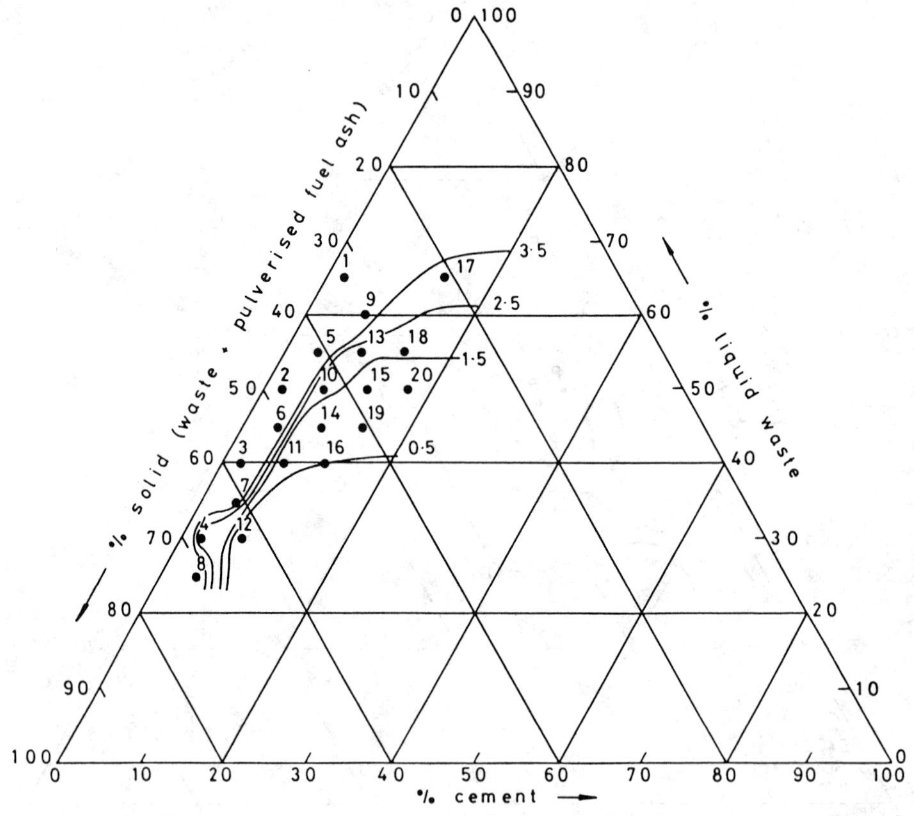

FIG. 5—*Day 7 penetrometer measurements plotted against the Stage 2 waste mix compositions.*

In Stage 2, only solidified waste products containing over 12% cement gave compressive strength values capable of bearing the load of a typical 10-tonne truck (approximately 2×10^6 Nm^{-2}) (Fig. 7). Based on the laboratory results, this would suggest that at lower cement/higher liquid waste compositions a vehicle traversing across the top of a solidified waste landfill could cause localized crushing and powdering of the landfill surface.

Hydraulic Conductivity

Two methods were used to measure hydraulic conductivity of water through the solidified wastes, but neither method produced satisfactory results. Consequently, a detailed appraisal of the performance of each waste mix was not possible. The main problem encountered was the poor reproducibility of results obtained. These were in the range 10^{-3} to 10^{-5} cm/s^{-1}. However, they are within the range of published values claimed for solidification processes (Table 2). Further development work is in hand to improve the precision and reproducibility of hydraulic conductivity measurements, particularly for "weak" solidified wastes.

Leaching Tests

Eight metals were investigated, each present at a significant concentration in at least one of the original wastes (Table 3). With the exception of arsenic, the retention of metals was

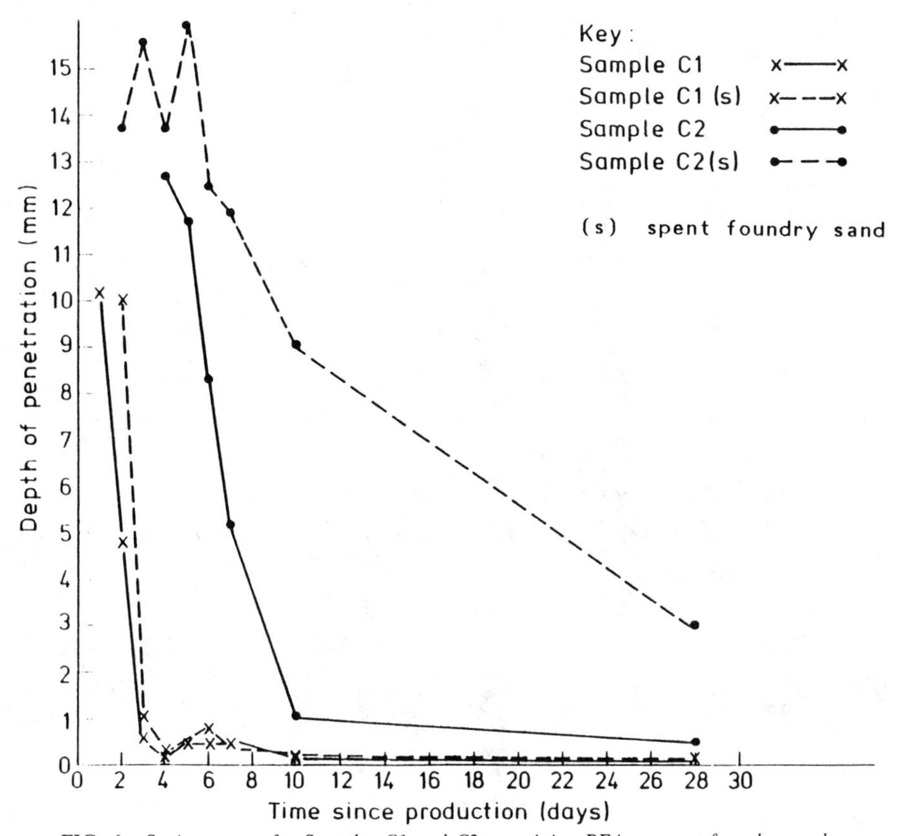

FIG. 6—*Setting curves for Samples C1 and C2 containing PFA or spent foundry sand.*

superior in the Stage 2 waste mixes than in those from the Stage 1, poorly managed waste mixes. Less than 1% of the total content of each metal was leached after an accelerated leaching test was conducted using shakers. In reality, this would probably mimic decades of leaching in the field. In the Stage 1 samples, considerably higher leaching rates were obtained, with chromium, copper, nickel, and zinc all demonstrating at least a 50% loss from some samples. Once again those samples with low cement proportion (less than 7%) or high liquid ratio (greater than 55%) or both exhibited the poorest retention. A factor of two difference for copper, chromium, and zinc and a factor of five for lead and nickel was found in the Stage 2 samples when the higher cement/lower liquid and lower cement/higher liquid waste mixes were compared. In the earlier Stage 1 work, variations up to an order of magnitude between these groups of samples were obtained.

Three factors relating to the waste pretreatment procedures may have contributed to the improved retention of metals in the Stage 2 samples:

(*a*) greater degree of mixing during preparation,
(*b*) small waste particles sizes, and
(*c*) possibly also, improved cyanide destruction.

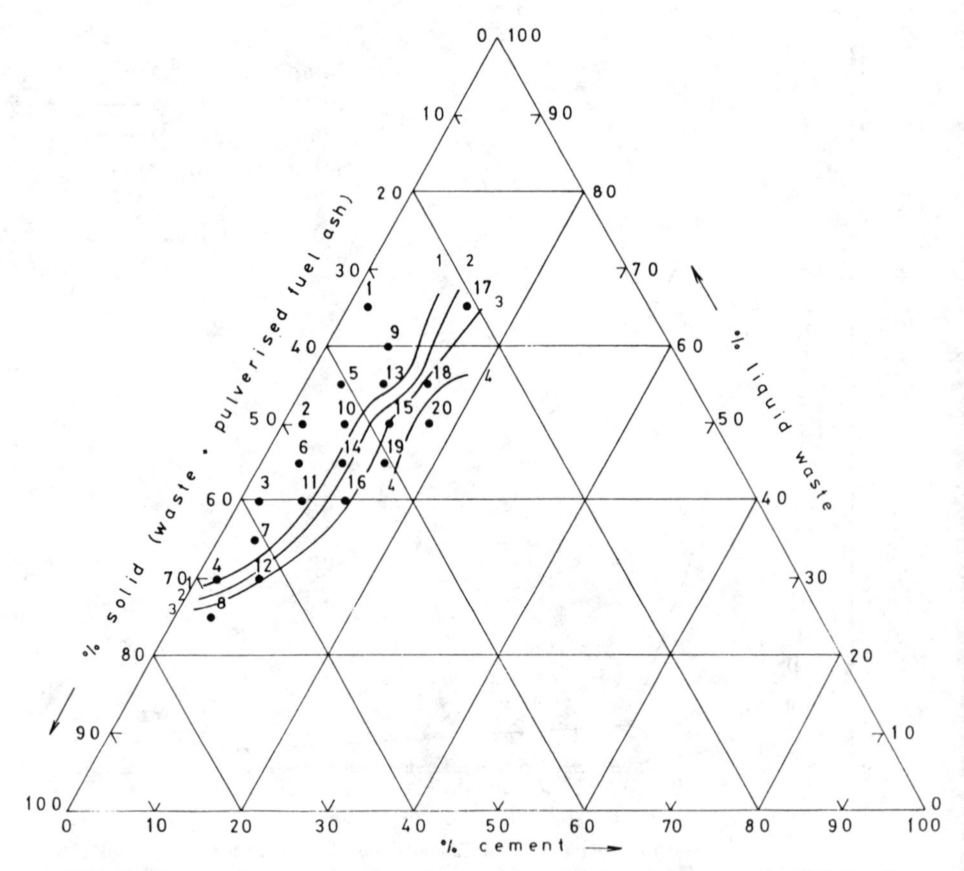

FIG. 7—*Compressive strength of Stage 2 waste mixes plotted against composition* (10^6 Nm^{-2}).

On this last point, a separate study provided no evidence of excess residual cyanide in Stage 1 samples complexing heavy metals and thereby promoting their enhanced mobilization.[5]

The leaching of arsenic was considered only in Stage 2 and was the only metal found to be leached in significant amounts. An unpublished study[6] which investigated the incorporation of arsenic-bearing wastes into concretes suggested that arsenic retention was strongly influenced by the proportion of cement and PFA used. It is surprising, therefore, that there is little evidence in our results to support this view. However, the findings presented in Table 4 suggest a possible relationship between liquid waste content and arsenic release. It is intended to undertake further work to ascertain if other factors such as leachate pH or the nature of the arsenic species in the original waste influence the observed mobility of arsenic.

Supernatants

Solidified waste mixes containing over 35% liquid on standing were found to produce clearly identifiable supernatant in laboratory samples. The presence and subsequent reten-

[5] Belcham, H., Young, P. J., and Heasman, L. A., U.K. Atomic Energy Research Establishment, 1985, unpublished report.

[6] Cook, J. D. and McGugan, P., U.K. Atomic Energy Research Establishment, 1975, unpublished report.

TABLE 2—*Average permeabilities of clay, concrete, and solidified wastes.*

Material	Hydraulic Conductivity Values, m/s^{-1}	Reference
Clay	10^{-5}	[*13*]
Concrete: good quality	10^{-8}	[a]
average quality	10^{-6}	[a]
poor quality	10^{-4}	[a]
Proprietary solidification materials: (published estimates)		
Poz-O-Tec	10^{-3} to 10^{-6}	[*9*]
Enviroclean	10^{-3} to 10^{-5}	[*9*]
Terra Crete	10^{-5}	[*9*]
Chemfix	10^{-3} to 10^{-4}	[*9*]
Sealosafe	10^{-5}	[*14*]
Stablex	10^{-3}	[*15*]
Environmental Technology Corporation	10^{-4}	[*9*]
Environmental Safety Centre solidified waste samples—Stage 2	10^{-3} to 10^{-5}	

[a] Atkinson, A., U.K. Atomic Energy Research Establishment, 1984, personal communication.

TABLE 3—*Composition of heavy metals in individual waste types.*

	Oxidized Solid Cyanide Waste[a]	Oxidized Liquid Cyanide Waste[a]	Lead Slag	Arsenical Slag	Acid Waste[a]	Alkali Waste[a]
Cadmium mg · kg^{-1}	<0.024	<0.025	550	1.1	0.137	<0.16
Chromium mg · kg^{-1}	0.24	<0.08	11 400	50	0.55	<0.16
Copper mg · kg^{-1}	0.53	0.5	3 180	4 080	1165	<0.16
Iron mg · kg^{-1}	3.7	1.1	199 000	166 000	109	12.3
Nickel mg · kg^{-1}	<0.08	<0.08	5 190	3 120	1.52	<0.39
Lead mg · kg^{-1}	<0.4	<0.4	12 800	14 700	0.69	0.78
Zinc mg · kg^{-1}	0.65	3.2	342 000	80 900	2.34	3.12
Arsenic mg · kg^{-1}	ND[b]	ND[b]	<3	6 930	ND[b]	ND[b]

[a] Metal compositions in these wastes are converted from mg l^{-1} to mg · kg^{-1} by multiplying by the density values, that is,

oxidized solid cyanide	1.23 (mg · kg^{-1})
oxidized liquid cyanide	1.19
lead slag	3.01
arsenical slag	3.15
acid waste	1.45
alkali waste	1.28

[b] ND = not detected.

TABLE 4—*Percentage of arsenic released in the leachate from each Stage 2 mix.*

Sample No.	Cement Content, %	Liquid Waste Content, %	Arsenic Release, %[a]
F1	2	65	17.7
F2	2	50	18.9
F3	2	40	2.8
F4	2	30	9.2
F5	4	55	24.3
F6	4	45	15.4
F7	4	35	4.5
F8	4	25	6.0
F9	7	60	19.2
F10	7	50	10.4
F11	7	40	10.1
F12	7	30	1.4
F13	9	55	17.8
F14	9	45	5.0
F15	12	50	28.8
F16	12	40	1.8
F17	14	65	13.8
F18	14	55	5.6
F19	14	45	4.6
F20	17	50	...

[a] After only one bed volume of water, that is, the volume of water required to just saturate the waste.

tion of the supernatant was strongly related to the proportion of liquid in the initial mix (Fig. 8). As is the case with most concretes, more water than is required for the hydration reactions is added to aid mixing. As a consequence, an increase in the proportion of cement used, from 2% to 18%, had only a small effect on the additional amount of liquid (less than 10%) which could be incorporated without increasing the quantity of supernatant produced or the retention time. Further work is needed to establish the correlation between laboratory results and supernatant production from landfilled solidified wastes.

Implications of the Results on the Control of Solidification Operations

It is apparent that a better understanding of the chemical and physical processes occurring within waste mixes as they solidify is only beginning to emerge [*16–18*]. To date, most work has concentrated upon establishing more reliable "external" measurements to gauge the field performance of a solidified waste, notably setting rates, compressive strength, permeability, and leachate quality. This is considered to be a sensible and practical approach to follow until a better understanding of the relationship between external physical field performance and the internal physico-chemical solidification processes occurring within a mix is obtained.

In this study, the physical properties and leaching characteristics of waste mixes are generally improved by increasing cement and decreasing liquid waste proportions. Furthermore, it is evident that different proportions of cement, waste (liquid and solid), and PFA can be identified which produce solidified products with characteristics likely to be acceptable, marginally acceptable, or unacceptable to regulatory authorities. To illustrate this, the results from the second stage of the laboratory program were plotted onto triangular graph paper (see Figs. 5, 7, and 8). From this type of chart, those mixes which are likely

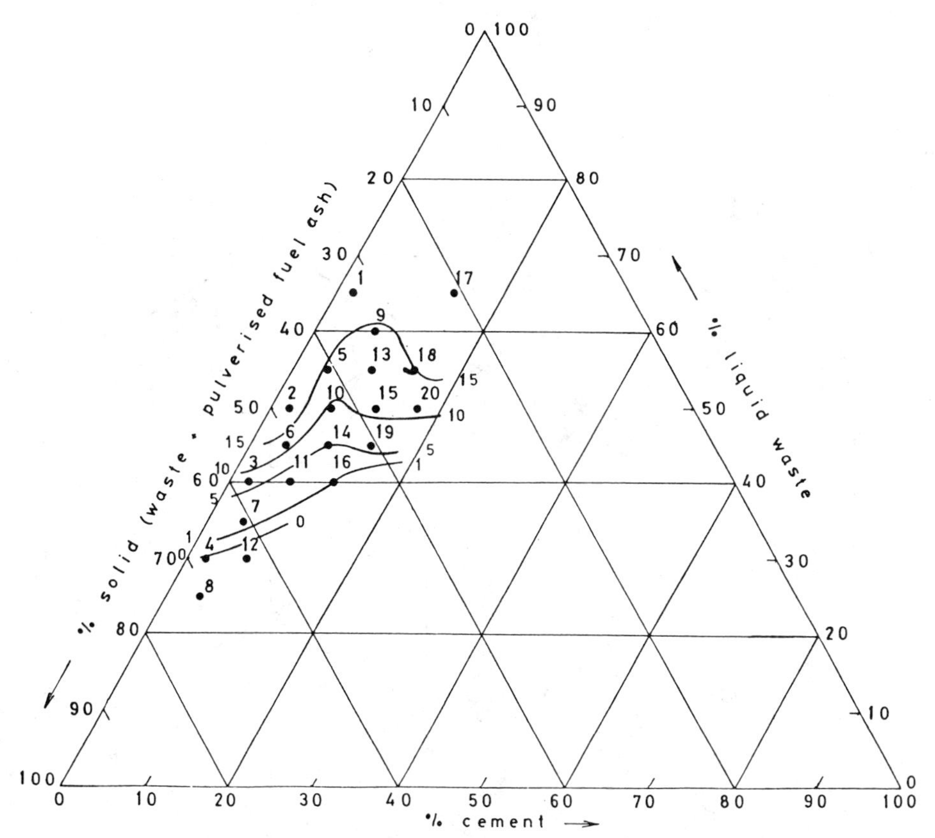

FIG. 8—*Retention of supernatent by Stage 2 waste mixes (days) (stored horizontally at between 15 and 20°C).*

(on the basis of laboratory results) to either pass or fail the selected performance criteria (for example, × 10^6 Nm^{-2} compressive strength after 28 days) can be easily identified.

In the United Kingdom, the setting of physical and leaching criteria to determine "acceptable" and "unacceptable" waste mixes is established independently by each local regulatory authority on a site-by-site basis. Elsewhere, the criteria could be set at regional or national level. The quality-control procedure introduced in this paper is intended to enable the quality of waste mixes to be assessed during production rather than after 28 days (or some other agreed length of time). This is achieved by extrapolating back the selected 28-day physical characteristics obtained from a range of known waste/cement proportions to predict the likely characteristics of mixes containing identical waste/cement proportions produced in future operations. Such a technique offers a significant advantage to both regulators and operators.

Implementation of the procedure is site-specific and involves three stages; the first two stages are concerned with analysis, and the third stage involves the production of a "quality-control chart" (as presented in Fig. 9).

The first stage involves conducting physical tests on sample mixes using the bulk wastes produced at the pretreatment plant. It is envisaged that 20 to 30 mixes would be prepared, representing a broad range of proportions of cement, waste, and PFA. These tests could

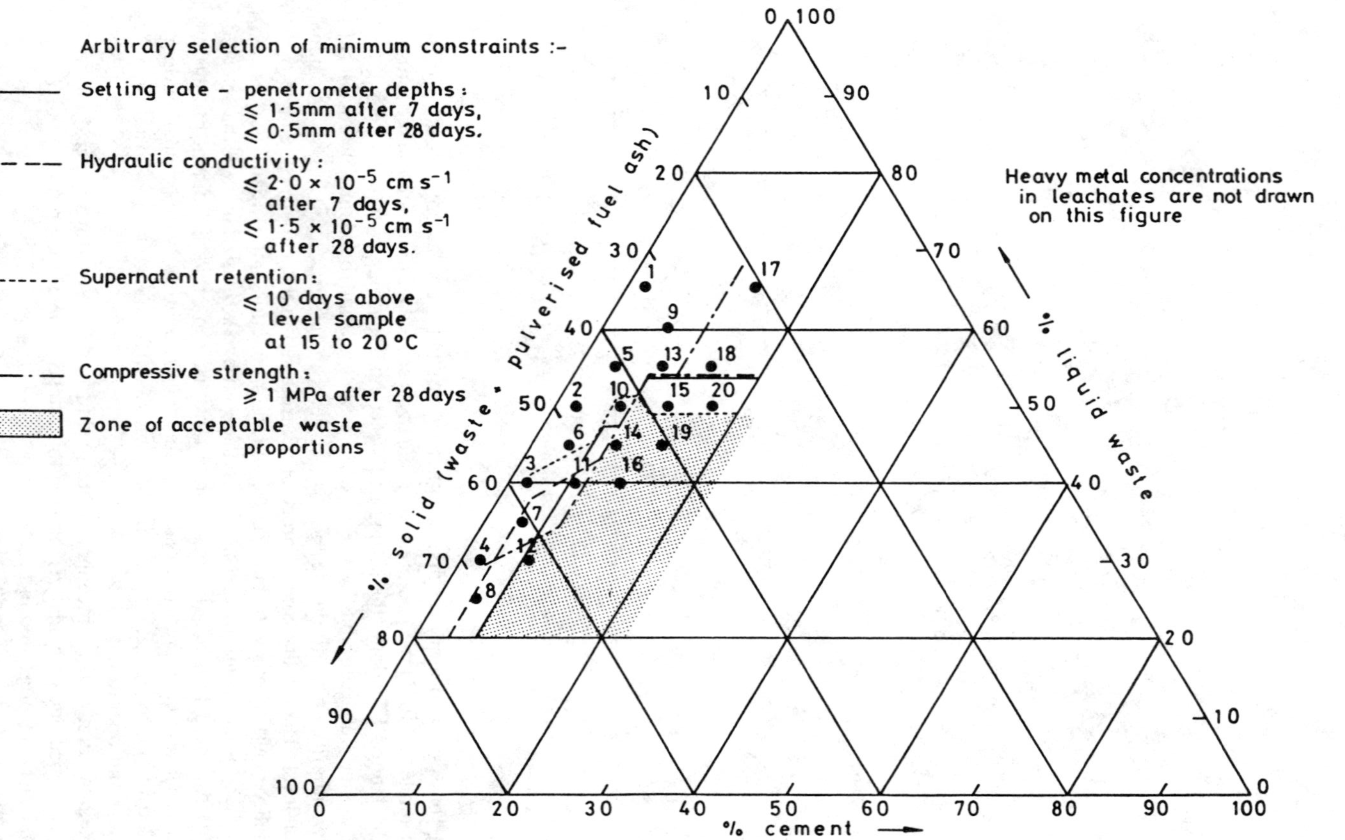

FIG. 9—*Example of a solidification control chart with a suggested "zone of acceptable waste proportions"* (*Stage 2 samples considered here*).

include setting rate, compressive strength, supernatant retention, hydraulic conductivity, and perhaps also heat of reaction and porosimetry. The results would then be plotted onto triangular diagrams (for example, Fig. 9) and any discernible trends noted. In Fig. 9, trends have been indicated using isometric lines.

The second stage involves carrying out leaching tests on those solidified samples which satisfy the physical criteria acceptable to the regulatory authority. Analysis would include heavy metal concentrations and other parameters required by the regulatory authority.

The third stage is the preparation of a control chart. Minimum values to which an acceptable mix must conform would be set by the regulatory authority. A control chart similar to Fig. 9 then would be drawn up using the results from the previous stages. This chart would indicate the compositions of those mixes falling above and below the minimum acceptable values for each physical criterion. Subsequently, the composition of all mixes produced by the solidification process would be required to fall within the zone of acceptability. Values for all of the physical tests would be above the minimum agreed with the regulatory authority and in which any leachate produced would be below prescribed concentrations. It is possible that a control chart could become a requirement of the operator's disposal license or permit. However, the control chart is not a permanent feature and would need to be reviewed and updated periodically using the results obtained from new sample mixes. Updating would be required in any of the following situations:

(*a*) when the composition of any major constituent in the bulk waste stream, or the source or characteristics of cement or PFA, are changed by more than predetermined values (for example, say, 10% for wastes and 2% for cement or PFA),
(*b*) significant modifications are made to the pretreatment processes or solidification plant,
(*c*) if new information becomes available and the licensing authority requires the minimum acceptable values for one or more of the physical tests to be altered, and
(*d*) every two or three years to ensure the existing control chart is still reliable, for example to ensure that regular minor changes to a treatment plant and waste streams do not have an adverse accumulative effect on product quality.

Once a treatment operation is under way, the regulatory authority should regularly visit the site. To ensure that the mix being produced is likely to give rise to an acceptable product once landfilled reference can be made to the plant process control log and the agreed quality-control chart. Proper monitoring would require an operator to divulge details of the proportions of constituents being used. Obviously, the procedure would be most effective where there is cooperation between the operator and the regulator. It is also suggested that an "observation cell" is set aside in the landfill where mixes whose compositions are in dispute (between the operator and authority) can be placed and left to set. If it transpires that the mix is shown to be unacceptable, it would then become the responsibility of the operator to ensure alternative disposal via an approved route. For example, this may involve reprocessing through the treatment plant.

To verify the operator's records on the compositions of mixes, it might be possible to install security-locked flow meters or weighing equipment on input feed lines, or the mixer itself could be mounted on a weighing device. In addition, it is recommended that every three to six months (or more regularly, if problems have been found) the regulatory authority should collect and analyze samples of the waste streams and mixes under preparation. This would ensure that the control chart was still valid for existing operations. Failure to link the results from these samples to that predicted by the control chart might lead the regulatory authority to suspect that the information on proportions, waste streams, or plant operations supplied by the operator was incorrect, and would warrant further investigation.

The conclusiveness of any quality-control procedure must also be considered very carefully before such a procedure can be implemented, although it is stressed that collaboration rather than confrontation between all parties would be the best way forward.

Conclusions

The results obtained in the current research program and elsewhere have indicated that relationships between varying proportions of cement and liquid waste used to produce solidified wastes can be related to some of their physical and leaching characteristics. This is especially useful when seeking to establish a quality-control procedure. The method presented in this paper is not foolproof, but is seen as a first step toward achieving safe, more controllable solidified waste production. It is presented as a basis for scientific discussion and as a stimulus for future developments in the quality control of waste solidification.

Acknowledgment

This work was funded by the Land Wastes Division of the U.K. Department of the Environment (DoE). The authors are grateful to the DoE for their advice and assistance throughout the research program.

References

[*1*] Côté, P. L. and Hamilton, D. P., "Leaching Comparison of Four Hazardous Waste Solidification Processes," *Proceedings*, 28th Purdue Industrial Waste Conference, Purdue University, West Lafayette, IN, 10–12 May 1983.

[*2*] "Hazardous Waste Management—An Overview," First Report of the Hazardous Waste Inspectorate, Department of the Environment, London, June 1985.

[*3*] "Getting to Grips with Waste Solidification," ENDS Report No. 120, Environmental Data Services, Jan. 1985.

[*4*] Myers, T. E., "A Simple Procedure for Acceptance Testing of Freshly Prepared Solidified Waste," *Hazardous and Industrial Solid Waste Testing: Fourth Symposium, ASTM STP 886*, J. K. Petros, W. J. Lacy, and R. A. Conway, Eds., American Society for Testing and Materials, Philadelphia, 1986, pp. 263–272.

[*5*] Baldwin, G., Rushbrook, P. E., and Dent, C. G., "A Method for the Reliable Prediction of the Quality of Cement-Based Solidified Hazardous Wastes at the Time of Preparation," *Proceedings*, International Congress on the Recent Advances in the Management of Hazardous and Toxic Wastes in the Process Industries, Vienna, Austria, 8–13 March 1987.

[*6*] Poon, C. S., Peters, C. J., Perry, R., Barnes, P., and Barker, A. P., "Mechanisms of Metal Stabilisation by Cement-Based Fixation Processes," *The Science of the Total Environment*, Vol. 41, 1985, pp. 55–71.

[*7*] Reeve, R. C., "Air-to-Water Permeability Ratio," *Soil Analysis*, Vol. 1, Chapter 41, C. A. Black et al., Eds., American Society Agronomy, Madison, WN, 1965, pp. 520–531.

[*8*] Wilson, D. C. and Young, P. J., "Testing of Hazardous Wastes to Assess Their Suitability for Landfill Disposal," Atomic Energy Research Establishment Report R10737, Her Majesty's Stationery Office, London, 1982.

[*9*] "Guide to the Disposal of Chemically Stabilized and Solidified Waste," U.S. EPA Report SW-872, U.S. Army Engineers Experimental Station, Vicksburg, MS, Sept. 1980.

[*10*] "Chemical Pretreatment of Hazardous Waste in Containers," European Patent Application EP 0 013 822 A2, 19 Dec. 1979 (date of filing).

[*11*] "Method of stabilizing organic waste," United States Patent 4 514 307, 30 April 1985.

[*12*] "Detoxification," United States Patent 4 116 705, 26 Sept. 1978.

[*13*] Todd, D. K., *Groundwater Hydrology*, Wiley, Chichester, U.K., 1959.

[*14*] Chappell, C. L. and Willetts, S. L., "Some Independent Assessment of the Sealosafe/Stablex Method of Toxic Waste Treatment," *Journal of Hazardous Materials*, Vol. 3, 1980, pp. 285–291.

[15] Leathers, J., "Cory Waste Management Expands its Industrial Waste Disposal Facilities," *NAWDC News,* Oct. 1984, pp. 24–25.
[16] Eaton, H. C., Tittlebaum, M. E., and Cartledge, F. K., "Techniques for Microscopic Studies of Incineration and Treatment of Hazardous Wastes," *Proceedings,* 11th Annual Environmental Protection Agency Research Symposium, Cincinnati, OH, 29 April–1 May 1985, pp. 135–142.
[17] Brown, T. M. and Bishop, P. L., "The Effect of Particle Size on the Leaching of Heavy Metals from Stabilized/Solidified Wastes," *Proceedings,* New Frontiers for Hazardous Waste Management International Conference, Pittsburgh, PA, 15–18 Sept. 1985, pp. 356–363.
[18] Poon, C. S., Clark, A. L., and Perry, R., "Permeability Study on the Cement-Based Solidification Process for the Disposal of Hazardous Wastes," *Cement and Concrete Research,* Vol. 16, 1986, pp. 161–172.

Chi S. Poon[1]

A Critical Review of Evaluation Procedures for Stabilization/Solidification Processes

REFERENCE: Poon, C. S., **"A Critical Review of Evaluation Procedures for Stabilization/Solidification Processes,"** *Environmental Aspects of Stabilization and Solidification of Hazardous and Radioactive Wastes, ASTM STP 1033,* P. L. Côté and T. M. Gilliam, Eds., American Society for Testing and Materials, Philadelphia, 1989, pp. 114–124.

ABSTRACT: The leaching procedures for evaluation of cement-based stabilization/solidification processes are considered. The difficulties in obtaining a consensus test method are discussed. The relative merits and shortcomings of the equilibrium and dynamic leaching test approaches are underlined, and various leaching models are considered.

Although mass transport and diffusion models have been generally used to describe the leaching behavior of soluble species, the use of these models to describe the cement-solidified heavy metal waste is questioned. This is based on theoretical consideration of the interaction between cementitious materials and waste and experimental evidence of the leaching characteristics. An alternative approach toward modeling the leaching behavior of the solidified waste is presented.

Because the performance of the solidified material is also dependent on the physical properties of the waste form, a physical test procedure in conjunction with an appropriate leach test is proposed as necessary to evaluate the quality of the solidified waste. A tentative approach to formulate such a combined test method is suggested.

KEY WORDS: stabilization/solidification, review, leaching tests, cement, hazardous waste, modeling

The use of cement-based stabilization/solidification processes for the disposal of waste requires the use of adequate test procedures to safeguard the environment from undesirable pollution. In the context of hazardous (that is, excluding radioactive) waste, to which the discussion of this present paper will largely be confined, most of the solidified waste will eventually be disposed of in a landfill environment. The most important factor in determining whether a particular stabilization/solidification process and its parameters are effective in treating a particular kind of waste is the reduction in the short- and long-term leachability of the waste. This is of course also subject to economic constraint and availability of resources.

Leaching can be defined as the process by which a component of waste is removed mechanically or chemically into solution from the solidified matrix by the passage of a solvent such as water. To devise an adequate leaching test procedure for assessing the compatibility of the stabilization/solidification process and the quality of the solidified waste, it is necessary first to understand the retention/immobilization mechanisms of the cement-based stabilization/solidification process. Two objectives can be identified for the use of

[1] Formerly, research fellow, Department of Metallurgy and Science of Materials, Oxford University, Parks Road, Oxford, 0X1 3PH, U.K.; presently, environmental protection officer, Environmental Protection Department, 28/F Southorn Centre, 130 Hennessy Road, Hong Kong.

such a process for waste disposal [1]. First, the waste component of the waste stream is chemically stabilized by the processing materials to induce chemical changes that would render them less hazardous or to insoluble and unavailable forms. Second, the added processing material would also be hardened over time and produce a readily handleable monolithic solid material with improved structural integrity and physical characteristics. This can also act as a physical barrier to reduce leaching.

The many factors which would affect the leaching potential of the solidified waste can be generally grouped into two categories: the external factors which are governed by the designated landfill environment, and the intrinsic factors which are inherent to the solidified waste form. The former category includes factors such as the amount of precipitation, ground water flow and geological conditions, temperature, and the chemical potential and acidity of the rain and ground water [2]. The latter includes the chemical specification of the pollutants in the cementitious matrix, solubility, alkalinity, redox potential, mechanical strength, permeability, porosity, tortuosity, and durability [3,4]. Some of these factors are interrelated. For example, permeability is a function of porosity, and durability is a function of permeability and porosity. In practical terms, only the intrinsic factors are controllable by the design and processing of the stabilization/solidification process, and therefore the following discussion will concentrate on the evaluation methods devised to assess such parameters.

These evaluation methods can be divided into physical and chemical properties test methods.

Physical Properties Test Procedures

It is believed that permeability and durability are the two physical properties of the solidified waste form that would significantly affect its leaching potential. Contrary to the general assumption that the only effective transport mechanism of pollutants in the solidified waste is molecular diffusion [5,6], the author believes that convective flow through the waste does contribute substantially to the flow regime. Actually, the philosophy behind the use of stabilization/solidification processes for hazardous waste disposal is to reduce the cost and burden of using secure landfill. Therefore, a dilute and disperse philosophy is more favored economically than the concentrate and contain method.

Assuming a simplified situation where flow is governed solely by Darcy's flow equation [7]

$$\frac{F}{A} = K\frac{h}{L} \tag{1}$$

where

F = flow,
A = cross-sectional area,
K = permeability constant,
h = hydraulic head, and
L = thickness of specimen.

A wide range of K values (10^{-4} – 10^{-7} cm^{-1}) has been reported and quoted by different organizations and companies marketing solidification processes [8,9]. However, the procedures from which these results were obtained were often either not reported or not performed with a standardized method. Furthermore, the condition of the specimens on which the measurements were made were not given. It should be noted that the permeability of most cementitious materials varies with different water/cement ratios, additives used, curing time, and conditions. To generalize and report the result of one particular stabilization/solidification process by quoting test results of a few unspecified samples is far from satisfactory. There is at present no standard procedure for permeability testing of solidified hazardous waste, and some of the difficulties of this have been discussed by Poon et al. [*10*]. However, it is generally known that the permeability of the cementitious material decreases with increasing amount of cement used in the processing [*10*].

Another factor that would significantly affect the leaching potential (especially long-term) of the solidified waste form is durability. Durability can be defined as the ability of the material to retain its mechanical and dimensional properties under different weathering conditions. It has been generally concluded that most solidified waste form (again, in the context of hazardous waste) fails when subjected to the usual wet/dry and freeze/thaw test procedures for concrete assessment [*8,9*]. This is probably due to the use of a much lower cement content in the processing and the deleterious effect of different waste components on cement hydration. However, a question which must be raised is whether or not any stabilization/solidification process aims to produce a product with concrete-like properties. If not, then these severe durability test procedures for concrete would not be relevant in assessing the properties of the solidified waste. Actually, it should be noted that in a normal landfilling situation, the major part of the disposed waste will not be subjected to these severe environmental conditions.

The third method and result that is most often performed and reported for solidified waste forms is the unconfined compression strength. Strength values of 1.4 to 14 MPa (200 to 2000 psi) have been reported for various processes [*8,9*]. These values are low for any practical application for structural material. In fact, in this present climate of waste disposal and environmental consciousness, it is unlikely that such solidified waste will be used as building materials except for small land reclamation in nonstructural applications. As with permeability, strength values also depend on the condition of the specimens and its curing conditions. Generally, strength increases with the amount of cement used in the processing.

Since permeability, durability, and strength of most cementitious material are interrelated, the simple strength measurement should be able to give some indication of the permeability or durability characteristics of the solidified waste. For a high water/cement ratio, hydrated cementitious material, permeability is largely governed by the capillary porosity (pore size >1320 Å [*11*]. This type of porosity is also the most important factor in affecting strength [*12*]. A typical relationship between strength and porosity is given in the equation [*13*]

$$\sigma = \sigma_0 (1 - \epsilon)^A \tag{2}$$

where

σ = strength,
σ_0 = strength at zero porosity,
ϵ = porosity, and
A = constant.

Therefore, permeability is generally inversely proportional to strength. However, due to the high gelatinous nature of certain solidified waste, unconfined strength values sometimes do not truly reflect the pore structure and permeability of the material. A shear-strength parameter based on a triaxial apparatus or peneotrometer is suggested to be more representative of the mechanical behavior.

The relationship between durability and strength is more complex. The former is a function of strength and has a relationship with porosity [*14*]. But in the context of low strength and relatively porous solidified materials, the strength element should be the more dominant factor. For most solidified heavy-metal waste form, it is to be expected that durability would increase with strength. Therefore, we have a solidified material for which both the permeability and durability can be addressed by a simple strength measurement.

Chemical Properties Test Procedures

The most important and frequently tested chemical property of the solidified waste form is the extraction potential of the waste component by various simulated leaching fluids. Many researchers and organizations prefer to use the term "leaching test." But it must be stressed that most of these test procedures do not and are not designed to simulate the actual leaching environment of the waste once disposed of.

There are a number of extraction methods developed for evaluating solidified wastes. Mahlock [*15*] suggested that there are four general factors to be considered when applying a particular test procedure for leach testing:

(*a*) "representiveness" of the actual field condition,
(*b*) "compatability" of using the same test for different kinds of waste,
(*c*) "reproducibility" for giving consistent results upon repeated application, and
(*d*) "stability" for giving consistent results as a function of the time of testing.

There are no accepted test procedures developed to date that would satisfy all four factors. For example, a column test would be more representative of the actual field environment, but the reproducibility of the test data is poor [*16*]. Since it is generally accepted that a reproducible result is more important for regulatory purposes for the disposal of highly toxic waste, the batch procedure is generally adopted as the method in evaluating the solidified material [*2*]. A vast number of batch test procedures has been developed during the last decade, ranging from simple elutriate test procedures to complex multiple extraction methods with carefully controlled test parameters [*17–19*]. A number of published reports validate and compare the suitability of these test procedures [*20–22*]. However, Mahlock [*15*] suggested that most of these methods should produce interrelated leach rates and proposed that a particular method adopted for testing is not critical, but that subsequent interpretation and application are important.

Recently, three batch extraction procedures have been considered potentially the most satisfactory test methods for evaluating solid waste. There is the EP extraction procedure of the United States Environmental Protection Agency (EPA) [*23*], the Canadian method of leaching [*24*], and the Multiple Extraction procedure developed by Harwell, United Kingdom [*25*]. The EPA testing procedures are developed primarily for the purpose of defining and classifying a waste as hazardous or otherwise. The U.K. Harwell procedure attempted to incorporate the landfill attenuation factors by using a low liquid solid ratio (defined by "bed" volume).[2] The merit of the EPA test is to provide a simple and quick method for assessing a waste, whereas the Harwell test takes into account the real environmental factors but suffers from the subjectivity of the definition of bed volume.

Since batch test methods often assume measuring the equilibrium concentrations of the leachate [*3,15*], the results of the tests do not provide the kinetic information necessary to

[2] Bed volume is the amount of liquid that would fully saturate a waste.

predict the leach rate of the system. The measured data are generally regarded as the "worst case" situation. The data are usually compared with the established criteria for contaminants under consideration. The merit of using this worst-case situation is thought to be that if there is no probable environmental impact resulting from the extracted concentration, the long-term adverse behavior of the disposed waste would not be significant. However, for the heavy metal wastes that have been solidified by a cement-based process and because of the high alkalinity of cement, the equilibrium concentrations are usually lower than the actual leached concentrations. This is because most heavy metals would be precipitated in the highly alkaline environment of the cement pore solution.

Leaching Models

There have been a number of attempts to study the kinetics of leaching in order to predict long-term leachability of the waste component. An expression based on Fick's diffusion theory is often used to describe the release of pollutants from the solidified waste [*26–28*]

$$\left(\frac{\Sigma An}{Ao}\right)\left(\frac{v}{s}\right) = 2\left(\frac{De}{\pi}\right)^{1/2} t^{1/2} \tag{3}$$

where

An = contaminant loss during leaching period n,
Ao = initial amount of contaminant present in the specimen,
v = volume of specimen,
s = surface area of specimen,
t = time, and
De = effective diffusion coefficient.

To formulate such an equation, a semi-infinite medium and a zero surface concentration are assumed during leaching.

When a kinetically controlled chemical dissolution process is also taken into account, a more complex equation can be formulated of the form [*26*]

$$\left(\Sigma\frac{An}{Ao}\right)\left(\frac{v}{s}\right) = (Dek)^{1/2}\left[\left(1 + \frac{1}{2k}\right)\text{erf}\,(kt)^{1/2} + \left(\frac{1}{\pi k}\right)^{1/2}\exp\,(-kt)\right] \tag{4}$$

where

k = dissolution rate constant and
$\text{erf}(u)$ = error function $= \dfrac{2}{\pi 1/2}\displaystyle\int_0^u \exp(-z^2)dz.$

A leachability index LX defined by the American Nuclear Society (ANS) has been used by a few researchers in hazardous waste [*27,28*] to provide a "figure of merit" that qualitatively describes leachability

$$LX = \frac{1}{7}\sum_{n=1}^{7} \log\,[\beta/De]n \tag{5}$$

where β is a constant and n is the leaching period.

The calculated effective diffusion coefficient from Eq 3 does not necessarily imply that diffusion is the only mechanism for leaching. This value represents the leaching behavior of the solidified waste over the time period t by assuming diffusion is the leaching mechanism. If this coefficient remains constant over the entire leaching period, one may then conclude diffusion is the primary transport mechanism responsible for leaching. However, it has been demonstrated that the De value changed over time during a batch experiment [*28*].

Nevertheless, based on the assumption that the solidified waste has a very low permeability compared with the surrounding medium and is subjected to a small hydraulic gradient, Côté and Hamilton [*5*] in their study used carefully controlled laboratory conditions with most boundary conditions of the diffusion model properly addressed. By using a leachate renewal period of

$$t_n = n^2\, t_1 \tag{6}$$

where

t_1 = initial extraction time,
t_n = extraction time at extraction n, and
n = number of extraction,

they produced leaching test results that lie on a straight-line plot of $\Sigma(An/Ao)$ versus $t^{1/2}$ for cadmium and lead, implying a diffusion mechanism. The leachate concentrations of arsenic, chromium, and phenol, however, did not fit the model, indicating that mechanisms other than diffusion were also affecting the leaching. These discrepancies between actual leaching from cement and the diffusion model were also observed by Zamorani et al., who found that the kinetics of calcium release from a cementitous material were complex and that a true diffusion coefficient could not be properly defined [*29*].

The uncertainties in using a diffusion model to describe the leaching behavior is further illustrated by Piciulo et al. [*30*]. In their leaching experiment on organic acids stabilized by five different cementitious grouts, the leach indices being calculated varied with time and the plot of cumulative release did not fall on the straight line of square root of time.

The successful use of this diffusion model is limited, therefore, because of the inherent assumption made in utilizing the model. Although convective flow due to the hydraulic gradient would be small during the initial stage of leaching, the dissolution process imposed by the leaching fluid and other environmental effects will undoubtedly make the solidified material more vulnerable to infiltration, making convection flow possible. In fact it is known that the strength of ordinary portland cement paste decreases at the rate of 1 to 2% per percentage loss of lime [*31*]. Furthermore, to obtain a simple analytical solution such as Eq 3 for the diffusion model, it is assumed that a constant leaching environment is maintained during the leaching period [*32*]. Contrary to this assumption it is generally found that the leaching environment is continually modified by the leaching process [*33*]. Thus a simple diffusion model is probably suitable to describe the short-term leaching behavior of the solidified waste, but is inappropriate to extrapolate the long-term leaching characteristics.

Another more satisfactory approach which has been adopted to model the long-term leaching characteristics is based on an empirical method using the expression [*32*]

$$\left(\Sigma\frac{An}{Ao}\right) = k_1 + k_2 t^{1/2} + k_3 t \tag{7}$$

where

k_1 = constant representing the immediate dissolution,
k_2 = constant representing the diffusion controlled transport mechanism, and
k_3 = constant representing the long-term kinetically controlled dissolution.

The equation is used to produce the best fit to the long-term experimental data, and the k_1, k_2, and k_3 can be used to extrapolate the leaching characteristics.

However, the equation itself does not incorporate the measurable physical and chemical properties of the solidified waste. Extensive effort and time, therefore, are needed to produce experimental data for each waste component in each stabilization/solidification process, which also vary with the different processing parameters.

Based on the theory that even a perfectly simulated leaching test would not provide a time based prediction of the leaching behavior at field conditions, Côté et al. [*34*] proposed the use of a modeling method to measure the intrinsic properties of the solidified waste that would affect leaching. These workers quantified the matrix-related properties (for example, porosity and durability), which indicate the way in which water could come into contact with the waste, and the contaminated-related properties (for example, solubility of contaminants in different pH and *Eh* conditions), which describe the behavior of contaminants in the solidified matrix and leaching water. A hypothetical leaching model was developed to describe the kinetics of contaminant leaching under various regimes which can be adopted in different field disposal conditions. A more fundamental approach toward modeling the chemistry and microstructure of the hydrated cementitious matrix is also being developed [*35,36*] which may be useful in formulating a more accurate leach model.

Immobilization and Leaching Mechanisms

Regarding the chemical mechanisms by which the waste components are stabilized by the cement-based stabilization/solidification process, there are still uncertainties with respect to the actual immobilization mechanisms. It is generally known that most heavy metals are satisfactorily immobilized by the stabilizing agents and that inorganic anonic and organic wastes are not well retained [*1,16*]. The study initiated by the interest in using cement and related materials for radioactive-waste disposal demonstrated that cement in general has poor sorptive capacity for alkaline metals such as cesium and strontium [*37*]. This may be due to the high solubility of their hydroxides. The study on mercury [*38*] (a heavy metal with some covalent characteristics) showed a combination of chemical and physical effects active in the immobilization and waste component by the cementitious matrix.

For most heavy-metal waste streams—regarded as those most suitable for cement-based stabilization/solidification process—the mode of interaction between the metals and the stabilizing materials will influence clearly the leachability of the metals. Extraction studies [*39–41*] of certain stabilized heavy-metal wastes showed that the amount present in solution is often lower than the calculated value based upon the theoretical solubility products. A variety of mechanisms have been postulated to account for this, involving absorption by cement hydrates, substitution, and solid solution in the hydrate structure and formation of various insoluble compounds [*39–45*]. However, many of these inferred claims relate to semi-quantitative observations, and interpretations leave many of the fundamentals to be resolved.

But one general conclusion agreed between these studies is that for a cement-stabilized heavy-metal waste, at pH above 7.0 little heavy-metal concentration would be detected in solution. Below pH 7, however, heavy metals start to solubilize. The solubility can be a function of its hydroxides, carbonates, silicates, or other complex forms depending on the actual chemical interaction between the individual metal and cement matrix.

The pH of a hydrated cement solution decreases with the addition of acidity [*38,46*]. Figure 1 plots the decrease in pH upon added acidity. The actual profile of the curve depends on the lime content of cement, water/cement ratio, and curing time. The most significant feature of this profile is the drastic drop of pH value roughly in the middle range of the

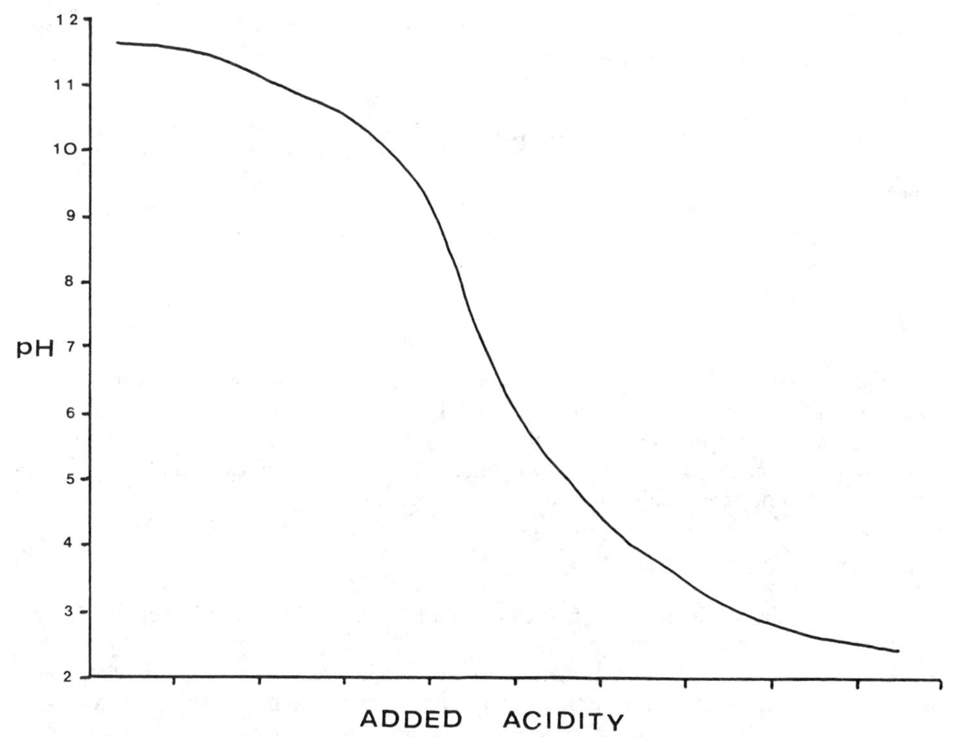

FIG. 1—*Schematic of variation of pH values of hydrated cement paste upon addition of acidity.*

curve. The initial alkaline environment was suddenly changed to acidic. This is significant as far as heavy-metal leaching is concerned. The acid-neutralizing capacity of anhydrous portland cement is known to be about 20 meq/g. This value varies with different types of cement and when additives such as flyash, silicates, and clay are used. A simple titration procedure is thus required to determine the overall neutralizing capacity of the solidified waste. If this value and the acidity and flow rate of the potential leaching fluid are known, the duration by which the solidified waste become acidic can be estimated.

Methodology to Develop a Realistic Assessing Procedure

So far we have shown that the leaching potential of a cement-based solidified waste is a function of its physical and chemical properties. These can be modeled by mechanical strength and extractability. We have also identified that for some heavy-metal waste, there is a potential breakthrough point at which a large part of the waste component would solubilize as the leaching water turned acidic.

Therefore, it is suggested that for cement-based solidified wastes (especially for heavy-metal wastes), the acid-neutralizing capacity of the waste form should be determined first by titration. Also, the time ot the potential breakthrough point should be determined.

The leach profile of the waste components before this breakthrough point is more difficult to assess because most of the short-term extraction test data are not extrapolatable to predict the long-term leaching in field condition. However, taking into account the discussed parameters, strength and extractability, that would mostly affect leaching, the leach profile can be formulated by the empirical relationship

$$\text{Cumulative release} = \left[\frac{1}{\sigma}, \epsilon, t, I\right] \tag{8}$$

where

σ = strength,
ϵ = extractability,
t = time, and
I = infiltration.

σ and ϵ are the characteristics of the solidified waste form, and they vary with types of stabilizing agent used, water/cement ratio, and the chemical interaction between and effect of waste components on cement hydration.

σ can be determined from a shear strength test. But due to chemical and physical weathering, σ should decrease with time with the relationship [*47*]

$$\sigma = \sigma_0 (1 - kt^{1/2}) \tag{9}$$

where σ_0 is initial strength and k is a constant which can be determined separately by a weathering experiment.

The extractability term ϵ can be defined as the weight fraction of the waste component extracted from the solidified waste form by a specific extraction procedure. The exact kind of test procedure is not so critical, but a multiple extraction test with a real solvent (such as rain or groundwater) is preferred since it gives a better representation of the long-term leaching behavior. The most important factor is the reproducibility, and some of the well-developed extraction procedures thus can be used.

More studies are needed to establish and expand Eq 8—the relationship between leachability, strength, and extractability to formulate a satisfactory model to predict the leaching characteristics of the waste form. This should involve some pilot scale leaching studies in field condition with solidified waste form of varying σ and ϵ values.

This prediction obtained for any waste component concerned can be used to interpret the suitability of the stabilization/solidification process and its processing parameters for the particular kind of waste. We must also calculate the cumulative amount of material leached from the waste up to the time when breakthrough would take place, as discussed earlier. If the calculated cumulative release is less than the total amount of waste component in the solidified mass, a breakthrough should occur which may have a significant detrimental effect on the environment. However, when the cumulative amount leached is greater than the original amount present, the leaching of the waste component has been safely delayed and diluted into the environment, causing no harmful effects.

References

[*1*] "Survey of Solidification/Stabilization Technology for Hazardous Industrial Wastes," EPA-600/2-79-056. Environmental Laboratory U.S. Army Engineer Waterways Experiment Station, U.S. Environmental Protection Agency, Cincinnati, OH, 1979.
[*2*] Anderson, M. A., Ham, R. K., Stegmann, R., and Stanforth, R. in *Toxic and Hazardous Waste Disposal*, Vol. 2, R. B. Pojasek, Ed., Ann Arbor Science, Ann Arbor, MI, 1979, pp. 145–167.
[*3*] Josephson, J., *Environmental Science and Technology*, Vol. 16, No. 4, 1982, pp. 219A–223A.

[4] Lowenbach, W. A. in *Toxic and Hazardous Waste Disposal,* Vol. 2, R. B. Pojasek, Ed., Ann Arbor Science, Ann Arbor, MI, 1979, pp. 89–142.
[5] Côté, P. L. and Hamilton, D. P., "Leachability Comparison of Four Hazardous Waste Solidification Processes," presented at the 38th Annual Purdue Industrial Waste Conference, 10–12 May 1983.
[6] Côté, P. L. and Isabel, D., "Application of a Dynamic Leaching Test to Solidified Hazardous Wastes," *Hazardous and Industrial Waste Management and Testing, Third Symposium, ASTM STP 851,* L. P. Jackson, A. R. Rohlik, and R. A. Conway, Eds., American Society for Testing and Materials, Philadelphia, 1984, pp. 48–60.
[7] Neville, A. M., *Properties of Concrete,* 2nd ed., Pitman, London, 1973, p. 388.
[8] Thompson, D. W. and Malone, P. G. in *Toxic and Hazardous Waste Disposal,* Vol. 2, R. B. Pojasek, Ed., Ann Arbor Science, Ann Arbor, MI, 1979, pp. 35–50.
[9] "Guide to the Disposal of Chemically Stabilized and Solidified Waste," EPA-SW-872, U.S. Environmental Protection Agency, Cincinnati, OH, 1981.
[10] Poon, C. S., Clark, A. I., Perry, R., Barker, A. P., and Barnes, P., *Cement and Concrete Research,* Vol. 16, 1986, pp. 161–173.
[11] Metha, P. K. and Manmohan, D., "Pore Size Distribution and Permeability of Hardened Cement," *Proceedings,* 7th Congress of Chemistry of Cement, Paris, Vol. 3, 1980, pp. V11/1–V11/5.
[12] Powers, T. C. and Brownyard, T. L., *Bulletin 22,* Portland Cement Association, Stokie, IL, 1948.
[13] Bye, G. C., *Portland Cement,* Pergamon Press, Oxford, U.K., 1983, p. 116.
[14] Celleja, J., "Durability," *Proceedings,* 7th Congress of Chemistry of Cement, Paris, Vol. 1, 1980, pp. V11-2/1–V11-2/48.
[15] Mahlock, J. L. in *Toxic and Hazardous Waste Disposal,* Vol. 2, R. B. Pojasek, Ed., Ann Arbor Science, Ann Arbor, MI, 1979, pp. 187–199.
[16] Poon, C. S., Peters, C. J., and Perry, R., *Effluent and Water Treatment Journal,* Vol. 23, No. 11, 1983, pp. 451–459.
[17] Thompson, D. W., "Elutriate Test Evaluation of Chemically Stabilized Waste Materials," EPA-600/2-79-154, U.S. Environmental Protection Agency, Cincinnati, OH, 1979.
[18] Fisher, S. in *Toxic and Hazardous Waste Disposal,* Vol. 2, R. B. Pojasek, Ed., Ann Arbor Science, Ann Arbor, MI, 1979, pp. 169–186.
[19] Perket, C. L. and Webster, W. C., "Literature Review of Batch Laboratory Leaching and Extraction Procedures," *Hazardous Solid Waste Testing: First Conference, ASTM STP 760,* R. A. Conway and B. C. Malloy, Eds., American Society for Testing and Materials, Philadelphia, 1981, pp. 7–27.
[20] Jackson, K., Benedik, J., and Jackson, L., "Comparison of Three Solid Waste Batch Leach Testing Methods and a Column Leach Test Method," *Hazardous Solid Waste Testing: First Conference, ASTM STP 760,* R. A. Conway and B. C. Malloy, Eds., American Society for Testing and Materials, Philadelphia, 1981, pp. 83–98.
[21] Welsh, S. K., Gagnon, J. E., and Elnabarawy, M. T., "Comparison of Three Acid Extraction-Leaching Test Protocals," *Hazardous Solid Waste Testing: First Conference, ASTM STP 760,* R. A. Conway and B. C. Malloy, Eds., American Society for Testing and Materials, Philadelphia, 1981, pp. 28–39.
[22] Lowenbach, W., "Compilation and Evaluation of Leaching Test Methods," EPA-600/2-78-095, U.S. Environmental Protection Agency, Cincinnati, OH, 1978.
[23] "Test Methods for Evaluating Solid Waste," EPA-SW-846, U.S. Environmental Protection Agency, 1980.
[24] Côté, P. L. and Constable, T. W., "Development of a Canadian Data Base on Waste Leachability," *Hazardous and Industrial Solid Waste Testing: Second Symposium, ASTM STP 805,* R. A. Conway and W. P. Gulledge, Eds., American Society for Testing and Materials, Philadelphia, 1983, pp. 53–66.
[25] Young, P. J. and Wilson, D. C., "Testing of Hazardous Wastes to Assess Their Suitability for Landfill Disposal," AERE Harwell R10737, Her Majesty's Stationery Office, London, 1982.
[26] Moore, J. G., Godbee, H. W., and Kibbey, A. H., *Nuclear Technology,* Vol. 32, 1977, pp. 39–52.
[27] Brown, T. W., Bishop, P. L., and Gress, D. L., "Use of an Upflow Column Leaching Test to Study the Release Patterns of Heavy Metals from Stabilized/Solidified Heavy Metal Sludges," *Proceedings of the Third International Symposium on Industrial and Hazardous Waste,* Alexandria, Egypt, 24–27 June 1985.
[28] Bishop, P. L., "Prediction of Heavy Metal Leaching Rates from Stabilized/Solidified Hazardous Waste," *Proceedings of 8th Mid-Atlantic Industrial Waste Conference,* G. D. Boardman, Ed., Atlanta, 1986, pp. 236–252.

[29] Zamorani, E., Serrini, G., and Blanchard, H., *Cement and Concrete Research,* Vol. 16, 1986, pp. 394–398.
[30] Piciulo, R. L., Davies, M. S., and Adams, J. W., "Leachability of Decontamination Reagents from Cement Waste Forms," *Scientific Basis for Nuclear Waste Management,* Materials Research Society Symposium Proceedings, Vol. 44, 1985.
[31] Lee, F. M., *The Chemistry of Cement and Concrete,* Edward Arnold, London, 1976, p. 345.
[32] Côté, P. L., "Evaluation of the Long-Term Stability of Solidified Wastes," presented at the Workshop on Environmental Assessment of Waste Stabilization/Solidification, Vegreville, AB, Canada, 1983.
[33] Poon, C. S., Clark, A. I., Peters, C. J., and Perry, R., *Waste Management and Research,* Vol. 3, 1985, pp. 127–142.
[34] Côté, P. L., Bridle, T. R., and Benedek, A., "An Approach for Evaluating Long Term Leachability from Measurement of Intrinsic Waste Properties," presented at the 3rd International Symposium on Industrial and Hazardous Waste, Alexandria, Egypt, 1985.
[35] Berner, U. R., "Modelling Porewater Chemistry in Hydrated Portland Cement," presented at Materials Research Society Fall Meeting, Boston, 1986.
[36] Jennings, H. M. and Johnson, S. K., *Journal of American Ceramic Society,* Vol. 69, No. 11, 1986, pp. 790–795.
[37] Komarneni, S. and Roy, D. M., *Cement and Concrete Research,* Vol. 11, 1981, pp. 789–794.
[38] Poon, C. S., Peters, C. J., Perry, R., Barnes, P., and Barker, A. P., *The Science of Total Environment,* Vol. 41, 1985, pp. 55–71.
[39] Shively, W., Brown, T., Bishop, P., and Gress, D., "Heavy Metal Binding Mechanisms in the Solidification/Stabilization Hazardous Waste Treatment Process," presented at the 57th Water Pollution Central Federation Conference, New Orleans, 1984.
[40] Falcone, J. S., Jr., Spencer, R. W., Reifsnyder, R. H., and Katsanis, E. P. L., "Chemical Interactions of Soluble Silicates in the Management of Hazardous Wastes," *Hazardous and Industrial Waste Management and Testing: Third Symposium, ASTM STP 851,* L. P. Jackson, A. R. Rohlik, and R. A. Conway, Eds., American Society for Testing and Materials, Philadelphia, 1984, pp. 213–229.
[41] Tashiro, C., Takahaski, H., Kanaya, M., Hirakida, Y., and Yoshida, R., *Cement and Concrete Research,* Vol. 7, 1977, pp. 283–290.
[42] Malone, P. G. and Larson, R. J., "Scientific Basic of Hazardous Waste Immobilization," *Hazardous and Industrial Solid Waste Testing: Second Symposium, ASTM STP 805,* R. A. Conway and W. P. Gulledge, Eds., American Society for Testing and Materials, Philadelphia, 1983, pp. 168–177.
[43] Poon, C. S., Clark, A. I., and Perry, R., *Environmental Technology Letters,* Vol. 7, 1986, pp. 461–467.
[44] Salas, R. K. in *Toxic and Hazardous Waste Disposal,* Vol. 1, R. B. Pojasek, Ed., Ann Arbor Science, Ann Arbor, MI, 1979, p. 321.
[45] Schofield, J. T. in *Toxic and Hazardous Waste Disposal,* Vol. 1, R. B. Pojasek, Ed., Ann Arbor Science, Ann Arbor, MI, 1979, p. 297.
[46] Bishop, P. L. and Brown, T. M., "Alkalinity Release and the Leaching of Heavy Metals from Stabilized/Solidified Wastes," *Proceedings of 5th International Conference on Chemistry for Protection of the Environment,* Lenven, Belgium, 1985.
[47] Prudol, S., *Cement and Concrete Research,* Vol. 7, 1977, pp. 77–84.

Hans A. van der Sloot,[1] Gerard J. de Groot,[1] and Jan Wijkstra[1]

Leaching Characteristics of Construction Materials and Stabilization Products Containing Waste Materials

REFERENCE: van der Sloot, H. A., de Groot, G. J., and Wijkstra, J., **"Leaching Characteristics of Construction Materials and Stabilization Products Containing Waste Materials,"** *Environmental Aspects of Stabilization and Solidification of Hazardous and Radioactive Wastes, ASTM STP 1033,* P. L. Côté and T. M. Gilliam, Eds., American Society for Testing and Materials, Philadelphia, 1989, pp. 125–149.

ABSTRACT: In this work, several construction materials prepared with an admixture of waste material and various stabilized waste products have been subjected to leaching studies in order for the leaching behavior of trace elements from these materials to be characterized. Static and dynamic leach tests have been applied, in which the specimen to be studied is fully immersed in demineralized water. Often the leaching process is governed by outward diffusion of constituents dissolved in the pore water. In some cases, dissolution from the surface of the product is most prominent, and in some materials short-term surface release has been observed. Groups of materials can be distinguished with similarities in the leaching behavior of trace elements.

The alkalinity of the material, the open porosity of the product, and the surface-to-volume ratio prove to be important factors in controlling the release of potential hazardous elements from materials containing waste products. In these studies, leach parameters of trace elements are related to those of sodium. Since the interaction of sodium with the solid phase is usually small, sodium can be used as an indicator for the tortuosity of the product. Elements leached from cement-based waste products are mainly anionic species, like MoO_4^{2-}, BO_3^{3-}, VO_4^{3-}, F^-, and SO_4^{2-}, whereas leaching of metals such as copper, cadmium, zinc, and lead is limited due to the high pH in the pore solution. The leaching experiments have been verified by scanning electron microscopy for major components on field samples and by measuring depth profiles in waste products for trace constituents using an apparatus developed for this purpose.

KEY WORDS: hazardous elements, waste materials, construction materials, leaching process, leach test, diffusion, element speciation, mobility measurement

In recent years, the utilization of waste materials as an admixture in construction materials and the stabilization of various waste materials to minimize adverse effects on the environment have received considerable attention. Waste disposal without measures to avoid soil and ground-water pollution has become unacceptable. A number of waste materials have properties that allow utilization in construction materials either as a structural component or as a filler [*1–3*]. In this field, the variability of waste material quality forms one of the main limitations [*3*]. Stabilization of wastes to achieve a physical and chemical confinement is widely used for those materials, which cannot be used or disposed of in pure form because of their composition or properties. Technical, economical, and environmental factors are involved in the choice between different options for a given waste material. For a proper

[1] Section head of environmental research, scientific co-worker, and senior technician, respectively, Netherlands Energy Research Foundation (ECN), P.O. Box 1, 1755 ZG Petten, the Netherlands.

judgment of the environmental aspects of waste utilization/stabilization, adequate tools are lacking. Current regulations based on a single extraction test [*4*] are not adequate for evaluating the wide variety of conditions encountered in practice.

In order for the environmental consequences of a given application to be assessed, information is needed on the leaching behavior as a function of time. This calls for a new approach in which waste materials and products are characterized by intrinsic leach properties [*5–7*]. At present, quantitative information on the factors controlling the leaching process is limited. For instance, the speciation of potentially hazardous elements within a waste product, which speciation is crucial for the release of these elements, has not received sufficient attention. As soon as these relations are better defined, technical modifications can be made to minimize the release of potentially harmful elements to the environment.

In this work, a tank leach test for waste products and the processes governing the leaching behavior of a variety of stabilized waste materials and construction materials containing waste materials are discussed. The possibility to arrive at intrinsic leach properties for groups of similar materials is addressed. In this respect, a recently modified procedure for the measurement of trace element mobilities in porous media has been applied [*8*]. With this method, detailed information on the mobility of potentially hazardous elements within the pore system of waste material or product can be obtained. A verification of the leaching behavior of waste forms in laboratory experiments is achieved by studying concentration profiles in the interior of a waste form after a known exposure time.

Procedure

Materials

The variety of stabilized waste materials and construction materials to be discussed in this work were obtained from different sources. In Table 1, the sample codes and origin of the samples are indicated. In Table 2, the physical properties of the materials are given [*3,9–12*]. The compressive strength of the structural lightweight concrete is already reached after 24 h, since the material is prepared in an autoclave.

Methods

An overview of the different test procedures applied in this work and their interrelations is given in Fig. 1. An attempt is made to relate leach data obtained from tank leaching studies, concentration profile analyses of exposed products, and mobility measurements with radiotracers giving information on chemical speciation within the product matrix.

Analytical Methods—Solids: The waste materials and stabilized products prepared from waste materials have been analyzed for major and trace elements by instrumental neutron activation analysis (arsenic, selenium, antimony, tungsten, uranium, and vanadium), atomic absorption spectrophotometry (sodium, calcium, and magnesium), induced-coupled plasma emission spectrometry (molybdenum), anodic stripping voltammetry (cadmium, copper, and zinc), and ion-chromatography (F^- and SO_4^{2-}). All waste products studied were crushed in a Retch model BB2/A jaw mill crusher (Retch GmbH, Germany Democratic Republic) using tungsten carbide plates. In the case of typically inhomogeneous materials, such as stabilized domestic waste incinerator residues (IA), care was taken to obtain representative samples for analysis. This product was ground in a tungsten carbide rotary-disc mill (Labor Scheiben-schwingmuhle TS100, Siebtechnik, German Democratic Republic). Contamination

TABLE 1—*Sample codes and origin of materials.*

Specimen Code	Material Composition	Origin	Reference
PCA	pulverized coal ash/cement (9:1)	INTRON B.V. (Maastricht)	[*10,11*]
PCA13	pulverized coal ash/cement (9:1)	KEMA (Arnhem)	
PCA/PG	pulverized coal ash/ phosphogypsum/lime (16:3:1)	INTRON B.V. (Maastricht)	[*10,11*]
PG/PCA	phosphogypsum/pulverized coal ash/cement (10.2:5.5:1)	INTRON B.V. (Maastricht)	[*10,11*]
PGH	phosphogypsum hemihydrate	fertilizer industry	
BFS	blast furnace slag/lime (99:1)	INTRON B.V. (Maastricht)	[*10,11*]
IA	domestic waste incinerator slag/ cement (5.7:1)	INTRON B.V. (Maastricht)	[*10,11*]
MPCA	concrete mortar (coal ash/ cement 1:4)	RPA (Heemstede)	[*3*]
LWPCA	structural lightweight concrete (45% coal ash by weight)	GASCON Nederland B.V. (Harderwijk)	[*9*]
PS	phosphate slag	National Waterworks (Lower River Branch, Dordrecht)	[*12*]

by the crushing and grinding operation cannot be avoided. Depending on the construction material of the grinding equipment and the hardness of the material, increased concentrations of tungsten, cobalt, and tantalum can be expected. To obtain representative tungsten data, part of the samples were crushed in a stainless-steel rotary-disk mill.

Liquids: Contact solutions from tank leaching experiments and extracts from availability tests have been analyzed directly by atomic absorption spectrometry (sodium, calcium, and magnesium), ion-chromatography (F^- and SO_4^{2-}), and anodic stripping voltammetry (cadmium, copper, and zinc), and, after preconcentration on active carbon, by neutron activation analysis (arsenic, antimony, selenium, molybdenum, tungsten, uranium, and vanadium). Liquid samples were acidified at pH 2 with 2N HNO_3 pro analysi (p.a.), except for subsamples for analysis of major anions, which were stored without pretreatment. Storage time between sampling and analysis was kept to a minimum.

TABLE 2—*Properties of materials studied.*

	Stabilized Waste Products							Concrete		Slag
Parameter	PCA	PCA13	PCA/PG	PG/PCA	PGH	BFS	IA	MPCA	LWPCA	PS
size, cm	ϕ10.2 × 11.8	4 × 4 × 16	ϕ10.2 × 11.8	ϕ10.2 × 11.8	10 × 10 × 10	ϕ9.7 × 11.8	ϕ10.2 × 11.8	4 × 4 × 16	10 × 10 × 10	[a]
surface area, cm^2	541	288	541	541	600	507	541	288	600	2990
compressive strength, N/mm^2, at 56 days	6.4	ND[c]	5.0	4.5	2–6	3.3	23.7	58–67	15–25[b]	ND
apparent density, kg/m^3	1.43	1.37	1.35	1.18	1.32	1.90	1.90	2.27	1.18	3.8
porosity	0.43	0.41	0.38	0.52	...	0.25	0.14	0.07	0.51	<0.05

[a] Irregular.
[b] Reached after 24 h.
[c] ND = not determined.

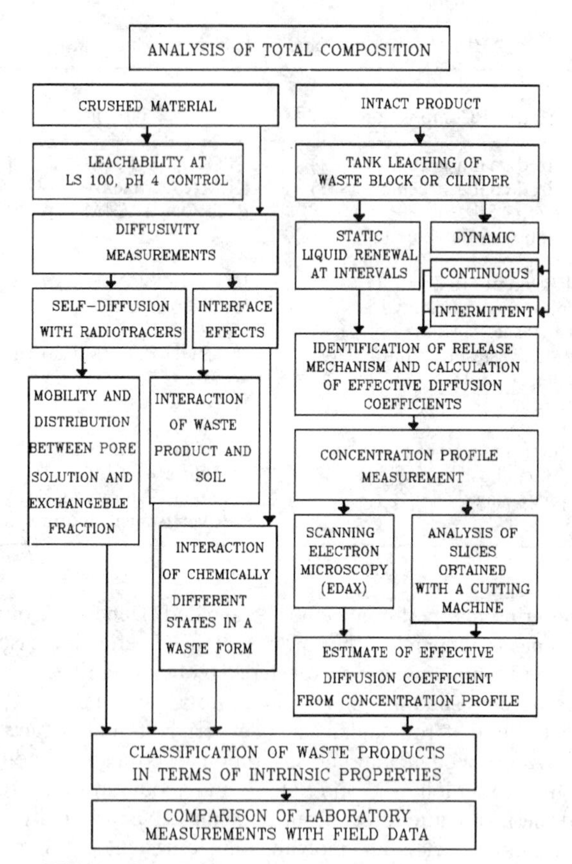

FIG. 1—*Schematic overview of waste product leaching.*

Tank Leaching Experiments—In order for the release of trace elements from solid waste forms to be measured, the products are placed on a polyethylene tripod in prerinsed polyethylene containers to maintain direct contact of all sides of the specimen with the leachant during the experiment. The leachant is demineralized water using five times the volume of the specimen to be leached. Exposure to demineralized water is applied for mutual comparison of materials. In addition, the concentration of trace elements in demineralized water is small or negligible compared with the concentrations usually measured after a certain contact time, thus allowing the measurement of leach factors for a wide range of trace elements without interference from a variable composition of a natural leachant (for example, ground water or surface water). Experiments in other media have been carried out, but will not be discussed here. The tank leaching experiments are carried out at room temperature. At certain time intervals, the contact solution is replaced. The solution is immediately filtered over a 0.45-μm membrane filter to remove suspended particulates, and pH and conductivity are measured. Part of the sample is acidified for the analysis of trace elements, and part of it is analyzed as such for major anions (F^-, Cl^-, and SO_4^{2-}). From the amount of each element leached in the successive leaching cycles, the flux (J_i) in mmoles $\cdot$ m^{-2} $\cdot$ s^{-1} is calculated and plotted on a log-scale against log t_i (t_i = contact time, s). The flux J_i is obtained from

$$J_i = \sum_{i=1}^{n} (C_i - C_0)/A \cdot t_i$$

where

C_i = concentration in the contact solution (mmoles · L^{-1}) in the ith interval,
C_0 = original concentration in the extractant (mmoles · L^{-1}),
V = volume of the extractant (L),
A = exposed surface area of the specimen (m^2),
t_i = overall contact time after the ith cycle of liquid renewal, and
n = number of cycles.

To compare the leaching characteristics of waste products under other conditions of exposure, the experimental setup as shown in Fig. 2 is used. In Fig. 2*a*, the conditions as described above are represented. In Fig. 2*b*, a dynamic leaching with a continuous inflow of fresh water is shown. In comparison with the previous case, the concentrations to be measured in the contact solution are lower to a degree, which is related to the flow rate in the system. Superficial flow over the waste products in an intermittent mode is represented in Case C. The experiment is conducted with representative specimens from the same batch. The same flow rate is maintained at 0.014 L/s in the two dynamic leaching systems to allow a direct comparison of these different exposure conditions. Similarly as described above the release of trace elements is plotted as log (J_i) against log t_i.

Element Speciation—Depending on the element under consideration, a substantial fraction of the total concentration of that element in the waste material is tied up in silicate minerals or other very insoluble mineral phases. This fraction does not contribute significantly to the leaching process. In order for the magnitude of this fraction to be assessed, a representative portion of the specimen under study is crushed until 95% of the weight is below 125 μm in size. A 10-g sample of the crushed material is leached in a liquid/solid ratio of 100 in demineralized water, while the pH is kept at 4 for 4 h. After filtration through a 0.45-μm

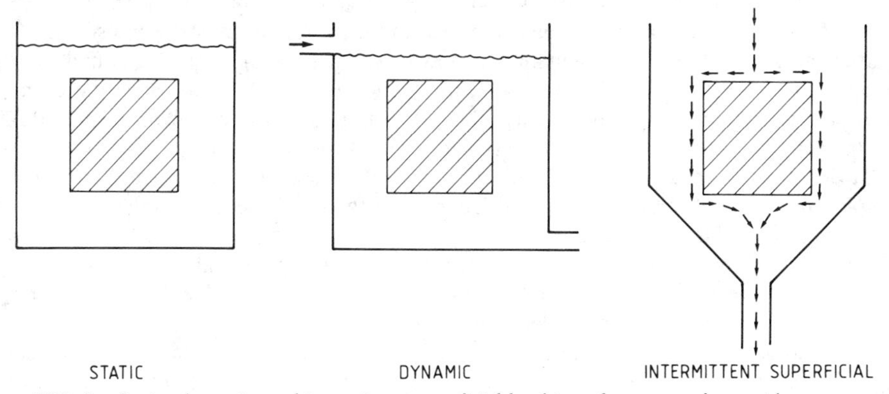

FIG. 2—*Static, dynamic, and intermittent superficial leaching of waste products with water.*

membrane filter, the extracts are acidified to pH 2 with 2N HNO_3 p.a. The leachable fraction (f) is obtained from

$$f = 100.C/S_0$$

where C is the concentration measured in the extract (mmoles · L^{-1}), and S_0 is the total element concentration in the waste product (mmoles · m^{-3}).

Measurement of Concentration Profiles—For the measurement of the solid-phase concentration profiles of elements in a direction perpendicular to the surface of a waste product, two approaches have been followed. One is to prepare a cross-section of the waste product for scanning electron microscopy (SEM-EDAX), which gives a picture of the segment and a distribution of major elements (for example, calcium, magnesium, aluminum, silicon, iron, and sulfur) over the segment. The other is to cut thin slices (>0.2 mm) from an exposed waste product using the homemade apparatus shown in Fig. 3. The apparatus was developed for dry cutting of waste products and subsequent collection of the dust generated on a filter. The amount of material collected in this way is sufficient for a full analysis of a wide range of trace elements. By replacing the filter after each cutting, a concentration profile can be obtained by analyzing the dust collected on the subsequent filters. The apparatus has been constructed from a wet stone saw (Universal-Trennmachine Type CON, Herbert-Arnold), with a diamond saw blade (Type D301), equipped with a housing and connected to a High-Volume Sampler (General Metalworks Type 2000H). The sawdust is collected on Gelman Glass Fiber Filter (Type A/E, 20.3 by 25.4 cm). From the diffusion or depletion depth (X_c) measurement by SEM-EDAX or by concentration profile analysis, the effective diffusion coefficient can be obtained from $X_c = \sqrt{2D_e t}$, in which t is the known exposure time.

Mobility Measurements—The measurement of mobility and interaction of potentially hazardous elements within waste materials and products is carried out in diffusion tubes. In Fig. 4, the principle of the method is schematically shown. A porous solid, part of which has been labeled with a radiotracer, is placed between pistons, and the diffusion in the non-labeled segment is measured after a certain exposure time. It is necessary to measure the mobility of sodium-22 in any solid to be studied, since sodium can be assumed to be relatively inert in the porous material and as such is an indicator for the pore structure and consequently the tortuosity of the material. Both water content and particle size distribution have an influence on the tortuosity. It is mandatory to have sodium-mobility data when diffusivity data of different materials are to be compared. In Fig. 5, the preparation and slicing of the diffusion tube contents are schematically shown. About 1.6 g of the material under investigation is mixed with 0.4 mL of a carrier-free radiotracer solution, of which the chemical form is known and the specific activity allows detection in the untreated segment at 0.01% of the activity per slice in the labeled segment. The consistency of the paste should be such that no liquid is spilled upon introduction in the diffusion tube. In practice, this always implies unsaturated conditions. A small quantity of the mix is kept as a reference.

The diffusion tube is filled with untreated, wetted material to a length of 25 mm with a consistency similar to that of the labeled material (water content: 25% weight/weight [w/w]). An interface marker thulium-170 ($T1/2 = 128\ d$, $E_\gamma = 84$ keV) of high specific activity is introduced [*8*], the labeled material is added up to a total length of 50 mm, and the second piston is put in position. The tube is stored in a water saturated environment to avoid drying out during the experiment. Experiments have been carried out without previous equilibration of tracer and solid to be able to study rapid interactions and exchange phenomena. By combining different materials in the diffusion tube, for example, waste material

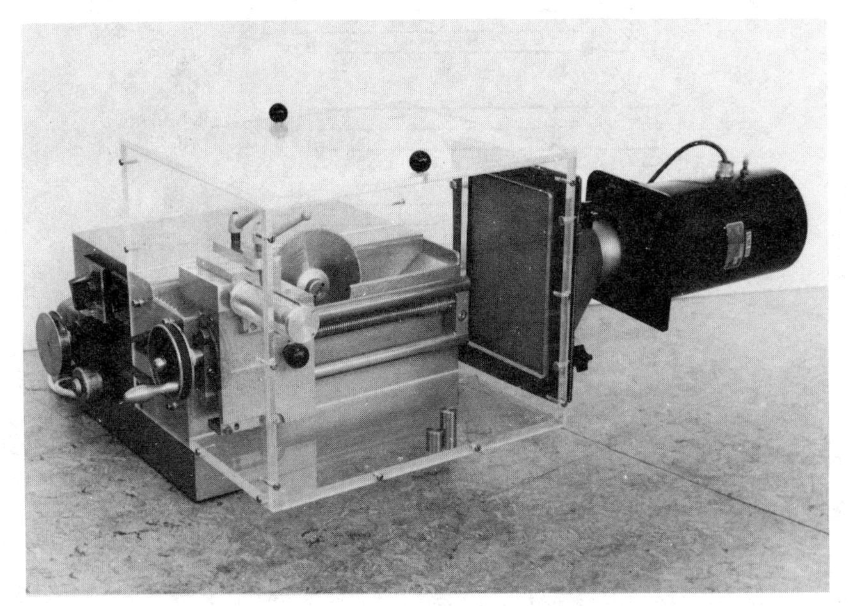

FIG. 3—*Homemade apparatus for concentration profile analysis by dry cutting of waste products.*

and soil, the interactions at the interface due to differences in chemical interaction of a diffusing species with both solids can be studied. By introducing the labeled species in a specific chemical form, the influence of speciation on migration can be assessed.

After a suitable storage time, which depending on the mobility of the element may vary from one day (sodium) to several month (or if necessary years), the combined segments are cut into slices of 0.3 to 1 mm starting from the nonlabeled side to avoid contamination by radioactive material adhering to the wall of the tube. The ideal exposure time is obtained from the product $D_e \cdot t = 2 \times 10^{-4} m^2$, in which D_e is the effective diffusion coefficient of the element and t is the exposure time. This relation is based on the fact that diffusing species should not reach the end of the diffusion tube, because the uncertainty in the calculation of D_e will be larger.

The slices are transferred to pre-weighed counting tubes, dried at 85°C, weighed, and counted in a sodium iodide crystal connected to a two-channel analyzer (radiotracer and interface marker). The mass of the slices and the total mass of the tube contents are used to calculate the axial length of each slice. Through the porous solid, flow of water is not

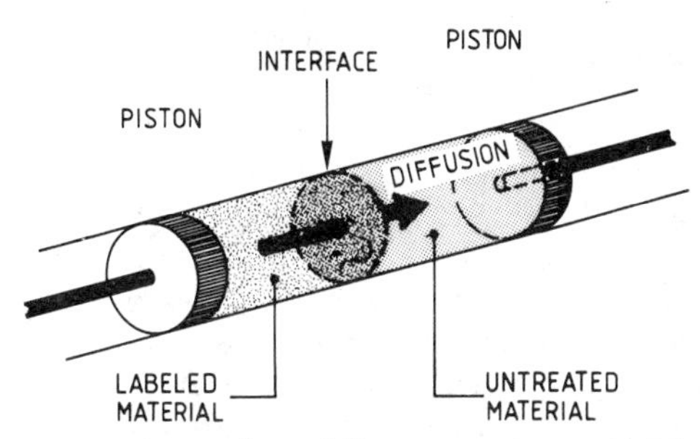

FIG. 4—*Experimental setup for mobility measurements with labeled material.*

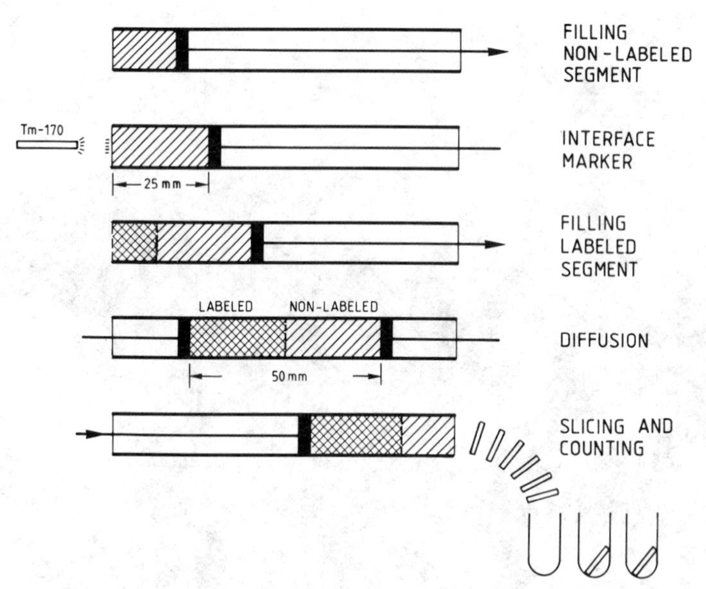

FIG. 5—*Assembling procedure for diffusion tubes used in mobility measurements of potentially hazardous elements in waste materials and waste products.*

possible, and consequently, the transport of species is diffusion controlled. Apart from molecular diffusion in the aqueous phase, adsorption phenomena and exchange reactions are important. Consequently, an effective diffusion coefficient (D_e, m^2/s) is measured. The solutions of Fick's second law, which can be applied to calculate effective diffusion coefficients from the measured activity profiles in the diffusion tubes are [*13,14*]

$$C = \frac{C_t}{A\sqrt{\pi D_e \cdot t}} e^{-x^2/(4D_e t)} \tag{1}$$

and

$$C/C_0 = 0.5 + 2/\pi \sum_{n=1} 1/n \cdot e^{-D_e \cdot t \cdot (n\pi/h)^2} \cdot \cos[n\pi(1 - x/h)] \cdot \sin[n\pi/2] \tag{2}$$

where

C_t = total amount of diffusing substance (counts $\cdot$ $mm^{-1} \cdot s^{-1}$),
C = concentration at position x (counts $\cdot$ $mm^{-1} \cdot s^{-1}$),
C_0 = concentration at interface $x = 0$ (counts $\cdot$ $m^{-1} \cdot s^{-1}$),
A = cross-sectional area through which diffusion occurs (mm^2),
D_e = effective diffusion coefficient ($m^2 \cdot s^{-1}$),
t = diffusion time (s),
x = distance from interface (mm), and
h = distance from interface to inactive end (mm).

The first equation is valid when the mobile component is originally located in a small layer on the boundary of the matrix in which the mobility is to be determined. The concentration C_0 of the component at the interface decreases with diffusion time. In our experimental setup, normally this is not the case, but it does provide a simple first approximation of the diffusion coefficient. In general, the obtained diffusion coefficient with this method is about

5 to 10% higher than the value obtained by Eq 2 which is more appropriate for our experimental setup. The second equation is the solution for linear, bounded diffusion [14]. The concentration C at the interface is constant in time. This relation is also valid in cases in which the mobile component has already reached the matrix boundary.

Calculation of D_e-values from Eq 1 is simple because plotting ℓnC against x^2 results in a straight line with slope $-1/4D_e t$. Although this model has limited applicability due to its boundary conditions, it does give a direct indication of the presence of more than one pD_e value. The second equation requires an iteration process to optimize the value. Since the plot of ℓnC against x^2 is convenient, it is used to obtain a good fit of calculated and original data points. From a constant offset between calculated and measured data points, a fractionation of the originally added activity in different chemical or physical forms can be deduced.

Since we are not equilibrating the solid prior to the diffusion measurement, dynamic processes can be observed, if the kinetics of the exchange are of the same order as the rate of diffusion. In most cases, however, the exchange/interaction is much faster than diffusive transport.

Leaching Model

Leaching of potentially hazardous elements from waste products is governed by a number of factors of both chemical and physical nature. In the field the biological factor may become relevant as well. A number of factors have been listed in Table 3. In this work, the biological aspects are not addressed, which does not imply that these aspects are not important. Quantifying their effects seems hard to achieve.

In Fig. 6, several phenomena occurring at the solid/solution interface and in the waste product are shown. The overall leach rate measured in tank leaching studies is composed of a number of more and less important superimposed effects. Depending on the concentration in porewater and contact solution, diffusion out of the product or into the product may occur. Dissolution from the surface may take place either continuously, if the whole material is relatively soluble, or only shortly, if a surface deposit of soluble components is

TABLE 3—*Factors influencing the leaching of trace elements from waste products.*

Chemical	Physical	Biological
alkalinity	porosity	colonization
chemical speciation in the product matrix	pore structure	product degradation by boring organisms
chemical interactions in the pores and at the surface	continuous or intermittent contact with water	pore clogging by biological substances
changes in the chemical environmental (pH, redox) in the product with time	surface-to-volume ratio of the waste product	changes in the chemical environment due to biological activity (redox)
surface dissolution	temperature in relation to diffusion rate and with respect to durability (freeze/thaw)	
chemical speciation in the pore water	density differences in the product matrix (for example, gravel in concrete)	
reaction kinetics		
chemical composition of the leachant (ground water, surface water)		

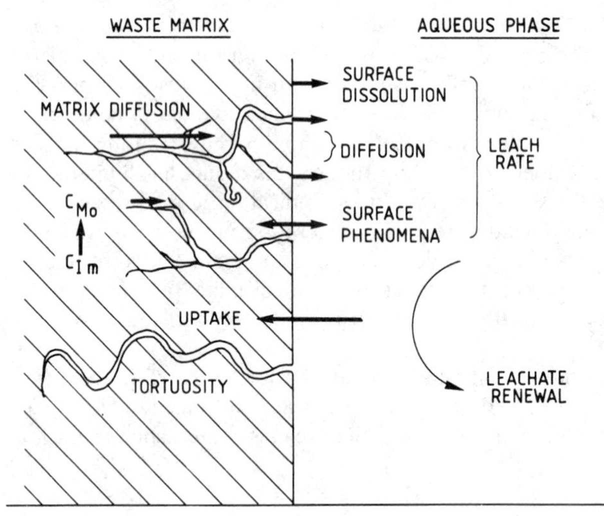

FIG. 6—*Phenomena at the solid/solution interface determining the leach rate from a waste product.*

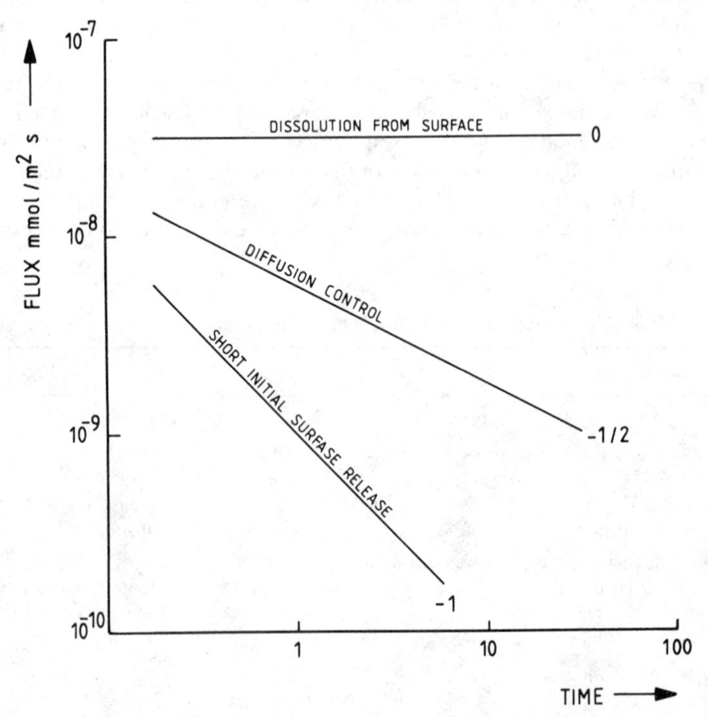

FIG. 7—*Cumulative element flux from a waste product as a function of time showing release governed by dissolution from the surface (slope 0), diffusion controlled release (slope −½,), and short initial surface release (slope −1).*

leached upon the first contact with water. In tank leaching studies, these phenomena lead to different slopes in the log-log plot of cumulative ion flux J_i against the contact time. If the product dimensions are not a limiting factor, a simple one-dimensional diffusion model can be applied, which in the case of diffusion-controlled leaching would lead to a slope $-½$ in this plot [*15*].

In Fig. 7, the effects of surface dissolution and short initial surface release on the ion flux through the solid/solution interface are indicated. This implies that from these measurements, information can be obtained as to the predominant release mechanism. Combinations of these mechanisms can also be found, resulting in a slope between 0 and $-½$ or between -1 and $-½$. If sufficient data points can be obtained within the time period set by the boundary conditions of the model, the relative importance of the different mechanisms can be isolated. The constant flux caused by surface dissolution can be quantified if present. The occurrence of short-term release of components can be recognized immediately and corrected for by subtraction of the initial peak concentrations. The diffusion component is modeled by a simple one dimensional diffusion described earlier [*15*]. If the diffusion depth of diffusing substance is small compared to the dimensions of the product, this approach is valid. An effective diffusion coefficient can be calculated if the slope in the log-log plot of element flux against t is close to -0.5. Since the fraction of the total concentration tied up in the silicate matrix and in poorly soluble mineral phases is not contributing to the concentration gradient that forms the driving force for the diffusive transport, this fraction has to be corrected for [*16*]. The value of D_e ($m^2 \cdot s^{-1}$) is obtained from

$$D_{e,x} = \pi t \left(\frac{J_i}{fS_0}\right)^2$$

where

J_i = ion flux through the interface in mmoles $\cdot$ $m^{-2}s^{-1}$ (Eq 1),
f = fraction available for leaching (-),
t = contact time (s), and
S_0 = total concentration in the solid (mmoles $\cdot$ m^{-3}).

In fact, for each chemical form, an individual D_e can be defined. Only the fastest leaching components are measured in tank leaching experiments.

If the D_e values for different chemical forms are very close, distinction in a tank leaching experiment may not be possible. In concentration profile measurement, a distinction in chemical forms is feasible. In Fig. 8, the concentration profile in a waste product is given after a certain exposure time. Here, three fractions can be identified. Fraction 3 represents the immobile fraction. From the profile a fraction 1 and 2 can be distinguished with definitely different effective diffusion coefficients, which can be estimated from $X_c = \sqrt{2D_e t}$. For mutual comparison of products the mobility of sodium is used, since sodium is assumed to be relatively inert in the product matrix and as such could reflect the influence of tortuosity on the effective diffusion coefficient. The tortuosity (T) can be obtained from the sodium mobility in the matrix ($D_{e,\mathrm{Na}}$)

$$T^2 = \frac{D_{\mathrm{Na}}}{D_{e,\mathrm{Na}}}$$

where D_{Na} is the free mobility of sodium in water.

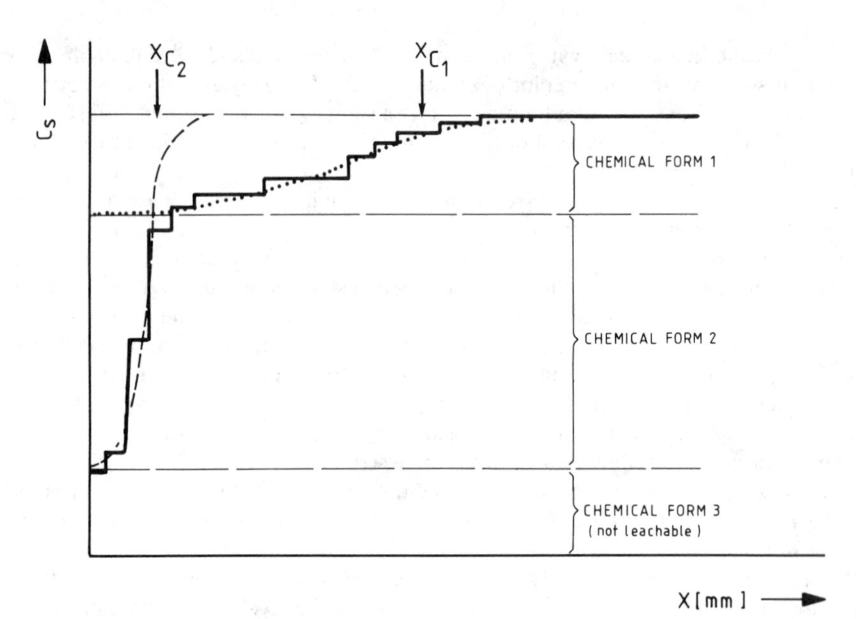

FIG. 8—*Concentration profile in a waste product as a function of distance from the surface; showing three chemical forms with different leach rates as derived from the depletion depth* X.

For a proper comparison, the difference in free mobility between sodium and other elements has to be taken into account [*17*]. From the difference in mobility of sodium and other elements in the product matrix, corrected for the difference in free mobility, a retention factor (R) can be isolated. A value of R can be obtained from

$$R = \frac{D_x}{D_{e,x} \cdot T^2}$$

where D_x is free mobility of component x and $D_{e,x}$ is the mobility of x in the product matrix. The factor R is dependent on pH in the pore solution and further changes in the chemical environment in the pores. In a comparison of laboratory and field conditions, the laboratory temperature and that in the field seldom match. For a temperature difference from 22 to 4°C, a reduction in mobility by a factor of 2 to 2.5 has to be taken into account.

Results and Discussion

Tank Leaching Studies

In the tank leaching studies, on a variety of waste products a number of major and trace elements has been measured. In Tables 4 to 6, the elemental composition of the products, the leachable fraction, and the calculated effective diffusion coefficient are given. The materials have been classified in the following categories: stabilized waste products, concrete, and slag. To give an impression of the alkalinity of the materials, the pH measured in the contact solution obtained in the first liquid renewal cycles is also listed.

The element concentrations in the various products cover a fairly wide range. Phosphogypsum is typically high in fluorine and phosphate slag is relatively high in uranium, whereas incinerator slag is usually high in metals (for example, copper, cadmium, zinc). From Table

TABLE 4—*Elemental composition of waste products studied.*[a]

	Stabilized Waste Products							Concrete		Slag
Element[b]	PCA	PCA13	PCA/PG	PG/PCA	PGH	BFS	IA	MPCA	LWPCA	PS
Na	0.30	0.28	0.26	0.16	0.28	0.49	2.68	0.16	0.45	0.78
Ca	5.90	5.44	7.2	16.1	26.9	17.8	11.2	11.0	14.9	33.5
Mg	0.35	0.61	0.39	0.16	0.03	6.31	0.49	0.27	1.14	0.45
F	180	120	980	3400	12 800	360	290	40	27	130
SO_4^{2-}	0.96	0.82	8.5	34.6	64	2.49	1.28	0.90	1.07	0.98
Cd	1.4	1.15	1.9	3.0	8.5	0.02	6.6	0.3	1.2	0.1
Cu	87	128	76	35	3.2	16	1620	25	117	4.2
Zn	250	253	224	117	4.1	22	1520	78	300	16
As	NA[c]	36	NA	NA	0.65	NA	NA	15	23.7	0.6
Se	10.4	4.6	9.3	4.1	NA	1.8	<1	9.5	1.6	<1
Sb	8.4	8.5	7.4	3.1	NA	0.05	17.2	2.5	10.5	<0.1
Mo	27	24	25	11	19	15	<10	2.5	6.0	<10
W	11.6	18	10.3	4.2	NA	0.5	<2	1.7	19	<1
U	12.6	NA	11.7	4.9	53	15	<5	3.6	11	128
V	253	159	215	99	5	213	48	68	226	21

[a] For specimen codes, see Table 1.
[b] Sodium, calcium, magnesium, and SO_4^{2-} in wt %, other elements in mg/kg.
[c] NA = not analyzed.

5 it becomes clear that sodium in pulverized coal ash and incinerator ash is largely tied up in the matrix, while it is largely available in phosphogypsum and blast furnace slag. In concrete, the calcium availability is high. The availability of magnesium increases progressively with phosphogypsum content. The availability of cadmium is generally high compared to the metals copper and zinc and the anions arsenic, antimony, selenium, uranium, and vanadium.

In products containing pulverized coal ash, the availability of molybdenum and tungsten is high, unless the alkalinity of the material is high. The difference in pD_e for sodium reflects the difference in tortuosity of the products. Obviously, the dense phosphate slag exhibits the highest tortuosity factor. As to be expected, the concrete products have a higher tortuosity factor than the stabilized products. For several elements, the differences in pD_e for

TABLE 5—*Leachable fraction of waste products (expressed as percent of total).*[a]

	Stabilized Waste Products							Concrete	
Element	PCA	PCA13	PCA/PG	PG/PCA	PGH	BFS	IA	MPCA	LWPCA
Na	13	15	14	45	100	45	5.6	16	41
Ca	10	81	45	23	26	20	50	83	85
Mg	38	32	56	63	89	14	22	40	39
F	18	78	80	20	8.6	18	14	70	>100
SO_4^{2-}	45	66	90	35	23	45	63	72	>100
Cd	ND[b]	26	ND	28	34	ND	5	65	33
Cu	2	8.5	1	2	17	0.8	2	1	6
Zn	3	6.1	11	18	40	10	14	9.5	36
As	ND	0.05	ND	ND	4.6	ND	ND	0.5	0.8
Se	6	3.7	5	4	ND	18	ND	2	13
Sb	3	1.5	7	4		13	1	2	4
Mo	45	13	56	56	2.7	3	ND	6	24
W	13	0.8	30	25	ND	63	ND	0.5	0.1
U	ND	ND	4	2	1.1	ND	ND	0.5	ND
V	NA[c]	1.1	NA	NA	8	NA	NA	2	0.9

[a] For specimen codes, see Table 1.
[b] ND = not detected.
[c] NA = not analyzed.

TABLE 6—*Effective diffusion coefficients from tank leaching* (pD_e = $-log De$; De *in* m^2/s) *and pH in contact solution.*[a]

Element	Stabilized Waste Products: PCA	PCA13	PCA/PG	PG/PCA	PGH	BFS	IA	Concrete: MPCA	LWPCA	Slag PS
Na	9.8	10.5	10.1	9.9	10.5	9.9	10.5	10.7	11.1	14.2
Ca	14.3	13.9	11.3	10.9	10.8	12.4	13.6	15.1	13.8	16.8
Mg	15.3	14.3	13.6	13.2	11.2	14.3	15.9	15.9	16.5	15.2
F	12.5	13.2	12.2	11.7	11.6	11.2	12.5	12.4	13.3	17.1
SO_4^{2-}	11.1	11.8	11.8	11.2	10.7	10.5	11.5	14.0	12.6	14.2
Cd	>16	>16	>16	16.2	12.8	>16	15.7	12.6	14.3	14.6
Cu	12.9	NA[b]	12.6	12.4	13.3	10.8	13.1	NA	13.6	15.3
Zn	13.4	NA	14.4	14.0	13.9	12.0	16.6	12.9	14.7	15.2
As	NA	10.0	NA	NA	10.9	NA	NA	10.6	11.3	16.6
Se	11.5	10.5	11.9	12.2	11.4	14.1	NA	11.7	11.1	>16
Sb	11.4	10.4	10.9	11.0	NA	10.8	11.8	11.1	12.1	>16
Mo	11.2	10.2	10.5	10.1	10.2	12.3	NA	12.2	11.5	16.6
W	11.7	9.2	11.9	11.2	NA	10.7	NA	NA	10.1	NA
U	NA	NA	13.1	12.5	9.9	NA	NA	NA	NA	17.7
V	NA	9.9	NA	NA	11.7	NA	NA	10.6	11.2	16.4
pH	11.3	9.8	8.6	8.2	7.4	11.2	11.5	11.5	10.8	9.0

[a] For specimen codes, see Table 1.
[b] NA = not analyzed.

different products cannot be explained by differences in tortuosity alone. In these cases, the alkalinity of the materials plays a dominant role. If the pD_e values are corrected for tortuosity by means of the sodium mobility, the retention factor R remains. A plot of R against pH in the contact solution shows that several elements are strongly influenced by pH (Fig. 9). The data on phosphate slag have not been corrected for availability. In the log-log plot of element flux against time, the slopes are often close to 0.8 to 0.9, indicating a considerable contribution of initial surface release. The data presented here have been corrected for this surface release. In the case of stabilized phosphogypsum, the other extreme has been observed in that a slope between $-\frac{1}{2}$ and 0 is found, indicating the importance of surface dissolution. In general, the mobility of anionic species in pulverized coal ash products is substantially higher than that of the metals, which implies that more attention has to be given to this group of elements in particular, since information on the behavior of these anionic species is rather limited in comparison with heavy metals. The availability test is relatively mild to simulate conditions which would reflect a long-term leachability. Therefore, the pH is not allowed to rise over 4. The high dilution (liquid/solid ratio 100) is used to minimize the effect of solubility limitations. An impression of the mild attack of the matrix can be inferred from the leachability of aluminum, silicon, and iron under these conditions. For the stabilized products, the availability of aluminum, silicon, and iron is generally less than 5%, 2%, and 6%, respectively. To compare the leaching from waste products under static and dynamic conditions, the sodium flux was measured from identical pulverized coal ash/phosphogypsum/lime blocks exposed to water, as indicated in Fig. 2. The flux is plotted against the contact time (Fig. 10). In the case of intermittent contact, only the wet periods—cycles of two to three days wet and two to three days dry—have been taken into account. After a couple of days of initial effects, the slopes in the flux-time plot is close to $-\frac{1}{2}$ in all cases. In view of the totally different exposure conditions, the agreement is good, indicating that in the three exposure conditions, diffusion is the rate-determining step in leaching of sodium. The pD_e values calculated from these data are for the static tank-leaching, the dynamic tank leaching, and the intermittent contact, 10.6, 10.8, and 10.5, respectively. Since in the intermittent contact with water, the pH in the contact

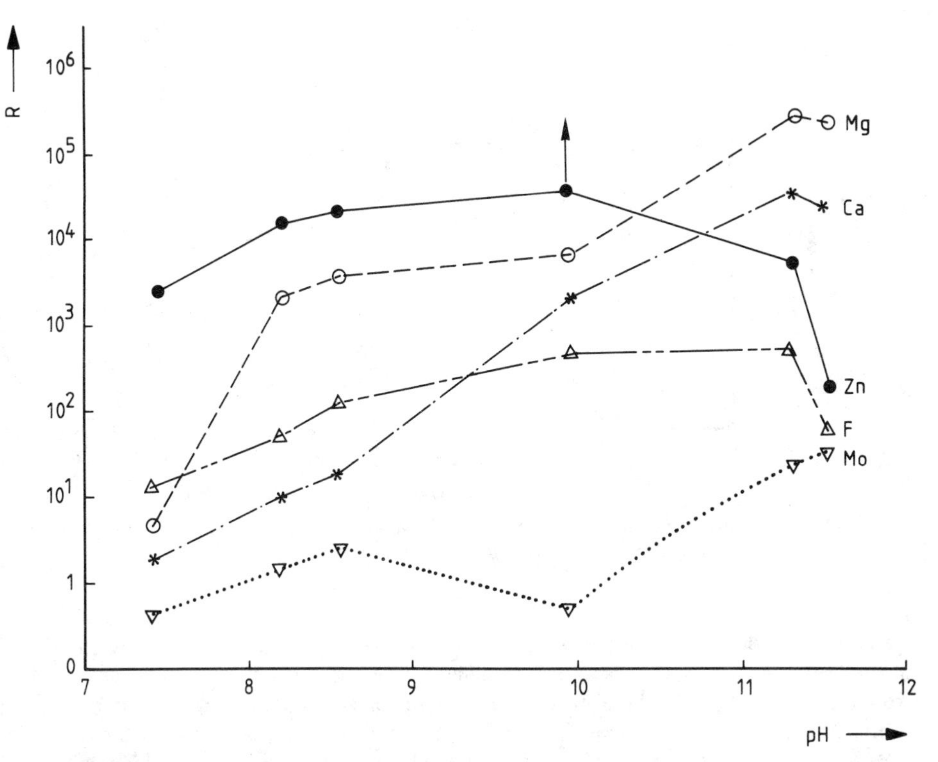

FIG. 9—*The retention factor* R *as a function of pH for magnesium, calcium, zinc, fluorine, and molybdenum in stabilized phosphogypsum/pulverized coal ash products and concrete containing pulverized coal ash.*

solution may be different due to the more intense contact during dry periods with carbon dioxide from the air, these findings cannot be translated directly to other elements, particularly not to elements sensitive to pH changes (such as magnesium).

Concentration Profile Measurement

Scanning Electron Microscopy—The concentration profile analysis by SEM-EDAX has been carried out on a cilindrical concrete specimen drilled in the wall of the cooling channel of the High Flux Reactor at Petten, which at the time of sampling had been exposed to fresh water from the North Holland Canal for 25 years. In Fig. 11, the SEM picture of a section of the specimen is shown together with the EDAX images of calcium, magnesium, silicon, sulfur, and manganese. The exposed surface is on the right-hand side of each picture. The plain of the picture is perpendicular to the exposed concrete surface. The sand grains clearly show up in the silicon image. The exchange of calcium for magnesium is reflected by the difference in X-ray intensity, in which calcium is depleted in the surface layer (1.7 mm) and magnesium is increased. Sulfur is depleted to a greater depth than calcium. In a very thin surface layer, manganese is detectable by SEM-EDAX ($>0.5\%$). Manganese is normally not present in concrete in concentrations detectable by SEM. Neutron activation measurements on material from the interior of the specimen resulted in 1100 to 1200 mg manganese/kg. A black coverage was found to contain 4.7% of manganese and a deeper layer 0.63% of manganese. The origin of the manganese, which clearly stems from the

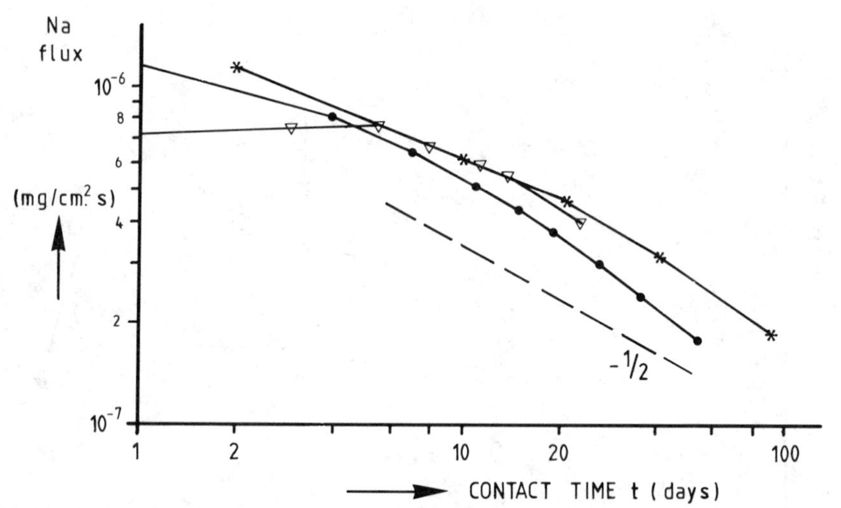

FIG. 10—*Sodium flux from a stabilized pulverized coal ash/phosphogypsum product in static mode* (*), *dynamic mode* (●), *and in intermittent superficial mode* (△).

waterphase, is the annual biodegradation of algea in autumn resulting in anoxic conditions in the bottom-sediments of the North Holland Canal and the subsequent release of Mn^{2+} from the sediments [*18*]. The dissolved manganese concentration in the canal increases and in the cooling channel manganese is retained in the alkaline surface layer of the concrete. This is a typical example of trace element uptake by a natural mechanism, which one would not consider in performing laboratory experiments.

Cutting Apparatus—By means of the homemade cutting apparatus, a segment of a stabilized pulverized/phosphogypsum/lime product was cut into thin slices, and the resulting dust of successive cuttings was collected on glass fiber filters, which were analyzed by neutron activation analysis and atomic absorption spectrometry for calcium, magnesium, sodium, scandium, lanthanum, zinc, arsenic, and antimony. The results of these analysis are presented in Fig. 12. The concentrations of the respective elements have been plotted in relation to the thickness of the successive slices sampled with the cutting apparatus. Towards the surface of the product, thinner slices were taken than deeper in the product. Since the loss of material in the surface layer due to dissolution of $CaSO_4$ ($CaSO_3$) leads to an apparent concentration increase of inert trace elements (scandium and lanthanum), all concentrations have been normalized to the average scandium concentration in the product (10.1 mg scandium/kg).

The resulting concentration profile can be used to estimate the depletion depth. From an exposure period of 10 months, the depletion depth of sodium cannot be estimated since the average mobility of sodium is highest of all elements measured and would correspond to a depletion depth >40 mm. The precision of the analytical techniques has to be very good to be able to draw conclusions from these measurements. In the case of zinc, values over 300 mg zinc/kg are suspect. The calcium profile exhibits two depletion zones, which indicates the presence of two leachable chemical forms, each with their own diffusion coefficient. The magnitude of each fraction is estimated as indicated in Fig. 8. This method of analysis applied to products exposed in the field will give information on the chemical speciation within the waste product as well as on phenomena that are not apparent in laboratory experiments. A spin-off of this work is the verification of chloride-penetration in relation

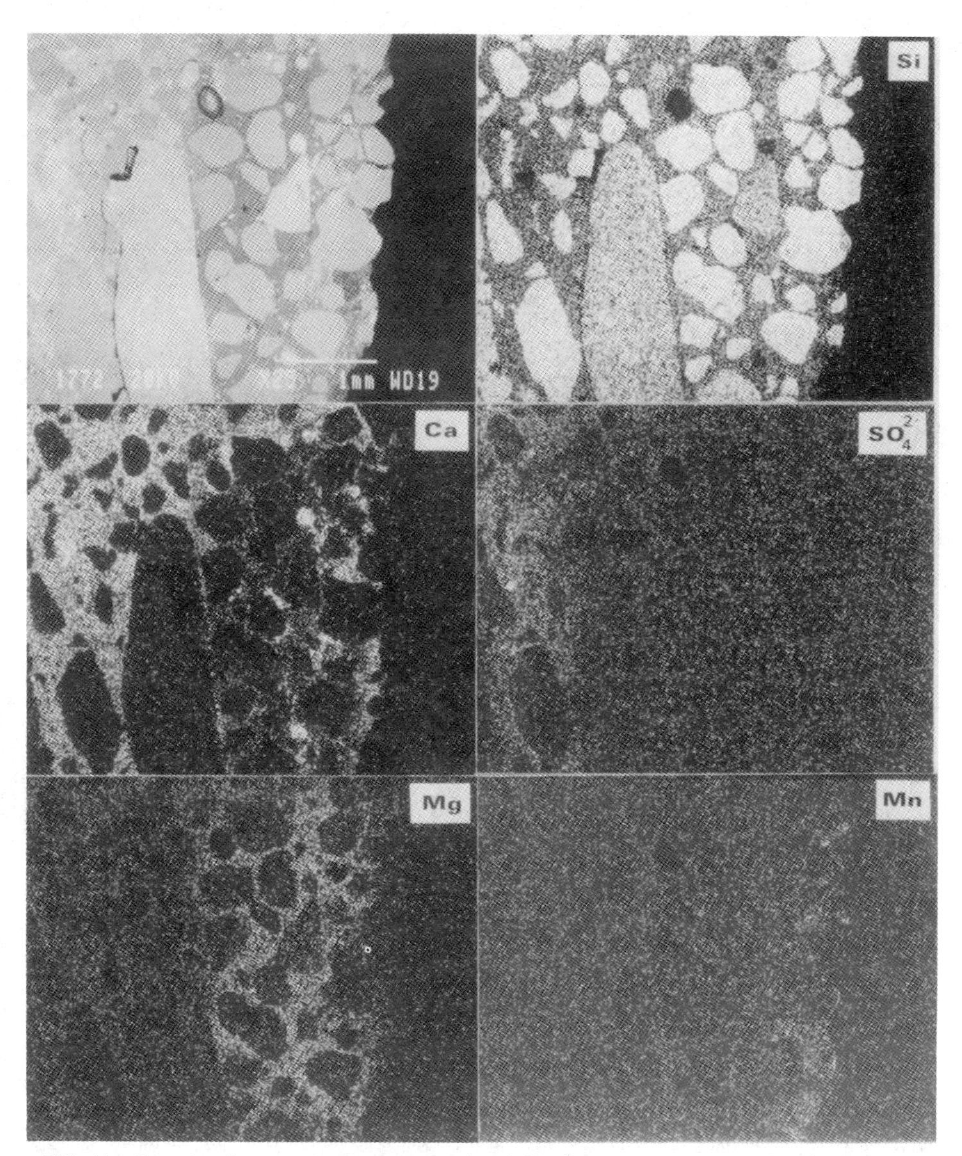

FIG. 11—*Scanning electron microscope picture* (top left) *of a concrete specimen taken from a cooling channel and exposed to leaching with fresh water for 25 years and the element concentration profiles (silicon, calcium, sulfur, magnesium, and manganese) as obtained with SEM-EDAX.*

to corrosion of steel reinforcement [*19*] and study of damage to construction materials by acid rain [*20*].

Mobility Measurements

The mobility measurements were carried out on the pulverized coal ash/phosphogypsum/lime product (PCA/PG). For that purpose, part of the product was crushed, homogenized, and sieved to obtain material passing a 150-μm sieve. The result of mobility measurements of sodium, molybdenum, arsenic, and zinc, using respectively sodium-22 (E_γ = 1275 keV,

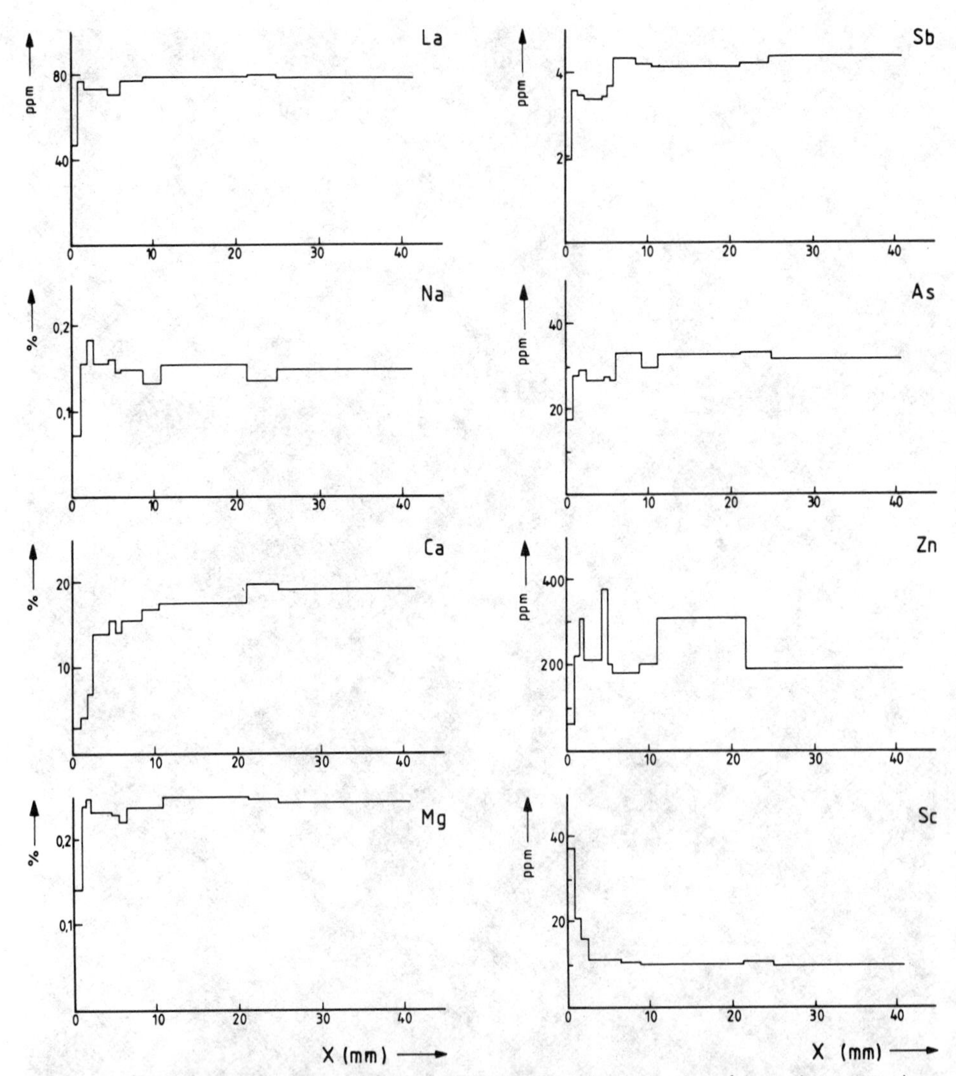

FIG. 12—*Concentration profiles of lanthanum, antimony, sodium, arsenic, calcium, zinc, magnesium, and scandium in a phosphogypsum/pulverized coal ash/cement product exposed to leaching by demineralized water for ten months. Data have been normalized to scandium (10.1 mg scandium/kg). The scandium data have not been corrected.*

T1/2 = 2.6*y*, Na^+), molybdenum-99/technetium-99^m (E_γ = 141 keV, T1/2 = 66*h*, molybdate), arsenic-74 (E_γ = 596 keV, T1/2 = 17.8*d*, arsenate), and zinc-65 (E_γ = 1115 keV, T1/2 = 244*d*, Zn^{2+}) as radiotracers, are given in Fig. 13. On the left-hand side of the graphs, the measured activity profile in the diffusion tube after a given exposure time is shown. On the right-hand side, the plots of ℓn (activity) against x^2 are given, from which the effective diffusion coefficients and the chemical fractionation are calculated. As expected, the mobility of sodium is highest. The mobility of molybdenum in the PCA/PG mix is relatively high, contrary to that of arsenic and zinc. Only a minor fraction of the latter elements is mobile; the remainder is retained fairly effectively in the product matrix. The amount of detail obtained in this type of mobility measurements is substantial. The technique offers a great potential for validation of measures to reduce the release of potentially hazardous trace elements from waste products. In addition, a better understanding of the chemical processes in the waste product will lead to the development of more efficient additives to control leachability.

Comparison of Diffusivity Measurements

Several interesting comparisons can be made of the methods mentioned above. In Table 7, the results of tank leaching data on the PCA/PG and data on concentration profile analysis using the cutting apparatus (Measurement of Concentration Profile subsection) are presented. Except for antimony, the agreement in pD_e and availability is good when the totally different methods are considered. If the difference in availability of antimony is taken into account, the data on it are comparable (pD_e = 11 in the tank leaching becomes 12.4 at an availability level of 20%).

A comparison of tank leaching data with SEM-EDAX measurements is possible on mortar prisms (Table 8) and the exposed concrete specimens from the cooling channel (subsection noted above). In Table 8, pD_e values for calcium, magnesium, manganese, sodium, and SO_4^{2-} derived from the depletion or penetration depth in the SEM data and those obtained from tank leaching studies on mortar prisms are given. An estimate of the fractionation is hard to give from the SEM data. The data on calcium and SO_4^{2-} are in the same order of magnitude. The discrepancy in the magnesium data may be explained by the higher alkalinity of the fresh product studied in the tank leaching experiment compared to that of the weathered surface of the concrete specimen from the field. For the stabilized PCA/PG, a comparison between tank leaching and mobility measurements is made (Table 9). The pD_e's calculated from tank leaching and mobility measurement are in the same order of magnitude. The tortuosities in the intact product and the crushed material used in mobility measurements is not the same. So the difference in sodium mobility has to be taken into account when comparing mobilities of other elements. The fractionation in both methods is not the same, as indicated in Fig. 14. So in addition to the fractionation of the tracer, information on the fractionation (availability) of the stable element is needed. The amount of detail in the mobility measurement is substantially greater than can be obtained in tank leaching. The methods are complementary because tank leaching can give information on the predominant leaching mechanism, whereas the mobility measurements yield information on processes occuring within the product matrix.

Main Factors Controlling Product Leachability

Alkalinity—The alkalinity of a stabilized product, which causes a high pH in the pore solution of the waste product matrix, proves to be a significant factor in controlling leachability. The leachability of metals is drastically reduced at high pH due to the formation of insoluble hydroxides or adsorption phenomena. The leachability of oxyanionic species is at a maximum at neutral pH and in most cases decreases at high pH [*21*]. The difference in leachability at neutral pH and at high pH can be orders of magnitude. So if there is a choice, an increase in alkalinity resulting in a contact solution pH higher than 9 or 10 is advantageous from environmental point of view.

Surface-to-Volume Ratio—The surface-to-volume ratio of a waste product greatly influences the release of potentially harmful elements to the environment. A smaller surface-to-volume ratio results in a smaller environmental impact. For cubic forms the surface to

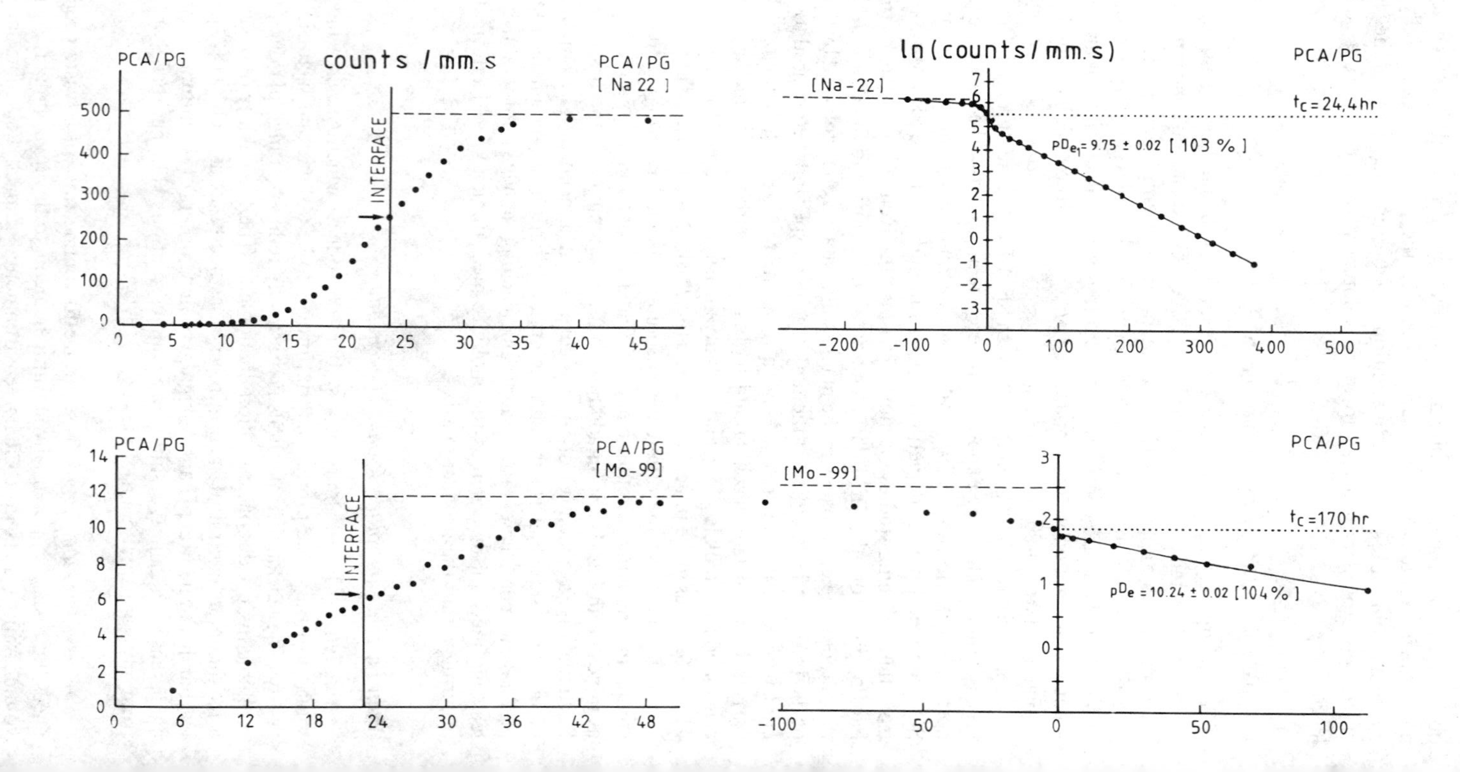
PCA/PG
counts / mm.s
PCA/PG
[Na 22]
INTERFACE
PCA/PG
PCA/PG
[Mo-99]
INTERFACE
ln (counts / mm.s)
PCA/PG
[Na-22]
t_c = 24,4 hr
pD_{e_1} = 9.75 ± 0.02 [103 %]
PCA/PG
[Mo-99]
t_c = 170 hr
pD_e = 10.24 ± 0.02 [104 %]

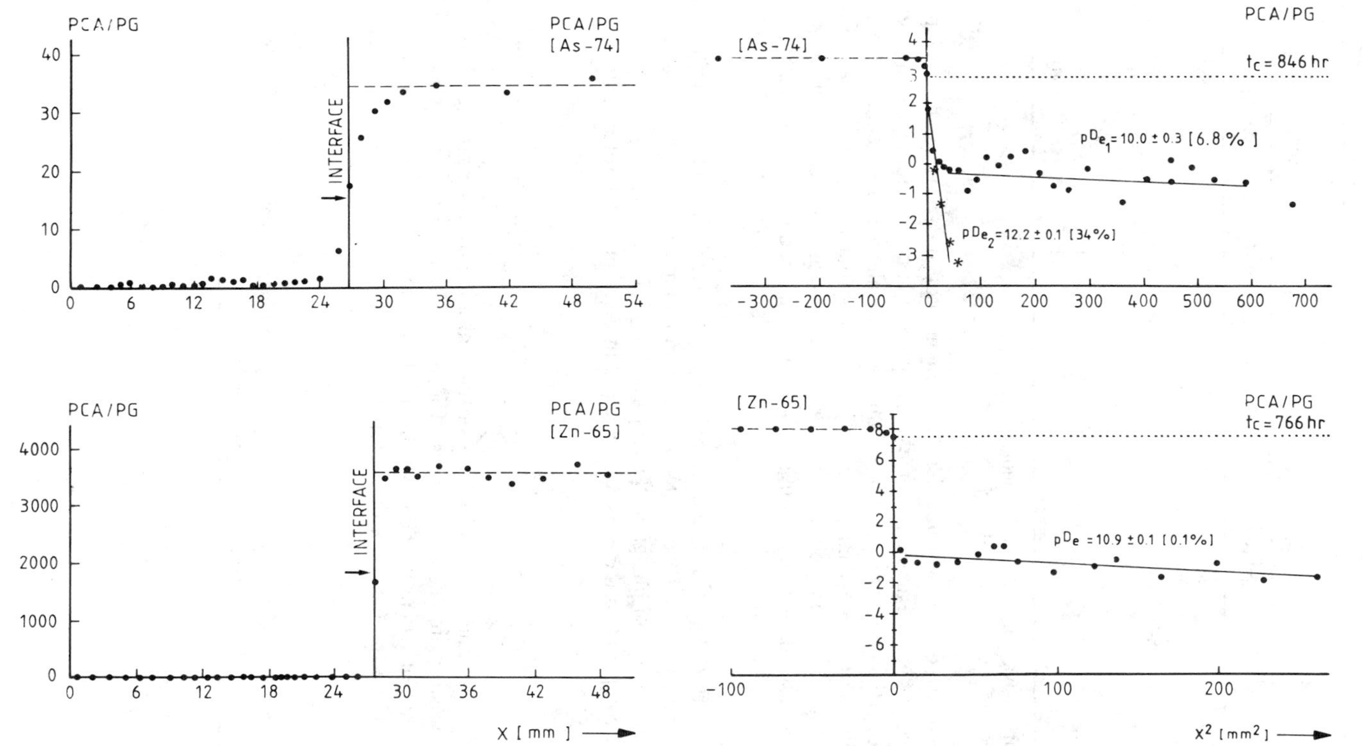

FIG. 13—*Mobility measurements of sodium, molybdenum, arsenic, and zinc in crushed PCA/PG. On the* left *the activity profile is given, on the* right *the* pD_e *values derived from the slopes* ($-1/4D_e \cdot t$) *of the lines in the plot of* ℓn (*counts* $\cdot$ $mm^{-1} \cdot s^{-1}$) *against* x^2.

TABLE 7—*Comparison of leach data from tank leaching and concentration profile analysis for a stabilized coal ash/phosphogypsum/lime product.*

	Tank Leaching		Concentration Profile in Exposed Products		
Element	pD_e	Availability, %	X_c, mm	pD_e	Availability, %
Na	9.9	45	>40	<10.5	20
Ca	10.9	23	20	11.1	20
			3	12.7	50
Mg	13.2	63	1	13.7	50
As	ND[b]	ND	7	12.0	20
Sb	11.0	4	6	12.2	20
Zn	14.0	18	1	13.7	(60)[a]

[a] Large uncertainty in availability due to low mobility.
[b] ND = not detected.

volume ratio is equal to $6/A$, in which A denotes the length of the axis (m). The leaching percentage relative to the total amount of an element present in a waste form is for a given exposure time proportional to the surface-to-volume ratio. So all measures leading to products with a smaller surface-to-volume ratio lead to a proportional decrease in leaching percentage; the long-term quantities released are not decreased, however. Granular materials prepared from waste materials with a surface-to-volume ratio greater than 200 m^{-1} are unfavorable compared with a monolith of the same material.

Tortuosity—The tortuosity, which is a measure for the extended path length for a diffusing substance to reach the surface of the waste product, can be influenced to some extent by mixing, compression, or use of fine-grained materials. The gain in reducing the release rate is relatively small compared to the effects of changes in alkalinity, state of oxidation, or surface-to-volume ratio.

Conclusions

1. Judgment of environmental consequences of the utilization of waste products or materials containing waste materials on the basis of a single extraction test is impossible in view of the wide range of actual situations which must be dealt with. Methods based on the determination of intrinsic properties of the materials concerned are more versatile in many respects.

TABLE 8—Comparison of leach data from tank leaching and concentration profile analysis by SEM-EDAX for concrete.

	Tank Leaching		Concentration Profile Analysis by SEM-EDAX	
Element	pD_e	Availability, %	X_c, mm	pD_e
Ca	15.1	83	1.7	14.7
Mg	15.9	40	1.7	14.7
SO_4^{2-}	14.0	72	2.9	14.2
Mn	NA[a]	NA	0.2	16.6
Na	10.7	16	NA	NA

[a] NA = not analyzed.

TABLE 9—*Comparison of tank leaching data with mobility measurements on crushed material for a stabilized coal ash/phosphogypsum/cement product.*

	Tank Leaching		Mobility Measurement	
Element	pD_e	Availability, %	pD_e	SP, %
Na	10.1	14	9.7	103
As	ND[b]	ND	10.0	7
			>12.0	>30
Mo	10.5	56	10.2	104
Cd	≧14	<5	13.3	4
Zn	14.4	11	10.9	0.1
			>13	>90

[a] SP = fractionation of tracer, matrix-incorporated elements not included.
[b] ND = not detected.

2. Chemical speciation of potentially hazardous elements within a waste product and the interaction of these elements within the pore system with matrix components are crucial for the release rate to the environment.

3. Release of potentially hazardous components from waste products is reduced by a high product alkalinity, a small surface-to-volume ratio, and a high tortuosity (low open porosity). The effect of the latter is small compared with that of the former measures.

4. Concentration profile measurement within a waste product allows a verification of results obtained in leach tests and yields information on the chemical speciation of elements within the product.

5. Comparison of different measurements of effective diffusion coefficients is in agreement within a factor of 2 to 5, in the case of very low mobilities sometimes only within one order of magnitude. In view of the widely different methods of determination, these results are encouraging for a further systematic approach of leaching from waste products.

6. Mobility measurements using radiotracers give a considerable amount of detailed information on element behavior in a waste material or waste product. The precision is good, and the versatility and simplicity of the method allow detailed testing of factors influencing mobility of hazardous components in waste materials, from which measures to control leachability of undesired components can be derived.

7. More information on different ways of contact with water are needed, particularly in

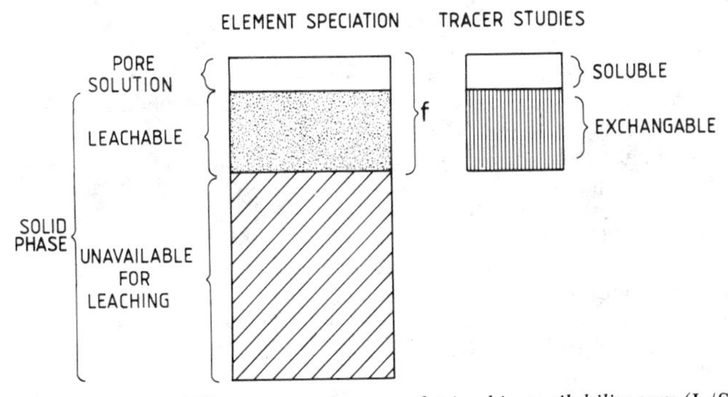

FIG. 14—*Element speciation within waste products as obtained in availability tests* (L/S *100, pH* = *4*) *and in mobility studies with radiotracers.*

relation to pH, to allow utilization of intrinsic leach parameters in a wide range of environmental conditions.

8. The different methods applied here are complementary rather than exchangeable, since each approach has its specific advantages. The tank leaching studies give information on the predominant release mechanism. The concentration profile measurements give an indication of the speciation within the waste material and uptake from the environment. The mobility measurements yield information on the mobility of specific chemical forms, in which an element may appear in the pore solution, and on the interaction of elements within the matrix.

9. Within groups of similar materials, the effective diffusion coefficients appeared to be mainly influenced by product alkalinity. Classification of waste materials and waste products in terms of leaching behavior therefore seems within reach.

Acknowledgment

Part of this work has been carried out with financial support from the European Community (EC) under contract EN3F-0032-NL and the National Coal Research Programme (NOK) through the Management Office for Energy Research (PEO) under Contract No. 20.51-013.10. The companies and organizations that supplied us with representative waste specimen are thanked for their contribution. Thanks are due to H. G. M. Eggenkamp and G. Hamburg for their contributions in the experimental work.

References

[*1*] Report DOE/METC-85/6018, *Proceedings,* 7th International Ash Utilization Symposium, Orlando, FL, 1985.

[*2*] *Proceedings,* 2nd International Conference on the Use of Fly Ash, Silica Fume, Slag, and Natural Pozzolans in Concrete, V. Mohan Malhotra, ed., Madrid, 1986.

[*3*] van der Sloot, H. A., Weijers, E. G., Hoede, D., and Wijkstra, J., "Physical and Chemical Characterization of Pulverized Coal Ash with Respect to Cement-Bound Applications," ECN-178, Netherlands Energy Research Foundation, Petten, the Netherlands, 1985.

[*4*] *Hazardous Waste Proposal Guidelines and Regulations,* U.S. Environmental Protection Agency, Section 250.13, paragraphs C through E, 1978.

[*5*] Côté, P. L., Bridle, T., and Benedek, A., "An Approach for Evaluating Long-Term Leachability from Measurement of Intrinsic Waste Properties," *Proceedings,* 3rd International Symposium on Industrial and Hazardous Waste, Alexandria, Egypt, 1985.

[*6*] Côté, P. L., "Contaminant Leaching from Cement-Based Waste Forms Under Acidic Conditions," thesis, McMaster University, Hamilton, Ontario, Canada, 1986.

[*7*] van der Sloot, H. A., Piepers, O., and Kok, A., "A Standard Leaching Test for Combustion Residues," Technical Report, BEOP-31, Bureau Energy Research Projects, Utrecht, the Netherlands, 1984.

[*8*] van der Sloot, H. A., de Groot, G. J., Eggenkamp, H. G. M., Tielen, J. A. L. W., and Wijkstra, J., "Versatile Method for the Measurement of Trace Element Mobilities in Waste Materials, Soil and Bottom Sediments, to be published.

[*9*] van der Sloot, H. A., Wijkstra, J., and Hoede, D., "Milieutechnisch Onderzoek van GASCON: Een Constructief Lichtgewicht Beton," Technical Report ECN-85-46, Netherlands Energy Research Foundation, Petten, the Netherlands, 1985.

[*10*] Anthonissen, I. H., "Onderzoek Naar de Mobiliteit van Metalen in Toepassingen van Bulkafvalstoffen," Technical Report RIVM/3924/141019, Rijks Institunt voor Volkesgezandheid en Milienhygiene, Bilthoven, the Netherlands, 1984.

[*11*] van der Sloot, H. A. and Wijkstra, J., "Short- and Long-Term Effects in the Leaching of Trace Elements from Stabilized Waste Products," *Proceedings,* International Ocean Disposal Symposium, Corvallis, OR, 1984.

[*12*] van der Sloot, H. A., Wijkstra, J., and Hoede, D., "Milieutechnisch Onderzoek aan Fosforslak," Technical Report ECN 86-37, Netherlands Energy Research Foundation, Petten, the Netherlands, 1986.

[13] Torstenfelt, B., "Migration of Fission Products Strontium, Technetium, Iodine and Cesium in Clay," *Radio Chimica Acta,* Vol. 39, 1986, pp. 97–104.
[14] Schreiner, F., Friend, S., and Friedman, A. M., "Measurement of Radionuclide Diffusion in Ocean Floor Sediments and Clay," Nuclear Technology, Vol. 59, 1982, pp. 429–438.
[15] Duedall, I. W., Buyer, J. S., Heaton, M. G., Oakley, S. A., Okubo, A., Dayal, R., Tatro, M., Roethel, F. J., Wilke, R. J., and Hershey, J. P., "Diffusion of Calcium and Sulfate Ions in Stabilized Coal Wastes," *Wastes in the Oceans,* Vol. 1, I. W. Duedall, B. H. Ketchum, P. K. Park, and D. R. Kester, Eds., Wiley, New York, 1983, pp. 375–395.
[16] van der Sloot, H. A., Wijkstra, J., van Stigt, C. A., and Hoede, D., "Leaching of Trace Elements from Coal Ash and Coal Ash Products," *Wastes in the Ocean,* Vol. 4, I. W. Duedall, D. R. Kester, and K. H. Park, Eds., Wiley, New York, 1985, Chapter 19.
[17] Li, Y. H. and Gregory, S., "Diffusion of Ions in Seawater and in Deep Sea Sediments," *Geochimica et Cosmochimica Acta,* Vol. 38, 1974, pp. 703–714.
[18] Sundby, B. and Silverberg, N., "Pathways of Manganese in an Open Estuarine System," *Geochimica et Cosmochimica Acta,* Vol. 45, 1981, pp. 293–307.
[19] Page, C. L., Short, N. R., and El Tarras, A., "Diffusion of Chloride Ions in Hardened Cement Pastes," *Cement and Concrete Research,* Vol. 11, 1981, pp. 395–406.
[20] Lal Grour, K. and Holden, G. C., "Pollutant Effects on Stone Monuments," *Environmental Science and Technology,* Vol. 15, 1981, pp. 386–390.
[21] de Groot, G. J., this publication, pp. 170–183.

William E. Shively[1] *and Mike A. Crawford*[2]

Extraction Procedure Toxicity and Toxicity Characteristic Leaching Procedure Extractions of Industrial and Solidified Hazardous Waste

REFERENCE: Shively, W. E. and Crawford, M. A., **"Extraction Procedure Toxicity and Toxicity Characteristic Leaching Procedure Extractions of Industrial and Solidified Hazardous Waste,"** *Environmental Aspects of Stabilization and Solidification of Hazardous and Radioactive Wastes, ASTM STP 1033,* P. L. Côté and T. M. Gilliam, Eds., American Society for Testing and Materials, Philadelphia, 1989, pp. 150–169.

ABSTRACT: The 1984 amendments to the Resource Conservation and Recovery Act (RCRA) require that the Environmental Protection Agency (EPA) restrict the land disposal of certain hazardous wastes. The 7 Nov. 1986 RCRA amendments promulgated specific treatment standards and effective dates for the first phase of the land disposal restrictions (dioxin- and solvent-containing hazardous wastes). EPA also promulgated the Toxicity Characteristics Leaching Procedure (TCLP) for determining compliance for these wastes with applicable treatment standards.

The purpose of this study was to evaluate the performance of commercial hazardous waste treatment, storage, and disposal facilities (TSDFs). Five facilities were sampled representing a wide range of treatment processes. Sampling was performed to monitor all influent, effluent, and treatment residues associated with each process. Leaching evaluations of industrial waste treatment residues and solidified/stabilized wastes are presented in this paper. Extraction procedure (EP) toxicity tests for metals and toxicity characteristic leaching procedure (TCLP) analyses for metals, volatile organic compounds (VOCs), and semi-volatile organics are reported. The results demonstrate effective treatment resulting in nonhazardous residues in the majority of the cases. In several instances, the TCLP results for treatment residues are not within the recently promulgated regulations.

The waste composition and leachate results before and after treatment are compared. Metal hydroxide sludges from hydroxide precipitation processes were tested. A waste oil, a paint sludge, a biological treatment pond sludge, and a metal hydroxide plating sludge were tested before and after fly ash solidification. Plating wastes mixed with acids and neutralized with lime were tested after sulfide precipitation. Mixed plating sludges were tested before and after solidification with excess lime. Two ash residues produced by different solvent incinerators were tested. Compositional analyses showed varying quantities of organic compounds.

KEY WORDS: leachate, Toxicity Characteristic Leaching Procedure (TCLP), extraction procedure (EP) toxicity, hazardous wastes, incinerator residues, metal hydroxide sludges, solidification, organics, metals, precipitation, Resource Conservation and Recovery Act (RCRA)

The Resource Conservation and Recovery Act (RCRA) directed the U.S. Environmental Protection Agency (EPA) to identify and establish regulations for wastes which pose a threat to human health and the environment. EPA established regulations to accomplish these objectives based on the characteristics of ignitability, corrosivity, reactivity, and extraction

[1] Project manager, CH2M Hill, 50 Staniford St., Boston, MA 02114.
[2] Regional office manager, Peer Consultants, P. C., 125 Cambridge Park Dr., Cambridge, MA 02140.

procedure (EP) toxicity. Maximum concentrations for eight metals and six pesticides in waste extracts were established as part of the EP toxicity test in an effort to establish regulations to prevent groundwater contamination.

The Hazardous and Solid Waste Amendments of 1984 (HWSA) amended RCRA and restricted the land disposal of untreated hazardous wastes and treatment residues. Treatment standards for restricted wastes are being established by EPA. Test procedures for characterizing the leaching potential of hazardous wastes have been evaluated. EPA promulgated the 7 Nov. 1986 land disposal regulations, which identified treatment standards and set effective dates for specific dioxin and solvent wastes. The Toxicity Characteristic Leaching Procedure (TCLP) was also adopted to determine whether the wastes meet applicable technology-based treatment standards. The TCLP is designed to determine the leachability of metals, pesticides, semi-volatile organic compounds, and volatile organic compounds (VOCs). Loss of the VOC's during the test is prevented by using a zero head-space extractor (ZHE). One of two leaching solutions is selected for the TCLP test based on waste alkalinity. Both solutions contain acetic acid diluted in deionized water; one solution is more concentrated. The waste alkalinity is tested in a preliminary step, and highly alkaline wastes are extracted with the more concentrated solution.

This paper summarizes the results of EP Toxicity and TCLP extractions of treatment residues and solidified wastes produced by full scale hazardous waste treatment facilities. Performance tests at five commercial treatment, storage, and disposal facilities (TSDFs) and the treatment of a variety of hazardous wastes are discussed here. Influent, effluent, and residue streams were sampled and analyzed for VOCs, semi-volatile organics, and metals. Both EP toxicity and TCLP tests were conducted on untreated solid wastes and treatment residues.

Sampling and Analysis Program

The focus of this study was the identification and sampling of well-operated full-scale hazardous waste treatment facilities. Sample locations were selected to characterize raw wastes, treated effluents, process residues, and air emissions, and to evaluate the performance of each unit operation. Most operations were batch processes, and for this reason, grab samples were collected at the beginning and end of each operation. For semi-continuous operations, like filtration or carbon absorption, discrete grab samples were composited to account for possible performance variations over time. Samples were typically collected from tank trucks, drums, storage tanks, and reactor vessels using composite liquid waste samplers (COLIWASAs) to ensure collection of representative samples including all phases. Samples were split into fractions in the field, stored in appropriate containers, preserved as appropriate for required analyses (Table 1), and then securely packed and forwarded to the analytical laboratory.

Results and Discussions

The wastes treated represent a broad cross section of different types with unique physical and chemical characteristics. A brief description of the wastes, the treatment processes employed, and the analytical results for each facility is presented.

Facility A—Cyanide Oxidation and Metals Hydroxide Precipitation

In addition to treating corrosive and metal-bearing wastes, this facility routinely treats metal-cyanide complexes. Wastes received and treated during the three sampling periods were predominately cyanide-bearing wastes from the electroplating industry (RCRA waste

TABLE 1—*Analytical methods and references.*

Parameter	Method No.
1. Volatile organics	Method 8240 [*2*]
chloromethane	trans-1,3-dichloropropene
bromomethane	trichloroethene
vinyl chloride	dibromochloromethane
chloroethane	1,1,2-trichloroethane
methylene chloride	benzene
acetone	cis-1,3-dichloropropene
carbon disulfide	2-chloroethylvinylether
1,1-dichloroethene	bromoform
1,1-dichloroethane	2-hexanone
trans-1,2-dichloroethene	4-methyl-2-pentanone
chloroform	tetrachloroethene
1,2-dichloroethene	1,1,2,2-tetrachloroethane
2-butanone	toluene
1,1,1-trichloroethane	chlorobenzene
carbon tetrachloride	ethylbenzene
vinyl acetate	styrene
1,2-dichloropropane	
2. Extractable organics	Method 8270 [*2*]
2,4,6-trichlorophenol	bis(2-chloroethoxy)methane
p-chloro-*m*-cresol	hexachlorobutadiene
2-chlorophenol	hexachlorocyclopentadiene
2,4-dichlorophenol	isophorone
2,4-dimethylphenol	napthalene
2-nitrophenol	nitrobenzene
4-nitrophenol	*N*-nitrosodimethylamine
2,4-dinitrophenol	*N*-nitrododi-*n*-propylamine
4,6-dinitro-*o*-cresol	*N*-nitrosodi-*n*-propylamine
pentachlorophenol	bis(2-ethylhexyl)phthalate
phenol	butyl benzyl phthalate
acenaphthene	di-*n*-butylphthalate
benzidine	di-*n*-octylphthalate
1,2,4-trichlorobenzene	diethylphthalate
hexachlorobenzene	dimethyl phthalate
hexachloroethane	benzo(a)anthracene
bis(2-chloroethyl)ether	benzo(a)pyrene
2-chloronaphthalene	3,4-benzofluoranthene
1,2-dichlorobenzene	benzo(k)fluoranthene
1,3-dichlorobenzene	chrysene
1,4-dichlorobenzene	acenaphthylene
3,3-dichlorobenzidine	anthracene
2,4-dinitrotoluene	benzo(ghi)perylene
2,6-dinitrotoluene	fluorene
1,2-diphenylhydrazine	phenanthrene
fluoranthene	dibenzo(a,h)anthracene
4-chlorophenyl phenyl ether	ideno(1,2,3,-cd) pyrene
4-bromophenyl phenyl ether	pyrene
bi(2-chloroispropy)ether	2,3,7,8-tetrachlorodibenzo-*p*-dioxin
3. Metals	
cadmium	Method 7130 [*2*]
hexavalent chromium	Method 218.5 [*3*]
total chromium	Method 7190 [*2*]
copper	Method 7210 [*2*]
iron	Method 236.1 [*3*]
lead	Method 7420 [*2*]
nickel	Method 7520 [*2*]

TABLE 1—*Continued.*

Parameter	Method No.
silver	Method 7760 [*2*]
tin	Method 282.1 [*3*]
zinc	Method 7950 [*2*]
4. Residue	
total solids	Method 160.3 [*3*]
total volatile solids	Method 160.4 [*3*]
total suspended solids	Method 160.2 [*3*]
total dissolved solids	Method 160.1 [*3*]
total volatile suspended solids	Method 160.4 [*3*]
total volatile dissolved solids	Method 160.4 [*3*]
5. Cyanide—total and amenable	Method 9010
6. Total organic carbon	Method 415.2 [*3*]
7. Alkalinity	Method 310.1 [*3*]
8. EP toxicity	Method 1310
9. TCLP	Method 13XX [*4*]
10. BTU content	ASTM D 2015 [*4*]

codes F006, F008, and F009). Cyanide content is limited to about 15% due to the excessive time required to dilute more concentrated wastes to avoid excessive exothermic reactions. Influent total organic carbon (TOC) is typically not permitted to exceed about 15 000 mg/L because above this limit the facility has identified historical limitations achieving the treated effluent limit of 1500 mg/L set by the local municipal pretreatment regulations. Laboratory bench-scale studies are conducted before waste acceptance to ensure the treatability of each waste.

The treatment process at Facility A is depicted in Fig. 1. Incoming cyanide wastes are pretreated in a single stage alkaline chlorination process. Unit processes for treatment of oxidized cyanide wastes and non-cyanide aqueous wastes include storage, blending, pH adjustment (with waste acids), chemical reduction, lime precipitation of heavy metals, filtration, neutralization, and carbon absorption (GAC).

Analytical results for the three sampling periods observed are presented in Table 2. Generally, the aqueous influent samples contained high concentrations of suspended solids, varying amounts of heavy metals, and high concentrations of organics. The most concentrated of the three samples (Period 3) contained 180 mg/L cadmium, 1500 mg/L copper, 500 mg/L nickel, 1700 mg/L zinc, and 19 000 mg/L TOC. After precipitation and filtration, the filtrate contained 50 mg/L cadmium, 380 mg/L copper, 22 mg/L nickel, 46 mg/L zinc, and 5000 mg/L TOC. Following GAC treatment, the effluent concentrations were 1.4 mg/L cadmium, 0.56 mg/L copper, 1.8 mg/L nickel, 0.7 mg/L zinc, and 1400 mg/L TOC. These effluent concentrations were all within the pretreatment regulations imposed by the local municipal sewerage authority.

Two waste acids were used for pH adjustment. Waste acid Sample 1 was identified as a concentrated hydrochloric acid. Heavy metals present in the acid, and therefore introduced to the aqueous waste, included 15 000 mg/L zinc, 5800 mg/L copper, and lesser concentrations of cadmium, chromium, lead, and nickel. Waste acid Sample 2 was identified as mixed sulfuric and chromic acid; chromium was present at 24 000 mg/L.

A dry clay-like filter press sludge was produced from the precipitated solids, and this was disposed in a landfill. High concentrations of metals were reported in each sample. In

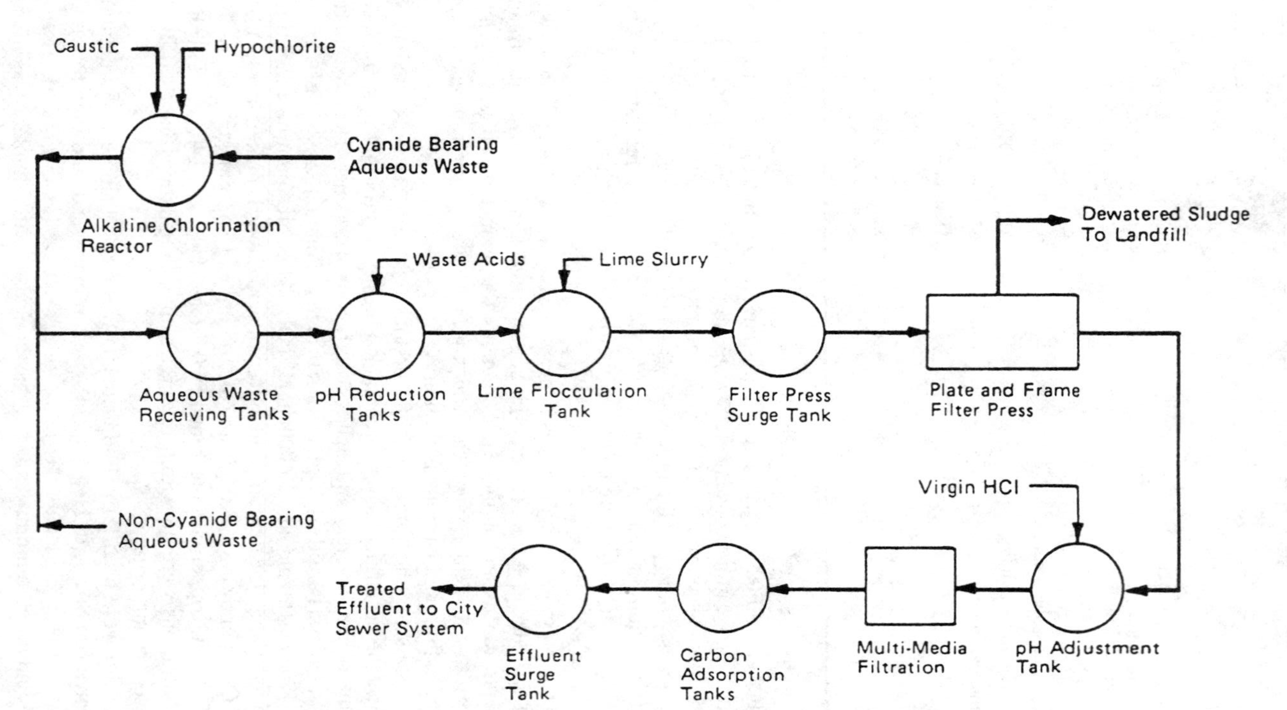

FIG. 1—*Waste treatment process, Facility A.*

TABLE 2—*Heavy metals treatment, Facility A.*

	Mixed Aqueous Influent, mg/L	Filter Press Sludge, mg/kg	Filtrate, mg/L	GAC Effluent, mg/L
		PERIOD 1		
cadmium	5.7	12 000	1.2	0.15
hexavalent chromium	<20	NS[a]	<10	<5
total chromium	12	10 000	2.1	0.09
copper	200	50 000	160	0.32
lead	1.5	1 300	1.4	0.56
nickel	7.3	9 300	2.2	2.2
zinc	400	35 000	24	1.4
total organic carbon	5 600	5 200	NS	1200
		PERIOD 2		
cadmium	3.9	6 700	7.9	0.51
hexavalent chromium	<5	NS	NS	<5
total chromium	12	23 000	1.9	0.12
copper	150	43 000	170	0.87
lead	1.1	790	.82	0.74
nickel	4.3	6 400	2.3	2.0
zinc	60	14 000	16	1.6
total organic carbon	1 200	NS	2400	1500
		PERIOD 3		
cadmium	180	2 900	50	1.4
hexavalent chromium	<5	NS	NS	<5
total chromium	18	2 200	0.07	0.09
copper	1 500	25 000	380	0.56
lead	3.8	830	.67	0.63
nickel	500	1 400	22	1.8
zinc	1 700	4 000	46	0.70
total organic carbon	19 000	NS	5000	1400

[a] NS = Not sampled.

addition, fairly high amounts of organics were concentrated in a hydroxide sludge. This was consistent for each of the samples and duplicate samples collected. Organic priority pollutants present in the sludge in the highest concentrations were 290 mg/kg toluene and 110 mg/kg 1,1,1-trichloroethane. The sludge samples were extracted using the EP toxicity and TCLP tests. Table 3 summarizes the analytical results and presents the EPA treatment standards for organics promulgated as part of the 7 Nov. 1986 land disposal restrictions for solvent wastes.

The treatment process successfully removed the metals and toxic organic compounds from the aqueous waste. The toxic organic compounds were not detected in the effluent from the treatment facility. Detection limits for the organic priority pollutants were generally 0.5 mg/L, and the metals were removed to below the pretreatment limits. The composite sludge samples met the maximum EP toxicity threshold criteria for the metals tested. More importantly, VOCs in the TCLP extractions of the composite sludge exceeded the 7 Nov. 1986 regulations for solvent wastes (RCRA waste codes F001, F002, F003, F004, and F005). If comparable best demonstrated available technology (BDAT) regulations were established for the electroplating wastewater sludges and spent plating baths treated during this sampling episode (F006, F008, and F009), the residues from the treatment process, as measured by the TCLP test, would be not acceptable for landfill disposal. Either Facility A would not be able to accept these wastes for treatment or additional treatment processes would be

TABLE 3—*TCLP and EP toxicity analytical results for the composite metal hydroxide sludge, Facility A.*

Pollutant	Influent Concentration Range,[a] mg/L	Sludge Compositional Analysis,[b] mg/kg	TCLP Results, mg/L	Land Disposal Regulations for TCLP, mg/L	EP Toxicity Results, mg/L	Maximum EP Toxicity Threshold, mg/L
1,1,1-trichloroethane	3.1	230	3.1	1.05	NA[c]	NA
methylene chloride	34	50	2.8	0.20	NA	NA
trichloroethylene	1.0	95	1.7	0.062	NA	NA
tetrachloroethylene	1.3	200	0.66	0.079	NA	NA
toluene	1.6	690	7.2	1.12	NA	NA
arsenic	NS[d]	NS	NS	NA	NS	5.0
barium	NS	NS	NS	NA	0.13	100.0
cadmium	63	7 200	NS	NA	0.26	1.0
chromium	14	11 200	NS	NA	0.38	5.0
lead	2.1	970	NS	NA	<0.10	5.0
mercury	NS	NS	NS	NA	NS	0.2
selenium	NS	NS	NS	NA	NS	1.0
silver	NS	NS	NS	NA	NS	5.0

[a] Average of three composite influent samples.
[b] Composite of three filter press sludges.
[c] NA = Not applicable.
[d] NS = Not sampled.

required. Organic compounds could be removed prior to precipitation, or the sludge could be stabilized to prevent solvents from leaching.

Facility B—Cyanide Oxidation and Metal Sulfide Precipitation

Facility B uses unit processes for the treatment of both liquid and solid hazardous wastes. Liquid wastes are primarily from surface-finishing operations such as the steel, jewelry, printed circuit board, automotive, and electroplating industries. These wastes are predominately acids, alkalines, acids and alkalines with heavy metals, chromic acids, and cyanide wastes (RCRA waste codes D002 and D003). This facility also accepts a variety of hazardous solid wastes that may contain cyanide, hexavalent chromium, or metals (RCRA waste codes F006, F007, F008, and F009). Results presented in this paper address only observed liquid-waste processing operations.

Liquid-waste treatment (Fig. 2) entails four basic processes: cyanide oxidation, chromium reduction, metals precipitation, and solids separation. Cyanide oxidation uses a waste sodium hypochlorite solution in a single-stage batch reactor. Chromium reduction incorporates waste acid containing ferrous iron to reduce hexavalent chromium in a single-stage batch reactor. Upon the successful completion of the cyanide oxidation or chromium reduction, the pH is neutralized with lime slurry or waste acid. Heavy metals precipitate as insoluble hydroxides, and a soluble sulfide waste is then added to precipitate any remaining dissolved metals as a polishing step. The contents of the reactors are their pumped to two precoat vacuum filters for solids separation. Diatomaceous earth or volcanic ash is purchased for the precoat material. The solids are transported to a landfill for disposal, and the filtrate from the precoat filter is discharged to the municipal sewerage system.

During the sampling program, a total of 12 separate batches were treated. Five of these batches were similar in nature; the primary waste was a sludge (total solids content of about 8 to 10%) from one generator. Similar proportions of lime slurry, waste chromic acid, ferrous-iron acid, waste nitric acid, sulfide waste, and rinsewaters were blended with the primary waste. The composite mixture was sampled from the reaction tank before it was

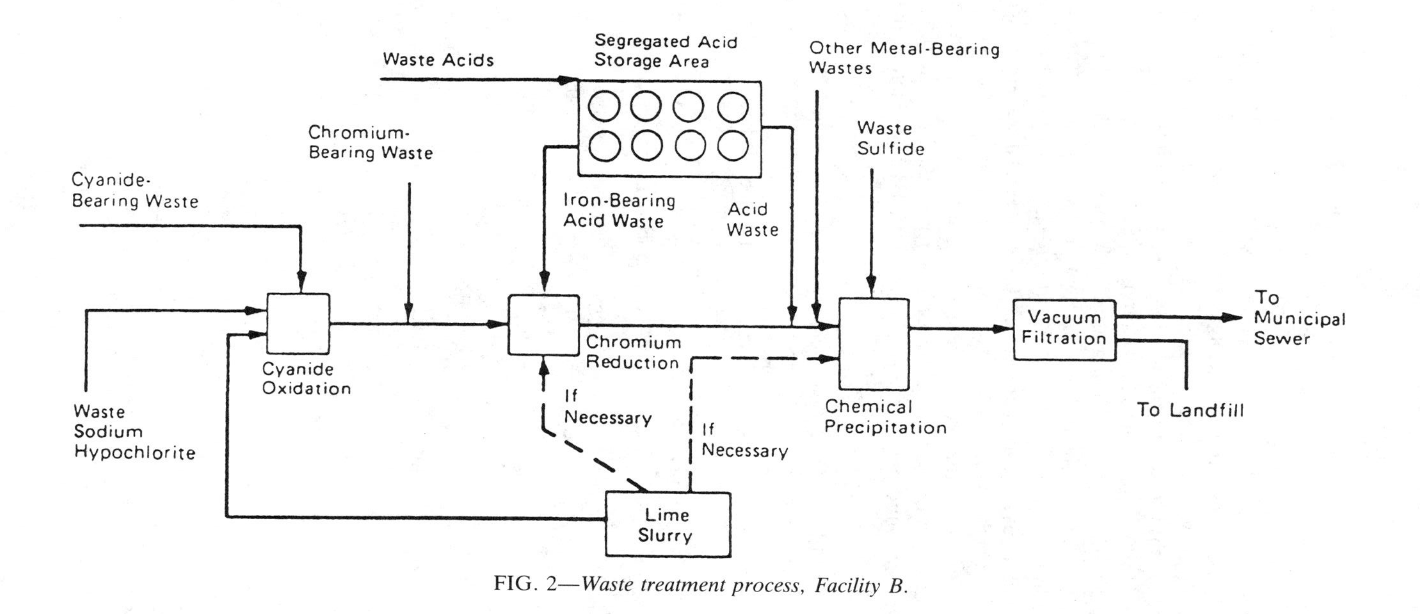

FIG. 2—*Waste treatment process, Facility B.*

filtered by two precoat vacuum filters. The results of all five batches were comparable, and the results of one batch are presented (Table 4) as Waste No. 1. Table 5 presents the results from the treatment of a different waste type, Waste No. 2. The results were similar for the treatment of the three batches of Waste No. 2. Waste No. 1 at Facility B was generated from operations associated with the electroplating and printed circuit board industries. The waste was pumped into a reactor vessel containing lime slurry. Due to the absence of cyanide, no initial oxidation was necessary. A waste chromic acid was added to the reactor, and this waste contained high concentrations of hexavalent chromium (78 400 mg/L). A waste acid containing ferrous-iron was added to reduce the chromium based on the stoichometric requirements. The mixture was agitated, and additional waste nitric acid was added to reduce the pH to about 9.0 for optimum sulfide precipitation. Sulfide-containing waste was blended to precipitate soluble metal ions remaining after hydroxide precipitation.

The precipitated waste was then filtered by two precoat vacuum filters. Both total and soluble metal fractions were analyzed from the filtrate stream to gage the efficiency of both the precipitation and solids separation processes. Only the total metal analyses for the filtered effluent are presented because there was no difference in the results. The excellent treatment efficiency indicates that both the precipitation and filtration processes were effective.

An interesting observation was the presence of several VOC's in the 0.5 mg/L range in the sludge. These compounds were apparently concentrated as a result of the precipitation

TABLE 4—*Analytical results for Waste No. 1, Facility B.*[a]

Pollutant	Primary Waste	Reaction Tank Composite	Filtered Effluent	Filter Cake	TCLP	EP Toxicity
Volume	18.0 m^3	36.8 m^3[b]	36.8 m^3	NA[c]	NA	NA
		VOLATILE ORGANIC COMPOUNDS				
1,1-dichloroethane	NS[d]	0.340[e]	<0.01	0.340[e]	0.022	NA
1,1,1-trichloroethane	NS	0.220[e]	<0.01	0.470[e]	0.017	NA
trichloroethylene	NS	0.100[e]	<0.01	0.130[e]	<0.01	NA
1,1,2,2-tetrachloroethylene	NS	0.074[e]	<0.01	0.150[e]	<0.01	NA
toluene	NS	0.140[e]	<0.01	0.230[e]	0.017	NA
ethylbenzene	NS	0.200[e]	<0.01	0.390[e]	0.017	NA
		METALS				
arsenic	<1[e]	<1[e]	<0.1	<1[e]	<0.01	<0.50
barium	<10[e]	<10[e]	<1	30[e]	0.18	<0.02
cadmium	19[e]	10[e]	<0.5	20[e]	<0.02	0.008
chromium	760[e]	2310[e]	0.12	16 300[e]	<0.05	<0.01
lead	125[e]	108[e]	<0.01	375	<0.10	0.08
mercury	<1[e]	<1[e]	<0.01	<1[e]	<0.002	<0.002
selenium	<10[e]	<10[e]	<1	<10	<0.01	<0.01
silver	<2[e]	<2[e]	<0.2[e]	<2[e]	<0.02	0.01
hexavalent chromium	0.016[e]	0.1[e]	0.121	I[e,f]	NS	<0.005
copper	36[e]	72[e]	0.16	330[e]	NS	0.05
nickel	14[e]	426[e]	0.40	1700[e]	NS	0.10
zinc	190[e]	171[e]	0.115	375[e]	NS	0.01

[a] All units in mg/L unless noted otherwise.
[b] Includes roughly 7.57 m^3 of dilute waste from floor drains, 0.57 m^3 of waste nitric acid, 18.0 m^3 of a sludge slurry from an electroplating operation, 0.38 m^3 of waste chromic acid, 4.54 m^3 of a ferrous-iron acid, 5.68 m^3 of lime slurry, and 0.04 m^3 of a sulfide-containing waste.
[c] NA = Not applicable.
[d] NS = Not sampled.
[e] Units are in mg/kg of solids on a wet weight basis.
[f] Analytical color interference.

TABLE 5—*Analytical Results for Waste No. 2, Facility B.*[a]

Pollutant	Primary Waste	Reaction Tank Composite	Filtered Effluent	Filter Cake	Filter Cake TCLP	Filter Cake EP Toxicity
Volume	19.7 m^3	39.4 m^3[b]	39.4 m^3	NA[c]	NA	NA
		VOLATILE ORGANIC COMPOUNDS				
ethylbenzene	NS[d]	<0.01	<0.01	0.011[e]	<0.002	NA
		METALS				
arsenic	<1[e]	<1[e]	<0.1	1[e]	<0.01	<0.05
barium	<10[e]	<10[e]	<2	<10[e]	<0.1	<0.02
cadmium	<5[e]	<5[e]	<0.5	<5[e]	<0.02	0.007
chromium	1200[e]	2550[e]	0.10	10 000[e]	<0.05	<0.01
lead	<10[e]	<10[e]	<0.01	42[e]	<0.10	0.08
mercury	<1[e]	<1[e]	<0.1	<1[e]	<0.002	<0.002
selenium	<10[e]	<10[e]	<1	<10[e]	<0.01	<0.01
silver	<2[e]	<2[e]	<0.2	<2[e]	<0.02	0.09
hexavalent chromium	0.012[e]	734[e]	I[f]	1.78	NS	<0.005
copper	10[e]	150[e]	<0.12	432[e]	NS	0.01
nickel	<3[e]	590[e]	<0.33	1600[e]	NS	0.06
zinc	<2[e]	4[e]	<0.10	68[e]	NS	0.01

[a] All units in mg/L unless noted otherwise.
[b] Includes roughly 19.7 m^3 of metal-bearing slurry waste, 4.31 m^3 of wastewater necessary to dilute the waste prior to pumping, 6.43 m^3 of lime, 0.38 m^3 of waste chromic acid, 4.88 m^3 of ferrous-iron acid, 3.48 m^3 of waste nitric acid, and 0.19 m^3 of a sulfide waste.
[c] NA = Not applicable.
[d] NS = Not sampled.
[e] Units are in mg/kg of solids on a wet weight basis.
[f] Analytical color interference.

processes. In contrast to Facility A, the VOCs in the waste influent at Facility B were present at much lower concentrations. The subsequent extracts from these residues likewise contained significantly lower organic concentrations, and the concentrations in the extract were within the limits recently promulgated for solvent land disposal restrictions. Waste No. 2 was treated in a similar manner. This waste was very viscous, and dilution with rinsewater was needed to pump the waste from the transport vehicle to the reaction tank. Lime slurry, waste chromatic acid, ferrous-iron acid, nitric acid, and sulfide-containing waste were added as reagents to this batch. The treatment results from this waste were similar to the results from Waste No. 1. High total chromium content waste (1200 mg/kg) was blended with a small volume of chromic acid requiring chemical reduction. A ferrous-iron acid was added as the reducing agent, and waste acids were used for gross pH adjustment. The composite waste was precipitated with lime and then sulfide. Treatment removals were excellent, and the concentration of metals in the EP toxicity extract from the precipitated solids were below the established EP toxicity regulations.

Facility C—Fly Ash Solidification

Facility C operated a secure landfill, and wastes were solidified and stabilized before landfilling. The fly ash was light tan in color, alkaline, and received in fine powder form. It was received from several power plants burning western coal and was stored in piles in a corner of the landfill cell. A front-end loader placed two or three buckets of fly ash on a concrete mixing pad (Fig. 3). Wastes were dumped from roll-off containers and tank trucks onto the fly ash on the pad. The front-end loader added more fly ash on top and then mixed the materials by lifting, turning, and spreading them over the pad. The solidified material

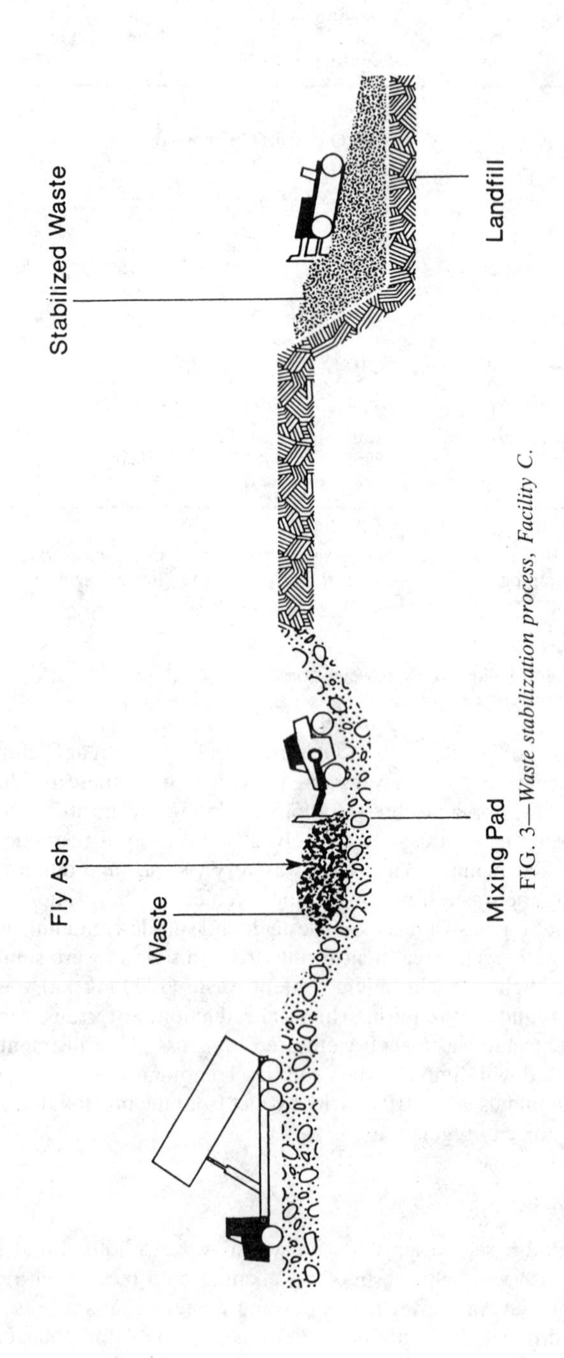

FIG. 3—*Waste stabilization process, Facility C.*

was piled in the landfill cell, and a bulldozer spread it evenly over the surface. Solidification of 16 m^3 (21 yd^3) of sludge and transport to the landfill required approximately 1 h. Fly-ash dosages were estimated by the volume of the loader's bucket, and in addition to our samples, the facility took process-control samples after mixing. All wastes exhibited no free liquids by the paint filter test after mixing with fly ash. Four different wastes were solidified and stabilized during the sampling program. Fly-ash samples were taken and characterized during the treatment of two wastes. These results are discussed with the results for Wastes Nos. 2 and 4.

The first waste was an opaque brown oil waste (oil-water emulsion) which contained 14% oil and grease, 2.2% TOC, and 72% volatile solids. Compositional organics included 180 mg/kg 1,1,1-trichloroethane and 71 mg/kg butylbenzylphthalate; the matrix elevated analytical detection limits to approximately 10 mg/kg. Compositional metal analyses measured 15 mg/kg cadmium, 150 mg/kg copper, 20 mg/kg lead, and 15 mg/kg zinc. The waste was liquid, failed the paint filter test, and required solidification before landfilling. Extraction tests were not conducted on this untreated waste.

The oil waste was difficult to solidify, requiring 1.2 parts fly ash per part of waste and 1.5 times the typical mixing time. The fly-ash dose was much higher than the typical waste solidifications observed at this facility. Bearing strength increased from 1758 kg/m^2 (0.18 tons/ft^2) when sampled to over 48 828 kg/m^2 (5.0 tons/ft^2) after six days. Other materials solidified at this facility did not exhibit increased strength with time. This phenomenon could be caused by the higher fly-ash does and pozzolan reactions of fly ash with water present in the emulsion. Extraction tests were not conducted on this fly ash.

The alkalinity of the solidified material was 6300 mg/kg as calcium carbonate. The material contained 89% total solids and 14% volatile solids. Compositional analysis showed 1.5 mg/kg 1,1,1-tricholoethane, 1.1 mg/kg 4-methyl-2-pentanone, 12 mg/kg toluene, 14 mg/kg ethyl benzene, and 25 mg/kg total xylenes. All metals in the extracts from EP toxicity and TCLP tests were below 0.1 mg/L, except 0.51 mg/L barium and 0.32 mg/L chromium in the TCLP extract. The only organics measured above detection limits of 0.1 mg/L in the TCLP extract were 0.43 mg/L toluene and 0.12 mg/L 1,1,1 trichloroethane. Though requiring a high fly-ash dose and additional mixing time, the process effectively solidified this waste and reduced metal and organic leaching to below regulatory limits for land disposal.

The second waste was a grey paint sludge mixed with trash including plastic buckets, plastic sheeting, and gloves. It failed the paint filter test for free liquids. The waste was nonhomogeneous containing unmixed oil and green and blue sludge pockets. Total solids were 41%; volatile solids were 33%. The EP toxicity test extracted 0.13 mg/L cadmium, and TCLP extracted 0.35 mg/L barium and 0.22 mg/L lead. Compositional organics included 0.67 mg/kg methylene chloride, 9.9 mg/kg acetone, 4.3 mg/kg 2-butanone (methylethylketone), 2.4 mg/kg, 4-methyl-2-pentanone, 3.8 mg/kg toluene, 3.7 mg/kg ethylbenzene, and 9.0 mg/kg total xylenes. TCLP zero head space extraction showed 0.28 mg/L methylene chloride and 1.3 mg/L toluene.

The process required 0.35 parts fly ash per part of waste, and bearing strength was 2148 kg/m^2 (0.22 tons/ft^2) after mixing. From the fly ash, EP toxicity extracted 1.9 mg/L barium, 0.14 mg/L chromium, and 10.1 mg/L lead; TCLP was not conducted. The fly ash would be classified as hazardous waste because of the high lead concentration in the EP toxicity extract. The solidified paint sludge was still nonhomogeneous, exhibiting pockets of green and blue sludge, but no free oil.

Alkalinity increased from 3200 to 43 000 mg/kg as calcium carbonate by addition of fly ash with 90 000 mg/kg alkalinity. Total solids were 75%, volatile solids were 21%, and TOC was 7.0 mg/L. EP toxicity extracted 1.3 mg/L barium, and TCLP extracted 0.54 mg/L barium. All other metals were below 0.1 mg/L in both extracts. The only organics present above 0.1 mg/L in the TCLP extract were 0.76 mg/L methylene chloride, 5.2 mg/L toluene, and 0.11 mg/L 1,1,1 trichloroethane. The main reasons for solidifying this waste were to increase bearing strength and eliminate the free oil present. TCLP extracted toluene at 5.2 mg/L after solidification, indicating organic stabilization was not completely effective. Metal leaching from the fly ash was reduced after solidification, and only barium (0.51 mg/L) was extracted at low concentrations.

The third waste was a thick, black, pourable biological treatment-pond sludge, and it failed the paint filter test for free liquids. Total solids were 15%, volatile solids were 11%, and TOC was 2.7 mg/kg. Compositional organics included 20 mg/kg vinyl chloride, 4.7 mg/kg 1,1-dichloroethene, 170 mg/kg trans-1,2 dichloroethene, 8.8 mg/kg 1,1,1-trichloroethane, and 23 mg/kg trichloroethene. The TCLP extract contained 6.1 mg/L carbon disulfide, 11 mg/L chloroform, 21 mg/L methylene chloride, 20 mg/L methyl ethyl ketone, 2.0 mg/L tetrachloroethene, 130 mg/L toluene, 0.91 mg/L barium, and 0.35 mg/L chromium. Metal concentrations in the EP toxicity extract were below 0.10 mg/L.

This waste was difficult to solidify, requiring 1.1 parts of fly ash per part of waste and over 2 h for mixing. The bearing strength was 2246 kg/m^2 (0.23 tons/ft^2) after mixing. Fly-ash extractions were not conducted. Alkalinity increased from 1500 mg/kg as calcium carbonate to 39 000 mg/kg with the addition of 204 000 mg/kg alkalinity in the fly ash.

The solidified biological treatment sludge was 70% solids, 5.5% volatile solids, and TOC was not measured. TCLP extract contained 0.69 mg/L barium; all other organics and metals were below 0.10 mg/L. Organic compositional analysis were not performed on the solidified waste within specified holding times, and the results are not reported here. The only metals present in the EP toxicity extract above 0.10 mg/L were 0.83 mg/L barium and 3.0 mg/L selenium. The only requirement for solidification was to eliminate free liquids before landfilling. Selenium leaching could be caused by the fly ash or the nonhomogeneous mixture. With the exception of the one high selenium concentration, stabilization was effective for both organic and inorganic constituents. However, this result also reflects are relatively low concentrations present in the sludge and the high fly-ash dose added.

The fourth waste was a blue-green metal hydroxide sludge. It contained pockets of rust-colored material and no free liquids as defined by the paint filter test. It had a stiff clay-like consistency with some hard lumps. Total solids were 16%, volatile solids were 2.6%, and TOC was 0.47 mg/L. Compositional organic analysis showed only 1.2 mg/kg methylene chloride. TCLP organic analysis was not conducted because the low compositional concentrations indicated that regulatory levels would not likely be exceeded. TCLP extraction for metals showed 0.26 mg/L barium, 400 mg/L cadmium, and 99 mg/L chromium. Metals were extracted at concentrations of 0.16 mg/L barium, 180 mg/L cadmium, 4.7 mg/L chromium, and 0.18 mg/L lead by the EP toxicity test.

The solidified material was light brown, moist, pliable, and cohesive, containing green and rust-colored lump. Bearing strength was only 586 kg/m^2 (0.06 tons/ft^2) after mixing 0.25 parts of fly ash per part of waste. The alkalinity was reduced from 540 000 mg/kg as calcium carbonate to 98 000 mg/kg by the addition of fly ash with only 12 000 mg/kg alkalinity. EP toxicity extraction from the fly ash contained 2.8 mg/L barium, 0.20 mg/L chromium, and 3.0 mg/L selenium.

The solidified metal hydroxide sludge contained 48% total solids, 40% volatile solids; TOC was not measured. TCLP organic analysis was not conducted due to the low waste compositional concentrations. Extractions contained 0.33 mg/L barium, 4.1 mg/L cadmium, 0.26 mg/L chromium, and 0.15 mg/L lead in the TCLP test and 0.37 mg/L barium, 34 mg/L cadmium and 0.43 mg/L chromium in the EP toxicity test. The unsolidified sludge extractions exceeded regulatory levels for cadmium and chromium in spite of the waste's high alkalinity. Leaching was reduced for all metals after solidification; however, cadmium concentrations still exceeded EP toxicity limits for hazardous waste.

Facility D—Dual-Chamber Incineration

A dual-chamber (pyrolysis/combustion) incinerator is used to burn organic liquids and solids at Facility D (Fig. 4). All wastes are introduced into the lower pyrolysis chamber, and fuel oil is burned in the after burner (upper chamber) to maintain required temperatures. Most wastes are received in drums which are opened, and the liquids are pumped to blend tanks. The solids are loaded into fiber packs, and the drums are cleaned for landfill disposal.

Blended liquid wastes are injected with steam for atomization. Solid wastes in fiber packs are loaded into a ram feeder. Ash residue is continuously discharged from the lower chamber. A caustic scrubber is employed to control air emissions, and it permits the burning of wastes with up to 40% total chlorine. Scrubber wastewater treatment includes pH adjustment, precipitation, clarification, filtration, and sludge dewatering.

Wastes accepted for incineration at Facility D include chlorinated and non-chlorinated solvents (RCRA Waste Codes F001, F002, F003, F004, and F005), paint sludges, printing inks, and dyes, polymer wastes, agricultural products, pharmaceutical intermediaries, lab packs, contaminated soils, and other hazardous wastes (RCRA Waste Code D001). Wastes burned during the sampling program represented waste solvent mixtures regulated under RCRA. These wastes included liquid and solid epoxy resins, non-halogenated solvents, waste tetrachloroethene and wax, polymer wastes with methylene chloride, photopolymer washing solutions, solvent filter bags, and solid pigments. The composition of the average liquid feed, a typical solid waste, ash residue, and TCLP extracts of the ash are summarized in Table 6.

The water content of the liquid feed was 15.6%, and total solids were 16.3%. The non-pumpable liquids and solids associated with the epoxy resins and polymer wastes were viscous and very sticky. Separate samples of the individual solid wastes, pumpable liquids, and mixed-liquid feed tanks were taken. The concentrations of individual compounds in the liquid and solid samples of each waste were very similar. Many of the solid wastes were non-pumpable materials settled in the bottoms of liquid waste drums. Total solids ranged between 25.0 and 75.3%, and water content ranged between 0.50 and 32.9% in all solid wastes. These solids contained free liquids according to the paint filter test and could not be landfilled without treatment. The solid wastes all failed recently promulgated TCLP regulations for solvent wastes because the extracts contained VOCs at concentrations above the limits for solvent wastes. TCLP extracts contained methylene chloride, acetone, and tetrachlorethene at concentrations between 27.0 and 9800 mg/L. Metals in TCLP extracts of the solid wastes did not exceed current EP toxicity regulatory limits with the exception of (16.1 mg/L) zinc in one sample.

The ash residue from the lower chamber of the incineration was black, with granular particles up to a few inches in diameter. The material appeared damp, and was 60% total solids. It passed the paint filter test and contained no free liquids.

High metals concentrations are not surprising since these are nonvolatile and are expected to accumulate in the combustion ash. Many of the volatile compounds found in the ash showed very high removal efficiencies compared to the liquid and solid feed. The removal of semivolatile compounds was much lower, but the individual concentrations were much lower in the feed and ash. While VOC removal was higher, the total amount present in the ash was still greater than semivolatile compounds. The sum of all contaminants in ash still only represented a few percent of the total weight of the original waste materials.

The organic and metal compounds measured in the ash were also found in the solid wastes. Note that acetone, methylene chloride, tetrachoroethene, toluene, and total xylenes were in high concentrations in the solid wastes associated with epoxy resins, polymer wastes, and wax. These solids may have been difficult to completely burn in the lower pyrolizing chamber.

TCLP extracts showed high concentrations of methylene chloride, tetrachloroethene, and

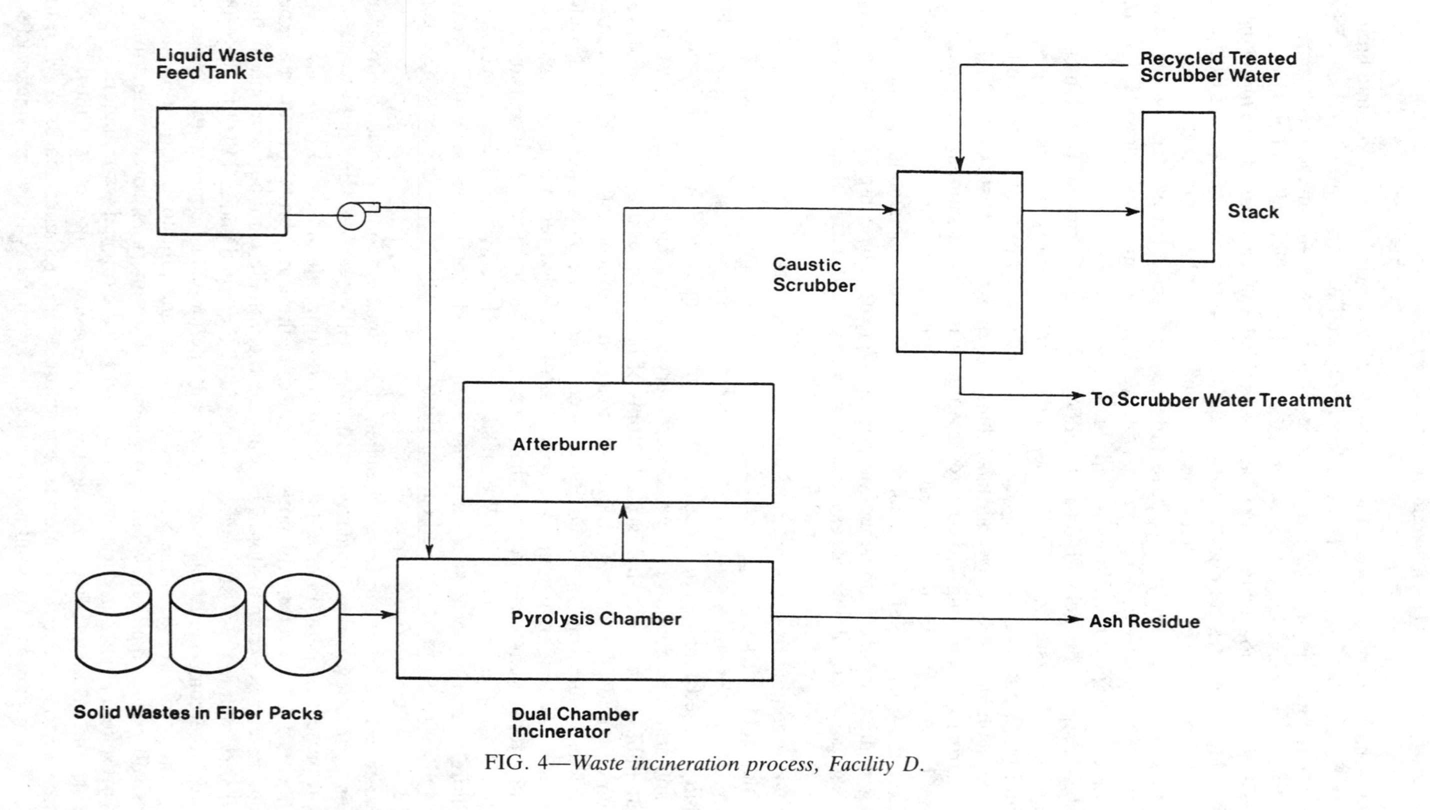

FIG. 4—*Waste incineration process, Facility D.*

TABLE 6—*Facility D analyses.*

VOC's	Average Liquid Waste, mg/L	Typical Solid Waste, mg/kg	Ash, mg/L	TCLP, mg/L
methylene chloride	47 200	34 000	38.0	8.80
acetone	ND[a]	210	20.0	ND
2-butanone	3 620	ND	2.00	ND
trichloroethene	1 700	8 600	ND	ND
tetrachloroethene	477 000	220 000	1 200	48.0
toluene	174 000	32 350	2.50	11.0
total xylenes	1 020	9 240	1.90	ND
4-methyl-2-pentanone	350	ND	2.30	ND
		SEMI-VOLATILES		
di-*n*-butylphthalate	1 375	170	160	0.04
diethyl phthalate	98.8	75.0	120	0.41
2,4-dimethylphenol	328	ND	23.0	0.87
2-4-dichlorophenol	715	ND	15.0	0.25
4-methylphenol	55.8	ND	2.50	0.16
phenol	121	ND	40.0	ND
napthalene	40.6	121	24.0	0.31
nitroaniline	ND	140	18.0	1.30
		METALS		
barium	120	41.0	150	0.02
chromium	19.0	12.0	71	0.01
copper	60.0	13.0	13 800	0.79
lead	40.0	20.0	30	ND
nickel	10.0	12.0	190	1.14
zinc	46.0	8.0	280	1.15

[a] ND = Not detected.

toluene. Other VOCs were not detected, and 2-nitroaniline (1.3 mg/L) was the only semi-volatile compound measured at concentrations above 1.0 mg/L. The three VOCs would restrict land disposal of the ash under the new regulations. All metal concentrations in both TCLP and EP toxicity extracts were below 2.0 mg/L, with the exception of copper (4.0 mg/L) in the EP toxicity extract.

Facility E—Rotary Kiln Incineration

Facility E operates a rotary kiln with after burner and a separate liquid injection incinerator (Fig. 5). Both incinerators use natural gas as auxiliary fuel and burn the same liquid wastes with air atomization. Liquid wastes are stored in mixed tanks which simultaneously feed both incinerators. Solid wastes are fed into the kiln by a hydraulic elevator which dumps drum contents into an air lock ram feeder. Empty drums are cleaned and reused. Average solid residence time in the kiln is between 40 and 60 min.

Each incinerator is followed by a water quench and a cyclone. Exhaust gases from the two systems enter a common three-stage packed tower scrubber system. Scrubber water is recycled, and blow down is neutralized before treatment. The neutralized blow down is treated at the facility's central wastewater treatment facility, and treated effluent is deep-well injected. Ash residue accumulates in the quench system, so the incinerator is typically shut down once a month for scheduled ash removal.

Wastes burned at Facility E typically are classified as spent solvents (RCRA Waste Codes F001 to F005) and miscellaneous hazardous wastes (RCRA Waste Codes D001 and D002). Chlorinated and non-chlorinated solvents, paint wastes, chemical intermediaries, and chem-

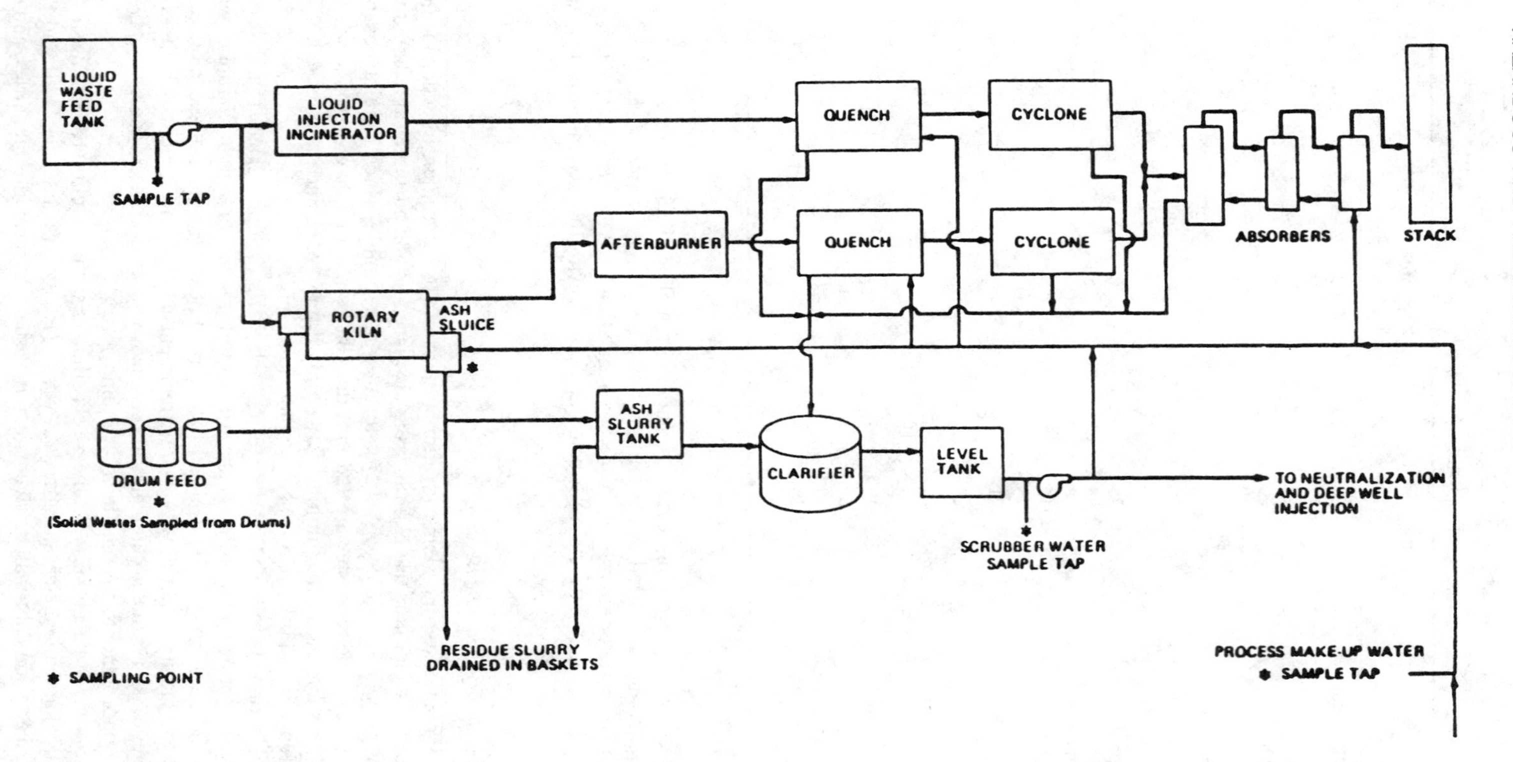

FIG. 5—*Waste incineration process, Facility E.*

ical production wastes are incinerated. Liquid wastes incinerated during the sampling program included waste solvents, nitrites, and chloroprene catalyst sludge. Solid wastes incinerated during sampling included paint sludges, chlorinated solids, coke solids, sludges from wastewater treatment, waste filter elements, and polymer wastes.

Separate samples of the mixed liquid wastes and individual solid wastes were collected. Results of the analyses of the average liquid waste, a typical solid waste, the ash residue, and the TCLP extraction of the ash are summarized in Table 7. The liquid waste contained similar compounds to the solid wastes, but they were generated by different processes. The liquid waste contained 7.3% total solids and 17.2% water. One typical solid waste burned during the sampling program is presented here, and it contained 62.5% total solids and 27.6% water. Most of the solid wastes contained free liquids, and TCLP extractions exceeded regulatory limits for VOCs. The solid wastes would require treatment before land disposal.

The ash residue was a brown granular material with the consistency of damp sand. It contained no free liquids, and the total solids content was 81.0%. The nonsolid material was largely water from the quenching process. Solids removed directly from the kiln were dry and had the consistency of pee gravel. Metals were concentrated in this ash as expected. Phenol (0.35 mg/L) was the only organic compound measured at concentrations above 0.10 mg/kg. In contrast to Facility D, the ash from Facility E contained only trace levels of a few organic compounds.

It is important to note that methylene chloride and acetone, which appear at trace levels in both the ash and TCLP extract, were measured at similar concentrations in the laboratory blanks. The concentrations of methylene chloride and acetone in these samples were likely

TABLE 7—*Facility E analyses.*

VOC's	Average Liquid Waste, mg/L	Typical Solid Waste, mg/kg	Ash, mg/L	TCLP, mg/L
methlene chloride	30.0	ND[a]	0.02	0.05
acetone	28.0	ND	0.02	0.04
carbon disulfide	55.0	ND	ND	ND
toluene	230 000	4900	ND	ND
benzene	47.3	90.7	ND	ND
ethylbenzene	80.3	75.0	ND	ND
styrene	18.2	340	ND	ND
total xylenes	51.2		ND	ND
	SEMI-VOLATILES			
benzyl alcohol	1300	54.0	ND	ND
1,2 dichlorobenzene	1880	357	0.10	ND
n-nitrosodiphenylamine	9700	497	ND	ND
napthalene	ND	490	0.10	ND
2-methylnaphthalene	ND	2900	ND	ND
acenaphthene	ND	56.0	ND	ND
phenol	ND	ND	0.36	ND
	METALS			
antimony	ND	116	14.5	0.03
barium	ND	21.3	75.2	0.39
chromium	2.17	66.7	361	0.10
copper	68.3	73.7	4580	115
lead	2.00	30.0	343	0.48
nickel	321	500	4180	34.1
zinc	12.8	195	1160	25.1

[a] ND = Not detected.

related to low level contamination. With the exception of these two compounds, no organic compounds were detectable in the TCLP extracts. These results are typical for other rotary kiln incinerator ashes and are a dramatic contrast to the results of TCLP extracts of the ash from the dual chamber incinerator at Facility D.

Conclusions

Treatment performance as measured by TCLP and EP toxicity extractions of the residues varied. In most cases, the two tests gave similar results, and leachate concentrations related to waste composition. The exceptions appear to be related to waste variability where field sampling and laboratory quality assurance were difficult to ensure. In the few cases where the two tests gave noncomparable results, waste composition was observed to be nonhomogeneous. Extraction results were related to waste composition, and nonhomogeneous wastes would be expected to generate leachate samples varying in composition.

The influent concentration of individual metal and organic compounds effected treatment performance. Unacceptably high VOC concentrations in the TCLP extract of the metal hydroxide sludge at Facility A (cyanide oxidation and metal hydroxide precipitation) were related to high influent concentrations. The precipitated sludges at Facility B (cyanide oxidation and metal hydroxide and metal sulfide precipitation) concentrated organics at much lower levels, and influent concentrations were much lower. The solidified paint sludge from Facility C (fly-ash solidification) leached toluene, which was present at high concentrations in the waste. Metals leaching from hydroxide sludges and solidified wastes were also present at high concentrations in the untreated wastes.

The incineration processes were more effected by operating parameters because feed concentrations were so high. Ash from Facility D (Dual-chamber incineration) leached VOCs, which were present in both liquid and solid feeds. Organics in the feed to Facility E (rotary kiln incinerator) were at just as high concentrations, but solid feed concentrations were lower. The VOCs leaching from Facility D ash are believed to be related to incinerator operations. Metals were extracted in high concentrations from the ash residue at both facilities.

The need to carefully evaluate each individual waste for treatment is indicated. The concentration of VOCs during hydroxide precipitation was not expected, and treatment was not completely effective. Oil wastes and biological treatment-pond sludge were successfully stabilized after the addition of alkaline fly ash (from power plants burning western coal) dosages. Waste acids were used in the treatment of other wastes at Facility B. Metals were extracted from some hydroxide sludges and incinerator ashes at concentrations characteristic of hazardous wastes. Metals were extracted from the sulfide sludge at much lower levels.

Facilities need to screen wastes to identify physical and chemical parameters that are not compatible with their treatment processes. The use of a generically named treatment process does not guarantee acceptable treatment. Treatment residues need to be tested to evaluate process operating parameters and ensure regulatory compliance. Hazardous waste treatment processes are difficult to operate, and problems can be encountered.

Acknowledgments

The evaluation of full-scale treatment processes was conducted for the EPA's Office of Research and Development, which the authors wish to thank for their cooperation with this project. The efforts of all EPA staff and co-workers are appreciated. The authors were employed by Metcalf and Eddy, Inc., at the time this project was conducted. Correspondence should be directed to William Shively.

References

[1] "Hazardous Waste Management System; Land Disposal Restrictions; Final Rule," *Federal Register,* Vol. 51, No. 216, 7 Nov. 1986, p. 40572.
[2] *Test Methods for Evaluating Solid Wastes, SW-846,* 2nd ed., U.S. Environmental Protection Agency, July 1982.
[3] *Methods for Chemical Analysis of Water and Waste,* EPA 600-4-79-020, U.S. Environmental Protection Agency, March 1983.
[4] "Toxicity Characteristic Leaching Procedure," 40 CFR, Appendix I to Part 268, *Federal Register,* Vol. 51, No. 216, p. 40643.

Gerard J. de Groot,[1] Jan Wijkstra,[1] Dirk Hoede,[1] and Hans A. van der Sloot[1]

Leaching Characteristics of Selected Elements from Coal Fly Ash as a Function of the Acidity of the Contact Solution and the Liquid/Solid Ratio

REFERENCE: de Groot, G. J., Wijkstra, J., Hoede, D., and van der Sloot, H. A., **"Leaching Characteristics of Selected Elements from Coal Fly Ash as a Function of the Acidity of the Contact Solution and the Liquid/Solid Ratio,"** *Environmental Aspects of Stabilization and Solidification of Hazardous and Radioactive Wastes, ASTM STP 1033,* P. L. Côté and T. M. Gilliam, Eds., American Society for Testing and Materials, Philadelphia, 1989, pp. 170–183.

ABSTRACT: To estimate the environmental consequences of the storage and utilization of coal fly ash, in the Netherlands a standard leaching test has been developed. In this test, leach conditions are used which aim at close approximation with field conditions. The standard test consists of three main parts: a column leaching test, a serial batch leaching test, and a test for the determination of maximum leachability. As a time scale the liquid/solid ratio (LS) is used, where LS is the ratio between the total quantity (L) of liquid used for leaching and the weight (kg) of the leached material. The relation between this relative time scale and the actual time scale depends on the time required for a given LS ratio to be reached in the actual situation.

In recent years, 50 ashes from a wide variety of sources were analyzed. Five different ashes were submitted to the complete standard leaching test described. All ashes show common leaching characteristics for several groups of elements. The elements present in the form of anionic species (for example, arsenic, antimony, selenium, molybdenum, tungsten, and vanadium) behave similarly. In contrast with literature information, limited solubility of anions at high pH (>11) has been observed. The metals lead, copper, cadmium, and zinc show minimum solubility at high pH, whereas major elements like aluminum and silicon show two minima in the pH range 7 to 9 and at pH values higher than 11. The latter was related to the formation of new mineral phases (for example, ettringite). To verify the pH dependence of the element concentrations leached, experiments with an artificial fly ash solution, containing all major elements normally found in fly ash extracts at pH 4 and LS ratio 5, have been performed. By stepwise increase of the pH, by adding calcium oxide, the relation between pH and element concentrations in the solution has been established. The results are in good agreement with the concentrations found in normal fly ash extracts. The experimental data of all batch experiments are summarized in three-dimensional graphs relating element concentration in solution, pH, and LS ratio. It shows that pH is the main factor in controlling the leachability.

The leaching behavior of trace elements in products containing fly ash has also been determined. After grinding, the materials were subjected to the standard test described. The leaching data retrieved from the products agreed well with the results from the pure ash/water system, indicating a consistent solubility control in solution. To what extent this information on ash/water mixtures can be used for other waste materials remains to be answered.

KEY WORDS: classification, elements, fly ash, leachates, leaching, quantitative chemical analysis, solid waste, systematics, trace amounts, waste products

[1] Scientific research worker, senior technician, senior technician, and section head of environmental research, respectively, Netherlands Energy Research Foundation (ECN), P.O. Box 1, 1755 ZG Petten, the Netherlands.

It becomes increasingly important to be able to assess the environmental risk involved in the storage and utilization of waste materials. A large variety of tests has been developed for this purpose. Due to the lack of agreement on test results, in the Netherlands in 1981 a working group was formed for the development of standard leaching procedures for combustion residues. The aim was to develop a test or series of tests in which the leaching conditions are in close relation with field conditions; that is, no complexing agents and strong acids should be used. The chemical composition of the leachate should be controlled solely by the material itself.

The standard leaching test developed consists of three main parts: a column test, a serial batch test, and a test to determine the maximum leachability. The liquid-to-solid (LS) ratio is used as a relative time scale, which can be transformed into the actual time scale if the hydrogeological conditions at the actual site are taken into account. Also in the standard leaching test, a partial characterization of residues is accounted for to shorten the test procedure when other members of the same group of residues already have been characterized in sufficient detail.

In recent years, at our institute (Netherlands Energy Research Foundation, ECN) we investigated, in cooperation with the research department of the Research Development Waste Materials (VCN), the leachability of 50 different pulverized coal ashes from utilities in the Netherlands, the Federal Republic of Germany, and Belgium. Five different ashes were analyzed according to the complete standard leaching test and the results were published [*1*]. The examination of a wide variety of ashes under a wide range of pH and LS conditions creates the possibility of identifying systematics in fly ash leaching behavior.

Experimental Procedures

The elemental analysis of ash and liquid samples resulting from leaching experiments has been performed by atomic absorption spectrometry, neutron activation analysis, and ion-chromatography.

The ashes are mainly analyzed by instrumental neutron activation analysis. For some elements, such as magnesium and copper, atomic absorption is preferred. In these cases, dissolution has been carried out according to Silberman and Fisher [2]. A more detailed description of the analytical methods used has been published elsewhere [*1*].

The standard leaching test consists of three main parts: a column test, a serial batch shake test, and a test for the determination of the availability of elements for leaching under natural conditions. No extractions with strong acids or complexing agents are applied to maintain compatibility with field conditions. The LS ratio (L/kg) is used as a time scale.

The column test is used to assess the LS range from 0.1 to 10. A column is filled with the material under investigation and percolated by acidulated demineralized water (pH = 4) to assess short- and medium-term leaching (<50 years). In the serial batch test, the same quantity of material is extracted several times with fresh demineralized water to get an impression of long-term leaching behavioir (50 to 500 years). The maximum leachability is assessed by a shake experiment at a LS ratio of 100 for 5 h, maintaining the pH of the leachate at pH 4 during the experiment by adding nitric acid. In this test, the goal is the total element concentration that can be leached under natural conditions. It is assumed that elements tied up in silicate phases or minerals are not available for leaching at a time scale less than about 100 years. A detailed description of the standard leaching test has been described in Ref *3*.

To study the pH influence on leaching, we subjected all 50 fly ashes to batch shake experiments at LS ratios of 1, 5, 50, and 500 with demineralized water with an initial pH of 4. After 24 h extraction time, the pH, imposed on the contact solution by components dissolving from the ash, and conductivity of the leachate are measured and the solutions acidified with nitric acid to pH 2 before elemental analysis. Five ashes were analyzed according to the standard leaching test. The results have been reported in Ref *1*. The data obtained by the column test were in good agreement with the results from batch shake experiments, which was also reported by Jackson [*4*]. We used batch shake experiments for this extensive study of 50 ashes.

The two-dimensional plots were prepared by the graphic software package DISSPLA running on a Cyber main frame. To visualize the relation between elemental concentration, pH of the leachate, and the LS ratio used, three-dimensional plots were prepared by a UNIMAP program, using a VAX minicomputer.

Results and Discussion

Older studies on a limited number of coal ashes already indicated that a relation exists between leaching behavior and alkalinity of ash expressed as the ratio $(CaO+MgO)/(SO_3+0.04Al_2O_3)$ [*1,3*]. Without pH control, the 50 ashes exhibit a pH range at low LS ranging from 4 to 12.5. At higher LS, this range is reduced to 4.5 to 11.5. This large variety of ashes allows investigation of the effect of pH on leaching without artificial adjustment of the pH condition.

The results from the leaching experiments show common characteristics for different groups of elements. Figures 1 to 2 illustrate the elemental concentrations leached as a function of pH, determined by the ash itself, at an LS ratio of 5 for some anionic and metallic species, respectively. Other, mostly major elements, show different leaching patterns, as presented in Fig. 3.

Trace elements such as arsenic, antimony, selenium, molybdenum, tungsten, and vanadium show a characteristic maximum at neutral pH and a decrease in concentration towards lower and higher pH. The decrease at high pH is in contrast with leaching results reported by Theis [*5*] and Liem [*6*]. They concluded from a very small amount of data, that concentrations of arsenic [*5,6*] and molybdenum and selenium [*6*] increase under alkaline conditions. These incorrect conclusions illustrate the risk of interpreting a limited amount of data. Possible explanations for the decrease at high pH are the formation of insoluble compounds with calcium [*7*] or precipitation/sorption as barium arsenate [*8*].

According to the potential-pH equilibrium diagrams published by Pourbaix [*9*], the decrease in concentration at low pH for elements such as arsenic, antimony, molybdenum, vanadium, and tungsten could be explained by the formation of solid phases, arsenic oxide, antimony oxide, molybdenum oxide, vanadium oxide, and tungsten oxide, respectively. The metals copper, lead, cadmium, and zinc show a minimum in solubility at high pH values, likely caused by the formation of low-solubility hydroxide compounds. This is in agreement with Theis [*5*] and Liem [*6*] and also supported by the data published by Pourbaix [*9*].

The results for magnesium show a decrease of the concentration above pH 8. This can be explained by the limited solubility of magnesium hydroxide. Aluminum and silicon maxima are at pH 10 and 12, respectively, and minima at pH 6.5 and pH 10, respectively. The minimum at pH 6 to 7 for aluminum is probably caused by gibbsite formation. The sharp decrease of the concentration above pH 11.5 is not to be expected, because of the good solubility of aluminates and silicates. For aluminum the minimum at pH 6 to 7 and the decrease at pH values above 11 have been related to pozzolanic activity, for example, ettringite formation (see below). The increase of calcium concentration and the decrease of sulfate concentration at high pH are interrelated by the limited solubility of calcium sulfate.

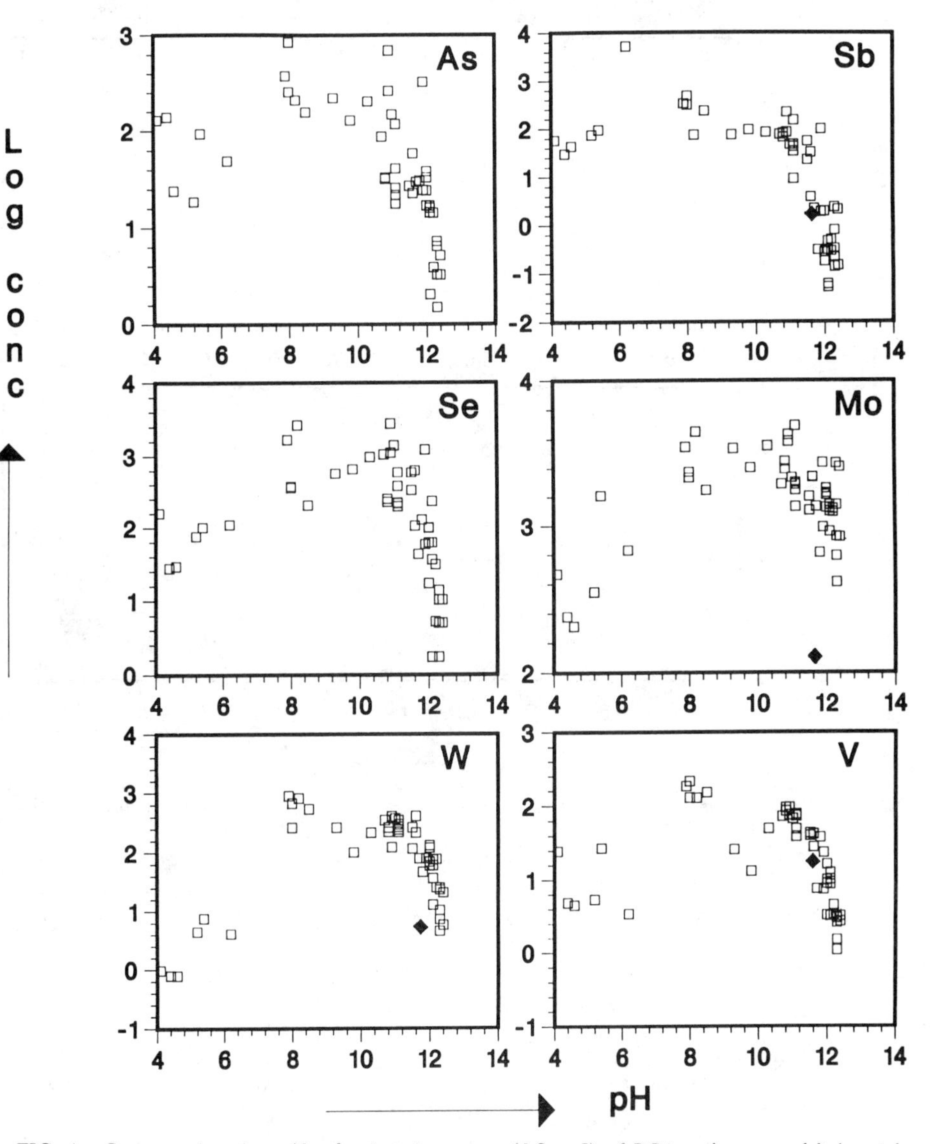

FIG. 1—*Concentrations in μg/L of anions in extracts (LS = 5) of PCA and structural lightweight concrete as function of pH:* □ *= fly ash;* ◆ *= structural lightweight concrete.*

These results show that despite differences in compositions in fly ash-water mixtures, pH is the main controlling factor, resulting in more or less fixed leaching patterns for different groups of elements caused by solubility criteria.

To verify the relation found between pH and the concentration in the leachate, experiments have been carried out in which an artificial fly ash extract was prepared containing all major elements with concentrations as found in normal fly ash extracts at pH 4 and LS 5 in demineralized water. By adding small amounts of calcium oxide, the pH has been increased stepwise from 4 to 12.5. At ten different pH values, the extract was analyzed to determine

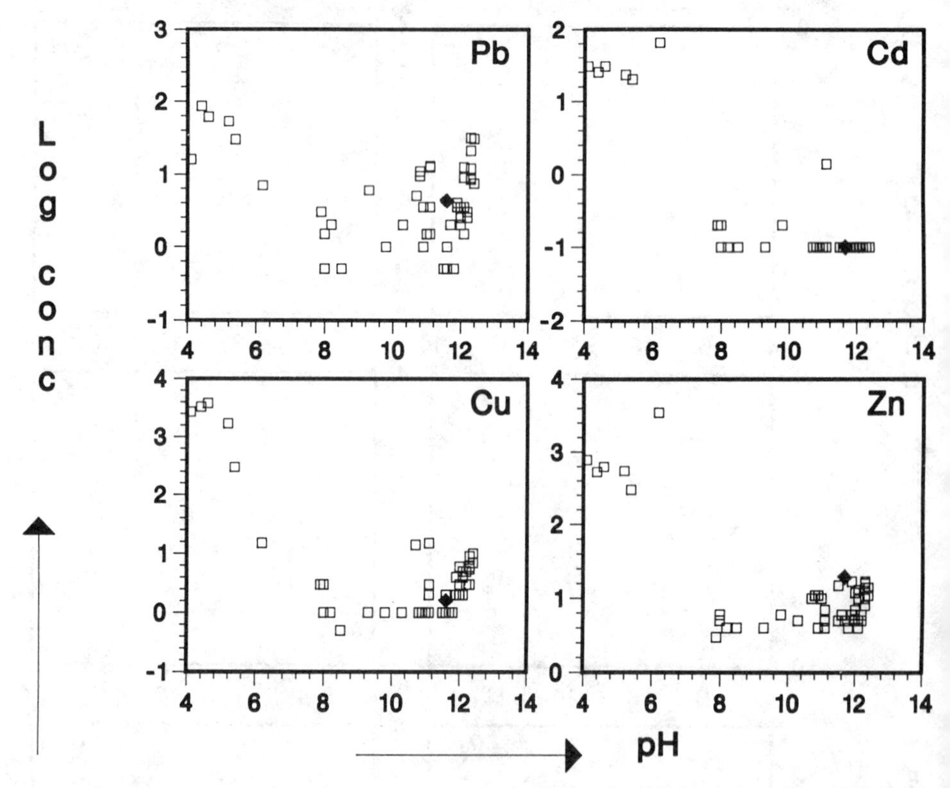

FIG. 2—*Concentrations in µg/L of metals in extracts (LS = 5) of PCA and structural lightweight concrete as function of pH:* □ = *fly ash;* ♦ = *structural lightweight concrete.*

whether the leaching results are mainly dependent on the composition of the extract or whether the presence of fly ash is essential. The starting composition of the artificial fly ash mixture for 2 L is

12 mg CaF_2	62 mg K_2CO_3
34 mg H_3BO_3	0.8 mg Sb
200 mg MgO	2.0 mg Cu
3600 mg $CaSO_4 \cdot 2H_2O$	1.0 mg W
560 mg $Al(NO_3)_3 \cdot 9H_2O$	
404 mg $Na_2SiO_3 \cdot 9H_2O$	pH = 4

In all fractions, the solid phase formed was also analyzed by X-ray diffraction. Fractions with pH greater than 8.5 showed increasing amounts of calcite ($CaCO_3$). Above pH 11, ettringite ($3CaO \cdot Al_2O_3 \cdot 3CaSO_4 \cdot 31H_2O$) was also detected. In Fig. 3, the concentrations found in the extracts for aluminum, silicon, calcium, sulfate, magnesium, and boron are shown as solid dots. They show a close resemblance with the results from normal fly ash leaching concentrations, indicating that pH dependence of the leaching characteristics is mainly determined by the composition of the extract. These findings are especially notable because concentrations for the anions and metals obtained from batch experiments performed on ground products containing fly ash, such as structural lightweight concrete (50% fly ash) and stabilized pulverized coal fly ash, are in good agreement with the results

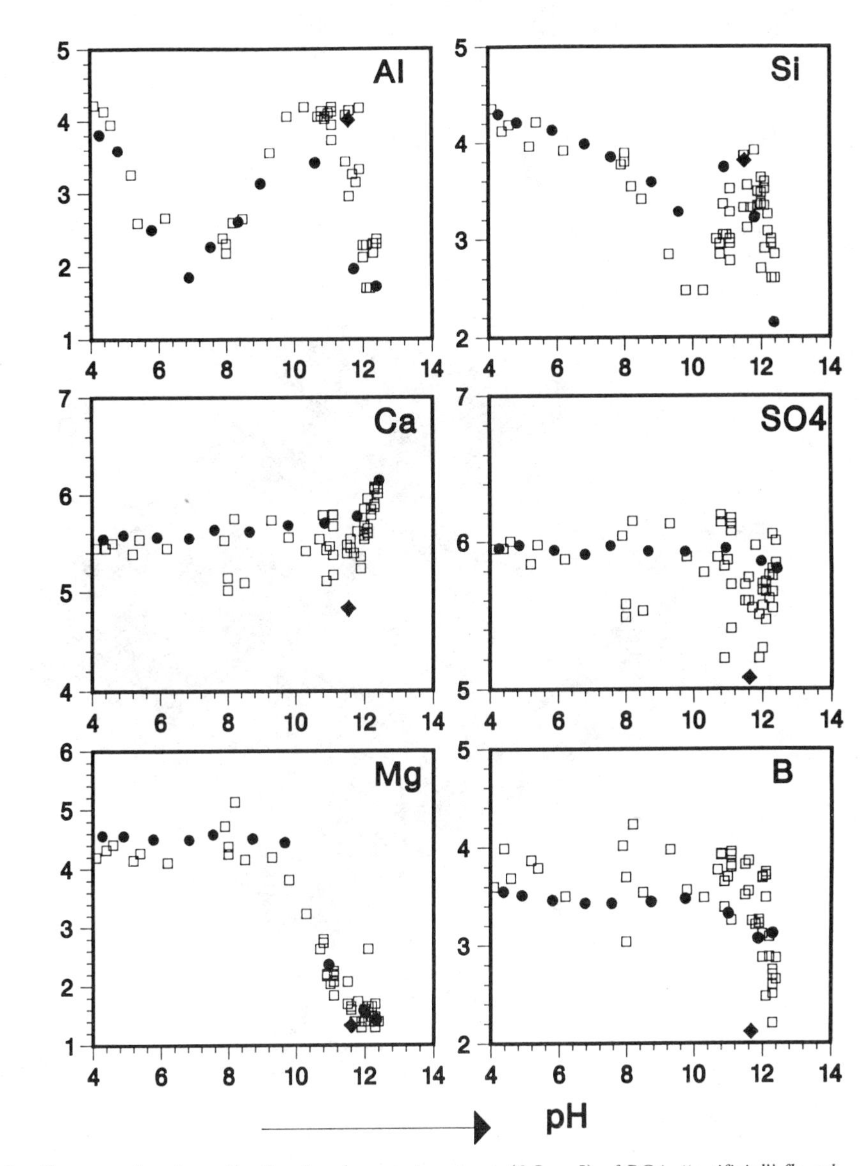

FIG. 3—*Concentrations in μg/L of major elements in extracts (LS = 5) of PCA, "artificial" fly ash, and structural lightweight concrete as function of pH: □ = fly ash; ● = artificial fly ash extract; and ♦ = structural lightweight concrete.*

described. In Figs. 1, 2, and 3, the concrete leaching data for LS 5 are indicated by diamonds.

Three-dimensional representation of some of the data is provided in Figs. 4 to 9. The relation with pH as well as the dependence of the LS ratio in the range of 1 to 1000 is shown. For the elements in regions where no solubility limitations occur, a typical dilution pattern is obtained, such as in sodium, potassium, bromine, chlorine, iodine, and sulfate. Also, at acid and neutral pH for magnesium, bromine, and calcium, at neutral pH for antimony and tungsten, and at low pH for copper, arsenic, and silicon, the concentration decreases roughly proportionally with the increase of LS ratio. From our earlier work [*1*]

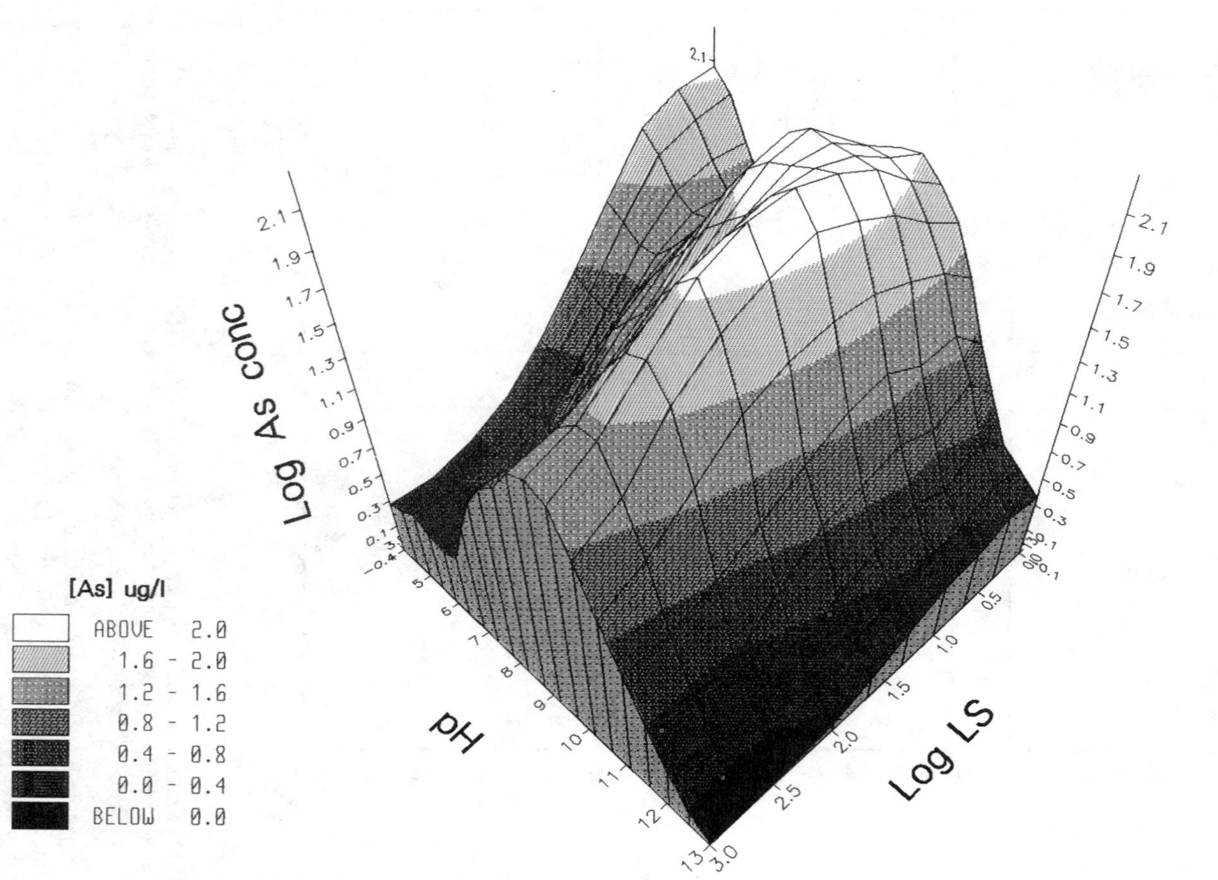

FIG. 4—*Concentrations in μg/L of arsenic in PCA extracts as function of pH and LS ratio.*

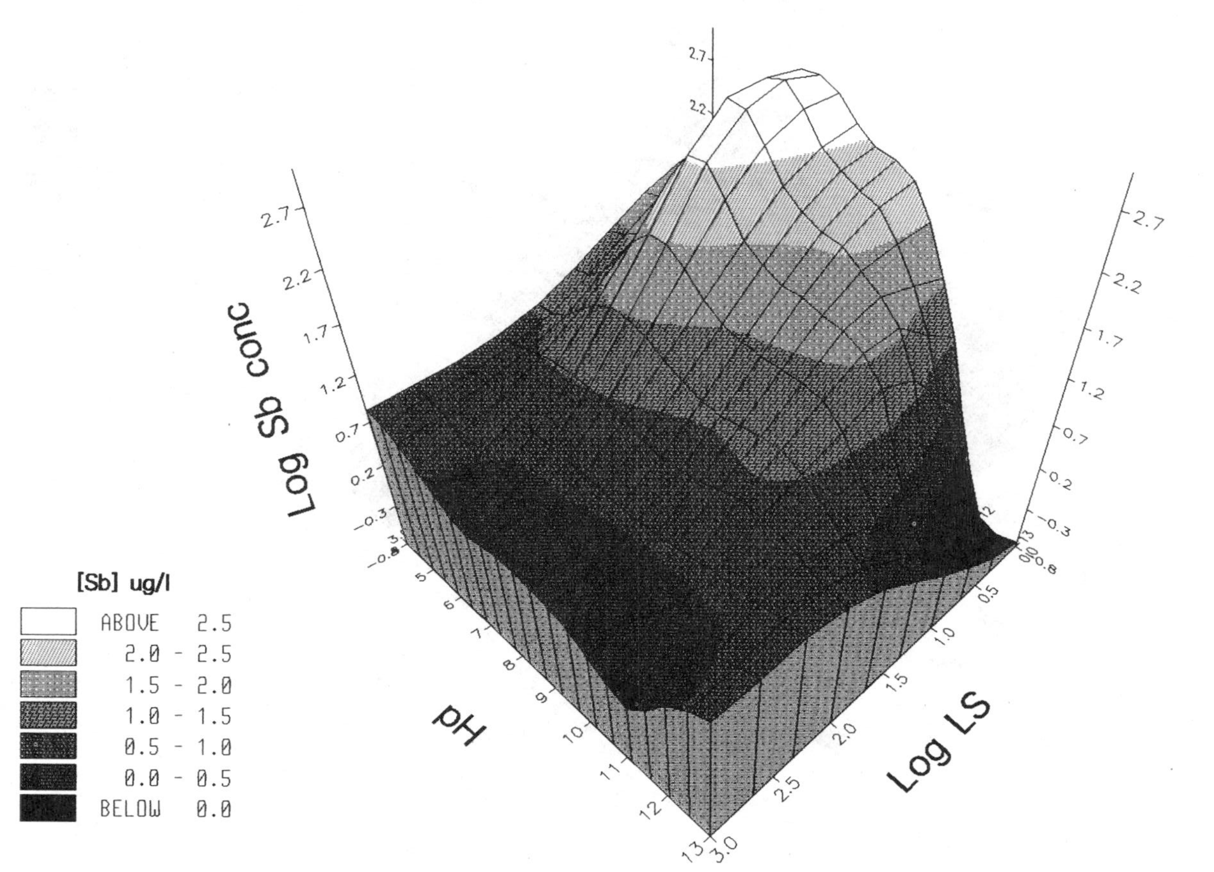

FIG. 5—*Concentrations in μg/L of antimony in PCA extracts as function of pH and LS ratio.*

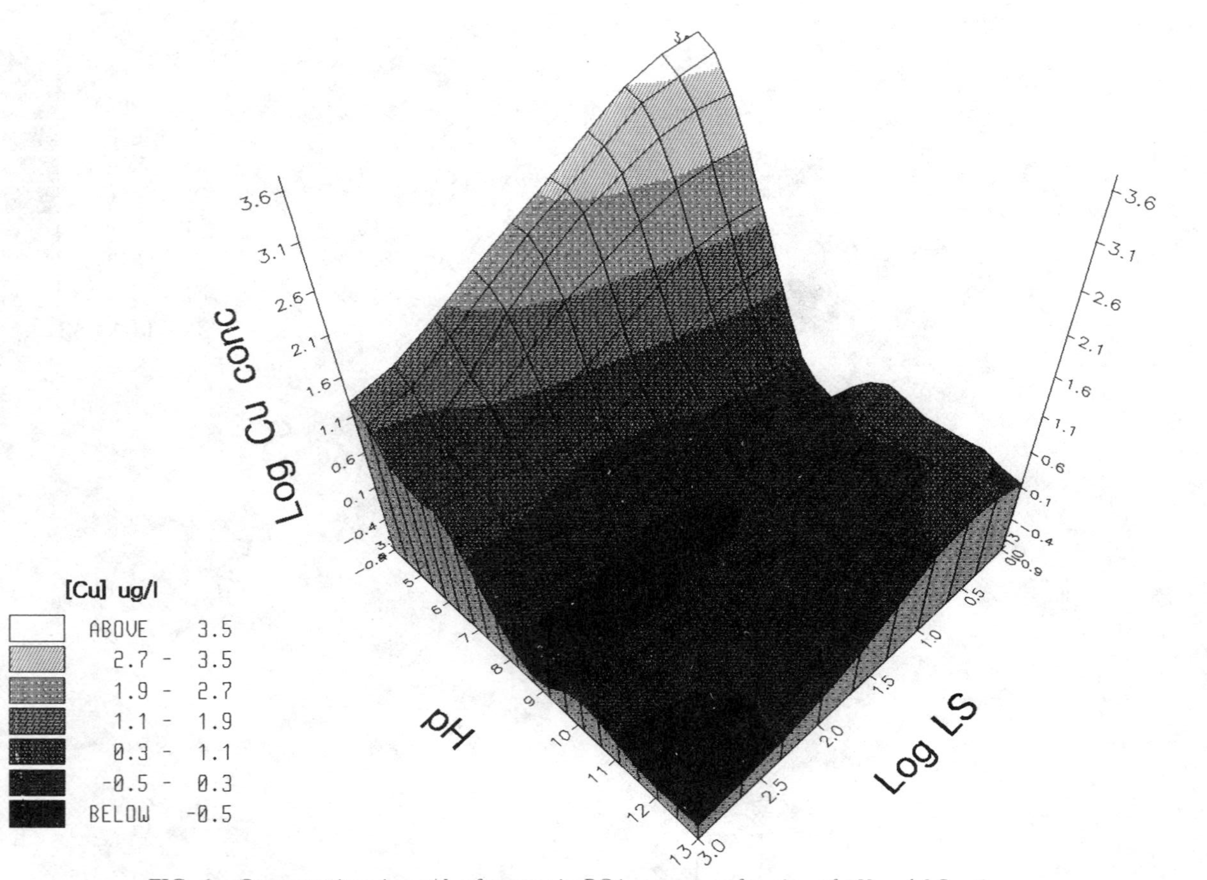

FIG. 6—*Concentrations in μg/L of copper in PCA extracts as function of pH and LS ratio.*

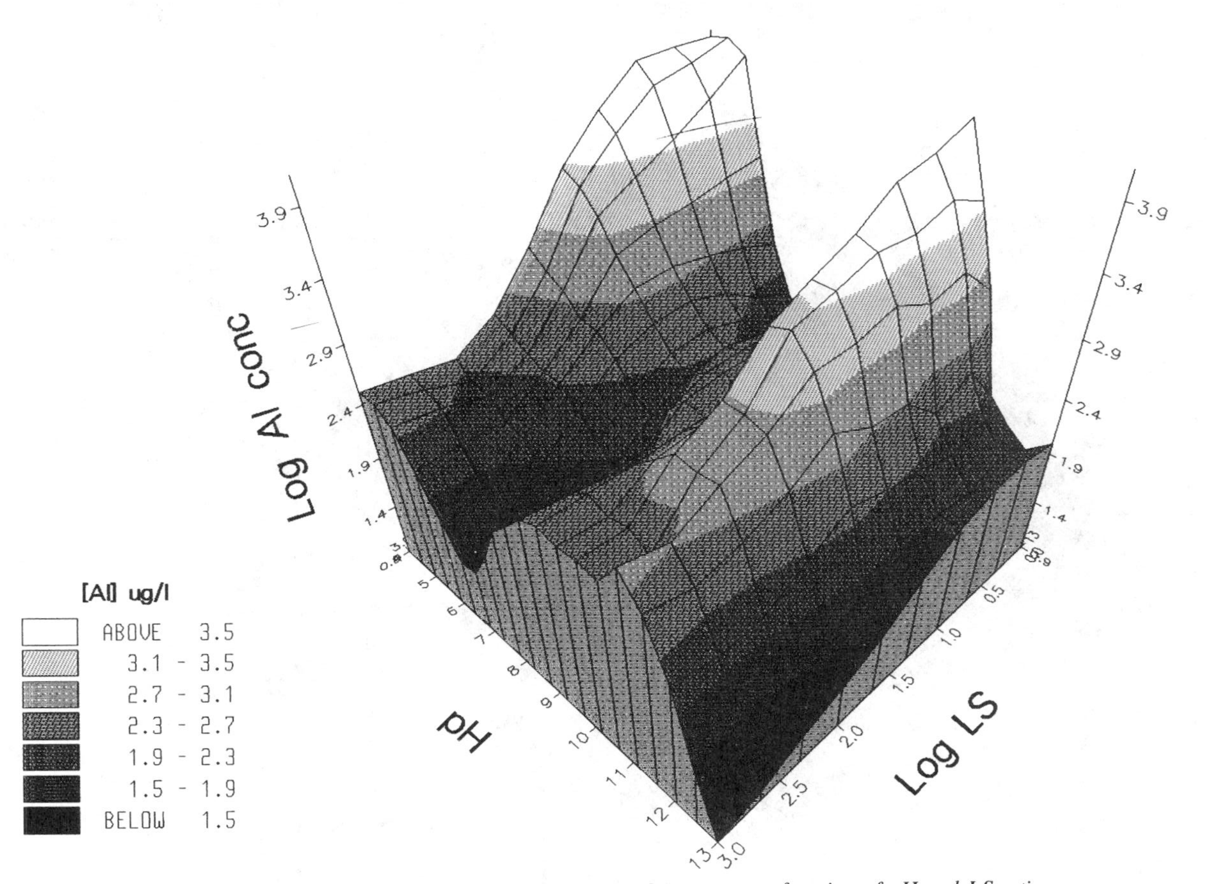

FIG. 7—*Concentrations in $\mu g/L$ of aluminum in PCA extracts as function of pH and LS ratio.*

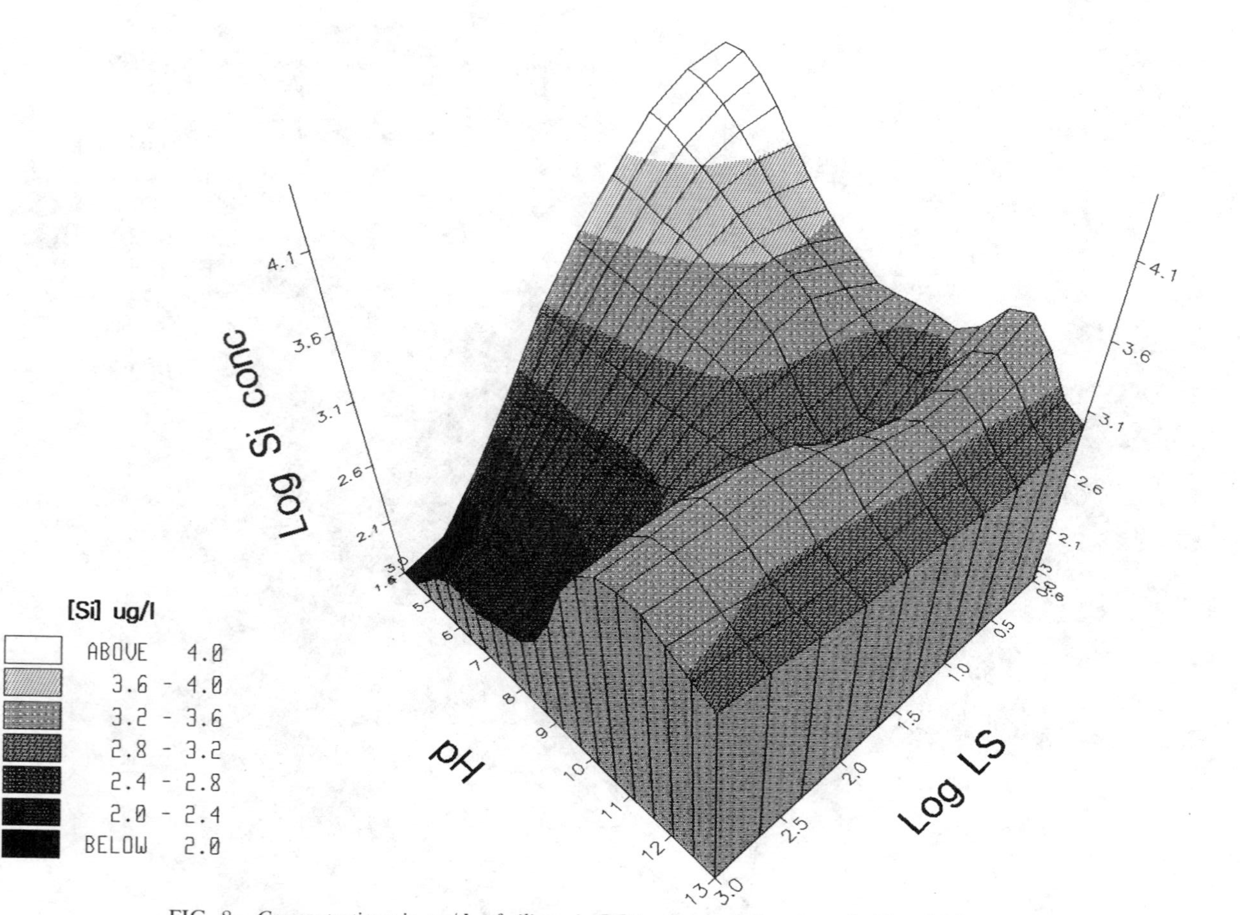

FIG. 8—*Concentrations in μg/L of silicon in PCA extracts as function of pH and LS ratio.*

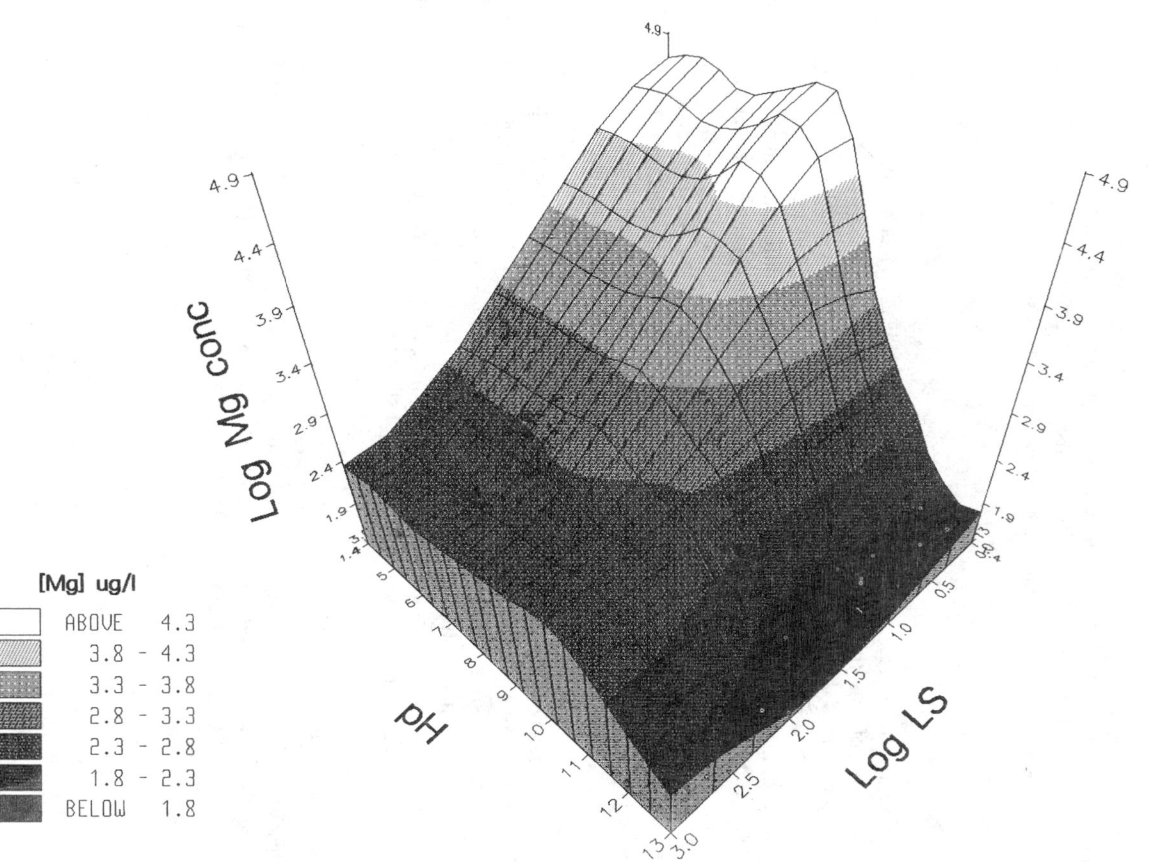

FIG. 9—*Concentrations in μg/L of magnesium in PCA extracts as function of pH and LS ratio.*

it appeared that the 24-h reaction time for the batch experiments is sufficient for most elements to reach equilibrium. Only at alkaline conditions and low LS ratios do the concentrations for vanadium, antimony, and arsenic still decrease after 24 h. This could be the explanation for the extremely low values for vanadium and antimony at high pH and low LS (see Fig. 5).

The relations found for LS 5 discussed before, are also confirmed by the three-dimensional graphs. Observations in Ref *1* from column experiments on resorption of anionic species are also in agreement with the three-dimensional data obtained by batch experiments. Further work is needed to identify the chemical forms, explaining the observed leaching behavior.

Conclusions

1. Fly ashes from coal combustion show systematic leaching behavior, allowing better understanding of this thermodynamically complex system. This could be recognized by examining results of the leach behavior of a large number of different fly ashes from a large number of sources, under well-defined conditions. With respect to leaching, differences in elemental composition of the ashes are far less important than differences in pH. Therefore, pH is the controlling factor in leachability. Elemental components can be grouped according to common leaching characteristics.

2. The three-dimensional relations allow an estimation of the element concentration in the fly ash/water system for any given LS ratio and pH condition. The uncertainty of this estimate is high in the steep gradients within these three-dimensional plots. The leaching time can be a significant factor only for a few elements for which the reaction kinetics are slow.

3. A database with leaching data of waste materials and waste products, containing data as those presented here, is an important tool for filling the voids in our knowledge. The matrices obtained are particularly useful as input for modeling studies and for revealing the systematics in leaching behavior of ash/water systems.

4. The presented relations are not limited to leaching of fly ash. Also, products containing fly ash, such as stabilized pulverized fly ash and structural lightweight concrete, fit well.

Acknowledgments

This work was carried out with financial support from the National Coal Research Programme through the Management Office for Energy Research (PEO) under Contract 4351.11.9.1 and Contract 20.53-021.10.

References

[*1*] van der Sloot, H. A., Wijkstra, J., et al., "Leaching of Trace Elements from Coal Solid Waste," ECN-120, Netherlands Energy Research Foundation, Petten, the Netherlands, 1982.

[*2*] Siberman, D. and Fisher, G. L., "Room Temperature Dissolution of Coal Fly Ash for Trace Metal Analysis by Atomic Absorption Spectrometry," *Analytica Chimica Acta*, Vol. 130, 1981, pp. 1–8.

[*3*] van der Sloot, H. A., Piepers, O., and Kok, A., "A Standard Leaching test for Combustion Residues," BEOP-31, Bureau for Energy Research Projects, Netherlands Energy Research Foundation, E.C.N., Petten, the Netherlands, 1984.

[*4*] Jackson, D. R., Garrett, B. C., and Bishop, T. A., "Comparison of Batch and Column Methods for Assessing Leachability of Hazardous Waste," Environmental Science and Technology, Vol. 18, 1984, pp. 668–673.

[*5*] Theis, T. L., "The Contamination of Groundwater by Heavy Metals from the Land Disposal of Fly Ash," Final Report C00-2727-1, Department of Civil Engineering, University of Notre Dame, Notre Dame, IN, 1979.

[6] Liem, H., Sandstrom, M., et al., "Studies on the Leaching and Weathering Processes of Coal Ash," Final Report Project KHM 105, Stockholm, Sweden, 1983, in Swedish.
[7] Sillen, L. G. and Martell, A. E., *Stability Constants of Metal-Ion Complexes,* The Chemical Society, London, 1971.
[8] Auerbach, S. I. and Reichle, S., ORNL-5620, Oak Ridge National Laboratory, Oak Ridge, TN, 1980, pp. 25–27.
[9] Pourbaix, M., *Atlas of Electrochemical Equilibria in Aqueous Solutions,* 2nd ed., National Association of Corrosion Engineers, Cebelcor, Brussels, Belgium, 1974.

Tahar El-Korchi,[1] David Gress,[2] Kenneth Baldwin,[2] and Paul Bishop[3]

Evaluating the Freeze-Thaw Durability of Portland Cement-Stabilized-Solidified Heavy Metal Waste Using Acoustic Measurements

REFERENCE: El-Korchi, T., Gress, D., Baldwin, K., and Bishop, P., **"Evaluating the Freeze-Thaw Durability of Portland Cement-Stabilized-Solidified Heavy Metal Waste Using Acoustic Measurements,"** *Environmental Aspects of Stabilization and Solidification of Hazardous and Radioactive Wastes, ASTM STP 1033,* P. L. Côté and T. M. Gilliam, Eds., American Society for Testing and Materials, Philadelphia, 1989, pp. 184–191.

ABSTRACT: The use of stress wave propagation to assess freeze-thaw resistance of portland cement solidified/stabilized waste is presented. The stress wave technique is sensitive to the internal structure of the specimens and would detect structural deterioration independent of weight loss or visual observations.

The freeze-thaw resistance of a cement-solidified cadmium waste and a control was evaluated. The control and cadmium wastes both showed poor freeze-thaw resistance. However, the addition of cadmium and seawater curing increased the resistance to more cycles of freezing and thawing. This is attributed to microstructural changes.

KEY WORDS: solidification/stabilization, heavy metal waste, hazardous waste, freeze-thaw resistance, acoustic testing, waste solidification, cement solidification

Stabilization/solidification (S/S) of toxic heavy metal waste with portland cement-based fixation processes is being considered as a technology available for treating selected banned waste prior to land disposal. The objective of S/S processes is to solidify liquid wastes and minimize their mobility to the surrounding environment. This is achieved through the desirable selection of additives that react with the contaminants to reduce their solubility and through the construction of a solid monolith which decreases the apparent surface area and accessibility of leachants to leaching sites. Any physical disintegration of the solid monolith will adversely affect the apparent permeability and hence increase the leachability upon contact with a corrosive fluid. Such physical disintegration could be caused by adverse climatic conditions such as freezing-thawing and wetting-drying.

These S/S wastes are designed to be enclosed under an earth cap of very low permeability to reduce water infiltration. Freeze-thaw damage must be given consideration for those situations which allow the S/S waste to freeze in a saturated state. At disposal sites where

[1] Formerly, graduate student, Department of Civil Engineering, University of New Hampshire, Durham, NH 03824; presently, assistant professor, Department of Civil Engineering, Worcester Polytechnic Institute, Worcester, MA 01609.

[2] Associate professor, Department of Civil Engineering, and associate professor, Department of Mechanical Engineering, respectively, University of New Hampshire, Durham, NH 03824.

[3] Formerly, professor, Department of Civil Engineering, University of New Hampshire, Durham, NH 03824; presently, professor of environmental engineering, University of Cincinnati, Cincinnati, OH 45268.

freezing and thawing occur, it is necessary to assess the freeze-thaw resistance as an integral part of S/S process evaluation.

The objective of this study is to utilize a nondestructive testing method to assess freeze-thaw durability of S/S wastes. The sonic method [*1*] was adopted to yield information relating to changes in the internal structure of the material. Weight measurements [*2–4*] are not always in direct agreement with the condition of the internal structure which contributes to the apparent permeability and hence leaching potential. Cadmium waste was chosen for this experimental evaluation due to its industrial importance and toxic behavior.

Experimental Procedures

Specimen Preparation

The cadmium sludge was synthesized by dissolving cadmium nitrate in deionized water at a concentration of 20 g/L. The cationic metal was precipitated to hydroxide with addition of sodium hydroxide to obtain a solution pH of 8.5. The sludges corrected for solid content were used as mix solution and were added to Type II portland cement at a liquid-to-cement ratio of 1:2. Mixing was conducted in accordance with ASTM Method for Mechanical Mixing of Hydraulic Cement Pastes and Mortars of Plastic Consistency (C 305-82).

The specimen beams used to evaluate freeze-thaw durability were prismatic, measuring 2.54 by 2.54 by 14 cm and were cast with mounting studs on the free ends. The stud used for attachment to the driving coil of an electromechanical shaker table was a threaded rod measuring 0.64 cm in diameter, with 7.8 threads per centimetre. The mounting stud for the transducer was 0.4 cm in diameter and 1.9 cm long, with one end rounded to have an exact fit to the base of the transducer. The specimen beams were cast in a well-greased mold and subjected to vigorous tamping and screeding to evacuate the entrapped air. The mold was placed in the humidity room after 2 h to prevent undesirable shrinkage during hardening of the samples. The samples were subsequently demolded after 24 h and cured for 28 days in approximately 100% relative humidity or in synthetic seawater at 20°C.

Freeze-Thaw Cycling

Both freezing and thawing cycles were carried out in water. Four beams were placed per container and submerged under 3 cm of water. The specimen beams were randomly placed in containers that were also randomly placed in the temperature cycling compartment of the freeze-thaw machine. The specimens were cooled to −17.8°C at a rate of 5°C/h and heated at a rate of 6.1°C/h to 20°C.

Thermocouples were cast in companion specimen beams, which were placed in the center and along the edge to monitor the temperature across the freeze-thaw compartment. This is to ensure that all specimens had reached the desired temperatures. The freeze-thaw apparatus was a standard Soiltest freeze-thaw machine used for concrete testing in accordance with ASTM Test for Resistance of Concrete to Rapid Freezing and Thawing (C 666-84). The apparatus is calibrated to control the temperature cycle using a 1200-Ω resistor cast into an air-entrained mortar specimen and placed in the center of the freeze-thaw compartment.

Derivation of Acoustic Parameters

The objective of the acoustic-stress wave testing effort was to utilize a nondestructive technique to evaluate the freeze-thaw durability of S/S waste in an effort to study the relationship between the sample microstructure and freeze-thaw durability. The dynamic modulus (E dynamic) [*1*] and quality factor (Q) [*1,5*] were chosen as the study parameters. The factor Q was adopted since it can be more sensitive to changes within the internal structure even if changes of natural frequency are not observed [*6*].

The natural frequency is routinely determined in concrete testing for calculating E dynamic using the ASTM Test for Fundamental Transverse, Longitudinal and Torsional Frequencies of Concrete Specimens (C 215-85). Assuming no change in weight and geometric properties, E dynamic is proportional to the square of the natural frequency $(f_n)^2$ of the specimen beams. The relative dynamic modulus is proportional to the square of the normalized natural frequency (NNF) and is used for evaluating concrete durability to freezing and thawing. The NNF parameter is determined before and after each freeze-thaw cycle, as shown in Eq 1. The NNF parameter is related to the durability factor typically reported as an indicator of freeze-thaw durability for concrete in accordance with ASTM C 666-84.

$$\text{NNF} = \frac{[f_{ni}]^2}{[f_{n0}]^2} \times 100 \tag{1}$$

where

NNF = square of normalized natural frequency,
f_{n0} = natural frequency at zero cycles, and
f_{ni} = natural frequency at the ith cycle.

The quality factor Q is a frequency domain measurement of the width of the spectral peak. The amplitude of the response is expressed in decibels (dB), as shown in Eq 2. A decrease in Q will indicate an increase in structural disintegrity and, in our particular case, increased freeze-thaw deterioration.

$$\text{Spectrum level (dB)} = 10 \log \frac{[V_{\text{rms}}\ f_i]}{[V_{\text{rms}}\ f_n]} \tag{2}$$

where $V_{\text{rms}}\ f_i$ is the root-mean-square voltage at the ith frequency and $V_{\text{rms}}\ f_n$ the rms voltage at natural frequency.

The spectrum level is plotted as a function of frequency as shown in Fig. 1. The spectral peak width is defined by -3dB frequencies (f_1,f_2) located on either side of resonance frequency. These frequencies correspond to the half-power frequencies as normalized at natural frequency. The quality factor Q is defined by Eq 3 [*5*].

$$Q = \frac{f_n}{f_2 - f_3} \tag{3}$$

The Q factor (which is also a measure of the sharpness of resonance) is related to the log decrement (δ), and in materials with low damping the relationship is defined by Eq 4 [*6*].

$$Q = \frac{\pi}{\delta} \tag{4}$$

Acoustic Testing

The continuous-wave technique involves exciting the specimen beams on one end and receiving on the other end. To begin this technique, the specimen beams are mounted

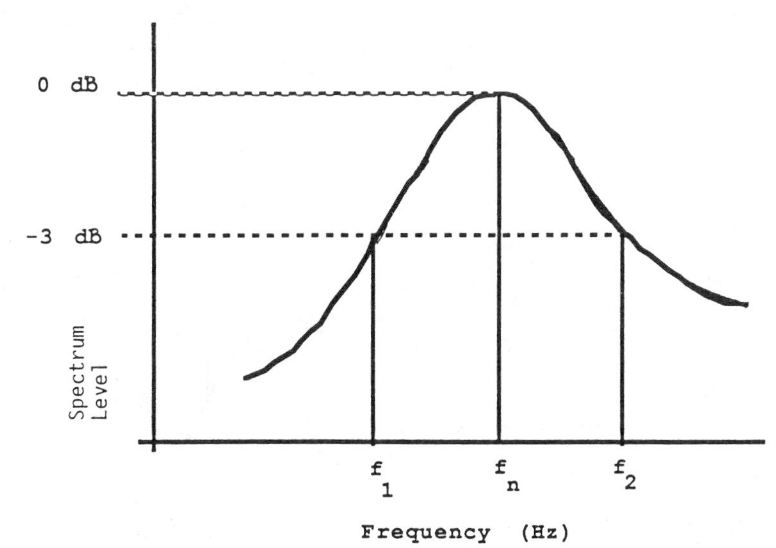

FIG. 1—*Generic spectrum level versus frequency plot, indicating the parameters necessary for determining* Q *factor, the sharpness of the resonance peak.*

directly to the electromechanical shaker table by way of the cast-in studs. The beams were mounted to the shaker table using a torque measuring 1.1 Nm.

The shaker table is driven by a sine-wave generator and amplifier, which provide the excitation to the specimen beams. The wave propagation is monitored on the free end with a Bruel & Kjaer 4374 accelerometer. The accelerometer output passes through a Bruel & Kjaer 2626 charge amplifier. Both the charge amplifier output and the sine-wave generator are monitored on a Nicolet 320 digital oscilloscope. While a constant drive voltage is maintained, the sine-wave frequency is varied and changes in received signal amplitude are recorded. The shaker table and the accelerometer both have flat frequency responses in the frequency band of interest. Figure 2 is a schematic drawing of the measurement system and test setup.

The first step of the test procedure is to sweep the frequency range between 2 and 5 kHz to determine f_n. This is followed with a constant drive voltage, and the frequency is sampled contingent upon f_n and the observed rate of amplitude decay with frequency during the experiment. Effectively, a conditional sampling scheme is manifested in the neighborhood of the resonant frequency. A plot of relative amplitude in decibels versus frequency in hertz is then constructed with the data curve fitted to enable the determination of the -3 dB frequencies, f_1 and f_2, required for the Q parameter calculations.

Results and Discussion

A total of 15 specimen beams were tested in this study. The control mix was prepared using portland cement and water. Four cadmium waste specimens were cured in 100% relative humidity and four in seawater. Three control specimens were cured in 100% relative humidity and four cured in seawater.

The average values for NNF and Q factor as a function of freeze-thaw cycles are shown in Figs. 3 and 4, respectively. The results show that the control as well as the cadmium samples all have very poor freeze-thaw resistance as compared to a well-designed portland cement concrete [7]. The results are as expected as there was intentionally no attempt to

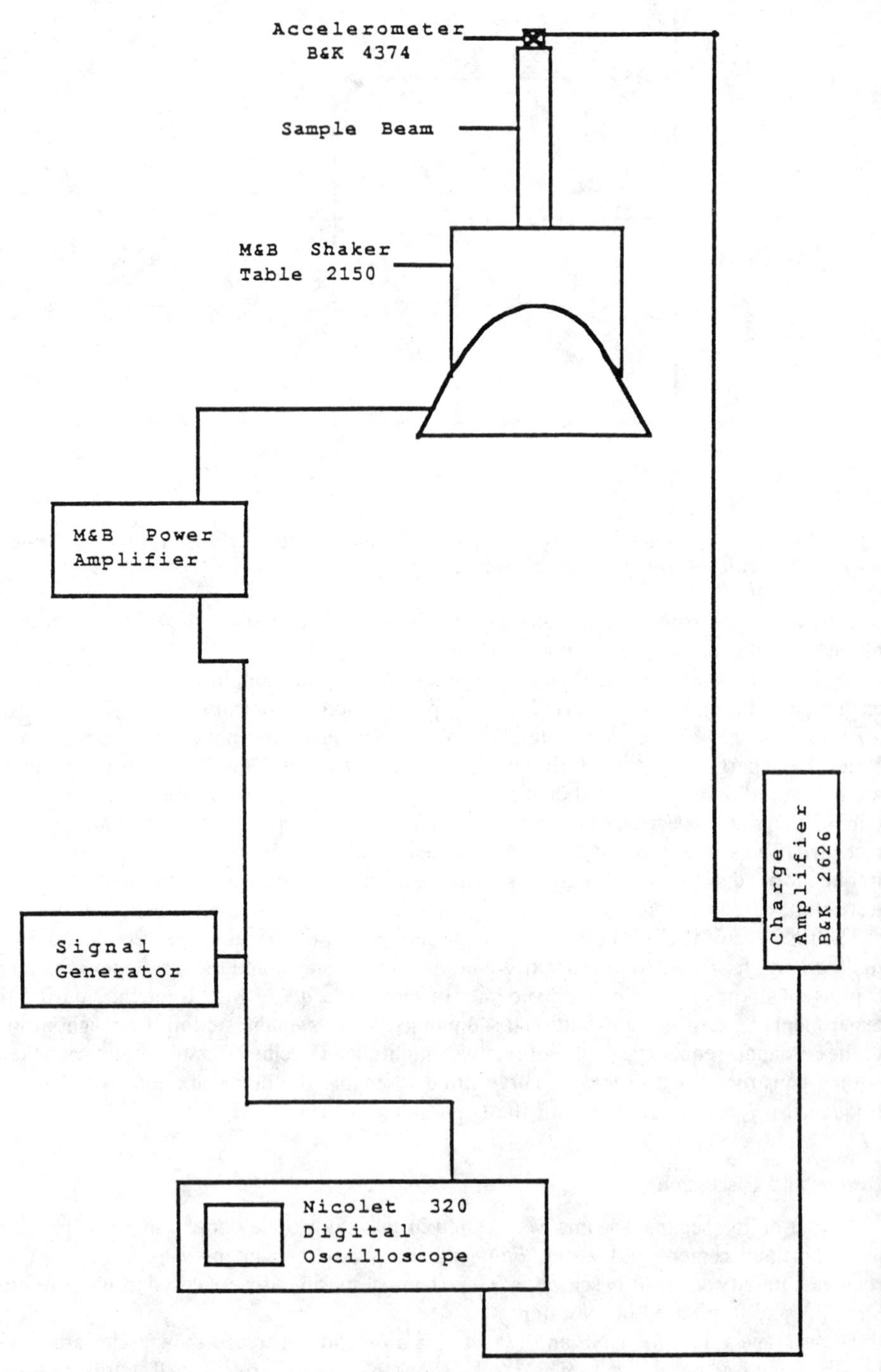

FIG. 2—*Measurement system schematic, indicating the arrangement of equipment necessary to obtain the spectrum level-frequency data.*

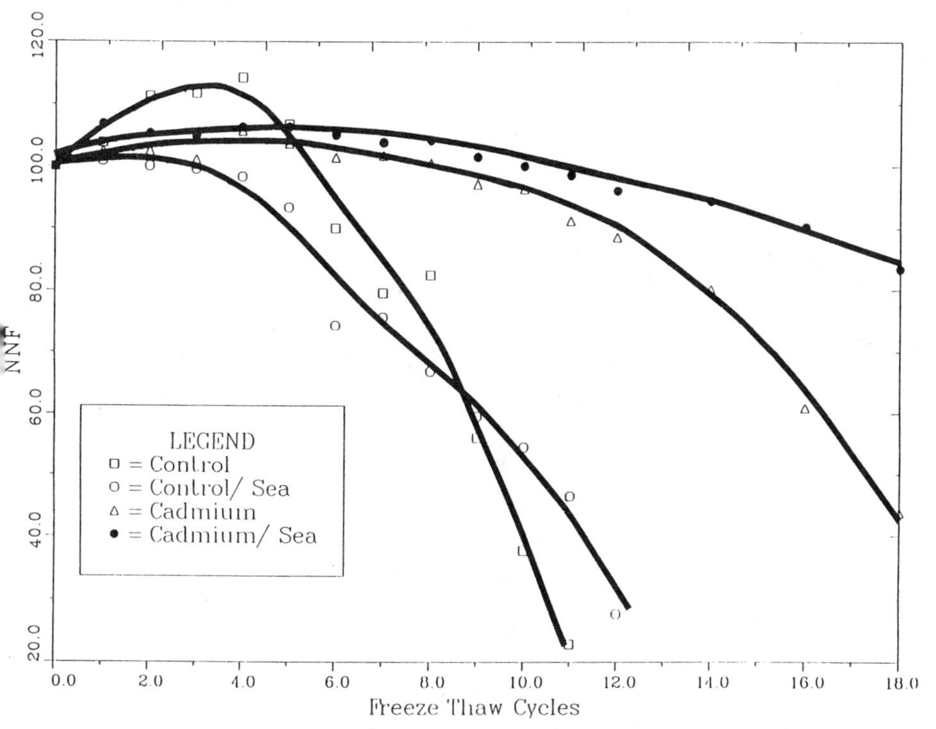

FIG. 3—*Freeze-thaw durability as monitored by NNF values.*

air-entrain the cementitious matrix. However, there is a significant difference between the resistance of the control paste and the cadmium waste.

The cadmium waste specimens show better resistance to endure more freeze-thaw cycles as compared to the control. Seawater curing changes the slopes of the curves, indicating a decreased rate of freeze-thaw deterioration and hence an improved performance. Based on the NNF plots in Fig. 3, the cadmium specimens cured in seawater undergo a less rapid change in their mechanical properties and the control specimens undergo the most drastic change. Based on the *Q* factor, the peak above the initial value is of interest, and it occurs in both mixtures and both treatments. The specimens cured in 100% relative humidity exhibit a higher initial peak, followed by a more pronounced decrease in integrity as compared to seawater treatment. This suggests that seawater cure produces a more desirable result especially with the cadmium waste.

One of the mechanisms of frost action in portland cement concrete, according to Powers [*8*], is attributed to the hydraulic pressure generated by the 9% volume increase when water freezes in a cavity. The resistance to the generated hydraulic pressure is controlled by the simultaneous interaction of several factors, namely the pore structure, continuity of pores and their pore size distribution, distance to entrapped or entrained air voids, salt concentration of pore fluid, degree of saturation, rate of cooling, and tensile strength of the cementitious matrix. Addition of cadmium to the cementitious matrix and subsequent curing in seawater affects the above-mentioned physio-chemical properties.

Cadmium and seawater curing alters the microstructure by increasing the total porosity and shifting the pore size distribution as observed by mercury intrusion porosimetry and scanning electron microscopy [*9*]. This is obviously due to dissolution of hydration products

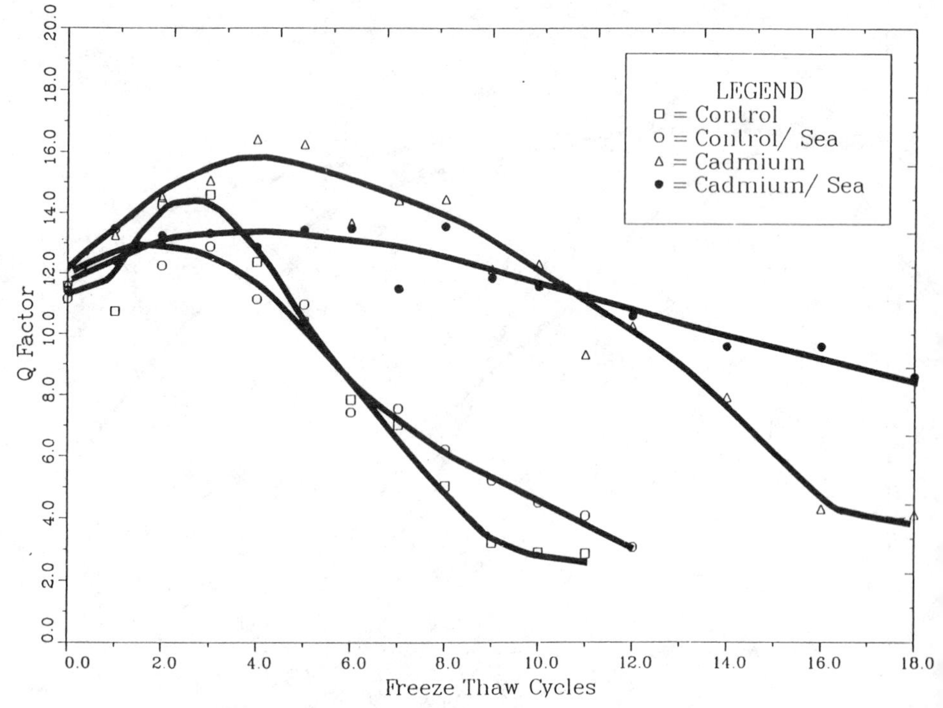

FIG. 4—*Freeze-thaw durability as monitored by* Q *factor.*

(that is, calcium hydroxide) and formation of new compounds (that is, magnesium hydroxide or brucite). For our particular freezing rate, the combination of total porosity, pore size distribution, formation of surface layer, and composition of pore solution enhances the freeze-thaw resistance in the cadmium specimens, even temporarily. All specimens will eventually fail as possible conditions of "critical saturation" (where the partially filled pores will not accommodate the frozen solution) will be attained.

The increase in the value of Q after initial freeze-thaw cycling is due to a decrease in internal friction caused by a decrease in void space. Note that all specimens appear to be initially improved by the freeze-thaw cycles. One way that the value of Q increases is through the creation of a more dense material through increased or continued hydration of cement. It is possible that during the freezing cycle, cement particle reorientation is caused due to the hydraulic pressures that develop. Continued cement hydration and hydration product formation would decrease the void space and hence decrease the damping and internal friction. This would minimize the energy loss associated with the transfer of p-waves, and hence a larger Q parameter value will result.

The increase in Q factor and the steepness of the slope of the curve in the seawater cured specimens was less dramatic than specimens cured in 100% relative humidity. For these specimens, initial freeze-thaw cycling does not alter the microstructure significantly such as to increase the rate of hydration [9]. The increased durability of these specimens is due to an altered microstructure and a possible freezing point depression caused by the added salt concentration in the pore fluid.

Conclusion

Acoustic stress-wave testing was successful in evaluating the freeze-thaw resistance of S/S wastes. Freeze-thaw deterioration was detected by acoustic analysis prior to visual observation.

The control and cadmium specimens both have very poor freeze-thaw resistance compared with a well-designed portland cement concrete. However, the addition of cadmium and seawater curing produces a temporarily improved matrix to freezing and thawing. This is attributed to microstructural changes in total porosity, pore size distribution, and surficial microstructure.

References

[1] Whitehyrst, E. A., *Evaluation of Concrete Properties From the Sonic Tests,* ACI Monograph No. 2, American Concrete Institute, Detroit, 1966, Chapter 2, pp. 23–38.

[2] Bartos, M. J. and Palermo, M. R., "Physical and Engineering Properties of Hazardous Wastes and Sludges," US EPA Report 600/2-77-139, Environmental Protection Agency, 1977.

[3] Malone, P. G. and Jones, L. W., "Solidification/Stabilization of Sludge and Ash from Waste Water Treatment Plants," US EPA 600/S2-85-058, Environmental Protection Agency, 1985.

[4] Hannak, P. and Liem, A. J., "Development of a Method for Measuring the Freeze-Thaw Resistance of Solidified/Stabilized Wastes," in press.

[5] Thomason, W. T., *Theory of Vibration with Applications,* 2nd ed., Prentice-Hall, Englewood Cliffs, NJ, 1981, p. 76.

[6] Kesler, C. E. and Higushi, Y., "Determination of Compressive Strength of Concrete by Using Its Sonic Properties," presented at the 56th Annual Meeting of the American Society for Testing and Materials, June 1953.

[7] "Guide to Durable Concrete," ACI 201.2R-77, American Concrete Institute, ACI Committee 201, Detroit, 1977.

[8] Powers, T. C., "Freezing Effects in Concrete," *Durability of Concrete,* ACI SP-47, American Concrete Institute, Detroit, 1975, pp. 1–12.

[9] El-Korchi, T., Ph D. thesis, University of New Hampshire, Durham, NH, 1987.

Carl P. Swanstrom[1]

Development of a Liquid Release Tester

REFERENCE: Swanstrom, C. P., **"Development of a Liquid Release Tester,"** *Environmental Aspects of Stabilization and Solidification of Hazardous and Radioactive Wastes, ASTM STP 1033,* P. L. Côté and T. M. Gilliam, Eds., American Society for Testing and Materials, Philadelphia, 1989, pp. 192–200.

ABSTRACT: The 1984 Amendments to the Resource Conservation and Recovery Act (RCRA) specify that bulk liquid wastes which have been stabilized/solidified for land disposal must not release the liquid under the mechanical pressures experienced in a landfill. How to determine this property has not yet been specified. In response to the need for a simple, yet highly effective test device to evaluate the tendency of a stabilized waste to release liquid, the author's company has developed a test apparatus and a test procedure. This simple tester distinguishes between stabilized wastes which will release liquids under pressure on a pass/fail basis.

The tester utilizes a 100-g sample that requires only minimal sample preparation. The test duration is 5 min. The only expendables are readily available paper filters, and the system requires only a balance and compressed air for operation.

The effects of test pressure, test duration, and sample cure time were evaluated. Results may be compared with other test procedures for evaluating stabilized waste.

KEY WORDS: RCRA, liquid waste, stabilized waste, solidified waste, landfill, liquid release, free liquids, bearing strength, Paint Filter Test

The 1984 Amendments to the Resource Conservation and Recovery Act (RCRA) specify that bulk liquids or wastes containing "free liquids" cannot be landfilled after 8 May 1985. The United States Environmental Protection Agency (USEPA) guidelines for implementing the Amendments state that the liquids cannot be treated by a material acting solely as an absorbent and that the test for the presence of "free liquids" is the Paint Filter Test. Temperature rise during mixing and increasing bearing strength with time have been used to demonstrate that a reaction has taken place and that the solidification is not just an absorption mechanism.

The 1984 Amendments also specify that bulk liquid wastes which have been stabilized (solidified) must not release the liquid under the mechanical pressures experienced in a landfill; regulations were to be promulgated by 8 Feb. 1986. The legislation does not specify how this property of the waste is to be determined, nor does it set specific limits for the allowable amount of liquid which can be released from the waste. No time, temperature, or pressure is specified. Congress appeared to have two concerns: first, that the liquid may form a leachate which might attack the cell liner and escape from the landfill, and second, that the movement of the liquid would tend to increase subsidence with attendant damage to the landfill cap.

On 24 Dec. 1986, the USEPA published "EPA Proposal Under RCRA to Prohibit Land Disposal of Most Free Liquid Hazardous Waste Held in Containers" (51FR46824). This proposed regulation "would require that if hazardous liquids or free liquids in containers

[1] Senior project manager, Chemical Waste Management, Inc., 2000 S. Batavia, Geneva, IL 60134.

are solidified by the use of an absorbent, the absorbent material must not be biodegradable and the absorbent/waste mixture must not release liquids when compressed under pressures experienced in landfills." Although this proposed regulation concerns absorbents used to solidify containerized wastes, it does give an indication of the type of test the USEPA desires. Assuming a common performance criteria basis, a liquid release testing procedure for containerized liquids solidified with absorbents should be the same for bulk wastes that have been chemically stabilized/solidified.

In response to the need for a simple, yet highly effective, method to evaluate the tendency of a stabilized waste to exude liquid under "typical" landfill pressures, Chemical Waste Management (CWM) has developed a simple test apparatus and a test procedure. The method distinguishes between samples on a pass/fail basis and yields an easily determined signal which is highly sensitive to liquid release. The tester has been used to develop stabilization formulations and as a quality control for stabilized and cured waste.

CWM proposed procedures require stabilized waste to pass through a hierarchy of tests for landfill disposal. No one test, of itself, is sufficient. CWM tests for sales samples determine

1. Ratio of stabilization agent to waste required to pass the Paint Filter Test.
2. Temperature rise during stabilization as an indication of chemical reaction.
3. Measurement of increase in bearing strength with time as an indication of chemical reaction.
4. Liquid release test to determine if liquid will be released under simulated landfill pressures.

The recipe developed during evaluation of the sales sample is used during stabilization operations. The Paint Filter Test is used for quality control during operations.

The CWM Stabilized Waste Liquid Release Tester was designed and built based upon knowledge of the properties of stabilized waste, the requirements of the RCRA amendments, and engineering experience. Since this is a new device and involves measurement of a stabilized waste property not previously measured, it has been necessary to evaluate test procedures and conditions and to relate these to conditions which would be experienced by the waste in a landfill.

Equipment

Several objectives were kept in mind when designing the Stabilized Waste Liquid Release Tester, namely,

1. Ability to apply a range of test pressures up to 690 kPa (100 psi).
2. Rapid test (15 min maximum).
3. Reasonable sample size (less than 250 g).
4. Require minimal support equipment and consumables.
5. Low cost.

For cleaning and maintenance, it was desirable to keep the sample holder a separate unit from the pressure-applying device.

The current design, including the sample holder, is shown in Fig. 1. The unit stands approximately 0.58 m (23 in.) tall, 0.37 m (14.7 in.) wide, and 0.15 m (6 in.) deep and weighs approximately 21 kg (46 lb). The pressure-applying cylinder is a standard double-acting pneumatic cylinder with a 0.051-m (2 in.) bore rated for 1723 kPa (250 psi). A standard manual control valve (handle on right side) is used to control the direction of travel of the

FIG. 1—*Liquid release tester.*

pressure-applying piston. A small industrial air regulator is used to adjust the operating pressure from 0 to 861 kPa (125 psi). The upper pressure gage displays the pressure applied to the sample, and the lower gage displays the supply pressure.

The sample holder assembly is shown in Fig. 2. The holder consists of a 5.1-cm (2 in.) vertical cylinder open at both ends with a flange on the bottom end that permits connection to a cup to retain the liquid and attachment of the detection system (a support plate, screen, and two filter papers) between the cylinder and the cup. The cylinder, cup, and detection system are held together by a hinged clamp with a wing-nut tightening device.

Pressure can be applied to the sample in the container by means of the ram connected to the double-acting pneumatic cylinder operated by compressed air. The pressure regulator is used to adjust the pressure to be exerted on the sample, and the lever-operated valve is used to actuate the ram movement. Since the piston in the air cylinder is the same diameter as the ram in the sample container, the pressure on the sample is the same as the pressure applied to the pneumatic cylinder.

The system used to detect any liquid that might be expressed from the sample consists of two thin filter papers and a plastic (or metal) screen approximately 5.5 cm in diameter. These are arranged so that the first filter paper comes in direct contact with the sample and is separated from the second filter paper by the screen. During the test, the filter paper in

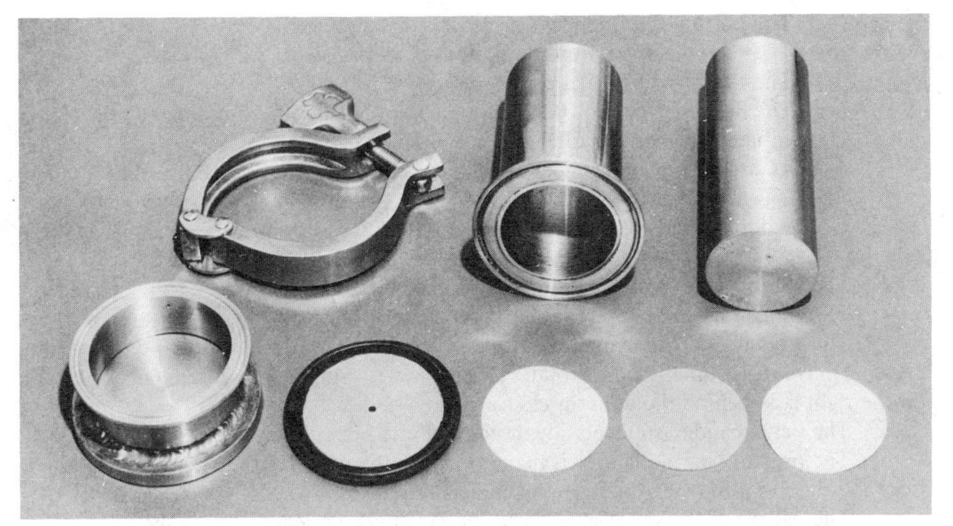

FIG. 2—*Sample holder assembly.*

contact with the sample prevents the sample from extruding through the screen and reaching the second or bottom filter paper. As pressure is applied to the sample, the thin filter paper in contact with the sample may intercept a small amount of liquid; however, if there is a significant amount of liquid released from the sample, it will penetrate the top filter paper and the screen separating the two papers and will cause a liquid stain on the bottom filter paper. A liquid stain detected on the bottom filter paper constitutes a failure of the test. The two filter arrangement is necessary to avoid "false positives" stemming from direct contact between the filter paper and the sample. New filter papers are used for each test. The separator screen can be cleaned and used again.

Detection of liquid on the second filter paper is easy; however, later work using a dyed filter paper has made detection even more positive.

Sample Preparation

The physical characteristics (density, strength, particle size, etc.) will vary greatly among stabilized wastes (or other materials that may be tested). At this stage of the procedural development, the only sample preparation required is that the sample pass through an eight-mesh screen. This is a compromise between limiting the maximum granule size based on the sample holder dimensions and minimizing sample preparation effort. Monolithic samples must be ground to less than eight-mesh. In the case of granular (powder) samples containing pieces larger than eight-mesh, the pieces should be crushed if possible. Pieces that cannot be crushed by normal laboratory methods may be removed from the sample. Although molded samples might be satisfactory, they would require many sample holders over a period of 7 to 14 days.

Tester Operation

The tester is operated in the following manner:

1. The clean container is assembled as follows (starting from the bottom).
   (*a*) The support plate with a single hole is placed on the bottom reservoir.

(*b*) The bottom filter paper is placed on the support plate.
(*c*) The screen is placed on the bottom filter paper.
(*d*) The top filter is placed on the screen.
(*e*) The flanged end of the 0.051-m-diameter (2 in.) by 0.152-m-long (6 in.) cylinder is placed on the top filter paper.
(*f*) The reservoir and cylinder are clamped together. Make sure the filter assembly is centered.

2. One hundred grams of the sample prepared as described previously is placed in the cylinder.
3. The solid piston is inserted into the cylinder.
4. The sample assembly is set on the tester base against the alignment stops.
5. The pressure to the cylinder is adjusted to 344.5 kPa (50 psi) (or other specified pressure). This is accomplished by rotating the yellow ring on the pressure regulator located on the right side behind the upper pressure gage.
6. The valve handle is moved downward to exert pressure on the solid piston.
7. The pressure on the sample is maintained for the specified time period as measured from the time that the air cylinder ram contacts the piston.
8. The valve handle is moved to the "up" position to release the pressure.
9. After the pressure is released, the sample assembly is removed from the tester. The solid piston is removed, the clamp is removed and the cylinder is carefully separated from the bottom reservoir.
10. The bottom filter is closely examined for discoloration or wet spots due to released liquid. Any discoloration or spotting of the bottom filter paper constitutes a failure of the sample to pass the test.

Variations of this procedure and detection arrangement are possible. For instance, a perforated plate could be used in place of the separation screen, or a fine screen could be used in place of the top filter. These variations were not evaluated during the work, however.

Generally, some liquid is absorbed from the sample and discolors the top filter paper which is in contact with the sample. It is important that this filter be a thin filter with minimum liquid retention.

Evaluation of Tester

Evaluation of the tester required the investigation of sensitivity to test conditions and consideration of how test conditions related to landfill conditions. The tester was initially designed to test stabilized hazardous wastes. Since waste streams are not typically consistent in their chemical and physical properties, all testing reported here was done using surrogate wastes. The surrogate wastes were produced by blending motor oil with water at concentrations of 10 and 30% by weight. The stabilization agent used for all tests was a cement-kiln dust produced in Colorado. The term *mix ratio* is used to indicate the weight ratio of cement-kiln dust to surrogate waste.

Test duration and pressure are the two major test variables. The test pressure selected was chosen based on the pressure that material may be subjected to near the bottom of a landfill. The pressure was assumed to be equal to the height of the landfill times the density of the waste. Typically, landfills are less than 20 m (66 ft), with densities ranging from 1280 kg/m^3 (80 lb/ft^3) to 1920 kg/m^3 (120 lb/ft^3), producing an average of 1600 kg/m^3. This results in a pressure of 320 kPa (46.4 psi), which was rounded to 345 kPa (50 psi). This somewhat arbitrarily selected pressure was used as the test pressure.

Wastes in landfills will be subjected to these pressures for many years; direct simulation of this condition is not feasible. Initial compression tests indicated that the waste density as indicated by piston travel increased significantly during the first 1 to 2 min, and only gradually increased after this initial compaction. On the basis of these observations a test duration of 5 min was selected.

Also of concern is the sample size and sample age. When formulas for stabilization of wastes at CWM were evaluated, sample mixtures of 250 to 500 g are usually prepared, therefore the required sample size should be less than 250 g. Using readily available hardware components resulted in a sample holder of approximately 0.21 L, which can hold approximately 100 to 200 g of uncompacted sample. Based on the above two criteria, a sample size of 100 g was selected.

The question of how long to cure the sample before testing is subject to considerable debate. Previous experience indicates that a minimum of 24 h is required between mixing and testing. In addition to cure time, the curing conditions are also a major factor affecting the cure rate. These effects are discussed in the results section.

The last parameter to be evaluated was the liquid release detection system. A simple, reliable and sensitive method of determining if any liquid had been forced from the sample under the test conditions was needed. The filter-paper staining method appears to meet the requirements of the test. Even for colorless liquids, a single wet spot on the bottom filter paper can readily be seen. Measurement of the increase in weight of the bottom filter paper due to the stain failed to produce consistent results. In general, it appeared that detectable stains increased the weight of the filter paper by 2 to 5 mg or 0.002% to 0.005% of the sample weight.

Effect of Test Duration

As mentioned previously, the goal was a 5-min test. The shorter the test cycle, the more tests that can be performed in a given time period. Since this test is envisioned for use only during the development of a treatment recipe and not as a control check during processing, a test time longer than 5 min can be tolerated.

The results of the test duration study using well-cured (greater than 16 weeks) samples are given in Table 1. For the samples tested, there appears to be no difference between a test of 5 or 200 min duration. At 1440 min (24 h), both mix ratios for the 10% oil waste failed, while the 30% oil waste passed. It must be pointed out that the failure means that as little as one drop of liquid (amounting to less than 0.005% of the sample weight) was released. From a practical standpoint, a 24-h test duration is not reasonable. Since there seems to be no major difference between the 200 min and 5 min duration, the latter seems like a reasonable test duration.

Effect of Test Pressure

The selected test pressure is 345 kPa (50 psi). The maximum test pressure evaluated was 620 kPa (90 psi) since many smaller air compressors cycle on at about 620 kPa, so 620 kPa is frequently the maximum continuous pressure available when using "shop air" to operate the tester. The results are given in Table 2 for samples tested at various pressures. The first two data sets show that a 50% increase in pressure from 345 to 515 kPa did not affect the results. Sample 305M, however, did fail when the pressure was almost doubled by raising it to 620 kPa. Since 620 kPa considerably exceeds expected landfill pressures, failure of the

TABLE 1—*Effect of test duration on liquid release.*[a]

Sample Code	Weight % Oil	Mix Ratio	Cure Time, days	Test Duration, min	Result
102M	10	0.8	117	5	pass
102M	10	0.8	117	200	pass
102M	10	0.8	117	1440	fail
103M	10	1.0	117	5	pass
103M	10	1.0	117	200	pass
103M	10	1.0	117	1440	fail
304M	30	0.8	120	5	pass
303M	30	0.8	120	1440	pass

[a] All tests performed at 345 kPa (50 psi).

sample at this pressure should not be a major concern. Sample 317M provides some insight into the sensitivity of the tester. After seven days cure, the sample passed at the reduced pressure of 170 kPa and failed at 345 kPa. After seven more days of curing, the sample passed the test at 345 kPa. The selected pressure of 345 kPa seems to be a reasonable value.

Effect of Cure Time

The age of the sample (cure time) is critical when the properties of stabilized wastes are measured. Most stabilization processes use cement, cement-kiln dust, fly ash, or lime, or a combination of two or more. Cure times for these types of formulations are in the range of weeks, not hours, at typical ambient temperatures. Some samples used in this study were cured at ambient temperature; others were cured at 37.7°C (100°F). All samples were cured in closed containers.

Table 3 presents the results obtained from samples cured at room temperature. Sample 108M failed after both 7 and 14 days curing. When the mix ratio was increased to 1.2 for Sample 109M, the 7-day cure still failed the test, but the 14-day passed. In the case of the 30% oil samples, mix ratios that failed after 14 days passed when the samples were cured

TABLE 2—*Effect of test pressure on liquid release.*[a]

Sample Code	Weight % Oil	Mix Ratio	Cure Time, days	Test Pressure, kPa	(psi)	Result
303M	30	0.8	110	345	(50)	pass
304M	30	0.8	120	515	(75)	pass
305M	30	0.8	120	620	(90)	fail
310M	30	1.1	53	345	(50)	pass
310M	30	1.1	53	515	(75)	pass
108M	10	1.1	14	170	(25)	pass
108M	10	1.1	14	345	(50)	fail
317M	30	1.2	7	170	(25)	pass
317M	30	1.2	7	345	(50)	fail
317M	30	1.2	14	345	(50)	pass

[a] All tests were 5 min in duration.

TABLE 3—*Effect of cure time on liquid release.*

Sample Code	Weight % Oil	Mix Ratio	Cure Time, days	Result
108M	10	1.1	7	fail
108M	10	1.1	14	fail
109M	10	1.2	7	fail
109M	10	1.2	14	pass
317M	30	1.2	7	fail
317M	30	1.2	14	pass
307M	30	1.1	7	fail
310M	30	1.1	53	pass
315M	30	1.0	7	fail
315M	30	1.0	14	fail
306M	30	1.0	112	pass
313M	30	0.8	7	fail
313M	30	0.8	14	fail
303M	30	0.8	110	pass

[a] All tests performed at 345 kPa and a test duration of 5 min.

for more than 100 days. Note that the chemical stabilization was still effective even when the waste contained 30% organic material (oil).

Comparison of Compression Tester Results

Currently, CWM uses the Paint Filter Test and bearing strength (as measured using a pocket penetrometer) as criteria for determining acceptable mix ratios for waste stabilization. Table 4 presents the testing results of several mix ratios for a generic waste with 10% oil

TABLE 4—*Comparison of liquid release test to other methods for evaluating stabilized waste.*[a]

Sample	Weight % Oil	Mix Ratio	Paint Filter Test		Bearing Strength[b] tons/ft²				Liquid Release Test,[c] days				
			5 min	1 h	1 h	1 day	7 days	14 days	1	2	3	7	14
1	10	0.7	o	o	...	...	0.8	1.3	o	o	o	o	o
2	10	0.8	o	o	...	...	1.8	2.2	o	o	o	o	o
3	10	0.9	o	P	...	0.5	2.4	3.0	o	o	o	P	P
4	10	1.1	P	P	0.5	1.8	>4.5	>4.5	o	o	P	P	P
5	10	1.2	P	P	0.5	2.9	>4.5	>4.5	o	o	P	P	P
6	10	1.3	P	P	0.5	4.1	>4.5	>4.5	o	P	P	P	P
7	30	0.7	o	o	...	...	0.5	0.5	o	o	o	o	o
8	30	0.8	o	o	...	...	0.5	1.3	o	o	o	o	o
9	30	0.9	P	P	0.5	0.5	1.3	1.9	o	o	o	o	o
10	30	1.0	P	P	0.5	0.5	1.9	3.5	o	o	o	o	P
11	30	1.1	P	P	0.5	1.0	3.0	3.5	o	o	o	P	P
12	30	1.2	P	P	0.5	1.5	4.1	4.2	o	o	P	P	P
13	30	1.3	P	P	0.5	1.8	>4.5	>4.5	o	P	P	P	P

[a] o = Fail; P = Pass; 1 ton/ft² = 95.7 kPa.

[b] Bearing strength measured using a Model CL-700 pocket penetrometer supplied by Soiltest, Inc. A dash indicates the value was less than the minimum detection level of 0.25 ton/ft².

[c] All tests performed at 50 psi (345 kPa) and 5 min duration.

TABLE 5—*Summary of method comparisons.*

	Minimum Mix Ratio as Determined by		
Weight % Oil in Stabilized Sample	Paint Filter Test	Bearing Strength	Liquid Release Tester
10	0.9	1.0[a]	0.9
30	0.9	1.1	1.1

[a] Interpolated value.

and one with 30% oil. Three different parameters were measured at each mix ratio: the ability to pass the Paint Filter Test, the bearing strength, and the ability to pass the liquid release test. The Paint Filter Test was performed at both 5 and 60 min after mixing. The samples were then maintained at 38°C (100°F) and analyzed over a 14-day period. Except where different requirements are imposed, 96 kPa (1 ton/ft^2) bearing strength is the minimum desired strength within 24 h for CWM operations.

We now have three different criterion for the minimum mix ratio: paint filter test, bearing strength, and liquid release test. How do they compare? The results are summarized in Table 5. For both wastes, the minimum mix ratio as determined with the liquid release tester is within the range of values determined by either the paint filter test or the bearing strength. It must be remembered that the paint filter test was performed 1 h after mixing, the bearing strength after 24 h, and the liquid release test after seven days. The liquid release tester is intended to be used in the laboratory for the development of stabilization formulations and is not intended for production quality control because of the minimum seven-day cure time.

Conclusions

The initial test data presented in this paper indicate that the test apparatus has achieved its goals. The equipment quickly produces consistent results that agree with other test methods used by CWM to evaluate stabilized wastes. The equipment is easy to use and relatively inexpensive to build and operate.

CWM received U.S. Patent Number 4,697,457 on the liquid release tester, 6 Oct. 1987. Any questions or comments should be referred to the author.[2]

[2] Author's telephone number at Chemical Waste Management is 312/513-4500.

Maria Neuwirth,[1] *Randy Mikula,*[2] *and Peter Hannak*[1]

Comparative Studies of Metal Containment in Solidified Matrices by Scanning and Transmission Electron Microscopy

REFERENCE: Neuwirth, M., Mikula, R., and Hannak, P., **"Comparative Studies of Metal Containment in Solidified Matrices by Scanning and Transmission Electron Microscopy,"** *Environmental Aspects of Stabilization and Solidification of Hazardous and Radioactive Wastes, ASTM STP 1033,* P. L. Côté and T. M. Gilliam, Eds., American Society for Testing and Materials, Philadelphia, 1989, pp. 201–213.

ABSTRACT: Solidification/stabilization processes are widely used for the disposal of metal-containing industrial/hazardous wastes. There is little understanding of the mechanisms of metal containment in cementitious matrices. Leaching studies characterize the mechanism by which contaminants are released from the matrix, but give little information on contaminant-matrix interaction and the partitioning of contaminants within different phases of the solid structure. Recent studies have shown that different mechanisms are involved in stabilizing metals such as zinc and mercury [*1,2*] or cadmium and lead [*3*].

In this study, selected concentrations of metal salt solutions (cadmium, chromium, and lead) were stabilized using portland cement as additive. Scanning electron microscopy (SEM), energy dispersive X-ray analysis (EDX), and backscattered electron detection were performed on fractured and polished surfaces of specimens. Transmission electron microscopy (TEM) and X-ray microanalysis were carried out on ground particles of approximately 1 μm in size from the same specimens. The results indicate that different containment mechanisms are responsible for stabilization of the metals used and that they provide additional information on potential release of contaminants. The study shows the importance of correlating SEM and TEM analyses as well as other pertinent techniques in order to understand more fully the mechanism of waste containment.

KEY WORDS: hazardous waste, solidification, stabilization, contaminants, leaching, transmission electron microscopy, scanning electron microscopy

Although solidification/stabilization (S/S) of hazardous wastes has increased over the past decade, supporting research activities have not kept abreast of the practical applications. Several essential theoretical and practical questions are still unanswered due to the complexity of the hazardous waste management research field.

Factors such as the variety of wastes and contaminants, the wide range of concentration levels of contaminants, and the use of several binder/additive systems with a selection of containment mechanisms contribute to the complexity of the research.

The methodology to address containment and leaching properties also has not always been readily available and, therefore, development or adaptation of appropriate methods was required. Some of the methods, models, and theoretical considerations used by the

[1] Section head of Electron Microscopy Services and Physical Chemical Processes, respectively, Alberta Environment, Alberta Environmental Centre, Bag 4000 Vegreville, AB, TOB4LO Canada.

[2] Research scientist, Energy, Mines, and Resources Canada, Coal Research Centre, Bag 1280 Devon, AB, TOC1EO Canada.

nuclear industry (mainly for glass-type waste forms) were adopted for characterization of toxic wastes.

Research efforts were focused on leaching studies since these results can be related to water standards. Regulatory goals were also served by these leaching tests, some of which characterize the equilibrium conditions while others the leaching kinetics. Mathematical modeling of the rate of leaching has not been successful because of the complex interactions occurring within the waste form that are also affected by the site specific conditions. Furthermore, dissolution problems related to the chemistry of binders have also not been addressed.

The concentration of contaminants in the leachates could be regarded as an integrated value and hence does not reflect the phase of origin within the waste form. In other words, contaminants are considered to be evenly distributed in a homogenous structure. This is an approximation, which may be acceptable in most glassy nuclear waste forms, but not in cementitious matrices. It is well known that the majority of containment methods for toxic wastes use cement or pozzolanic binders which yield products of hydration with several chemically and physically different phases [*4,5*]. A comprehensive review summarizes techniques and binder systems for S/S processes [*6*]. Research was also directed to inventory potential mechanisms of containment [*7*].

In order for S/S processes to be improved, the affinity of contaminants for any particular phase of a waste form has to be understood. It is believed that partitioning of contaminants among different phases is due to differences in the containment mechanisms, reflecting a variety of interactions of chemically different species and binder components. Control of simple parameters such as moisture content, pH, additive type, or ratio could result in a product with better characteristics. By improvement of the process, contaminants may be more tightly bound physically or chemically to binder phases and would not lend themselves to leaching. Since leaching methods and the analysis of leachates do not show directly the origin of contaminants within the waste form, identification of phases with reduced leaching availability requires additional tools which are capable of showing the surface of waste forms, internal structure, and changes in the structure, including binding or release of contaminants or both. Available techniques such as electron microscopy and spectroscopic methods used in surface characterization have been presented in the literature [*8*].

Summary of Research to Date

At the Institute of Waste Research, Amersfoort, the Netherlands, electron microscopy was used to study inorganic metal contaminants in four commercial S/S products although no identification of waste contaminant bonds was attempted [*9*].

Transmission electron microscopy energy dispersive X-ray (TEM-EDX) analysis and scanning electron microscopy (SEM) were used in the analysis of calcium hydroxide and calcium-silica-hydrate particles, separated from portland cement, with emphasis on the calcium:silicon ratios [*10*].

Binding properties of calcium sulfoaluminates (ettringite) to iron, chromium, and lead compounds were studied in synthetic minerals by X-ray diffraction analysis and SEM. Strong binding of ettringite to chromium and lead was reported [*11*]. Distribution of ethylene glycol and parabromophenol in hydrated portland cement was also studied by SEM, and the presence of brominated phenol in calcium silica hydrate gel phase was observed [*12*].

The importance of microstructure was demonstrated not only in metal stabilization but also in metal leaching for mercury and zinc from cement/silicate waste forms using SEM scanning electron microscopy and EDX spectroscopy. The progress of leaching was observed through structural changes. Leaching of contaminants was correlated with a reduction in calcium content of the waste form, and the effect of zinc on the internal structure was also observed [1,2].

Objectives and Approach

In the present study, methods associated with the use of electron microscopy were selected and applied to cement/silicate type S/S processes. The objectives of this research were as follows:

Stage 1: observation of contaminant distribution in cement/silicate waste forms using selected inorganic contaminants.
Stage 2: characterization of leaching properties of cement/silicate waste forms and identification of phases available or unavailable for leaching.

The novel elements of the approach include combined applications of TEM and SEM techniques to S/S waste forms, and direct observation and possible quantitation of waste concentration rather than inferring concentrations from changes in calcium:silicon ratios or leaching rates.

This paper presents the progress of Stage 1 to date.

Materials and Methods

Experimental Design

Table 1 summarizes the composition of the synthetic waste specimens. The contaminants chromium, cadmium, and lead were selected because of their frequent presence in industrial wastes and their toxic properties. Two binder systems were selected. Portland cement is extensively used as a "traditional" binder, while cement kiln dust is less expensive and a readily available by-product.

Type N sodium silicate (SiO_2/Na_2O 3.22) was applied in a ratio used previously [13]. Several levels of contaminants were used in order to study the effect of concentration on the containment and to comply with the minimum detection limits of contaminants.

Water-to-binder ratios were selected to provide sufficient moisture required for the hydration process. Hydroxide precipitates were prepared to simulate industrial precipitation and to alleviate problems related to volatilization of metal salts under the experimental conditions (high vacuum, heat) as well as to study potential differences between interactions of salts and hydroxides. Blanks containing no contaminants (A3, B3) were also prepared. Specimens were also molded for Stage 2 leaching experiments.

Specimen Preparation

Table 2 summarizes the source of raw materials. Synthetic waste solutions (Al to A4, B1 to B4) were prepared by dissolving salts of contaminants in distilled water. In some (A5, B5), the pH was adjusted with sodium hydroxide (6N) to precipitate hydroxides. Other specimens (A6 to A8, B6 to B8) were prepared by direct addition of metal hydroxide precipitate. Specimens were mixed in a Hobart mixer in accordance with the ASTM Method for Mechanical Mixing of Hydraulic Cement Pastes and Mortars of Plastic Consistency (C 305-82) and as described previously [13]. The specimens were molded in 4.5-cm-diameter by 7.4-cm-high plastic molds and cured under 98% relative humidity at room temperature for a minimum of eight days.

TABLE 1—*Composition of synthetic wastes and additives (%).*

Raw Materials Specimen	Normal Portland Cement	Cement Kiln Dust	Sodium Silicate (1:1)	Water	Contaminant: Solution, Cd, Cr, Pb, 0.01 *M*	Contaminant: Solution, Cd, Cr, Pb, 0.04 *M*	Contaminant: Solution, Cd, Cr, Pb, 0.4 *M*	Contaminant: Precipitate, Cd, Cr, Pb, Weight %
A1	72.0					28.0		
A2	72.1				27.9			
A3	72.0			28.0				
A4	70.4						29.6	
A5	71.9						28.9[a]	
A6	71.0			27.1				1.9 Cd
A7	70.7			27.2				2.0 Pb
A8	69.4			28.6				1.9 Cr
B1		60.9	6.4			32.7		
B2		60.7	6.6		32.7			
B3		61.0	6.8	32.2				
B4		68.7	7.0				34.3	
B5		58.2	5.1				36.7[a]	
B6		58.7	11.7	27.4				2.0 Cd
B7		58.7	11.7	27.4				2.2 Pb
B8		57.6	11.7	28.8				1.9 Cr

[a] Sludge—pH adjusted to 8.5.

SEM and Energy Dispersive Spectrometry

The electron microscope used in this study is a Hitachi X-650 equipped with both wavelength and energy dispersive spectrometers. The energy dispersive spectrometer [Tracor 30 mm^3 Si(Li) detector] was used to produce all of the X-ray spectra shown in the figures. The incident beam current was approximately 0.2 nA at 35 keV unless otherwise specified. In order to more clearly illustrate the presence of some of the less abundant elements, the X-ray spectra are presented on a logarithmic scale. The X-ray emission which is excited by the incident electron beam is characteristic for each of the elements present in the specimen and by focusing the incident electron beam it is possible to determine compositional differences as a function of region in the specimen. Ideally, with a flat specimen surface perpendicular to the incident electron beam, this X-ray information can be made quantitative. With the cement specimens used in this study, there is always a certain surface "roughness" which makes this quantification difficult, especially when examining micrometre-sized occlusions or concentrations of impurities.

TABLE 2—*Source of raw materials.*

Material	Source
portland cement, type II	Genstar, Edmonton, Alberta
cement kiln dust	Western Canada Fly Ash, Sherwood Park, Alberta
sodium-silicate (3.22)	National Silicate Toronto, Ontario
cadmium: $CdCl_2 \cdot 2\frac{1}{2} H_2O$	BDH (Analar. Grade)
chromium: $Cr(NO_3)_3 \cdot 9H_2O$	BDH (General Purpose Reagent)
lead: $Pb(NO_3)_2$	Fisher Scientific

Specimens were prepared for observation by fracturing and carbon coating. The carbon coating is required in order to prevent specimen charging effects and to minimize X-ray attenuation for the acquisition of compositional information. Although gold coatings are more efficient for the production of micrographs, they attenuate the X-rays to a much greater extent and therefore were not used in this study. The fracturing generally produces an angled surface which must then be tilted during observation in order to present a reasonably flat surface perpendicular to the electron beam. The fracturing was done at room temperature and found not to be observably different in the electron microscope from fractures done at liquid nitrogen temperature. This indicates that for the present study and impurity concentrations, fracturing is not preferential along the impurity inclusions. For quantitative work using EDX techniques, frozen hydrated observation would be required to prevent specimen changes in the electron microscope. Loss of hydrated water and migration of ions can be minimized by frozen hydrated observation, and although comparison of frozen hydrated (uncoated) and room temperature (conventionally prepared) specimens showed that these effects were unimportant in the present study, quantitative analysis would demand frozen hydrated observation.

By using the appropriate detector, the electron microscope can be used to determine relative elemental or mineral abundances, which make it possible to identify material of different composition. For instance, a backscattered electron (BSE) detector will produce a brighter image from particles composed of higher Z (atomic number) elements. Therefore, concentrations of the higher Z impurities such as cadmium, chromium, and lead will appear brighter than the calcium, silicon, and aluminum in the cement matrix. By correlating the observed brightness to a particular compound, it is then possible to determine qualitatively the relative amounts of impurities in a given cement specimen. Furthermore, these data can be collected and quantified over a large number of fields of view by the appropriate interface of an image analyzer to the electron microscope.

A more direct and accurate determination of impurity inclusions can be made by acquiring X-ray dot maps of elements of interest. This involves rastering the electron beam across the specimen and correlating X-ray emissions with the position of the electron beam. This can provide more accurate compositional information than the indirect method of measuring the brightness in a BSE detector, especially in the case of these cement specimens, in which the BSE image brightness can be affected by surface roughness as well as by compositional differences.

Scanning Transmission Electron Microscopy (STEM) and EDX Analysis

The molded and cured specimens were broken into small fragments, which were ground and suspended in hexane. The heavier particles were allowed to settle for a few minutes, after which a drop of hexane containing the lighter particles still in suspension was applied to a Formvar-coated, 200-mesh copper grid. The hexane was allowed to evaporate and random particles of approximately 1 μm were analyzed without coating in a Hitachi H600 electron microscope. Analyses were done at 75 keV for 300 s. Spectra were acquired from each specimen, and automatic files were set up for ratios (to silicon) and weight percent.

Results and Discussion

Specimen A5 (with 1.25% cadmium, 0.58% chromium, and 2.32% lead by weight) was chosen to illustrate the potential of scanning SEM to monitor directly the morphology of the cement matrix and the concentrations of various stabilized impurities. Localization of the impurities was difficult to establish in Specimens A1 to A4, which had significantly lower impurity concentrations. This could be because at the lower concentrations, the impurities have not saturated the matrix and begun concentrating at certain sites or possibly because the sites of concentration themselves are too small to be resolved under the present electron beam conditions. The low-energy X-ray peak labeled as lead (Mα 2.35 keV and Mβ 2.44 keV) in the X-ray spectra has an interference from sulfur (Kα 2.31 keV and Kβ 2.47 keV), and, therefore, the height of this peak is not necessarily representative of the lead concentration. The higher-energy lead peaks (Lα 10.55 keV and Lβ 12.62 keV) are much more useful in this study. The X-ray spectra of a 10 μm^2 area of Specimens A1, A2, and A5 have been expanded in the region between chromium and lead in order to better illustrate the concentration differences (Fig. 1). Although the presence of impurities can be illustrated in large-area X-ray spectra, at the relatively long acquisition times the concentrations or inclusions of impurities are not evident under these experimental conditions until the concentrations reach that in Sample A5.

Figure 2 shows a conventional scanning electron micrograph of Specimen A5 with two areas marked for X-ray analysis. The spectra corresponding to these areas are shown on a logarithmic scale in the same figure. Although the X-ray spectra clearly show that Region

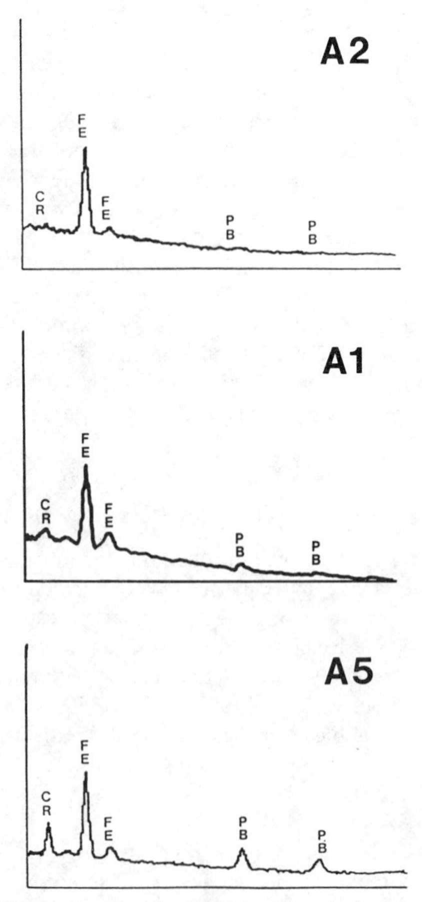

FIG. 1—*X-ray spectra of Specimens A2, A1 and A5 in order of increasing concentration (35 keV, 0.2 nA, 5000 s acquisition time).*

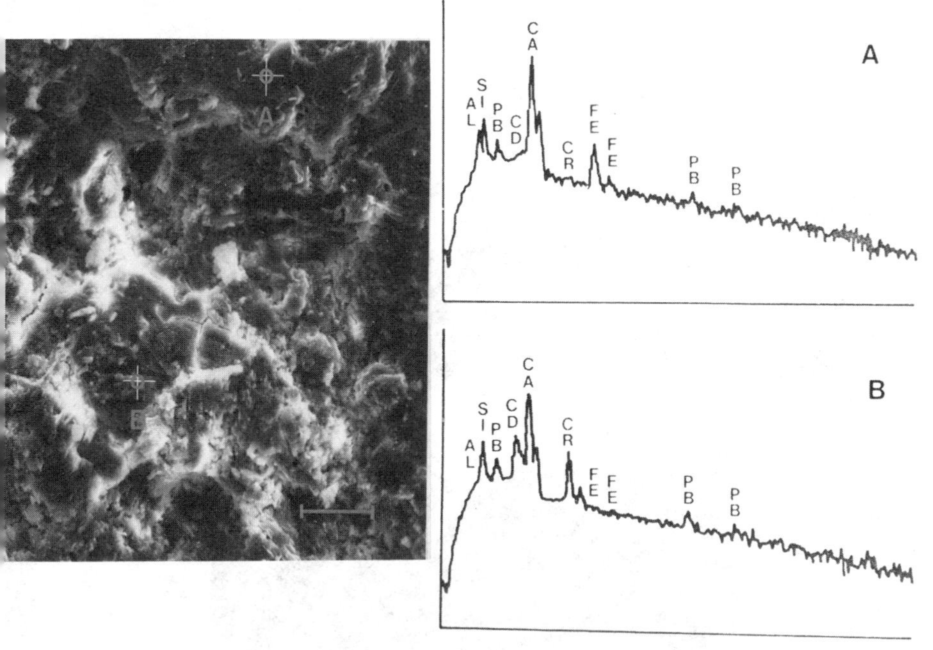

FIG. 2—*Specimen A5 secondary electron image and X-ray spectra taken at the points indicated (35 keV, 0.2 nA, 2000 s acquisition time, long intensity scale, bar = 30μm).*

B has a concentration of cadmium and chromium, it is not possible to determine that these areas represent different chemical compositions from their morphology alone. However, the BSE detector image in Fig. 3 clearly shows a brighter image in the region of the impurity (that is, the higher Z region). The X-ray dot map of the same field of view illustrates the utility of this technique in accurately delineating the region of impurity concentration. The dot maps in Figs. 3 and 4 show that the impurity inclusions are in the micrometre range and perhaps smaller and that the cadmium and chromium sometimes concentrate separately.

The resolution of the electron microscope image can be reduced to 30 to 60 Å with proper specimen preparation and coatings, but the X-ray resolution is in the micron or less range due to the penetration of the electron beam into the specimen. This present resolution limitation means that it is not possible to determine if the inclusions or concentrations of impurities are occurring after saturation of the impurity throughout the specimen or simply as a result of the growth of discrete sites to a size which can be resolved. Particles containing lower concentrations of impurity (A1 to A5 and B1 to B5) do not provide enough metal to be consistently detected by TEM-EDAX. In Specimens A6 to A8 and B6 to B8, the appropriate metal is also not detected in all particles examined. The results from the scanning transmission electron microscope (STEM) support the contention that the impurities concentrate at various sites regardless of the concentration, and it is the growth of the concentration at these discrete sites which becomes resolvable in the SEM. If the impurities were evenly distributed, then it would be expected that these elements would begin to appear consistently in the small 1-μm-diameter particles investigated in the STEM. A discrete concentration of impurity at every impurity level would explain the variation of impurity signal in the STEM since only a relatively small number of particles was investigated.

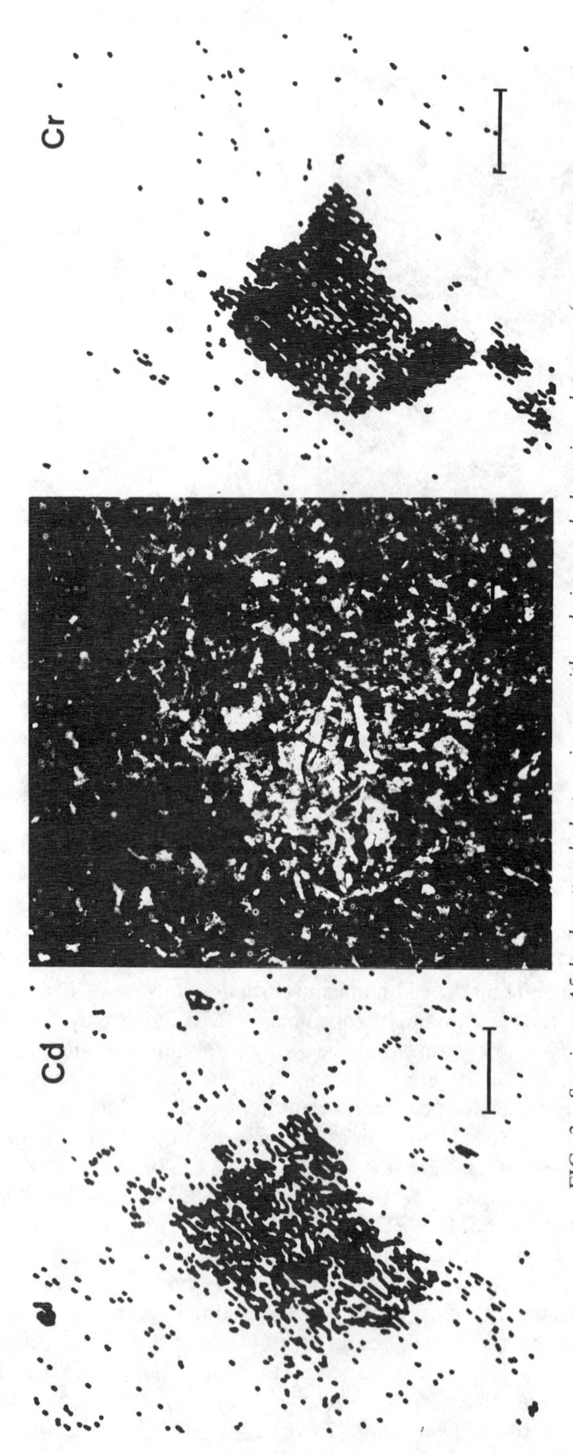

FIG. 3—*Specimen A5 backscattered electron image with cadmium and chromium dot maps (same field of view as Fig. 2, 35 keV, 0.2 nA, bar = 30 μm).*

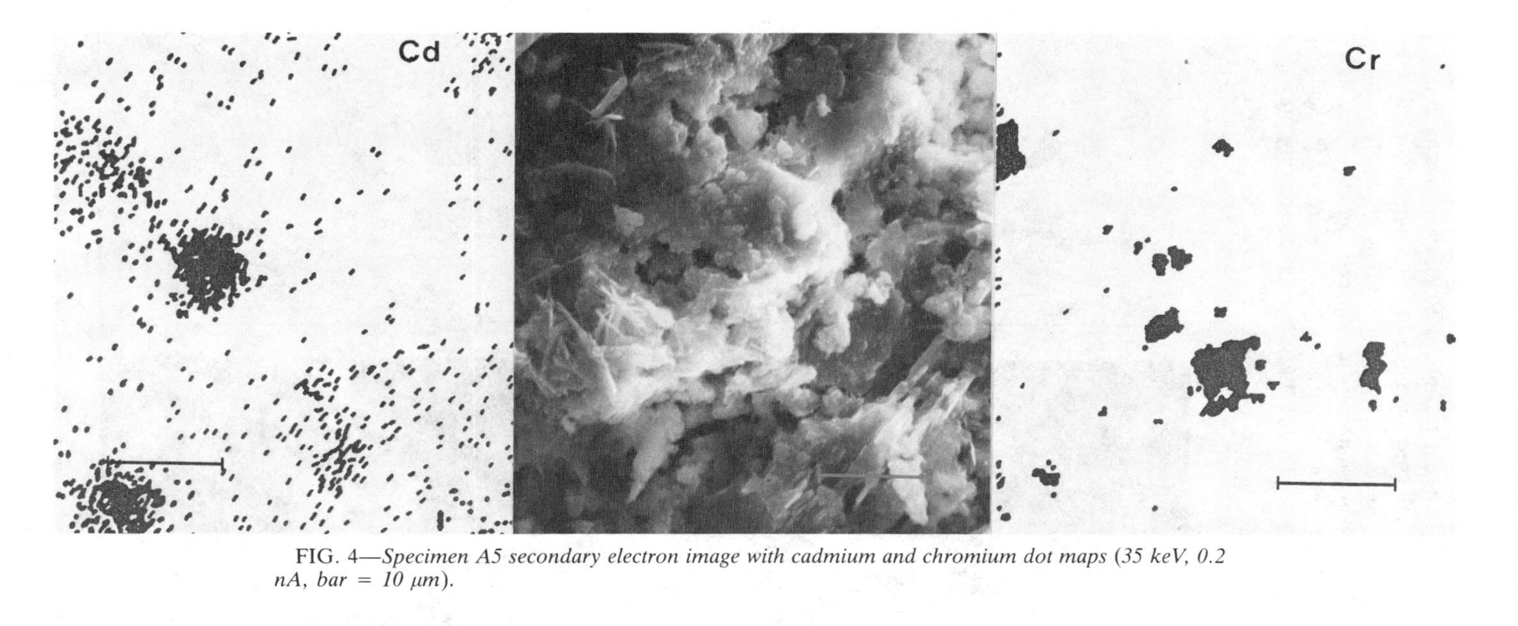

FIG. 4—*Specimen A5 secondary electron image with cadmium and chromium dot maps (35 keV, 0.2 nA, bar = 10 μm).*

Some of the qualitative observations which have been made about these samples are also relevant to the concentration of these impurities and to the mechanism of their containment. Generally, the cadmium and chromium appear to be concentrated together (see Fig. 3) and separate from the lead. This is in agreement with the observations of El Korchi et al. [3], who postulated different containment sites for lead and cadmium. Furthermore, it is also generally true that the aluminium and iron components are low in regions where the impurity level is high (Figs. 2 and 5). These observations at present are only qualitative, based on

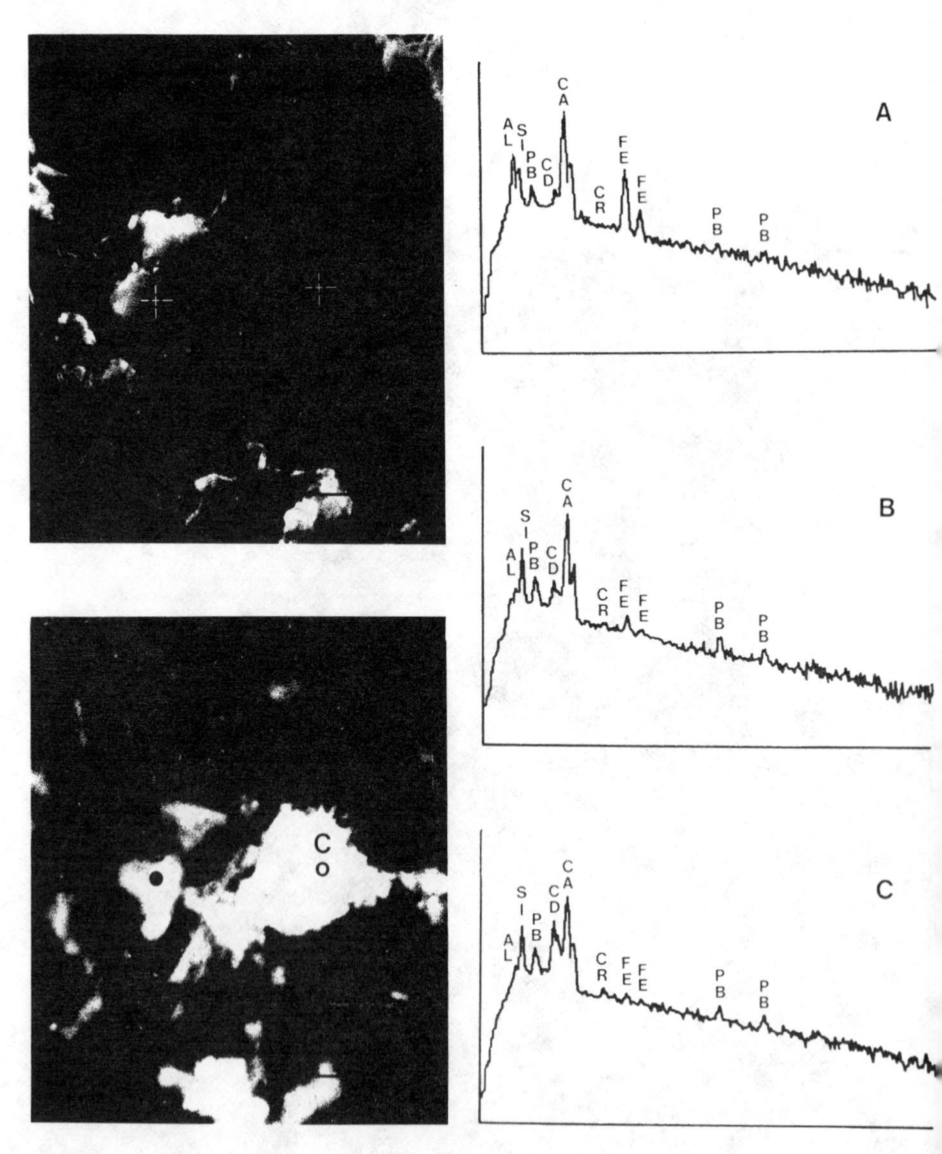

FIG. 5—*Specimen A5 secondary electron and BSE images with X-ray spectra acquired at the spots marked (35 keV, 0.2 nA, 2000 s acquisition time, long intensity scale, bar = 5 μm).*

the observation of a few hundred occlusions, but could have important implications in understanding the mechanisms of waste containment and leaching.

At the impurity concentrations found in Specimen A5, it will be possible to use automatic image analysis in order to quantify these observations, and to estimate the size distributions of the inclusions both before and after leaching. An example of this method is shown in Fig. 6, which shows Specimen A6 (1.9% cadmium) imaged with the conventional secondary electron signal as well as with the BSE signal. Either the backscattered signal or X-ray dot map information makes it possible to quantify the size distribution of the occluded cadmium impurity by using simple image analysis technique. In this case it is easy to differentiate the bright cadmium signal from the cement matrix and to determine that in this field of view, cadmium-rich areas constitute 2.8% of the matrix. By scanning many fields of view and folding in composition and density information, it would be possible to obtain statistically significant information about waste containment both before and after leaching. However, true quantitative determination of the cadmium impurity via EDX technique would require an appropriate quantitation routine (Phi-Rho-Z or Bence-Albee) and frozen hydrated observation in order to minimize specimen changes due to electron beam exposures or vacuum effects or both.

A total of 234 particles were analyzed by STEM EDX. Table 3 summarizes the calcium:silicon ratios for all A and B specimens analyzed. There is great variability in calcium:silicon ratios between particles as well as between specimens. The mean calcium:silicon ratio for the cement specimens (A) is 1.56. This value is in agreement with the calcium:silicon ratios of cements reported earlier [*10,14,15*]. The addition of metal contaminants to the cementitious matrix does not appear to change the calcium:silicon ratio appreciably from the blank (Specimen A3) or from the values reported earlier. Although there is variability in the calcium:silicon ratios, there is no correlation between ratios and waste concentrations

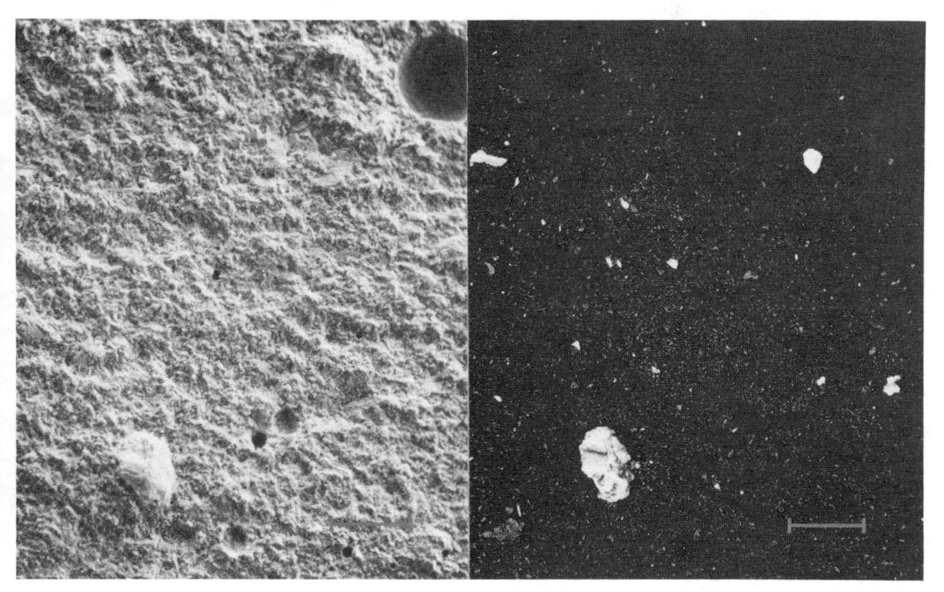

FIG. 6—*Specimen A6 secondary electron and BSE images showing cadmium inclusions (35 keV, 0.2 nA, bar = 100 μm).*

TABLE 3—*Mean calcium:silicon ratios of particles analyzed with STEM-EDAX.*

Specimen No.	Number of Particles Analyzed	Calcium:Silicon Ratio
A3 (blank)	23	1.64
A1	24	1.09
A2	22	1.46
A4	12	1.16
A5	29	2.01
A6	15	1.96
A7	11	1.41
A8	6	1.73
TOTAL	119	1.56
B3 (blank)	12	0.45
B1	17	0.62
B2	11	0.69
B4	10	1.41
B5	11	0.87
B6	16	0.81
B7	9	1.19
B8	6	0.75
TOTAL	80	0.87

and between ratios and waste forms prior to solidification. The mean calcium:silicon ratio for B3 (blank) is 0.45, while for the waste-containing B specimens it is 0.87, showing a significant difference. However, there is no correlation between the waste concentration and the calcium:silicon ratio, which varies from 0.62 to 1.41. From the present STEM data, it is not possible to characterize the particles analyzed due to the variability of the calcium:silicon ratios and the few numbers of particles analyzed from each specimen.

Conclusions

Microscopic techniques (SEM and TEM) combined with electron microprobe compositional analysis can be a powerful tool for investigating the containment mechanism of waste stabilized in cement matrices. At certain concentrations, the contaminants can be observed directly and their containment sites characterized. This direct observation of the waste metal inclusions in the cement matrix and the subsequent leaching behavior of the waste has a clear advantage over inferences about the containment site when studies are done via analysis of leachates. The investigations using electron microscopy techniques showed similarities in affinity of cadmium and chromium to particular containment sites, while lead preferred different containment sites. Although the present study showed the feasibility of techniques used, the chemical and morphological characterization of containment could not be targeted. Several limitations of the method were considered or identified. Low concentrations of contaminants (0.01 to 0.04 *M*) could not be discerned from that of contaminants present in the binders. Another limitation was the potential volatility or carry over of some metals/metal salts as well as the potential of a structural rearrangement in the matrix under high vacuum. Future work will address these limitations and involve quantification of elemental concentrations and their associations with each other and with certain components in the cement matrix. Automated image analysis is a tool which can be applied very efficiently to these systems, to quantify some of the hitherto qualitative statements which have been based on the relatively small numbers of sites examined manually.

Acknowledgment

We wish to thank Dr. Albert Liem and Dr. Malcolm Wilson for their continuing support in the research published in this paper. We also acknowledge Diana Chau, Rosemary Harris, and Arlene Oatway for their enthusiastic laboratory support. Special thanks to Dr. H. Eaton of Louisiana State University for introducing Dr. Neuwirth to the technical aspects of electron microscopy in hazardous waste management research.

References

[*1*] Poon, C. S., Peters, C. J., and Perry, R., "Mechanisms of Metal Stabilization by Cement-Based Fixation Processes," *The Science of the Total Environment,* Vol. 41, 1985, pp. 55–71.
[*2*] Poon, C. S., Clark, A. I., Peters, C. J., and Perry, R., "Mechanisms of Metal Fixation and Leaching by Cement-Based Fixation Processes," *Waste Management and Research,* Vol. 3, 1985, pp. 127–142.
[*3*] El Korchi, T., Melchinger, K., Gress, D., and Bishop, P., "Evaluating the Potential for Stabilizing Two Heavy Metals in Portland Cement Paste," *Proceedings,* American Institute of Chemical Engineers National Conference, Boston, MA, August 1986.
[*4*] Lea, F. M., *The Chemistry of Cement and Concrete,* 3rd ed., Edward Arnold, London, 1970.
[*5*] Pojasek, R. B., *Toxic and Hazardous Waste Disposal, Volume 1, Processes for Stabilization/Solidification,* Ann Arbor Science, Ann Arbor, MI, 1978.
[*6*] "Survey of Solidification/Stabilization Technology for Hazardous Industrial Wastes," EPA 600/2-76-056, U.S. Army Engineer Waterways Experiment Station, Vicksburg, MS, 1979.
[*7*] Malone, P. and Larson, R., "Scientific Basis of Hazardous Waste Immobilization," *Hazardous and Industrial Solid Waste Testing: Second Symposium, ASTM STP 805,* R. A. Conway and W. D. Gulledge, Eds., American Society for Testing and Materials, Philadelphia, 1983, pp. 168–177.
[*8*] Mendel, J. E. and Harker, A. B., "Workshop on the Leaching Mechanisms of Nuclear Waste Forms," PNL-4810, UC-70, Battelle Pacific Northwest Laboratory, Richland, WA, 1982.
[*9*] "Comparative Investigation on Four Immobilization Techniques," Publication No. 39, Institute for Waste Research, Amersfoort, the Netherlands, 1979.
[*10*] Lachowski, E. E. and Diamond, S., "Investigation of the Composition and Morphology of Individual Particles of Portland Cement Pastes," *Cement and Concrete Research,* Vol. 13, 1985, pp. 177–185.
[*11*] Kujala K., "Stabilization of Harmful Wastes and Muds," *Proceedings,* International Symposium on Environmental Geotechnology, H.-Y. Fang, Ed., ENVO Publishing Co., Allentown, PA, 1986, pp. 540–548.
[*12*] Eaton, H. C., Walsh, M. B., Tittlebaum, M. E., Cartledge, F. K., and Chalasani, D., "Microscopic Characterization of Solidification/Stabilization of Organic Hazardous Wastes," presented at Energy-Sources and Technology Conference and Exhibition, Dallas, Texas, 17–21 Feb. 1985. Publication of ASME 85-Pet-4.
[*13*] Reisfsnyder, R. H., "The Solidification of Liquid Waste with Sodium Silicate and Kiln Dust," Report R and D, 83-67T, The P.Q. Corp., Valley Forge, PA, 1983.
[*14*] Gard, J. A., Mohan, K., Taylor, H. F. W., and Cliff, G., "Analytical Electron Microscopy of Cement Pastes. I: Tricalcium Silicate Pastes," *Journal of the Ceramic Society,* Vol. 63, No. 5/6, 1980, pp. 336–337.
[*15*] Lachowski, E. E., Mohan, K., Taylor, H. F. W., and Moore, A. E., "Analytical Electron Microscopy of Cement Pastes. II: Pastes of Portland Cements and Clinkers," *Journal of the Ceramic Society,* Vol. 63, No. 7/8, 1980, pp. 447–452.

Barbara L. Forslund,[1] *Liane J. Shekter Smith,*[2] *and Wayne R. Bergstrom*[3]

A Physical Testing Program for Stabilized Metal Hydroxide Sludges

REFERENCE: Forslund, B. L., Shekter Smith, L. J., and Bergstrom, W. R., **"A Physical Testing Program for Stabilized Metal Hydroxide Sludges,"** *Environmental Aspects of Stabilization and Solidification of Hazardous and Radioactive Wastes, ASTM STP 1033,* P. L. Côté and T. M. Gilliam, Eds., American Society for Testing and Materials, Philadelphia, 1989, pp. 214–226.

ABSTRACT: A proposed method of closing two Resource Conservation and Recovery Act surface impoundments containing metal hydroxide (F006) sludges involved stabilization and on-site disposal. A sample testing program was developed as part of an evaluation of processes offered by various firms interested in performing the stabilization. For this application, the stabilized material was to have sufficient strength to support construction and cover loads, low permeability, long-term durability, and low settlement under anticipated loads.

The test procedures developed to determine those properties were based on standard soil and soil-cement test methods to the extent possible. The standard test methods proposed for use included the ASTM Test Methods for Unconfined Compressive Strength of Cohesive Soil (D 2166), for One-Dimensional Consolidation Properties of Soils (D 2435), Wetting-and-Drying Tests of Compacted Soil/Cement Mixtures (D 559), and Freezing-and-Thawing Tests of Compacted Soil/Cement Mixtures (D 560). A falling-head test was used to determine the coefficient of permeability. Appropriate molds and specimen preparation procedures were sent to each firm with a sample of the raw sludge in order that the stabilized sludge specimens would be prepared in the same manner.

Modifications to the proposed test procedures were required due to the condition of several sets of samples. The procedures for evaluating the freeze-thaw and wet-dry durability were altered by eliminating the brushing and weighing of samples between cycles. This sample handling was eliminated so that the relative effects of environmental exposure could be more readily compared. A few of the samples suffered significant expansion, shrinkage, cracking or some combination of these to a degree that made them unusable.

In general, the specimen preparation procedures and test methods developed worked well for the stabilized sludge samples. The unconfined compressive strength test results were highly variable and ranged from 0.5 kg/cm^2 (100 lb/ft^2) to over 35 kg/cm^2 (70 000 lb/ft^2). The compressibility under a load of 1.5 kg/cm^2 (3000 lb/ft^2) ranged from 0.5% to 5%. The permeability of the stabilized material was generally low and in the range of 10^{-6} to 10^{-7} cm/s. The specimens generally withstood more freeze/thaw cycles than wet/dry cycles; several of the specimens lost much of their strength and integrity upon saturation. No strong correlation between any of the physical properties was observed. However, specimens which included cement as an additive tended to fail at low strain, have higher unconfined compressive strength, and were less compressible.

KEY WORDS: stabilization, physical testing, hazardous waste, sludge, sludge testing, surface impoundments, closure, stabilized sludge, remedial action

[1] Project manager, NTH Consultants, 860 Springdale, Exton, PA 19341.

[2] Environmental engineer, Wilkins Wheaton Environmental Services, Inc., 169–171 Portage St., Kalamazoo, MI 49007.

[3] Senior project engineer, Wayne Disposal, Inc., 1349 Whitaker, Ypsilanti, MI 48197.

Current Federal regulations governing the disposal of hazardous waste under the Resource Conservation and Recovery Act of 1980 (RCRA) allow two methods of closing surface impoundments. One method requires complete removal of all wastes in the impoundments, removal of any liner system present, and removal of any contaminated subsoils, followed by disposal at a licensed hazardous-waste facility. For impoundments containing large volumes of waste or when removal of contaminated subsoils can pose serious practical problems, this method may be unfeasible or at least prohibitively expensive.

An alternative method of closing surface impoundments containing hazardous waste is to remove or solidify any liquids present, physically stabilize the remaining waste so that the resultant material can support a protective cover, and close the impoundments as a hazardous-waste landfill. This approach is being proposed at a site with two surface impoundments containing metal-hydroxide sludges. This paper presents a physical testing program developed as part of an evaluation of the use of commercially available stabilization processes at the site and discusses the effectiveness of the program.

Background

The surface impoundments were used for the disposal of sludges derived from the treatment of electroplating wastewaters until June 1981. Closure of the impoundments is regulated under the interim status provisions of RCRA (40 CFR 265). The sludges are listed under RCRA as Hazardous Waste No. F006 (40 CFR 261) and contain chromium, nickel, copper, and zinc concentrations exceeding 1000 ppm. Over 115 000 m^3 (150 000 yd^3) of sludge are contained in the two impoundments, which together cover an area of about 4 ha (10 acres). Two additional surface impoundments, which predate the RCRA-regulated impoundments, are also present at the site and contain 283 000 m^3 (370 000 yd^3) of similar material. Because of the large volume of sludge, on-site closure options were being explored as opposed to removal and off-site disposal.

As part of the preliminary efforts, an evaluation of the feasibility of using commercially available, on-site stabilization processes for the sludge was undertaken. If the use of such processes was determined to be feasible, the evaluation had the added goal of prequalifying potential stabilization contractors. The testing program developed as part of the evaluation included both physical and chemical analyses to determine the characteristics of the stabilized sludge. This paper focuses on the physical test procedures and the results of the physical testing program.

Objectives of the Physical Testing Program

The primary objective of the physical testing program was to provide an effective means of comparing the behavioral characteristics of products offered by various stabilization contractors. In order for this objective to be met, a representative sample of the raw sludge had to be obtained and given to the contractor, test specimens of the stabilized sludge had to be prepared in a consistent manner, and appropriate test methods had to be chosen and implemented.

Prior to the development of the physical testing program, desired general characteristics of the stabilized sludge were identified, assuming possible design features for the disposal area. Anticipated design features include a composite cover system to protect the waste from weather effects, and installation of a leachate collection system and perimeter slurry trench to enhance the protection offered by the natural clay at the site. Given these probable features, the most important physical characteristics of the stabilized wastes relate to its load-bearing capacity, that is, compressibility and strength.

The sole physical characteristic specified to the contractors was that the stabilized sludge have an unconfined compressive strength of at least[4] 1 kg/cm^2 (2000 lb/ft^3) after curing 14 days. Other desirable characteristics included low permeability, resistance to weathering, and low compressibility under anticipated loads. No acceptance criterion was established for these characteristics; qualitative comparison of results was used for the evaluation.

Sludge Sample Acquisition

A critical factor in the success of the testing program was to provide each contractor with a representative sample of raw sludge for trial stabilization. Since the degree of sludge variability was unknown at that time, a composite sample was formed by taking two full-depth cores of sludge at 26 locations throughout the impoundments.

The approximate depth of sludge at the sample location was determined by first pushing a probe of known length as far as possible and measuring the length remaining. The sludge samples were obtained using a sampler made of 5-cm (2-in.) diameter polyvinyl chloride (PVC) flush-threaded casing. A rubber piston attached to a chain was placed in the bottom of a 3-m (10 ft) section of sampler, as shown in Fig. 1. A handle was then clamped around the top of the casing to assist in inserting and withdrawing the sampler. The sampler was pushed into the sludge while the rubber piston was simultaneously drawn up inside the sampler. This created a slight vacuum within the casing and helped retain the sludge as the sampler was withdrawn.

A tripod was used to assist in pulling the piston and to keep the sampler vertical. Additional sections of casing could be added as needed to reach greater depths, being careful to maintain tension on the chain at all times. Sludge depths to 8.2 m (27 ft) below the surface were sampled this way.

The sampler was pushed until resistance was met; this depth was assumed to be the base of the sludge. The sampler was then withdrawn with the enclosed core of sludge. At eight locations, the first core was extruded using the rubber piston onto a flat, clean surface, and discrete samples at varying depths were obtained for chemical analysis to determine the variability within the sludge. The remainder of the sludge and all other cores retrieved were combined to form a composite sample. Approximately 190 L (50 gal) of sludge were obtained for the composite sample in this manner. The composite sample was thoroughly mixed using a cement mixer and placed into sealed buckets for distribution to the contractors. Each contractor received 13.2 L (3.5 gal) of raw sludge.

Some of the composite sample was retained for an extensive chemical analysis. This analysis and the variability testing showed that the sludge had the following characteristics:

1. Solids content of about 26% (composite sample).
2. A mean oil and grease content of approximately 23 000 ppm, with a standard deviation of 13 400 ppm for the individual specimens. Oil and grease content of the composite was 17 000 ppm.
3. Copper, zinc, total chromium, and nickel concentrations in excess of 1000 ppm in the composite specimen.
4. Total cyanide concentrations of 6.1 ppm in the composite specimen.
5. Xylene concentration of 2.7 ppm; no other volatile organic compounds in excess of 1 ppm in the composite specimen or individual specimens tested using gas chromatography/mass spectrometry.
6. No leachable inorganic constituents exceeding the extraction procedure toxicity criteria (40 CFR 261.24) in the composite specimen.

[4] Measurements generally were taken in English units during testing.

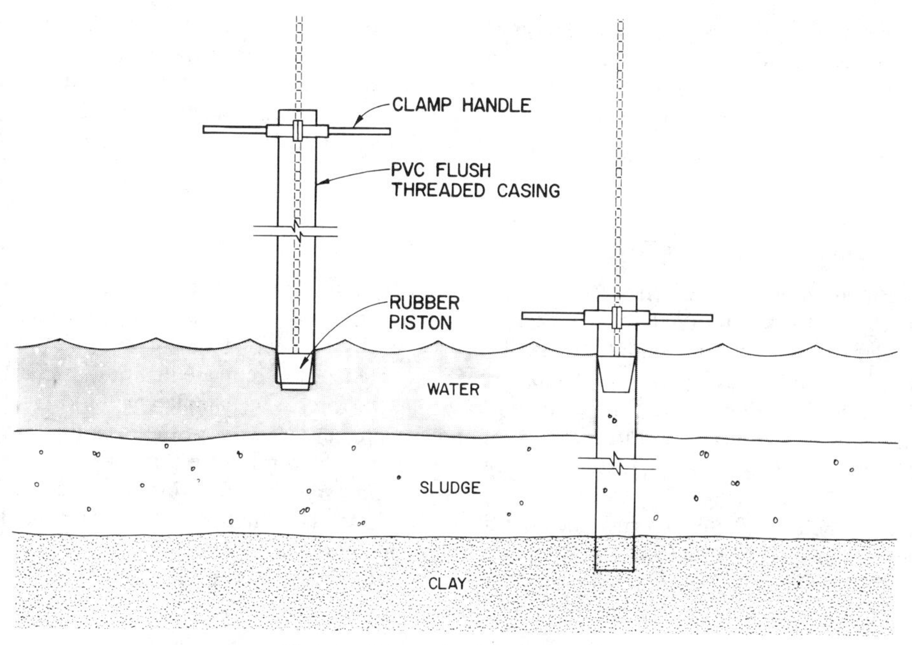

FIG. 1—*Schematic of sludge sampler.*

Physical Testing Program

Testing of stabilized sludge specimens was undertaken to determine certain basic characteristics of the stabilized material and to predict its physical behavior. The tests chosen included determination of unconfined compressive strength, consolidation, permeability, and freeze/thaw and wet/dry durability. The unconfined compression and consolidation tests would provide an understanding of how the material would behave during and shortly after stabilization when construction and cover loads would be applied. The wet/dry and freeze/thaw tests provide qualitative comparisons of the relative durability and integrity of the samples. The permeability tests would be used to confirm the anticipated low permeability of these types of materials.

The specific test procedures developed were derived from standard methods for testing soils, soil-cement mixtures and other construction materials. American Society for Testing and Materials (ASTM) test methods provided the basis for most of the procedures; however, ASTM currently does not have an applicable test method for measuring the coefficient of permeability for fine-grained materials. A falling-head test was used on the specimens, as is common geotechnical practice for clay soils.

Once the general test procedures were established, molds for stabilized specimen preparation were chosen using commonly available molds wherever possible. Other molds were fabricated as needed to be compatible with the testing equipment.

Specimen preparation procedures were made as simple as possible to assure uniformity. Instructions were prepared and sent to each contractor along with a complete set of specimen molds and 13.2 L (3.5 gal) of the composite sludge. The stabilized specimens were returned and subjected to the physical tests. The following sections present the details of the specimen preparation for each test and the specific testing procedures.

Along with the stabilized samples, the contractors were requested to provide information relating to general type of fixation process and additives used. Since most of this information is proprietary, details of the admixtures used were not supplied. However, the processes have been divided for purposes of this presentation into two categories: (1) one of the additives is known to be portland cement; and (2) either portland cement is not used as an additive or insufficient information was provided to determine if it had been used. Seven of the sixteen contractors participating in the evaluation are known to have used portland cement as an additive.

Unconfined Compression Test

Based on anticipated construction and cover loads, a minimum unconfined compressive strength of 1 kg/cm^2 (2000 lb/ft^3) after 14 days of curing was specified for the stabilized sludge. Test procedures outlined in ASTM Test for Unconfined Compressive Strength of Cohesive Soil (D 2166-66) were chosen. Stabilized specimens were made in 7.6 by 15.2-cm (3 by 6-in.) plastic cylinder molds (commonly used for preparing samples of soil-cement mixtures and grout). Opposite sides of the molds were cut vertically to the base to allow for future sample removal. The resultant seams and the mold rim were then taped closed to keep the mold rigid and closed during specimen preparation and curing. A 7.6-cm (3-in.) diameter by 0.6-cm (¼-in.) thick plastic disk was also placed in the bottom of the mold to help in specimen removal and protect the specimen base during removal.

The instructions for specimen preparation were as follows for six specimens, each a 7.6-cm (3-in.) diameter by 15.2-cm (6-in.) plastic cylinder:

1. Insert a 7.6-cm (3-in.) diameter × 0.6-cm (¼-in.) plastic disk into the bottom of each cylinder mold. Make sure it is seated onto the cylinder bottom.
2. Pour the material into the mold in four approximately equal layers. After adding each layer, use a spatula to rod the material in the center and around the edges of the mold to help eliminate air bubbles. Also, after adding each layer, lift the mold up from the table about 1 cm (½ in.) and let it drop 15 to 20 times to consolidate the material and further remove any air bubbles that still might be present.
3. After the top layer has been prepared in this manner, smooth off the top of the specimen with a spatula and set aside to cure.
4. Leave the specimen in the mold for shipping and place in a sealed plastic bag.

Six specimens were prepared in this manner, two of which were tested for unconfined compressive strength.

Most of the specimens returned were in very good condition. One set of specimens had expanded significantly, and these were not of uniform diameter. Shrinkage had occurred in some specimens; however, no cracking was apparent. In a few cases, the plastic disks had not been placed in the bottom of the molds, but had been placed on top of the stabilized material after the mold had been filled. Greater care was needed to remove these specimens without causing damage.

After the specimens were removed from the molds, the length and diameter of each specimen were measured. The test was generally run using a loading rate of 0.5 mm/min. Load readings were recorded at deflection intervals of 0.06 cm (0.025 in.).

Most of the specimens were tested 14 days after molding. A curing period of 14 days was chosen to assure adequate strength soon after stabilization in order to avoid contractor delays during construction. In three cases, specimen testing was delayed. For two pairs, this was due to receipt of the specimens after 14 days (29 to 35 days). In the last case, testing of the second specimen was delayed to examine the effect of extended curing time (62 days) on a material with initially inadequate strength.

The results of the strength testing were highly variable, with unconfined compressive strengths ranging from 0.5 kg/cm^2 (100 lb/ft^2) to over 35 kg/cm^2 (70 000 lb/ft^2). The distribution of strengths determined is shown in Fig. 2. Two specimens per mix received from each contractor were tested. Twelve of the twenty pairs tested had results which varied less than 15%, based on the difference between each result divided by the mean value of the pair. The largest variation was 41%. The extended curing time of 62 days resulted in a strength increase of 57% over the 14-day strength for a single pair.

Strain values at failure were relatively low when compared with most cohesive soils and in the range of 0.8 to 6.2%. The mean failure strain value for specimens containing cement was lower than for the rest of the specimens (1.8 and 2.9, respectively). In addition, variability in failure strains was less for the specimens containing cement.

The specimens generally exhibited brittle failures characteristic of cementatious materials. When the applied loading exceeded the sample strength, the specimen showed a distinct loss of integrity and sharp reduction in any additional load-bearing capacity.

Overall, the molds, specimen preparation, and testing procedures were successful for the unconfined compression tests. Average cross-sectional areas were used for specimens with nonuniform diameters; more rigid molds could help alleviate this problem, particularly if the specimens are shipped in the early stages of curing. It was not clear whether or not the plastic disk, when used, assisted in preventing damage to the base of the specimens, but it greatly simplified specimen removal from the molds. Because of the low strain at failure of these materials, more information regarding their behavior under load would be gained if more frequent load deflection readings were taken during the test. Standard reading intervals for typical soil testing were used in this program.

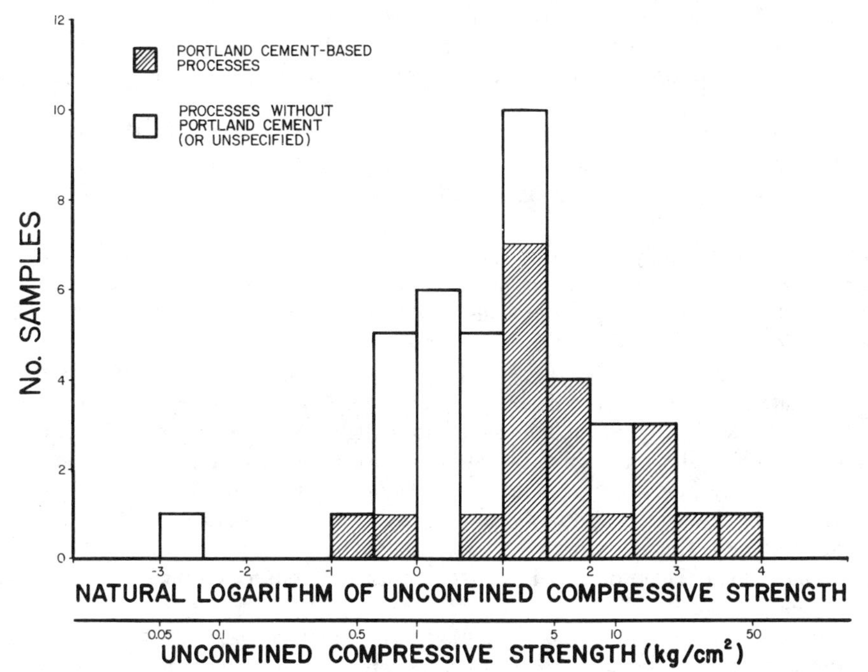

FIG. 2—*Distribution of unconfined compressive strengths.*

Consolidation Testing

The long-term functional integrity of a cover placed over a site with stabilized sludge can be detrimentally affected if the cover settles excessively. The compressibility of the stabilized waste when subjected to a confining load is a key property in determining the amount of such settlement which may occur. The compressibility of specially prepared specimens of stabilized sludge was investigated under conditions of one-dimensional consolidation.

The techniques used to examine compressibility behavior were in general accordance with the ASTM Test for One-Dimensional Consolidation Properties of Soils (D 2435-80). A fixed-ring, one-dimensional consolidometer was used to confine each specimen.

A vertical load was applied on the specimen after saturation with water and the amount of specimen consolidation was measured. After consolidation under one load was essentially complete, another load, twice the initial load, was applied, and the additional consolidation was measured. This sequence was repeated several times under increasing loads. Sequential unloading of the sample to 0.5 kg/cm^2 (100 lb/ft^2) was undertaken after consolidation under a load of 8 kg/cm^2 (1600 lb/ft^2). The sample was then reloaded in the same manner as initially loaded.

Specimens for this testing were prepared by rodding uncured stabilized sludge into specially fabricated ring molds. These resulted in specimens that closely fit into the consolidation ring, approximately 6.4 cm (2.5 in.) in diameter and 1.9 cm (0.75 in.) in thickness. After curing, a plastic plate was used to protect the top and bottom of each specimen during shipping. The instructions for preparing two 6.4-cm (2½-in.) inside diameter by 1.9-cm (¾-in.) plastic ring specimens were as follows:

1. Set the ring on a 8.3 by 8.3 by 0.6 cm (3¼ by 3¼ by ¼ in.) plastic plate.
2. Pour material into the ring and use a spatula to rod the material around the edges of the ring to be sure there are no trapped air bubbles on the ring.
3. Refill the ring, if needed, to just overflowing and set 8.3 by 8.3 by 0.6 cm (3¼ by 3¼ by ¼ in.) plate on top of the ring and set a moderately heavy object on top of the plate to hold it down.
4. After the specimen has had time to set, remove the weight from the top plate and secure the top and bottom plate to the ring using two rubber bands.
5. Set the complete assembly in a sealed plastic bag for shipping.

This approach generally resulted in specimens which were found to be satisfactorily intact for the intended testing. However, four of the total twenty specimens were found to be excessively cracked when received. This may have been a result of rough handling during shipping or of excessive shrinkage during curing.

The results of the consolidation testing are presented in Fig. 3. In general, the rate of consolidation was observed to decrease with the time under any given load. However, the consolidation versus log-time relationship did not often exhibit the classic "S" curve observed when similarly testing clay soils. Linearity on such a log-time graph also was not always observed. Rather, the decrease in the consolidation rate was slower than generally observed when testing clay soils, resulting in a slightly concave-downward graph of consolidation versus log time. However, the rates of consolidation were noted to be very slight after 24 h under load. Therefore, 24 h was chosen as the arbitrary termination time for each individual loading.

A typical curve of consolidation (strain) versus log-pressure is shown in Fig. 4. The shape of this curve is very similar to those which usually result from testing clay soils. For want

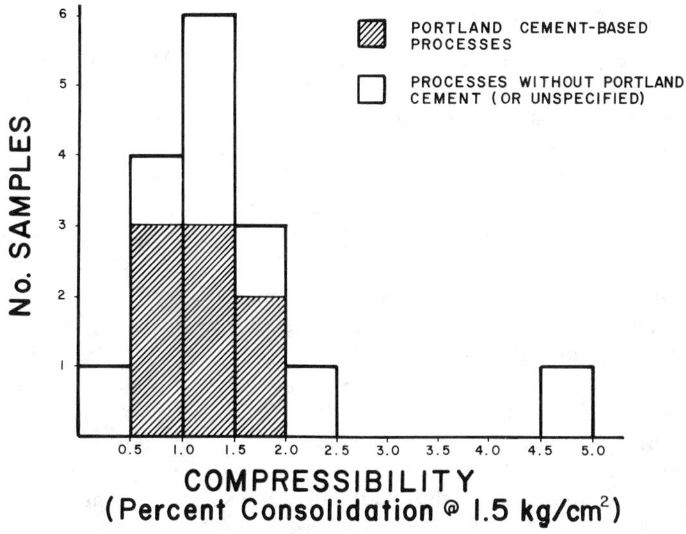

FIG. 3—*Distribution of consolidation test results.*

of better terminology, the various portions of this curve have been labeled with the terms commonly applied in soil mechanics. While the reasons for this settlement behavior may not be the same for stabilized sludge as for clay, the similarity in observed consolidation trends is noteworthy. This suggests that settlement predictive tools commonly used for clays may also be successfully applied to stabilized materials.

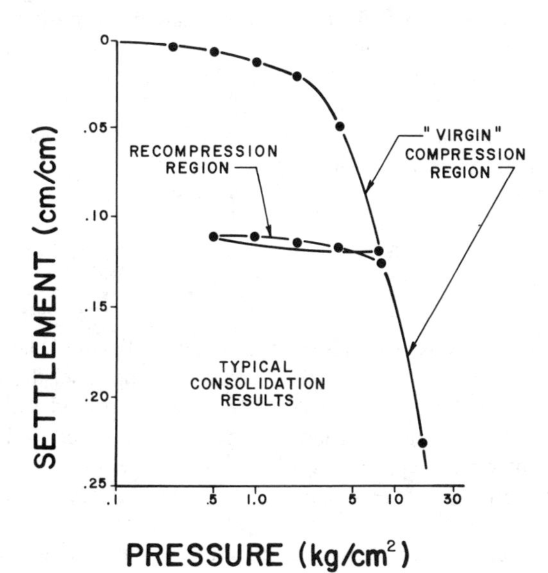

FIG. 4—*Typical consolidation curve.*

Permeability Testing

Based on published information and information supplied by the contractors, a low-permeability material was expected from the stabilization process. A low-permeability material is desirable because of reduced infiltration potential, thereby minimizing the potential for large volumes of leachate production if the protective cover is breached.

At the present time, there is no standardized ASTM test procedure for determining the permeability of fine-grained (low-permeability) materials. For this program, a falling-head test in a fixed-wall permeameter was chosen, largely because of equipment availability and time constraints.

Samples were prepared in 3.5-cm (1⅜-in.) inside diameter by 7.6-cm (3-in.) long brass molds. Instructions for preparing three brass tube specimens, each 3.5 cm (1⅜ in.) inside diameter by 7.6 cm (3 in.), is as follows:

1. Set the tube vertically on a flat surface.
2. Holding the tube down, fill it with the material.
3. Gently tap the sides of the tube until no more air bubbles rise to the surface.
4. If necessary, add more material to fill the tube and smooth out the top surface with a spatula.
5. Allow the specimen to set up completely, then place it in a sealed plastic bag for shipping.

It was intended that the specimen in the brass mold would be placed directly into the permeameter. However, after curing, a number of specimens did not maintain themselves fully in the mold. Some shrinkage had occurred so that the specimens could slip out of the mold. In order for these specimens to be tested, they were removed from the mold and then rolled and covered with a moist, bentonite paste. The specimen was replaced in the mold, allowing the bentonite to hydrate and swell, filling the annular space between the sample and the mold.

Back pressure was not used to achieve saturation. Rather, saturation was assumed to be complete when repeated measurements of the rate of head drop yielded consistent values.

The stabilized material has a permeability equivalent to a clayey silt to silty clay soil. The results of the permeability test were in the ranges given in Table 1. Because of the generally poor specimen conditions, the test results are reported as order of magnitude only. Use of bentonite paste worked well enough for the purposes of this evaluation; however, use of a flexible-wall, pressure-cell permeameter would be advisable for similar programs in the future.

Durability Tests

Resistance to weathering was also chosen an important behavioral characteristic for comparing the various stabilization products. Test procedures to model the weathering effects were adapted from ASTM Wetting-and-Drying Tests of Compacted Soil-Cement Mixtures (D 559-82) and ASTM Freezing-and-Thawing Tests of Compacted Soil-Cement Mixtures (D 560-82). Specimens were prepared according to the procedures described previously in the section of the unconfined compression test. Two specimens were designated for durability testing.

The wet/dry and freeze/thaw tests were modified from the standard ASTM procedures. Both tests were altered to use one specimen and eliminate brushing the specimens to determine material loss. The samples were rotated and cycled generally according to standard

TABLE 1—*Results of permeability test.*

Coefficient of Permeability, k, cm/s	Number of Occurrences
10^{-7}	7
10^{-6}	8
10^{-5}	1

procedures. The test was determined to be complete when the specimen had deteriorated to such a point that it could not support itself to be rotated again.

Figure 5 presents the results of the wet/dry and freeze/thaw testing. The value reported indicated the cycle at which disintegration occurred. Testing was concluded after 12 complete cycles regardless of specimen condition as per ASTM procedures. Results of the tests are somewhat variable. The specimens generally withstood more freeze/thaw cycles than wet/dry cycles. Some of the specimens lost much of their strength and integrity after a single wet/dry cycle; this lack of wet/dry durability may be an important limitation for certain applications.

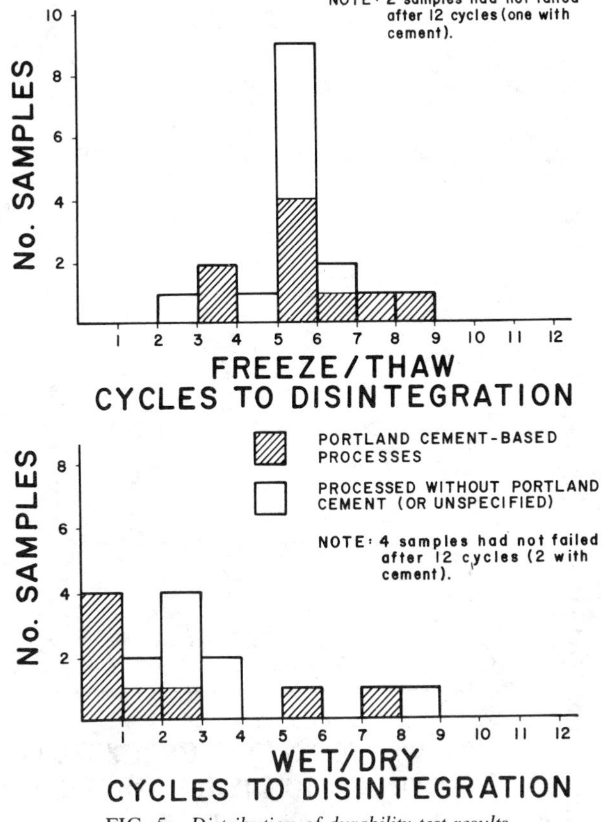

FIG. 5—*Distribution of durability test results.*

These tests represent much harsher conditions than the materials may be subjected to in the field. The stabilized material will be permanently covered and thus be protected from long-term weathering exposures. Also, the 24-h cycles of saturation/drying and freezing/thawing are more extreme than would normally be encountered. However, based on the test results, the anticipated long-term exposure resistance appears to be somewhat poor (and highly variable), indicating that the stabilized materials should be clearly protected from weathering exposure.

Relationship Between Physical Properties

Determination of a general correction between physical properties of the stabilized sludge would assist in streamlining similar testing programs in the future. For example, one might expect a general correlation between strength and durability. The data generated were reviewed to determine if any such correlations were apparent.

Figures 6 to 8 show the relationship between various physical properties. As can be seen on the graph comparing unconfined compressive strength and durability, specimens with higher strengths did not necessarily exhibit better long-term durability characteristics. Also, there is no apparent relationship between the results of the wet/dry and freeze/thaw testing.

Although no strong correlation between any of the physical properties was observed, the processes which included cement as an additive tended to exhibit certain qualities. These

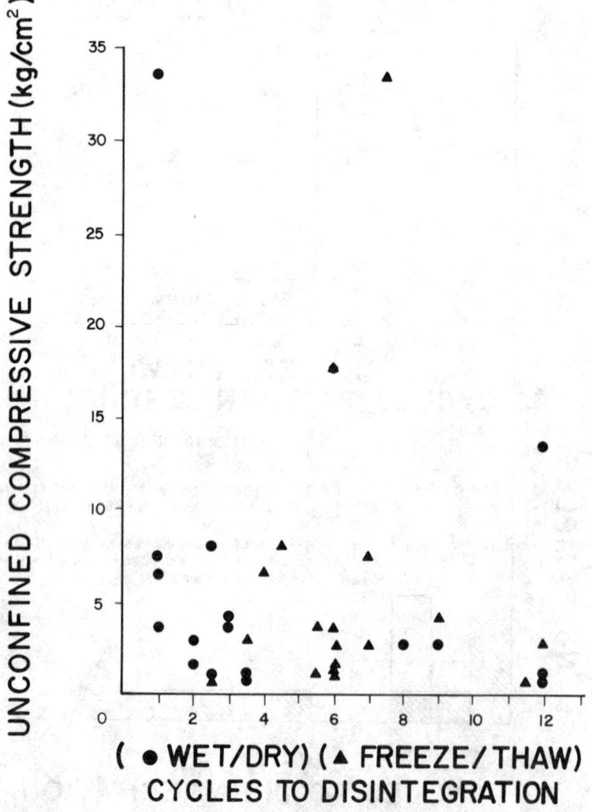

FIG. 6—*Comparison of strength and durability test results.*

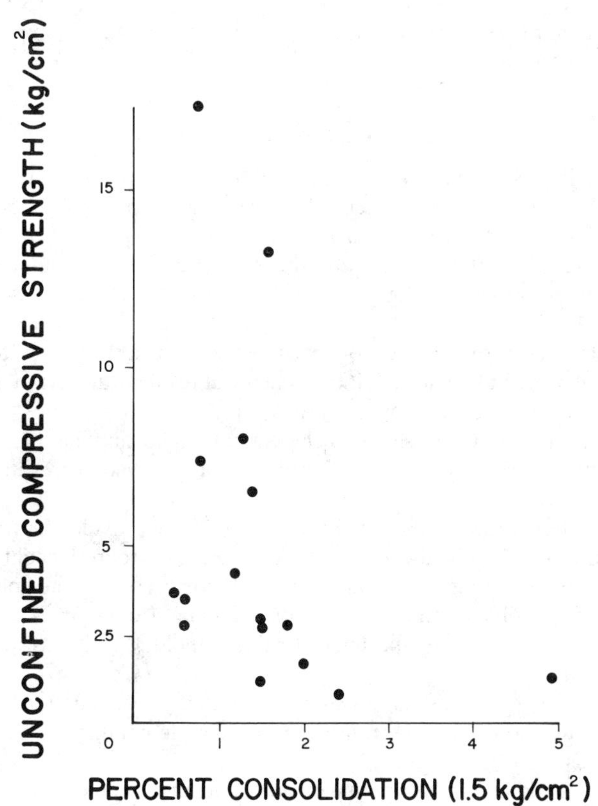

FIG. 7—*Comparison of strength and consolidation test results.*

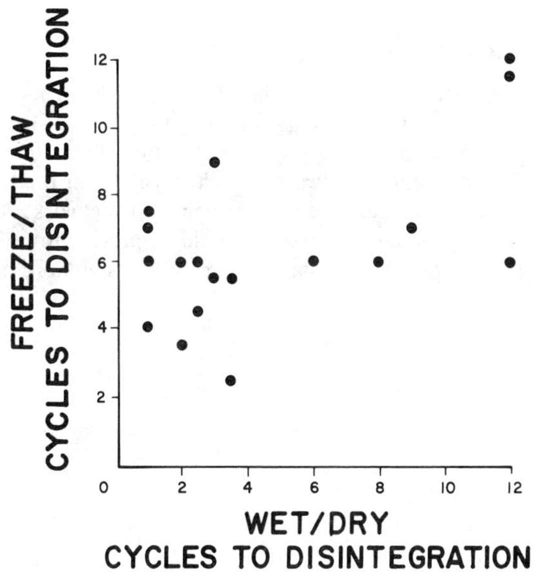

FIG. 8—*Comparison of wet/dry and freeze/thaw test results.*

samples, in general, tended to fail at low strain, have higher unconfined compressive strength, and were less compressible.

Possible Modifications to Physical Testing Program

The physical testing program as originally developed and implemented was successful in meeting its objectives; most of the problems that arose during the testing were handled with fairly minor modifications to procedures. However, certain aspects of the program could be improved by use of alternative test procedures or equipment or both. Some of these possible modifications are listed below:

1. Individual samples of raw sludge from each bucket sent to the contractor should have been retained. This would allow for additional chemical testing if necessary for verification of its uniformity with other contractor submittals.
2. Instructions to the contractor should include a note indicating that the sludge material should be well mixed prior to trial stabilization, because some segregation of material appeared after shipping.
3. If possible, more than 13.2 L (3.5 gal) of sludge should be provided for trial stabilization and sample preparation. This could also allow for evaluating the reproducibility of the results.
4. As an alternative to using the small rigid molds for preparing the consolidation test specimens, it may be possible to trim consolidation specimens from a larger sample. Cracking was not a problem with the larger specimens (cylinder molds), as it was with some of the consolidation specimens.
5. If a specimen can be trimmed successfully for use in the consolidation device, a falling-head permeability test could then be run using one sample confined in the consolidation cell at various loads.
6. A triaxial cell could be used for permeability testing of a sample prepared in a cylinder mold. Saturated strength or confined compression testing or both could also be performed in the triaxial cell.

Evaluation of Stabilization Processes

The physical testing program presented in this paper formed one portion of an evaluation to determine the feasibility of using stabilization processes for the sludges at the site. The overall evaluation will also be used to prequalify stabilization contractors. Additional factors considered in the evaluation included the results of chemical testing on the stabilized samples and the individual contractor's experience in performing on-site work. Additional specimen preparation and testing is planned as part of the bidding process to refine admixtures and to confirm both physical and chemical characteristics of the stabilized material.

Laboratory Evaluation

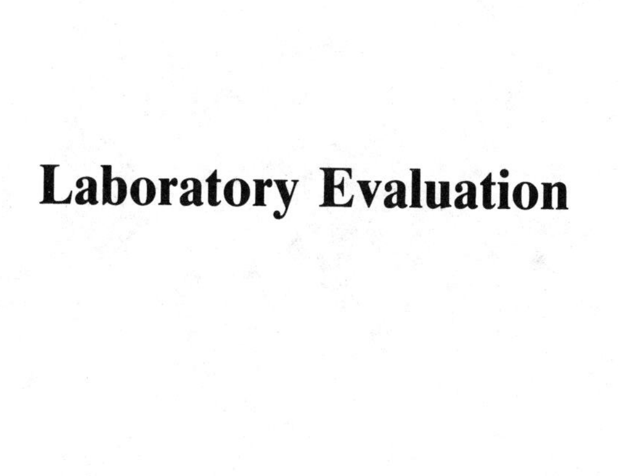

F. Medici,[1] *C. Merli,*[2] *G. Scoccia,*[1] *and R. Volpe*[1]

Experimental Evaluation of Limiting Factors for Leaching Mechanism in Solidified Hazardous Wastes

REFERENCE: Medici, F., Merli, C., Scoccia, G., and Volpe, R., **"Experimental Evaluation of Limiting Factors for Leaching Mechanism in Solidified Hazardous Wastes,"** *Environmental Aspects of Stabilization and Solidification of Hazardous and Radioactive Wastes, ASTM STP 1033,* P. L. Côté and T. M. Gilliam, Eds., American Society for Testing and Materials, Philadelphia, 1989, pp. 229–237.

ABSTRACT: Results are presented of a series of experimental tests performed to determine the influence of matrix characteristics on the leaching mechanism of lithium immobilized into cementitious matrices. A dynamic leaching test, using a fixed leachant renewal schedule proposed by the American Nuclear Society, was adopted. Experimental results showed the influence of the matrix characteristics on the different phases of the leaching. A hypothesis on the leaching mechanism from the cementitious matrices is proposed.

KEY WORDS: hazardous wastes, toxic metals, immobilization, cementitious matrix, dynamic test, leaching, release, permeability, capillary porosity

Different immobilization materials and methods which reduce the release of toxic elements, utilizing different matrices [*1*], have been studied from a number of sources [*1,2*]. From an economic point of view [*3*], cementitious matrices seem to be fairly promising. In this paper, the first results of research on the immobilization of toxic elements into cementitious matrices are reported. Apart from the comparison between the performances of different types of cementitious matrices, the first objective of this research was to investigate the leaching mechanism of toxic elements. A particular leaching test, proposed by the American Nuclear Society (ANS) [*4*] and modified by Côté and other authors [*5*], was used to simulate the release conditions.

Another objective of the research was to verify existing models for interpretation of leaching mechanism [*6–9*] and to define a model, both on empirical and physical basis, for evaluation of leaching mechanism from cementitious matrices.

There are, in fact, different theoretical models for interpretation of the experimental results of leaching tests, but the Jaffe-Ferrara [*6*] and Godbee-Joy [*7*] models appear to be the most suitable. The first is an empirical model which does not consider the chemical-physical parameters of the process; the second is a semiempirical model which permits the determination of the effective diffusion coefficient when the diffusion is the controlling factor. Moreover, a theoretical approach for evaluating long-term leachability from measurement of intrinsic waste properties has been proposed by Côté et al. [*8*].

[1] Department of Chemistry, Chemical Engineering, and Materials, University of L'Aquila, L'Aquila, Italy.

[2] Department of Chemical Engineering, Materials, Raw Materials, and Metallurgy, University of Roma "La Sapienza," Rome, Italy.

This paper describes the first results obtained from a series of experimental tests carried out on four different types of cementitious matrices, immobilizing lithium (added as lithium chloride). Lithium was chosen as the contaminant element because it is only physically fixed into cementitious matrices; in fact, among the univalent elements, lithium has the lowest atomic weight, and its hydroxide is quite soluble.

Experimental Procedures

Synthetic waste solutions were prepared from distilled water by addition of 0.4 *M* of lithium (as lithium chloride). Moreover, in order for possible interaction phenomena to be evaluated, mixed solutions by addition of lithium (as a solution of 0.2-*M* lithium chloride) and copper (as a solution of 0.2-*M* copper chloride), whose total concentration was 0.4 *M*, were prepared. Bivalent copper was chosen as additional contaminant element because its immobilization mechanism is different from that of lithium; in fact, bivalent copper hydroxide is not very soluble, and its solubility is highly influenced by pH of the medium.

Four types of cementitious matrices were used in the experiments:

(*a*) portland cement + sand,
(*b*) portland cement + silica fume + sand,
(*c*) portland cement + waterproofing + sand, and
(*d*) pozzolanic cement + sand.

All the mixtures were manufactured by using standardized sand (Torre del Lago). A water/cement ratio equal to 0.5 in weight and an aggregate/cement ratio equal to 3 in weight were adopted. Silica fume and waterproofing contents were 12 and 0.5%, respectively, in weight of cement.

For each cementitious matrix three mixes were prepared: a mix for the lithium immobilization, a mix for both toxic elements (lithium and copper), and a mix without toxic elements for the blank test.

Mixing of the constituents composing the various cement mixtures was carried out by a laboratory pan mixer with a vertical axis. Steel molds were employed in the preparation of specimens, and the mortar was manually compacted by means of a standardized rod so as to obtain the maximum density.

All the cubic specimens (4 by 4 by 4 cm) were cured for 28 days in a room conditioned at 20 ± 1°C and 98% relative humidity before the leaching experiments were initiated. Demolding took place 24 h after the casting for the specimens containing lithium and 72 h after the casting for the specimens containing lithium and copper, the latter to take into account the delay induced from this element on setting time of the cement [*10,11*].

A dynamic leaching test using a fixed leachant renewal schedule derived by the ANS [*4,5*] was adopted. The leaching test was conducted by immersing the specimens in distilled water, using a specimen surface area-to-leachant volume ratio of 0.064 cm^{-1}. The bottles were not agitated because it is assumed that the migration rate of the contaminants in the leachate is several orders of magnitude larger than in the solid matrix.

All the specimens were leached for 28 days, following a standard leachant renewal schedule proposed by Côté and Isabel [*5*]; the leachants were changed at 2, 7, 24, 48, 72, 96, 120, 192, 264, 336, 504, and 672 h. During all the leaching tests, temperature was maintained constant at 25°C.

At the end of each leaching period, the lithium content in each of the leachates was measured by atomic absorption (Perkin Elmer 3030) [*12*].

On the blank specimens of the four considered matrices, the coefficient of permeability was determined according to the British Standards Institution Methods of Testing Concrete (BS 1881/1970).

Experimental Results

The results of the dynamic leaching test conducted both on the specimens containing lithium and lithium plus copper are presented, in Tables 1 and 2, respectively. In these tables, for each matrix is reported: lithium concentration in the leachates for each leaching period (C), cumulative contaminant loss during each leaching period (Σa_n), and cumulative fraction leached for each leaching period (C_n) based on the initial amount of contaminant present in the specimen (A_0 = 44.4 mg).

The dynamic leaching test results are also reported in Figs. 1 and 2. The experimental data are presented by individual symbols, while the solid lines represent their best interpolation.

Figure 1 represents the results for the specimens containing lithium, Fig. 2 the results for the specimens containing both toxic elements (lithium and copper). All the tests were carried out in the laboratories of the Department of Chemistry, Chemical Engineering and Materials of the University of L'Aquila, L'Aquila, Italy.

Discussion

Experimental results show, as other authors agree [*5–9,13*], that two different phases exist in the toxic elements leaching. The first phase (initial washing period) is associated with an initial rapid release; the second phase (diffusion-controlled period) represents a long-term extending phase in which the leaching rate is lower.

In order for the cumulative contaminant loss during these two phases to be estimated, the trend of the cumulative contaminant loss (Σa_n) versus the square root of time ($\sqrt{t}$) was plotted according to the models of Godbee and Joy [*7*] and Côté et al. [*5,9*].

Cumulative contaminant loss during the first (F_1) and the second phase (F_2) were obtained by interpolating the trend of $\Sigma a_n = f(\sqrt{t})$ by two straight lines, each one characterizing the considered phases. The end of the washing period was fixed when the tangent to two

TABLE 1—*Results of dynamic leaching test (0.4-*M *lithium).*

	Matrix A			Matrix B			Matrix C			Matrix D		
t, h	C, mg/L	Σa_n, mg	C_n, %	C, mg/L	Σa_n, mg	C_n, %	C, mg/L	Σa_n, mg	C_n, %	C, mg/L	Σa_n, mg	C_n, %
2	1.150	1.725	3.88	1.124	1.686	3.79	2.020	3.030	6.82	0.719	1.078	2.42
7	0.745	2.842	6.40	0.864	2.982	6.71	1.528	5.322	11.98	0.544	1.894	4.26
24	1.303	4.796	10.80	1.406	5.091	11.47	2.041	8.383	18.88	1.018	3.421	7.70
48	0.779	5.964	13.43	0.843	6.355	14.31	1.457	10.568	23.80	0.664	4.417	9.95
72	0.596	6.858	15.44	0.383	6.929	15.60	0.472	11.276	25.39	0.344	4.933	11.11
96	0.370	7.413	16.69	0.203	7.234	16.29	0.310	11.741	26.44	0.199	5.231	11.78
120	0.305	7.870	17.72	0.144	7.405	16.67	0.178	12.008	27.05	0.178	5.498	12.38
192	0.694	8.911	20.06	0.370	7.960	17.93	0.421	12.639	28.47	0.319	5.976	13.46
264	0.493	9.650	21.73	0.246	8.329	18.76	0.220	12.969	29.21	0.272	6.384	14.38
336	0.613	10.570	23.81	0.139	8.537	19.23	0.199	13.268	29.88	0.242	6.747	15.19
504	0.655	11.552	26.02	0.093	8.677	19.54	0.400	13.868	31.23	0.446	7.416	16.70
672	0.276	11.966	26.95	0.122	8.860	19.95	0.339	14.376	32.38	0.370	7.971	17.95

TABLE 2—*Results of dynamic leaching test (0.4-*M *lithium and copper).*

	Matrix A			Matrix B			Matrix C			Matrix D		
t, h	C, mg/L	Σa_n, mg	C_n, %	C, mg/L	Σa_n, mg	C_n, %	C, mg/L	Σa_n, mg	C_n, %	C, mg/L	Σa_n, mg	C_n, %
2	0.435	0.653	2.94	0.336	0.504	2.27	0.245	0.368	1.66	0.205	0.308	1.38
7	0.370	1.208	5.44	0.301	0.956	4.30	0.245	0.736	3.31	0.193	0.597	2.67
24	0.686	2.237	10.07	0.506	1.715	7.73	0.524	1.522	6.86	0.430	1.242	5.59
48	0.435	2.890	13.02	0.279	2.134	9.61	0.359	2.061	9.28	0.328	1.734	7.81
72	0.223	3.225	14.53	0.155	2.366	10.66	0.197	2.356	10.61	0.168	1.986	8.94
96	0.159	3.464	15.60	0.114	2.537	11.43	0.122	2.539	11.44	0.110	2.151	9.69
120	0.128	3.656	16.47	0.095	2.679	12.07	0.101	2.690	12.11	0.100	2.301	10.36
192	0.406	4.265	19.21	0.160	2.919	13.15	0.200	2.990	13.47	0.243	2.666	12.01
264	0.231	4.611	20.77	0.114	3.090	13.92	0.146	3.209	14.45	0.199	2.964	13.35
336	0.279	5.029	22.65	0.090	3.225	14.53	0.116	3.383	15.24	0.148	3.186	14.35
504	0.197	5.324	23.98	0.152	3.453	15.55	0.278	3.800	17.12	0.281	3.607	16.25
672	0.244	5.690	25.63	0.112	3.621	16.31	0.215	4.123	18.57	0.216	3.931	17.70

consecutive points significantly moves from tangent between considered point and the point (0.0).

Calculated values F_1 and F_2 for all the different matrices and for the two series of experimental tests are presented in Tables 3 and 4, while the interpolations of the experimental results are shown in Figs. 3 and 4. Tables 3 and 4 also report, for each matrix, the cumulative contaminant loss (F_T), the cumulative fraction leached ($C_F = F_T/A_0$), and the coefficient of permeability (K).

Examination of Tables 3 and 4 shows how the amounts of contaminant leached during the washing phase (F_1) can represent considerable percentages of the cumulative contaminant losses (F_T). Moreover, the cumulative contaminant loss during the washing phase, whether in comparison with the complessive contaminant loss or in comparison with the immobilized contaminant amount, decreases as immobilized contaminant concentration decreases.

In addition, from an examination of Tables 3 and 4, we also discover how the values of F_2 increase as matrix permeability increases. Correlation between F_2 and permeability coefficient (K) is good for both the considered concentration values; in fact, for the matrices immobilizing lithium as a solution of 0.4-*M* lithium chloride, we have a linear correlation coefficient equal to 0.94, and for the matrices immobilizing lithium and copper we have a linear correlation coefficient equal to 0.93. Considering C_F as parameter to evaluate the performance of the matrix, it follows that Matrices B and D give the best results. Besides, compared examination of cumulative contaminant losses (Tables 1 and 2) does not allow the exclusion of possible interaction phenomena between lithium and copper.

Conclusions

Results of experimental tests to date show how the cumulative amount leached from the different matrices depends essentially on the concentration of the immobilized toxic element and on the matrix porosity. In particular, the experimentally observed relation of the matrix performance with porosity allows the formation of a hypothesis on the leaching mechanism of immobilized toxic. In fact, it is possible to assume that porosity and superficial macrocavity of the matrix constitute the controlling factors of the initial washing period, whereas capillary porosity controls the subsequent diffusive phase. Furthermore, it is necessary to investigate

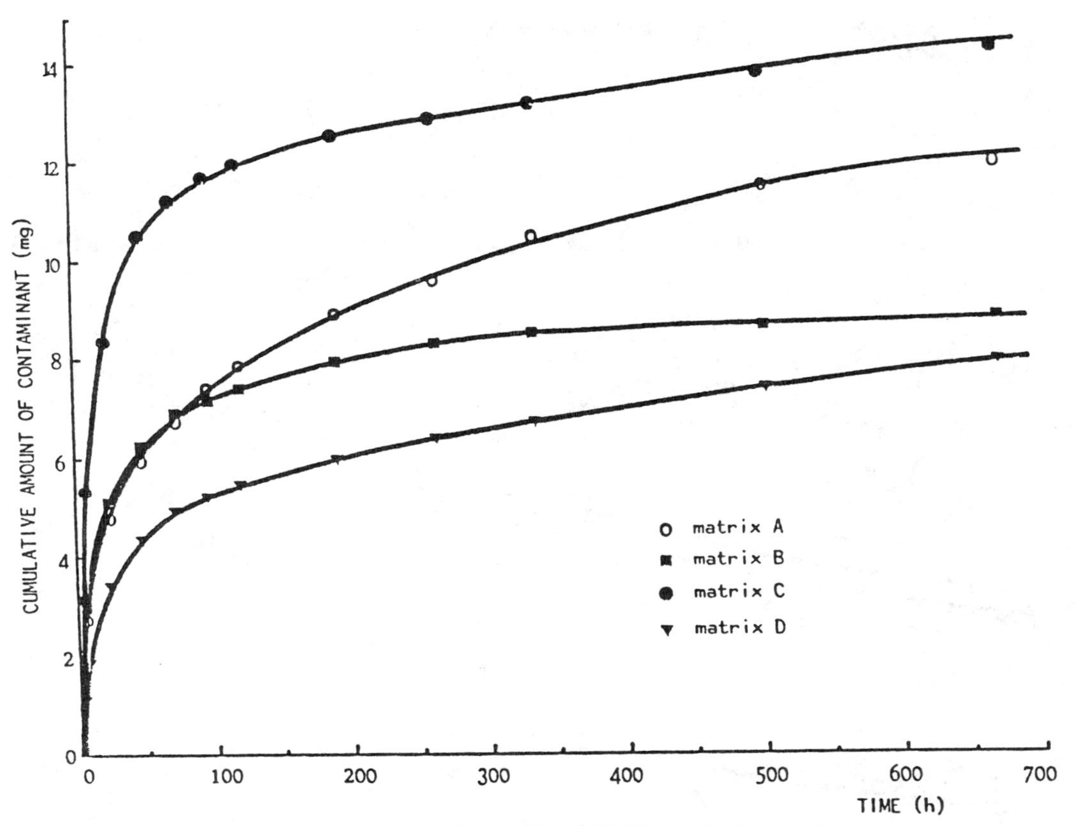

FIG. 1—*Dynamic leaching test (0.4-M lithium): obtained results.*

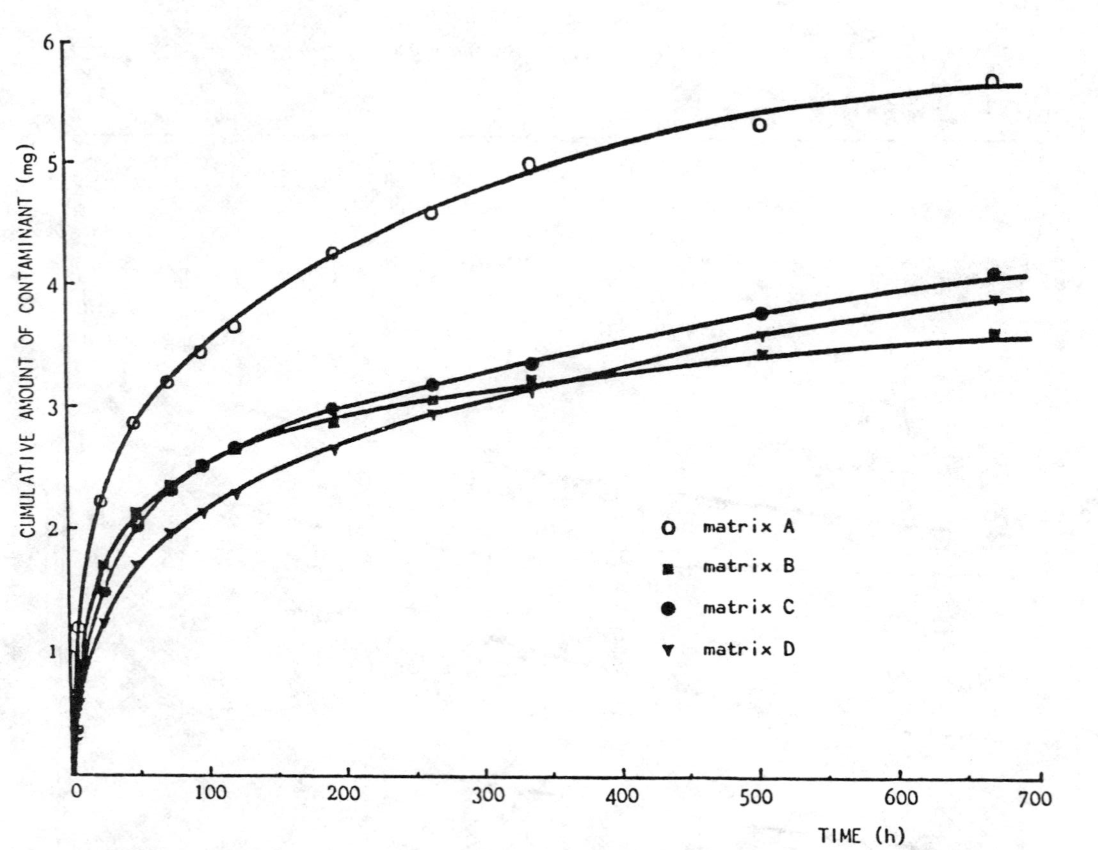

FIG. 2—*Dynamic leaching test (0.4-M lithium and copper): obtained results.*

TABLE 3—*Cumulative contaminant loss during different phases of leaching test (0.4-*M *lithium) and matrix permeability.*

Matrix	F_1, mg	F_2, mg	F_T, mg	C_F, %	K, cm/s
A	5.964	6.002	11.966	26.95	9.78×10^{-6}
B	6.355	2.505	8.860	19.95	6.52×10^{-7}
C	10.568	3.508	14.376	32.38	7.34×10^{-7}
D	4.417	3.554	7.971	17.95	4.82×10^{-6}

TABLE 4—*Cumulative contaminant loss during different phases of leaching test (0.4-*M *lithium and copper) and matrix permeability.*

Matrix	F_1, mg	F_2, mg	F_T, mg	C_F, %	K, cm/s
A	2.890	2.800	5.690	25.63	9.78×10^{-6}
B	2.134	1.487	3.621	16.31	6.52×10^{-7}
C	2.061	2.062	4.123	18.57	7.34×10^{-7}
D	1.734	2.197	3.931	17.70	4.82×10^{-6}

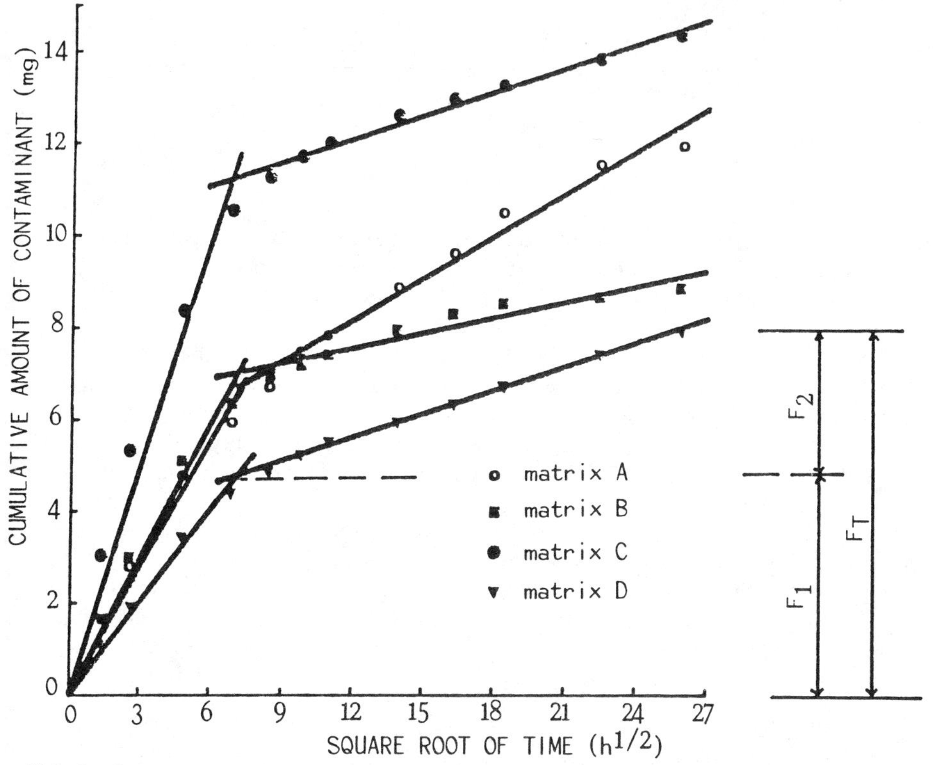

FIG. 3—*Determination of cumulative contaminant loss during washing and diffusive phase (0.4-*M *lithium).*

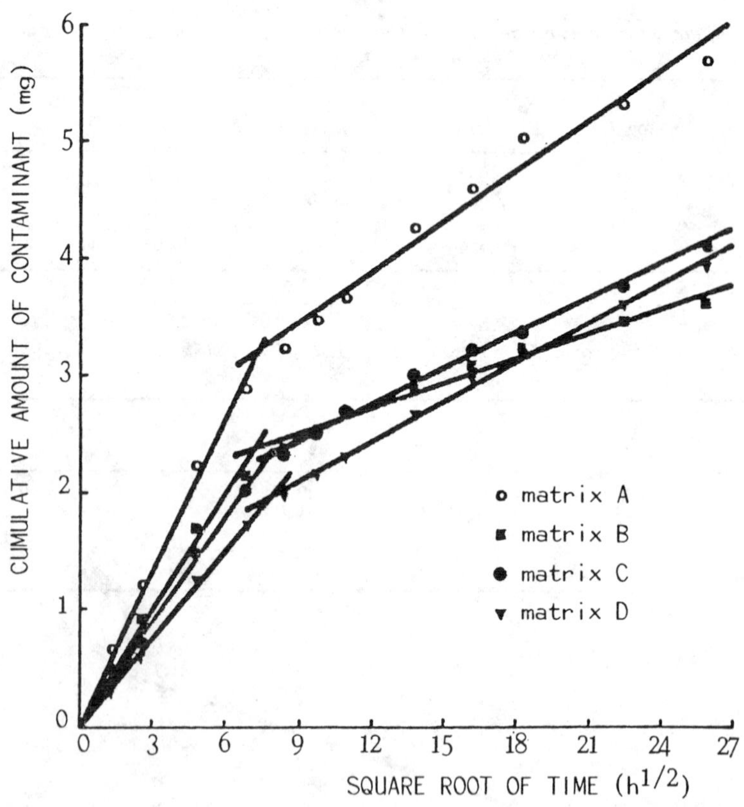

FIG. 4—*Determination of cumulative contaminant loss during washing and diffusive phase* (*0.4*-M *lithium and copper*).

the concomitant effect of the immobilized toxic concentration, which, according to knowledge on the diffusion mechanism, should have a significant influence both on modalities of release and cumulative contaminant losses during the two phases of the release (F_1 and F_2).

For this reason and also to evaluate the possible influence of the gel microporosity on the long-term leachability, experimental tests are being carried out. During the experiments, immobilized toxic concentration and matrix capillary porosity were varied by lowering water/cement ratio, by using superplasticizers and by adding micronized silica [*14–16*] to the mixtures. On the other hand, it is also necessary to take into account the chemical-physical parameters which influence the leaching mechanism, that is, cement-ion reactions [*17,18*], ion size, and ion solvation capacity.

Such problems need to be examined closely both from experimental and theoretical points of view [*17,18*]. In this way it will be possible to correlate the experimental results of the leaching tests with the microscopic and intrinsic properties of the matrix and to correctly interpret the leaching mechanism, in order to "design" immobilizing matrices as a function of the toxic element type.

Acknowledgments

Financial support for this research was provided by the Ministry of Education (MPI 40%, 1985). The authors would like to express their thanks to G. Ficara, Department of Chemistry,

Chemical Engineering, and Materials, University of L'Aquila, for carrying out the technical analyses.

References

[1] Pojasek R. B., *Toxic and Hazardous Waste Disposal,* Vol. 3, Ann Arbor Science, Ann Arbor, MI, 1982.
[2] *Guide to the Disposal of Chemically Stabilized and Solidified Waste,* Report SW-872, U.S. Army Corps of Engineers, U.S. Environmental Protection, Cincinnati, OH, Sept. 1980.
[3] Collivignarelli, C. and Urbini, G., *Ingegneria Ambientale,* Vol. 14, No. 11/12, Nov/Dec. 1985, pp. 644–654.
[4] *Measurement of the Leachability of Solidified Low-Level Radioactive Waste,* Working Group A.N.S. 16.1, American Nuclear Society, 1981.
[5] Côté, P. L. and Isabel, D. in *Hazardous and Industrial Waste Management and Testing: Third Symposium, ASTM STP 851,* American Society for Testing and Materials, Philadelphia, 1983, pp. 48–60.
[6] Joffe, P. R. and Ferrara, R. A., *Journal of Environmental Engineering,* Vol. 109, No. 4, April 1983, pp. 859–867.
[7] Godbee, H. W. and Joy, D. S., "Assessment of the Loss of Radioactive Isotopes from Waste Solids to the Environment, PART 1: Background and Theory," TM 4333, Oak Ridge National Laboratory, Oak Ridge, TN, 1974.
[8] Côté, P. L., Bridle, T. R., and Benedek, A. presented at the 3rd International Symposium on Industrial and Hazardous Waste, Alexandria, Egypt, June 1985.
[9] Côté, P. L., Bridle, T. R., and Hamilton, D. P. in *Proceedings,* National Conference on Management of Hazardous Waste and Environmental Emergencies, Hazardous Materials Control Research Institute, HMCRI, Silver Spring, MD, 1985, pp. 302–308.
[10] Koenne, W., *Zement, Kalk, Gips,* Vol. 14, No. 4, April 1961, pp. 158–160.
[11] Lieber, W. and Richartz, W., *Zement, Kalk, Gips,* Vol. 25, No. 9, Sept. 1972, pp. 403–409.
[12] *Standard Methods for the Examination of Water and Wastewater,* 15th ed., American Public Health Association, New York, 1982.
[13] Stone, J. A., *Nuclear and Chemical Waste Management,* Vol. 2, No. 2, Feb. 1981, pp. 113–118.
[14] Collepardi, M., *Scienza e Tecnologia del Calcestruzzo,* Edizioni Hoepli, Milan, 1980, pp. 178–180.
[15] Neville, A. M., *Properties of Concrete,* Pitman Publishers, London, 1975, pp. 329–335.
[16] Turriziani, R., *I Leganti e il Calcestruzzo,* Edizioni Sistema, Rome, 1972, pp. 213–227.
[17] Walsh, M. B., Eaton, H. C., Tittlebaum, M. E., Cartledge, F. K., and Chalasani, D., *Hazardous Waste and Hazardous Material,* Vol. 3, No. 1, Jan. 1986, pp. 111–123.
[18] Tittlebaum, M. E., Cartledge, F. K., Chalasani, D., Eaton, H. C., and Walsh, M. B. in *Proceedings,* International Conference on New Frontiers for Hazardous Waste Management, Pittsburgh, PA, 15–18 Sept. 1985, pp. 328–336.

Bette Kolvites[1] *and Paul Bishop*[2]

Column Leach Testing of Phenol and Trichloroethylene Stabilized/Solidified with Portland Cement

REFERENCE: Kolvites, B. and Bishop, P., **"Column Leach Testing of Phenol and Trichloroethylene Stabilized/Solidified with Portland Cement,"** *Environmental Aspects of Stabilization and Solidification of Hazardous Radioactive Wastes, ASTM STP 1033,* P. L. Côté and T. M. Gilliam, Eds., American Society for Testing and Materials, Philadelphia, 1989, pp. 238–250.

ABSTRACT: The ability of portland cement to stabilize/solidify organic wastes was evaluated by conducting column leaching tests on portland cement pastes containing phenol and trichloroethylene (TCE). Synthetic solidified waste was manufactured by mixing dilute aqueous solutions of phenol and TCE with Portland Type II cement at a 0.5 liquid-to-cement ratio. The liquid consisted of either 4000 mg/L phenol or 1100 mg/L TCE in purified water. The cement was cured for 3 days and 28 days at 100% humidity at room temperature. Thirty grams of particles approximately 5 to 10 mm in diameter were placed in an enclosed leaching apparatus. Purified water was pumped upward through the column at a rate of 0.08 to 0.18 mL/min. Every 48 h for eight days, samples of the leachate in the collection flask and samples of the vapor in the head spaces above the collection flask and leaching column were analyzed by gas chromatography. The concentrations of phenol and TCE found in these samples were used to calculate the total mass of the organic compound which was released from the solidified waste during eight days of leaching.

An empirical mathematical model was used to predict the long-term leaching characteristics of the stabilized phenol waste. The leachability indexes calculated for phenol in cement paste cured 3 days and 28 days were 7.7 and 8.2, respectively. This index rates the relative mobility of the contaminant on a uniform scale that varies from 5 (very mobile) to 15 (immobile). Similar leachability indexes could not be calculated for cement paste containing TCE due to unquantified losses of TCE by volatilization prior to the start of the leach tests and suspected decomposition of TCE in the highly alkaline environment. Gas chromatograms of the TCE waste leachates contained unidentified peaks which were thought to represent the products of TCE decomposition.

KEY WORDS: cement stabilization, leach test, phenol, solidification, stabilization, trichloroethylene

Stabilization/solidification (S/S) is a technology which has been developed for the treatment of hazardous industrial wastes. The stabilization process is designed to contain the hazardous constituents of the waste and prevent dissolution and loss of toxic materials into the environment. Prior to disposal in a landfill, the hazardous components are immobilized in a solid matrix either by chemical/physical interactions with a binding agent or by encap-

[1] Environmental engineer, Department of Civil Engineering, University of New Hampshire, Durham, NH 03824 and Roy F. Weston, Inc., Concord, NH 03301.

[2] Professor, Department of Civil Engineering, University of New Hampshire, Durham, NH 03824.

sulation with the matrix material. Cement-based stabilization techniques have been most applicable to inorganic wastes, particularly those containing heavy metals [*1*].

Waste sludges generated by the metal-finishing industry frequently contain organic solvents used for processing, cleaning, or degreasing. Surveys conducted by the Environmental Protection Agency (EPA) indicate that the ten most common toxic organic pollutants found in the waste streams of the metal finishing industry were: phenol, 1,1,1-trichloroethane, toluene, methylene chloride, trichloroethylene, tetrachloroethylene, methyl chloride, benzene, chloroform, and di-*n*-octyl phthalate [*2*]. The average daily maximum concentration of organics in the waste streams was found to be 802.0 mg/L, while the daily mean was found to be 11.3 mg/L [*2*].

Since research on the leachability of stabilized inorganic wastes has yielded favorable results, cement stabilization of metal finishing wastes has become increasingly popular as a method of ultimate disposal. However, little information is available on the leachability of organics from these stabilized wastes. It is therefore important to understand the fate of the organic compounds present in the metal sludges which have been stabilized and deposited in a landfill. This research project was designed to determine if two common organic compounds could be effectively immobilized by a cement-based stabilization process. Phenol and trichloroethylene (TCE) were chosen because they are considered representative of the two categories of organics which are typically found in metal-finishing wastes—aromatics and chlorinated hydrocarbons. Also, TCE is a very volatile compound with relatively low solubility in water, whereas phenol is non-volatile with a high solubility in water. Studies concerning the leachability of organic compounds from S/S waste can be found in the literature; however, there is a great variation in the organic compounds, the solidification matrix, and the leaching test methods which were employed. These studies also varied substantially in their conclusions. Some studies found S/S technology to be a feasible method of organic waste disposal [*3*], while others determined that it provided little treatment of the organic waste [*4*].

Continuous-flow column leaching tests were devised to simulate best the flow of groundwater through a landfill containing stabilized waste. The leaching apparatus was designed as an enclosed system to allow for monitoring of all volatilized components of the waste. Purified water was chosen as the leachant, and the solidified waste was broken into particles of approximately 5 to 10 mm (⅜ in.) diameter before being placed in the leaching columns. Gas chromatography was utilized for analysis of the leachate and vapor samples. Throughout the eight-day experiments, pH and alkalinity of the leachate were also monitored. The effects of curing time on the effectiveness of the stabilization were investigated by conducting leach tests on samples cured for 3 and 28 days.

The concentrations of organics mixed into the cement paste were relatively low (1100 to 4000 mg/L) because the objective of this research was to simulate the concentrations which might normally be found in waste treated by cement stabilization. The cement stabilization process has not generally been considered ammenable to treatment of organic solvents in large concentrations. This is mainly due to the ability of organics to prevent or delay hydration and setting of the cement paste. The concentrations chosen were low enough to be within the typical range of metal finishing waste, but high enough to be easily detected with gas chromatography.

Experimental Procedures

The stabilized waste used to conduct the leaching tests was manufactured in the laboratory using reagent grade organic chemicals. Four variations of cement paste containing organic compounds were prepared for this research: phenol cement paste cured 3 days, phenol cement paste cured 28 days, TCE cement paste cured 3 days, and TCE cement paste cured 28 days. All four cement pastes were made using Type II Portland cement with a 0.5 liquid-to-cement ratio. The liquid mixed with the cement was purified water containing either 4000 mg/L phenol or 1100 mg/L TCE. For each variation of cement paste which was prepared, a control cement paste which did not contain either TCE or phenol was also prepared.

The cement and water containing phenol were combined in a glass beaker and mixed with a glass stirring rod to apparent homogeneity. The cement paste was then poured into 10-cm (4-in.) long cylindrical molds made from 2.5-cm (1-in.) inside diameter copper tubing. The molds were lined with Parafilm[3] to ensure easy removal of the cement paste cylinders. The molds then were placed between two plates of glass, and sealed at the top and bottom with petroleum jelly to prevent evaporation or loss of water. The molds were then placed in a plastic Ziploc bag with a small beaker of water to maintain 100% relative humidity. The phenol cement paste used in Leach Test 1 was cured for 3 days, whereas the paste used in Leach Test 2 was cured for 28 days. We accounted for any possible contamination sources such as the petroleum jelly by conducting leach tests on control cement paste made by the same procedure. To determine if any phenol had adsorbed to the Parafilm liners of the cylindrical molds during curing, we placed each of the parafilm liners in a volatiles bottle with 40 mL methanol. After six weeks, the methanol was analyzed by gas chromatography for phenol.

The cement paste containing TCE was prepared by a slightly different method due to the volatile nature of this compound. For each leach test, the TCE cement paste was mixed and cured in a single 240-mL trifluoroethylene (Teflon) bottle (6 cm inside diameter and 8 cm deep). The bottle was unlined and sealed with an airtight Teflon lid for the duration of the curing period. Control cement paste was made following the same procedure, using a second Teflon bottle for mixing and curing. The TCE cement paste for Leach Test 3 was cured for 3 days, the TCE cement paste for Leach Test 4 for 28 days.

Once the cement pastes for Tests 1 through 4 had cured the designated length of time, they were removed from the molds and broken into pieces with a mortar and pestle. The pieces were sieved to collect particles 4.76 to 9.51 mm in diameter.

A diagram of the apparatus used to conduct the continuous flow leach tests is shown in Fig. 1. The leaching column consisted of a 60-mL Buchner-type funnel with a fritted-glass disk which had been modified by a glassblower. A ground-glass fitting was added to the top of the funnel so that an air-tight cover with a septum port could be placed on top. A side arm was added for effluent flow, and the neck of the funnel was tapered to a pipette tip to accommodate narrow bore tubing. A ground-glass stopper top was custom-made for the 500-mL collection flask. A septum port and influent and effluent glass tubing arms were added to the flask stopper. A 250-mL gas washing bottle filled with 200 mL methanol was connected to the effluent end of the collection flask.

The continuous-flow leach tests were conducted by placing 30 g of the stabilized waste particles in each column and pumping the leachate upward through the fritted-glass disk. Masterflex peristaltic pumps were used to maintain a constant flow of leachant through the column over an eight-day period. The flow rates were as follows: Test 1, 0.12 mL/min; Test 2, 0.08 mL/min; Tests 3 and 4, 0.18 mL/min. As the leachate filled the collection flask, the displaced air or vapor was forced through the gas washing bottle. Since both TCE and phenol are highly soluble in methanol, the gas washing bottle was expected to trap any TCE or phenol vapor passing through the system. The leachant was carried from a reservoir to the

[3] The use of tradename products in this research as described does not indicate any association with, sponsoring by, or recommendations for the manufacturer or manufacturer's product.

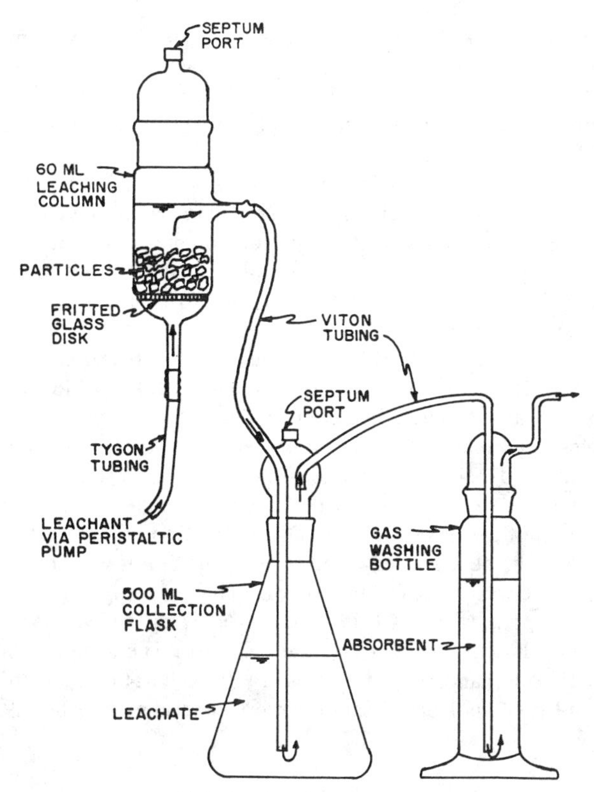

FIG. 1—*Leaching apparatus.*

leaching column through Tygon 3606 tubing. The tubing connecting the leaching column to the collection flask and the collection flask to the gas washing bottle was made from Viton when phenol waste was being tested and from glass with Teflon TFE connectors when TCE waste was being tested.

Five leaching columns were set up for each of leach tests 1 through 4. Three of the five columns contained cement paste with organics, and the remaining two columns contained control cement paste. The results from the replicate columns were averaged for the final analysis.

Every 48 h during the eight-day leach tests, four types of samples were collected from each of the leaching apparatuses. Vapor samples were collected from the head space above the leaching column and also from the head space above the collection flask. A 5-mL syringe was used to penetrate the septum, remove a sample, and immediately inject it into the gas chromatograph (GC) for analysis. A 40-mL sample of the leachate was taken from the collection flask and stored in a glass vial with a Teflon-lined lid. A second sample of the leachate (50 mL) was deposited in a polyethylene bottle for determination of pH and alkalinity. Both of these samples were immediately refrigerated at 4°C. Within one week of collection, the 40-mL leachate sample was analyzed on the GC. The pH and alkalinity determinations were conducted within two days of sample collection. Every 48 h, the methanol in the gas washing bottle was replaced with new methanol and a 40-mL sample of the spent methanol was collected. This sample was stored in a glass vial with a Teflon-lined lid at 4°C until it was analyzed on the GC (no later than two weeks after collection). The volume of leachate produced in each 48-h period was measured using a graduated cylinder.

A Perkin-Elmer Sigma 2000 Gas Chromatograph in conjunction with an LCI 100 Laboratory Computing Integrator was used for all organic analysis. A flame ionization detector and modified EPA methods of analysis were used [5]. A modification of EPA Method 604 was used for identification and quantification of phenol. TCE analysis was conducted using a modification of EPA Method 601.

The chromatographic column used for phenol analysis was 1% SP-1240-DA on 100/120 Supelcoport, 2 by 2-mm inside diameter glass. The phenol samples were not altered in any way before being injected into the column. Injection volumes were 2 μL for liquid samples and 5 ml for vapor samples. A 20-mg/L aqueous phenol standard was used for calibration of the GC program. Detection limits were determined to be 1.0 mg/L for liquid samples and 0.4 μg/L for vapor samples.

Analysis of samples containing TCE was conducted on a 60/80 Carbopack B/1% SP-1000, 8 ft by 1/8-in. outside diameter stainless steel column. A Tekmar Model LSC-2 automatic purge and trap concentrator was used in conjunction with the gas chromatograph to analyze the aqueous samples containing TCE. The parameters set on the purge/trap unit were those specified for EPA method 624 [6]. A sample size of 5 mL was used, and the trap material was Tenax. The method was calibrated with an aqueous solution containing 46.72 μg/L TCE. The detection limit for this method was determined to be 1.9 μg/L.

Vapor samples from the head spaces of the leaching apparatus were directly injected into the GC. The sample size was 5 mL. Calibration was achieved using a commercially available calibration gas of 10.2 ppm ± 2% TCE in nitrogen (56.67 μg/L). Methanol-based samples containing TCE were directly injected into the GC, also. The sample volume was 2 μL, and a 292 mg/L TCE in methanol solution was used as the calibration standard. The detection limits for vapor and methanol based samples were determined to be 1.9 μg/L and 1.0 mg/L, respectively.

Problems Encountered

TCE is an extremely volatile compound. After 63 to 80 min at 25°C, 90% of the TCE in a 1 ppm solution can evaporate [7]. Throughout this research, every attempt was made to minimize the loss of TCE by volatilization. Solutions containing TCE were made as rapidly as possible and the cement paste containing TCE was mixed, broken up, sieved, and weighed out as quickly as possible. Even with these precautions, however, a significant amount of volatilization was inevitable.

The design of the leach test apparatus was found to be inadequate for quantifying the release of TCE during the test because of the volatile nature of this compound. The gas washing bottles were incorporated into the design of the leaching apparatus to prevent any volatilized phenol or TCE from leaving the leaching system before being quantified. Phenol and TCE are both highly soluble in methanol, and their vapors were expected to be absorbed into the methanol in the gas washing bottles. TCE was so volatile, however, that it apparently volatilized out of the methanol more rapidly than it dissolved into it. Consequently, no significant quantity of TCE was found in the methanol in the gas washing bottles. Measurable concentrations of TCE were present in the headspaces of the leaching columns and collection flasks, indicating that volatilization losses did occur.

Volatilization of phenol during Leach Tests 1 and 2 was found to be minimal, making the gas washing bottles unnecessary. The concentration of phenol vapor in the head spaces of the leaching columns and the concentration of phenol in the methanol of the gas washing bottles were below the detection limits of the GC.

Results and Discussion

The results of this study indicate that generalizations about the effectiveness of S/S technology on all organic wastes are not possible. Phenol and TCE were found to behave very differently when subjected to the S/S process.

The cummulative mass of phenol which was released from the S/S waste during Leach Tests 1 and 2 was plotted on a graph as a function of the volume of leachant which flowed through the leaching column. These plots are shown in Figs. 2 and 3. Each data point on these plots represents the average of the results of three leaching columns. The standard deviations for the data are represented as error bars on the graphs.

The leachant flow rates were slightly different for Leach Tests 1 and 2 due to the difficulty of setting low flow rates on the types of pumps used for this research. The total volume of leachant which passed through the leaching columns during the eight-day test was 1346 mL for Leach Test 1 and 913 mL for Leach Test 2. Plotting the release of phenol as a function of leachant volume rather than time corrects for variations due to this discrepancy.

The ability of portland cement to serve as a matrix for solidification/stabilization of phenol appears to be limited based on the results of Leach Test 1. After the phenol cement paste was cured for three days and leached for eight days, 37 g of the original 40 g of phenol had leached out of the cement paste. Leach Test 2 provided more encouraging results: after 28 days of curing and 8 days of leaching, 23 g of phenol had leached out of the cement paste. The rate of release of the phenol during both leaching tests was greatest during the first 48-h leaching period and steadily declined for each successive leaching period. The original

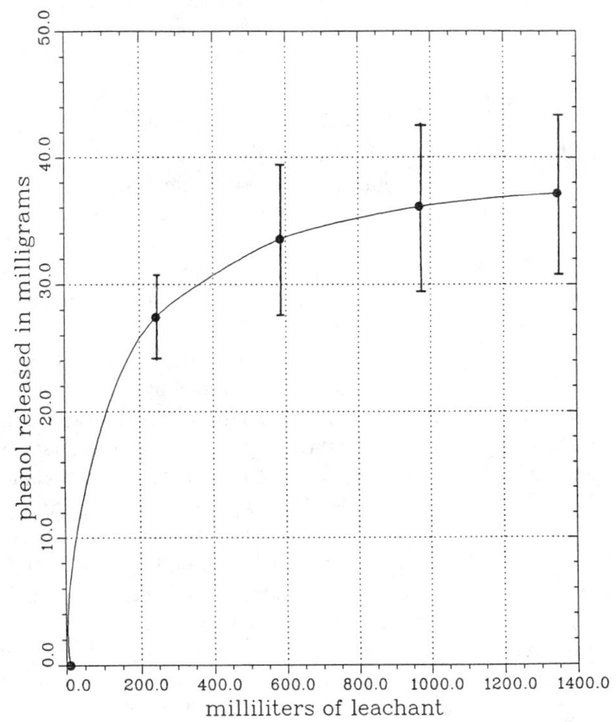

FIG. 2—*Phenol cement, cured three days, leachant flow rate = 0.12 mL/min.*

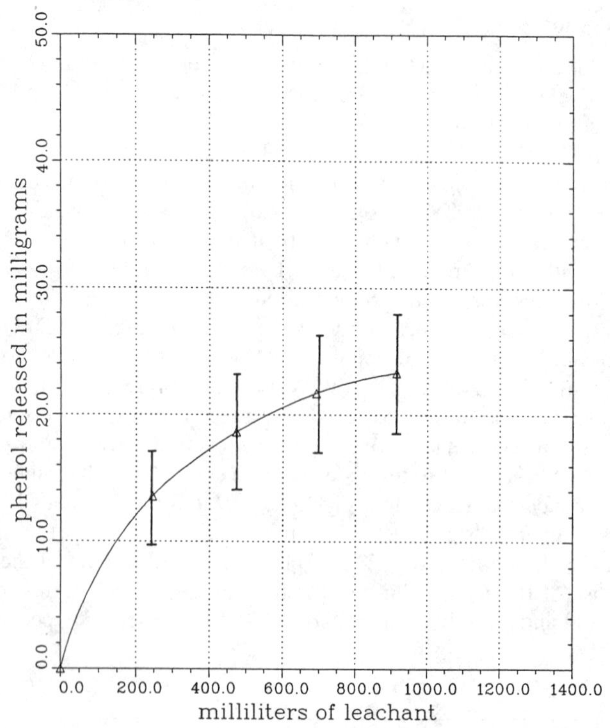

FIG. 3—*Phenol cement, cured 28 days, leachant flow rate = 0.08 mL/min.*

concentration of phenol present in the cement paste was not measured directly, but was calculated based on the concentration of phenol in the water used to make the cement paste. It was assumed that no losses of phenol occurred before the onset of the leach tests. Mass balances of the mass of phenol which was leached out of the cement paste are shown in Tables 1 and 2.

The volatile nature of TCE as opposed to phenol was readily evident in the results of the leach tests. In the case of the phenol cement paste, all detectable phenol which was released from the paste was released into the aqueous phase (dissolved in the leachate). The results of Leach Tests 3 and 4 were inconclusive with regard to the leachability of TCE; however, it was observed that TCE was released into both the aqueous and vapor phases. Losses of TCE due to volatilization prior to the start of the leach test, unquantifiable amounts of TCE vapor which left the leaching system during the test, and an unknown quantity of TCE which remained in the cement paste made it impossible to calculate a mass balance.

There was evidence that a portion of the TCE was transforming into another chemical compound during the leach tests. Gas chromatograms of leachate and vapor samples from Leach Tests 3 and 4 exhibited an unknown peak in addition to the peak representing TCE. The unidentified peak was not present in the control samples or the TCE solution used to make the cement paste. TCE has been known to react strongly with alkaline liquor or granulated potassium hydroxide to produce dichloroacetylene [*8*]. Both potassium and hydroxide ions are present in the highly alkaline cement paste ($pH > 12$), making this reaction possible during the mixing, curing, or leaching of the TCE cement paste.

The almost complete release of phenol from the three-day-old cement paste during eight days of leaching would tend to indicate that minimal chemical or physical interaction occurred

TABLE 1—*Mass balance: Leach Test 1.*

Leaching Period		Phenol Released, mg	Leachant Volume, mL
Days 1 and 2	leachate	27.42	245
	vapor	0.00	
Days 3 and 4	leachate	6.13	339
	vapor	0.00	
Days 5 and 6	leachate	2.54	385
	vapor	0.00	
Days 7 and 8	leachate	1.04	377
	vapor	0.00	
.otal phenol leached from 30 g of cement paste		37.13	
phenol adsorbed to liner of cement mold		0.05	
phenol mixed into 30 g of cement paste		40.00	
phenol retained in cement paste (calculated by difference)		2.74	

between the phenol and the cement constituents. The rate of release of the phenol therefore was more likely related to the diffusion of the phenol molecules through the pore spaces in the hardened cement paste. A change in the rate of leaching of phenol would be expected with increasing curing time if diffusion of the phenol to the outside of the cement paste was a limiting factor. However, it may not become a limiting factor until the pores reach a small enough diameter to restrict the movement of the phenol molecules. The reduction in the leaching rate of phenol between Leach Tests 1 and 2 may have been the result of some of

TABLE 2—*Mass balance: Leach Test No. 2.*

Leaching Period		Phenol Released, mg	Leachant Volume mL
Days 1 and 2	leachate	13.54	245
	vapor	0.00	
Days 3 and 4	leachate	5.08	228
	vapor	0.00	
Days 5 and 6	leachate	3.08	219
	vapor	0.00	
Days 7 and 8	leachate	1.60	192
	vapor	0.00	
total phenol leached from 30 g of cement paste		23.30	
phenol adsorbed to liner of cement mold		0.02	
phenol mixed into 30 g of cement paste		40.00	
phenol retained in cement paste (calculated by difference)		16.68	

the phenol being trapped in small-diameter pore spaces in the 28-day-old paste. Molecules of phenol may have been unavailable for leaching because their molecular diameter exceeded the size of the exit route from the cement paste or because pore spaces in the solid mass became discontinuous with increasing curing time.

A plot of the decrease in total pore volume of a cement paste as a function of curing time is shown in Fig. 4. A similar plot depicting the decrease in the mean diameter of the pores is shown in Fig. 5. The data points used to construct these graphs were obtained by averaging data developed by Diamond and Dolch [9] for cement pastes with water-cement ratios of 0.4 and 0.6 to approximate the total pore volume and mean pore diameter for a cement paste with a 0.5 water-cement ratio. Based on Figs. 4 and 5, between 3 and 28 days of curing, the total pore volume of a cement paste with a 0.5 water-to-cement ratio decreases by 21.4%. The mean pore diameter of the same cement paste decreases 87.7% between 3 days and 28 days of curing. Restrictive pore diameter or discontinuous pore structure as a result of this decrease in porosity is probably the primary reason for the decrease in the leaching rate observed between the 3- and 28-day phenol samples.

Total pore volume and pore diameter are only two factors which may alter the leaching rate with time. The diminishing supply of phenol in the cement paste as the leach tests proceeded was most likely an important factor in the decreasing rates of release of the organic compounds during the leach tests. The kinetics of any chemical reactions between the organic compounds and the cement constituents could have an effect similar to the diminishing supply. Another factor which could cause the rate of contaminant release to decrease is the formation of a surface film such as calcium carbonate. Calcium ions from dissolved calcium hydroxide in the cement paste can combine with carbonate ions from the water leachant to form calcium carbonate. No such deposit was observed on any of the specimens, but microscopic studies of the leached particles were not conducted.

The pH of the leachant was approximately 6 prior to each leach test, but increased to greater than 12 once it had passed through the leaching column. Dissolution of calcium

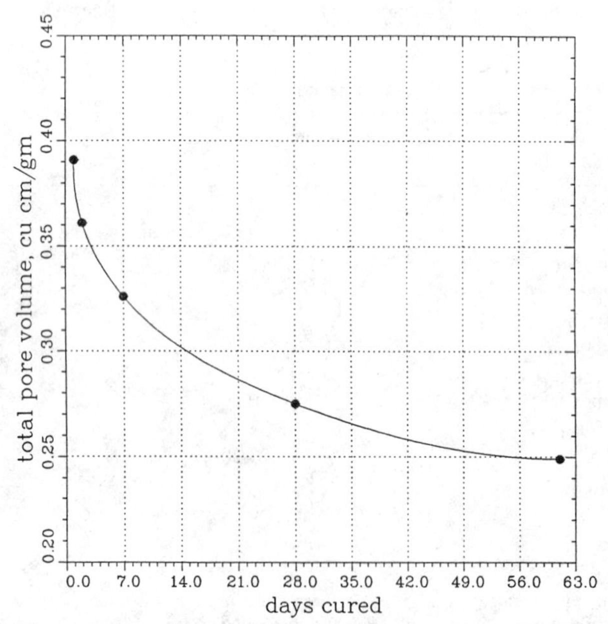

FIG. 4—*Pore volume versus curing time, 0.5 water-cement ratio (data from Ref 9).*

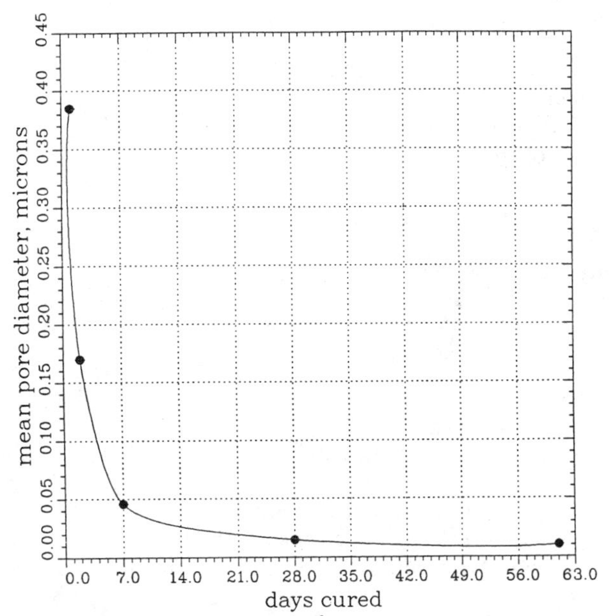

FIG. 5—*Pore size versus curing time, 0.5 water-cement ratio* (*data from Ref* 9).

hydroxide in the cement paste was assumed to be the major source of the high hydroxide ion concentrations and high alkalinities found in the leachates. Dissolution of calcium hydroxide in the cement paste would have the effect of increasing the porosity of the cement paste and therefore possibly increasing the leaching rate of contaminants in the pore structure. A decrease in the alkalinity of the leachates as the leach tests progressed may be an indication that the supply of calcium hydroxide in the cement paste was being depleted.

The dissolution of calcium hydroxide acting to increase the porosity and the formation of a calcium-carbonate surface film acting to decrease the porosity are two reactions which could occur simultaneously. The relative kinetics of each reaction would determine the overall effect on porosity.

All of the above factors probably influence the leaching rates of the organic compounds from the cement paste to a varying degree, but it would be very difficult to try to quantify individually each of their effects. Because of the complexity of the leaching mechanisms, empirical mathematical models are often used to evaluate the long-term leaching potential of contaminants from S/S waste.

The Godbee-Joy model is an empirical mathematical model which assumes that diffusional processes are the controlling factor in the leaching of contaminants from a solid matrix. To apply the model, the initial amount of the contaminant present in the stabilized waste and the surface to volume ratio of the solidified mass must be known or approximated. In order for this model to be effective, there must be no chemical or physical interaction of the contaminant with the chemical constituents of the solid matrix. The equation of the Godbee-Joy model takes the form [*10,11*]

$$\left(\frac{\Sigma a_n}{A_0}\right)\left(\frac{V}{S}\right) = 2\left(\frac{D_e}{\pi}\right)^{0.5} t_n^{\ 0.5}$$

where

a_n = contaminant loss during leaching period n, mg,
A_0 = initial amount of contaminant present in the specimen, mg,
V = volume of specimen, cm^3,
S = surface area of specimen, cm^2,
t_n = time to end of leaching period n, s, and
D_e = effective diffusivity coefficient, cm^2/s.

The leachability index (LX value) can be calculated using the equation [*11*]

$$LX_i = \frac{1}{i} \sum_{n=1}^{i} \log \left(\frac{1}{D_e} \right)$$

where i is the number of leaching periods.

One of the limitations of applying this model to cement stabilized waste is the changing porosity, and hence the changing surface-to-volume ratio, of the cement paste matrix with increased curing time. The surface to volume ratio of the 3- and 28-day cement pastes used in this research was not known, but was assumed to be the same as that of a solid sphere and assumed to remain constant for the purpose of applying the Godbee-Joy model. The porosity of the cement is then taken into account empirically as just one of the multitudes of factors which influence the diffusion rate of the contaminant through the solid matrix.

Because phenol appeared to have minimal interaction with the cement paste and because the initial amount of phenol present in the cement paste was known, enough data was available to apply the Godbee-Joy model to the release of phenol. This model was not considered appropriate for the TCE data, however, since an unknown quantity of TCE may have volatilized out of the waste before the onset of the leaching test and because there appeared to be chemical interactions occurring between the TCE and the cement paste. The original amount of TCE therefore was unknown, and diffusional processes were apparently not the limiting factor in the leaching rate.

When the Godbee-Joy model was applied to the data from Leach Tests 1 and 2, the effective diffusivity constants (D_e) listed in Table 3 were obtained. The ANS recommends using the D_e's obtained from seven leaching periods to calculate the dimensionless leachability index. Data from only four leaching periods were obtained from the leach tests, but the D_e's for these four periods were very consistant. The diffusivity constants varied from 10E-7.52 to 10E-7.86 for the 3-day paste, and from 10E-8.13 to 10E-8.26 for the 28-day paste. Four leaching periods were therefore considered sufficient for calculating the leachability index, refered to as LX_4.

The LX_4 values calculated for the 3- and the 28-day paste were 7.7 and 8.2, respectively. These values can be used to compare the relative mobility of different contaminants on a uniform scale which varies from 5 (D_e = 10E-5 cm^2/s, very mobile) to 15 (D_e = 10E-15 cm^2/s, immobile) [*11*]. The *LX* value is theoretically independent of the shape of the sample tested, the initial amount of contaminant present in the sample, and the flow rate of the leachant during the laboratory test.

The Godbee-Joy model does not take into account the effects of various factors, such as leachate characteristics, which could drastically affect the leachability and hence the *LX* value of a contaminant. Leachate characteristics such as pH, alkalinity, redox potential, or ionic strength could cause the phenol to leach out much more or much less rapidly. If the *LX* value were to be used to predict long-term leachability in a landfill, leachant repre-

TABLE 3—*Effective diffusivity constants and leachability indexes.*

Leaching Period	3-Day Phenol Cement Paste, D_e	28-Day Phenol Cement Paste, D_e
1 (48 h)	10E-7.52	10E-8.13
2 (96 h)	10E-7.64	10E-8.16
3 (144 h)	10E-7.76	10E-8.20
4 (192 h)	10E-7.86	10E-8.26
LX4 index	7.7	8.2

sentative of the actual field conditions should be used in the laboratory test. *LX* values generated from a more representative leach test would be much more valid for predictive purposes.

Summary

Phenol was released more rapidly from the 3-day cured cement paste than from the 28-day cured cement paste. At the end of an 8-day leach test, 93% of the phenol had been leached out of the 3-day paste, but only 58% of the phenol had been leached out of the 28-day paste.

The almost complete release of the phenol from the three-day-old cement paste is an indication that there was little interaction between the phenol and the cement constituents. The leaching of phenol from cement paste is therefore suspected to be controlled primarily by diffusional processes. The decrease in the amount of contaminant released after 28 days of curing is suspected to be related primarily to the decreases in the total pore volume and the mean pore diameter of the cement paste with increasing curing time.

Because of unquantifiable losses of TCE during the experiment, conclusions concerning the leachability of TCE from S/S waste are not possible. However, evidence of a chemical interaction between the TCE and cement constituents was observed. Gas chromatograms of the leachate and vapor samples contained unidentified peaks which were thought to represent the products of TCE decomposition.

An empirical mathematical model based on diffusional processes was applied to the phenol leach test data. The LX_4 values (leachability indexes based on four leaching periods) calculated for phenol in cement paste using the Godbee-Joy model were 7.7 and 8.2 for the 3-day and 28-day pastes, respectively. The LX_4 value is a rating of the relative mobility of different contaminants on a uniform scale that varies from 5 (very mobile) to 15 (immobile) and can be used to predict long-term leaching rates.

The Godbee-Joy model could not be applied to the TCE data because of unquantifiable losses of TCE during the leach tests.

References

[1] Tittlebaum, M. E., Cartledge, F. K., Chalasani, D., Eaton, H., and Walsh, M., "A Procedure for Characterizing Interactions of Organic with Cement: Effect of Organics on Solidification/Stabilization," *Proceedings,* International Conference on New Frontiers for Hazardous Waste Management, U.S. EPA, Washington, DC, 1985.

[2] *Survey of Solidification/Stabilization Technology for Hazardous Industrial Wastes,* U.S. Environmental Protection Agency, Washington, DC, 1979.

[3] Gilliam, T. M., Dole, L. R., and McDaniel, E. W. in *Proceedings,* International Symposium on Industrial and Hazardous Waste, Alexandria, Egypt, 1985.
[4] Escher, E. D. and Newton, J. W., "Waste Immobilization in Cement-Based Grouts," *Hazardous and Industrial Solid Waste Testing, ASTM STP 933,* American Society for Testing and Materials, Philadelphia, 1985, pp. 295–307.
[5] *Characterization of Hazardous Waste Sites: A Methods Manual, Vol. 3: Available Laboratory Analytical Methods,* U.S. Environmental Protection Agency, Washington, DC, 1984.
[6] *Liquid Sample Concentrator Operation Manual,* Tekmar Co., Cincinnati, OH, 1984.
[7] Verschueren, K., *Handbook of Environmental Data on Organic Chemicals,* 2nd ed., Van Nostrand Reinhold, New York, 1977.
[8] Ott, E. and Packendorff, K., *Berichte der Deutschen Chemischen Gesellschaft,* Vol. 64B, 1931, pp. 1324–1329.
[9] Diamond, S. and Dolch, W. L., *Journal of Colloid and Interface Science,* Vol. 38, No. 1, 1972.
[10] Godbee, H. W. and Joy, D. S., *Assessment of the Loss of Radioactive Isotopes from Waste Solids to the Environment, Part 1: Background and Theory,* Oak Ridge National Laboratory, Oak Ridge, TN, 1974.
[11] Côté, P. L. and Hamilton, D. P., "Leachability Comparison of Four Hazardous Waste Solidification Processes," *Proceedings,* 38th Annual Purdue Industrial Waste Conference, 1983.

Robert R. Landolt[1] *and Linda R. Bauer*[1]

A Study of Tritium Release from Encapsulated Titanium Tritide Accelerator Targets

REFERENCE: Landolt, R. R. and Bauer, L. R., "**A Study of Tritium Release from Encapsulated Titanium Tritide Accelerator Targets,**" *Environmental Aspects of Stabilization and Solidification of Hazardous and Radioactive Wastes, ASTM STP 1033,* P. L. Côté and T. M. Gilliam, Eds., American Society for Testing and Materials, Philadelphia, 1989, pp. 251–256.

ABSTRACT: Titanium tritide accelerator targets were encapsulated in several types of waste-disposal solidification media. The resulting monoliths were subjected to a standard leach testing procedure. A bare tritium target was also immersed in water and evaluated for tritium release. As anticipated, more tritium was released from the bare target than from the encapsulated targets, but not by a large factor. Of the encapsulating media tested, gypsum cement, coated and uncoated, performed slightly better than epoxy or microsilica cement. A gypsum cement-encapsulated target was also tested in air, and its tritium release was compared with that of the gypsum cement-encapsulated target which had been subjected to the leach test. Significantly more tritium was released from the specimen suspended in air. The results indicate that tritium can move through these barrier materials and may be released from an encapsulated target exposed to water or to air.

KEY WORDS: tritium waste disposal, tritium leach testing, waste encapsulation, tritium target disposal

Tritium targets are used in accelerators to produce 14-MeV neutrons for use in fusion research, cross-section measurements, shielding experiments, fissile material safeguards interrogation, borehole uranium exploration, activation analysis, and radiation therapy [*1,2*]. Such targets are typically titanium tritide adsorbed on a copper backing. They range in tritium activity from 10^5 MBq (2.7 Ci) [*3*] to as much as 3.7×10^7 MBq (1000 Ci) [*4*]. Often a considerable amount of tritium remains on the target after it has expended its usefulness. Consequently, upon removal from an accelerator, used targets may represent a significant radioactive waste disposal concern.

It has not been uncommon that the disposal of these targets has been in their original shipping containers, which offer minimal protection against water infiltration. The purpose of the present study was to evaluate the use of encapsulating media as a simple method of improving target disposal practices. To accomplish that objective, we encapsulated several titanium tritide targets in commonly used low-level waste solidification media and determined relative tritium release values by subjecting the encapsulated targets to the American Nuclear Society (ANS) Proposed American National Standard Measurement of the Leachability of Solidified Low-Level Radioactive Wastes (Draft) (ANS 16.1).

[1] Professor of health sciences and U.S. Department of Energy graduate fellow, respectively, School of Health Sciences, Purdue University, West Lafayette, IN 47907.

Procedures

Target Assay

Each tritium target was in the form of titanium tritide deposited on a copper disk 3.175 cm in diameter. Some of the targets were completely covered by tritium, whereas on other targets the active tritium area was 2.54 cm in diameter. The targets had activities between 1.85×10^5 and 2.25×10^5 MBq. Because the assay of tritium on the purchased targets was only an approximation, it was necessary to develop a hydrogen-3 (^3H) assay method in order to provide a mechanism for relating tritium activity on the foils to the leach test results. This was done by determining the relative ^3H activity on each target by measuring bremsstrahlung emission rates with a thin-window Geiger-Müeller (G-M) detector. All targets were assayed with the same counting system at a source-to-detector distance of 30.5 cm and a 60-s count time. A cobalt-60 (^{60}Co) reference source was counted concurrently with each target to permit correction for time-dependent variations in counter efficiency. Correction for coincidence loss was made on all counts using a combination of the point source method and the method of aliquots.

Leach Tests

Encapsulated tritium targets were subjected to the ANS 16.1 ten-interval, 90-day static leach test which was conducted at ambient temperatures. The simulated waste specimens were in the form of right circular cylinders 5 cm long and 4 cm in diameter, and all specimens were treated in the same manner with respect to the testing protocol. All leach testing was performed entirely in a glove box.

The exclusive source of leachant solution was distilled, deionized water having a total organic carbon of 0 to 2 ppm and an electrical conductivity of 5 to 10 μmho/cm. The target encapsulation cylinders had surface areas of 88 cm^2. Consequently, the required leachant volume used for each leachant period was 880 mL in order to meet the ANS 16.1 protocol that the quotient of leachant volume to surface area be approximately 10 cm. The leach test vessels used were 1-L wide-mouth, screw-top polyethylene bottles. A nylon filament was used to suspend each encapsulated target in its leach test vessel such that essentially all of the surface area of the test specimen was exposed to water. Leachant exchanges were conducted at elapsed times of 2 h, 7 h, and 1, 2, 3, 4, 5, 19, 47, and 90 days by withdrawing the specimen from the leachant and resubmerging it in an equal volume of fresh leachant solution in the next container. Each exchange took only a few seconds which precluded surface drying. Following each exchange, a 1-mL aliquot of the leachant solution was assayed for tritium by liquid scintillation counting.

Encapsulation Media

Gypsum Cement—A gypsum cement product consisting of a calcium sulfate hemihydrate binder in conjunction with a polymer was prepared in a cylindrical mold. A water-to-cement ratio of 0.60 was used. A tritium target was immediately inserted into the paste due to the exceedingly short initial set time of the gypsum cement. The target was oriented vertically within the cylinder to place the tritiated surface at the axial midplane of the specimen. A closed-end hook was also inserted into the paste to facilitate suspension of the monolith during leach testing. The specimen was covered and allowed to cure for 3 days. The mold was then removed and the specimen was allowed to cure for an additional 4 days.

Coated Gypsum Cement—A second gypsum cement-encapsulated target was prepared as described above. Following the cure period, however, a surface coating was applied to the vaste specimen. The surface coating corresponded to the U.S. Bureau of Reclamation standard CA-50 for cold-applied coal tar paint. Application of the coating was achieved by ompletely submerging the cement cylinder in the paint, allowing any excess paint to run ff, and then allowing the coating to dry thoroughly. Over a two-day period, a total of four oats were applied to the specimen to provide a surface coating of approximately 1-mm hickness.

Epoxy—A two-part epoxy system was also used to encapsulate a tritium target. Part A onsisted of polyamide resin, 16.76%; glycol ethers, 35.06%; pigment, 47.89%; and unpecified additives, 0.29%; by volume. Part B consisted of epoxy resin, 48.72%; silicon esin, 1.28%; and glycol ethers, 50%; by volume. The specimen was produced by handnixing a 1:1 volumetric ratio of Parts A and B. After an initial set time of approximately 0 h, the epoxy was sufficiently cured to support the weight of the target. Following insertion f the target, an additional four-day cure period was allowed before the mold was removed. further set time of 24 h ensured that the outer surface of the specimen was fully cured.

Microsilica Cement—A microsilica concrete additive was used to prepare a microsilica ement specimen having a water-to-cement ratio of 0.44 and a microsilica loading of 13.2% y weight. The paste was sufficiently cured to support the weight of the target approximately h after preparation. Following insertion of the target, the cement matrix was allowed a hree-day cure period, followed by an additional two-day cure period after the mold was emoved.

ritium Release from Dry Monolith

It was also of interest to determine how much tritium would escape from a dry monolith nto surrounding air with no leachant present. To estimate the fraction of tritium that would e released from a test specimen subjected to these conditions, a separate gypsum cementncapsulated target was suspended in an enclosed volume of air at ambient temperature for 0 days.

The test vessel was a 1-L glass bottle sealed with a ground-glass stopper. The stopper was nodified to have inlet and outlet ports to which tubing could be attached. Because the inlet nd outlet ports were unavoidably only 5 cm apart, it was necessary to insert a baffle in the ottle to prevent short-circuiting of the air streams.

To collect the airborne tritium at the designated sampling times, a vacuum pump was sed to push air past the encapsulated target and through two bubblers connected in series. he bubbler system consisted of fritted cylinders used in conjunction with 300-mL gas vashing bottles. Both bottles were filled with 100 mL of water from which 1-mL aliquots vere removed and assayed for tritium at intervals corresponding to the ANS 16.1 test ampling schedule.

irect Release to Water

It was hypothesized that the largest release of tritium from a target would occur when vater was permitted to come into direct contact with the active surface of the target. With hat in mind, an unprotected tritium target was exposed to water at ambient temperature or 90 days. The water was sampled for tritium activity at the intervals specified by the ANS 6.1 leach test. Due to the potential magnitude of tritium release anticipated during the xperiment and to minimize the volume of tritiated water generated, an essentially closed ystem containing a single volume of leachant solution was used.

The leachant vessel used was a 2-L round-bottom flask with center and side ground glass openings. A hollow 45/50 stopper provided the seal for the center opening. The stopper however, was modified with top and bottom openings of sufficient diameter to accommodate a nylon filament which was needed for lowering and suspending the target in the leachant. The closure for the side opening was a sampling adapter with a 24/40 joint at the bottom and an 8-mm septum port at the top. The septa used were composed of self-sealing rubber and were supplied with the adapter. After 1850 mL of water had been placed in the flask the target, held by a support fashioned out of gypsum cement, was lowered into the flask to just above the water level. After both the 45/50 and 24/40 openings were securely closed the target was completely submerged in the water.

To extract water samples at the designated intervals, a 1-mL syringe with a 17-gage needle was inserted through the septum. Each 1-mL aliquot was discharged from the syringe into 999 mL water diluent. From these solutions, 1-mL samples were withdrawn for liquid scintillation analysis.

Results and Discussion

Figure 1 illustrates the manner in which the cumulative activity released (CAR) increases with time for the four encapsulated targets and the bare target in water. The CAR values were normalized to the gypsum cement specimen to account for differences in initial tritium activity on the targets, as determined by the bremsstrahlung assay method. As can be seen from the figure, the gypsum cement specimens were slightly more effective at retaining the tritium than the other two media.

It is also noteworthy that, despite some initial improvement, the process of coating the gypsum cement apparently did not improve its retention capabilities by the end of 90 days. The reasons for this result are unclear. The coating was applied at room temperature and had been previously reported by Varghese et al. [5] as being quite successful in acting as a

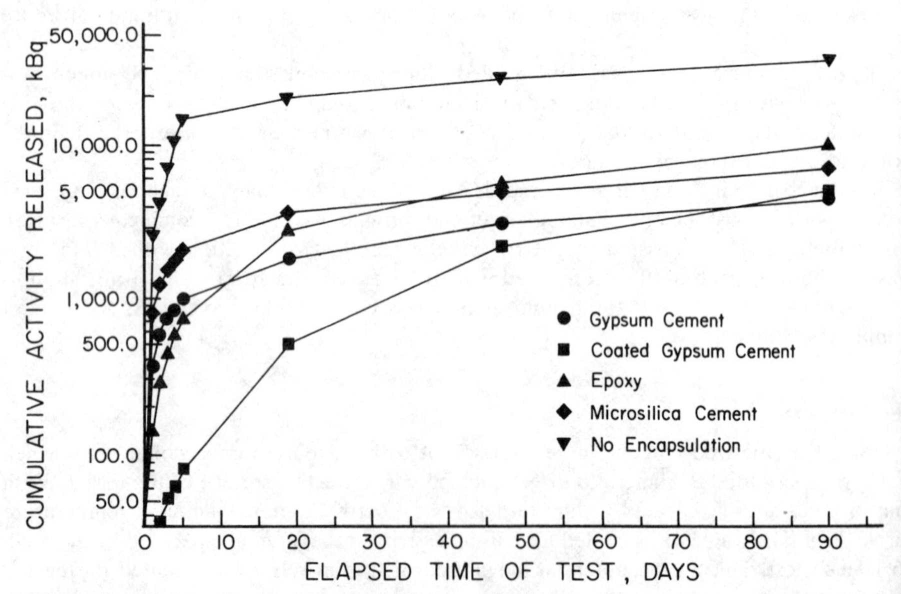

FIG. 1—*Cumulative tritium activity released from encapsulated and bare accelerator targets as function of leaching time.*

barrier to tritium migration from cement specimens. Possibly a much thicker coating was required. However, the need to thoroughly dry the cylinder between coating applications made an extended treatment period unacceptable for this study because the magnitude of tritium loss prior to commencement of the first leaching interval would not have been reflected in the leach test results.

It can also be noted from the figure that all four encapsulated specimens released less tritium than the bare target in water, but the difference was much less than had been anticipated. The fact that the encapsulated targets did not more successfully outperform the unprotected target suggests that the materials tested presented very little barrier to tritium movement. The encapsulating media may have retardation properties which were offset by an enhanced release of tritium from the target surface as the encapsulants cured. Perhaps the rate-determining factors for these competing processes are significantly affected by the temperature or the substances or both to which the titanium titride (TiT_2) is exposed. Iyengar et al. [6] have studied the effect of temperature on tritium release rates in air from a 1.48×10^4 MBq titanium tritide target. Their results were 1.95 to 2.47 Bq/s at 25°C, 34.02 Bq/s at 80°C, and 143.89 Bq/s at 200°C. If these values are upscaled to a 1.85×10^5 MBq target, they suggest that thermally driven or thermally catalyzed phenomena may be mechanisms of tritium release if sufficiently high centerline temperatures are encountered as the encapsulation media cure. This possibility may have been particularly important to this study in view of the fact that all of the targets were positioned along the axial midplane of the cylinders where the highest temperatures would be found.

Figure 2 shows that the gypsum cement-encapsulated target exposed to air exhibited a pattern of tritium release that very closely mimicked the release observed during the gypsum cement-in-water test. At the time the test was terminated, the cylinder tested in air had actually released significantly more tritium (7.88 versus 4.40 MBq) than the specimen tested in water. In addition, as shown in Fig. 2, if one expresses the results of the two tests on a normalized CAR basis, the difference in performances becomes even more distinct (9.62

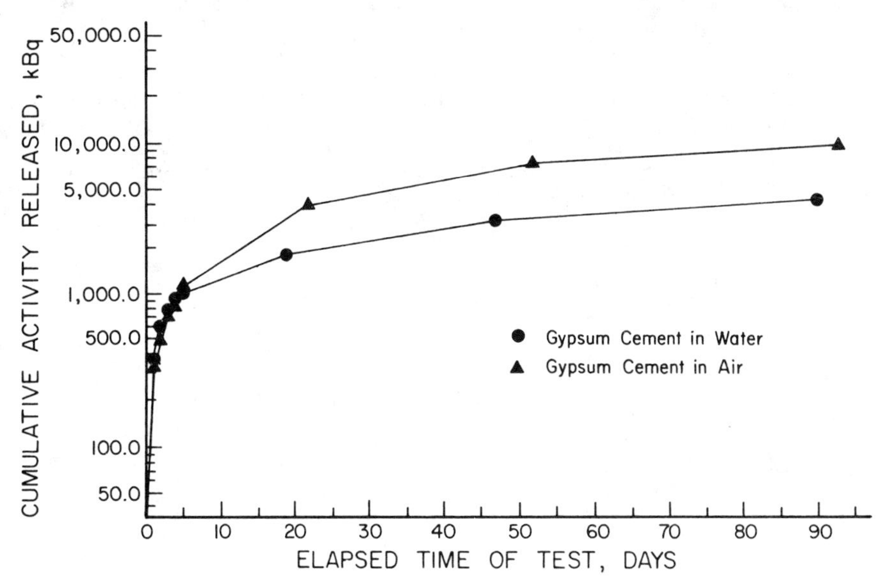

FIG. 2—*Cumulative tritium activity released from gypsum cement-encapsulated accelerator targets suspended in air and in water as function of time.*

versus 4.40 MBq). These results provide evidence that the gypsum cement-encapsulated targets have some inherent or baseline release rate which will be observed independent of the presence or absence of a leachant. In the above study, contact of the monolith with water in fact may have inhibited tritium emanation relative to the dry surface of the cylinder tested in air. However, it would seem unlikely that with much longer exposure periods, the release rate from the monolith in air would continue to exceed the release rate of the specimen tested in water.

As indicated by this research, the short-term roles of both an encapsulant and a leachant may be secondary to those related to the target itself in determining the rate of ^{3}H release from a plane source. The potential importance of such source-specific release mechanisms may not necessarily be revealed by standardized leach testing. It is therefore recommended that, when evaluating encapsulation effectiveness for highly mobile species (such as hydrogen-3, carbon-14, or iodine-129), there may be an additional need for complementary studies which examine non-leachant-based diffusion mechanisms.

Acknowledgments

Funding for this research was provided by the U.S. Department of Energy Thesis Research Program on Nuclear Waste Management under Agreement Number ARGONDO 055958, and by the Institute of Nuclear Power Operations under Agreement Number 052483.

References

[*1*] Barschall, H. H. in *Neutron Sources for Basic Physics and Applications,* S. Cierjacks, Ed., Pergamon Press, New York, 1983, pp. 57–80.

[*2*] Sauermann, P. F., Friedrich, W., Knieper, J., and Printz, H., "Fifteen Years of Experience in Handling Tritium Problems in Connection with Low-Energy Particle Accelerators," *Proceedings of Conference on Behavior of Tritium in the Environment,* International Atomic Energy Agency, Vienna, 1979, pp. 671–686.

[*3*] "Radiation Protection and Measurement for Low-Voltage Neutron Generators," NCRP Report No. 72, National Council on Radiation Protection and Measurements, 1983.

[*4*] Logan, C. M. and Heikkinen, D. W., *Nuclear Instruments and Methods,* Vol. 200, 1982, pp. 105–111.

[*5*] Varghese, C., Singh, I., Agarwal, R. P., Ramani, M. P. S., and Khan, A. A., "Handling of Tritium Contaminated Effluents and Wastes—A Final Report," BARC-1201, Bhabha Atomic Research Center, Bombay, India, 1983.

[*6*] Iyengar, T. S., Soman, S. D., and Ganguly, A. K. in *Tritium,* A. A. Moghissi, Ed., Messenger Graphics, Phoenix, 1973, pp. 764–772.

Ray Mark Bricka[1] and Donald O. Hill[2]

Metal Immobilization by Solidification of Hydroxide and Xanthate Sludges

REFERENCE: Bricka, R. M. and Hill, D. O., "**Metal Immobilization by Solidification of Hydroxide and Xanthate Sludges,**" *Environmental Aspects of Stabilization and Solidification of Hazardous and Radioactive Wastes, ASTM STP 1033,* P. L. Côté and T. M. Gilliam, Eds., American Society for Testing and Materials, Philadelphia, 1989, pp. 257–272.

ABSTRACT: The U.S. Department of the Army (DA) is responsible for the treatment and disposal of heavy metal wastes generated at DA facilities. Heavy metal wastewaters are generally treated by hydroxide precipitation. An alternative technique uses xanthates to precipitate the heavy metals. Both of these processes produce a sludge defined as hazardous by the Resource Conservation and Recovery Act (RCRA). One method proposed for treating such materials is solidification/stabilization (S/S). S/S renders a waste less toxic by chemically immobilizing the hazardous constitutes in a solid matrix.

This study evaluated the ability of cellulose and starch xanthate to immobilize heavy metals and compares the results with the metal immobilization capability of hydroxide precipitation. Solidified and unsolidified sludges resulting from starch xanthate, cellulose xanthate, and hydroxide precipitation processes were tested for their physical and chemical characteristics. Sludges were prepared using either cellulose xanthate, starch xanthate, or calcium hydroxide to treat a synthetic waste solution containing four heavy metals (cadmium, chromium, nickel, and mercury) in a laboratory-scale study. Sludges were solidified with portland Type I cement. Physical tests performed on the solidified and unsolidified xanthate and hydroxide sludges included unconfined compressive strength and the cone penetration tests. Chemical leaching characteristics of the unsolidified sludges were determined using the U.S. Environmental Protection Agency's extraction procedure test and a serial graded batch extraction procedure. Results from this study indicated that the solidified materials reduced the concentration of metals in the leachate when compared to the unsolidified sludges. The solidified sludges also developed unconfined compressive strengths as high as 2000 kPa.

KEY WORDS: hazardous waste, sludge, xanthate, solidification, metal waste, hydroxide sludge

The U.S. Environmental Protection Agency (EPA) estimates that there are over 13 000 generators of metal plating and finishing wastewaters. Treatment of these wastewaters is estimated to generate approximately 11 million metric tons/year of sludge requiring special handling and disposal. The generator of metal-contaminated wastewaters is faced with the dual problem of, first, removing the metal contamination from the wastewater and, second, disposing of the residual materials resulting from the wastewater treatment process. The U.S. Department of the Army (DA), as a generator of wastewater containing heavy metals, recognizes this problem and has initiated various research projects designed to reduce or more effectively treat and dispose of these wastes.

[1] Chemical engineer, U.S. Army Engineer Waterways Experiment Station, Vicksburg, MS 39180-0631.

[2] Chairman, Department of Chemical Engineering, Mississippi State University, Mississippi State, MS 39762.

Typically, wastewaters generated by metal plating and finishing operations are treated to remove metal contaminants using hydroxide precipitation. Hydroxide precipitation usually involves adding calcium hydroxide, calcium oxide, or sodium hydroxide to the wastewater being treated. The effectiveness of hydroxide precipitation is highly dependent on the solution's pH, and, in some cases, hydroxide precipitation does not remove sufficient contaminants necessary to meet today's stringent treatment standards. In addition to these difficulties, hydroxide precipitation produces large volumes of sludge classified as a hazardous waste under the Resource Conservation and Recovery Act (RCRA). This results in large sludge disposal costs. Further increases in cost results from the fact that hydroxide precipitated sludges are generally difficult to dewater, necessitating the handling and disposal of sludges with a high moisture content.

An alternative-treatment method, developed by the U.S. Department of Agriculture [*1,2*] uses insoluble starch xanthates for the removal of heavy metals from industrial wastewater. Closer investigation reveals that metal precipitation is not limited to starch, but an entire family of xanthates can potentially be used for wastewater treatment. Xanthate precipitation offers several advantages over the hydroxide precipitation. Advantages include better metal removal capability, less sensitivity to pH fluctuations, improved sludge dewatering properties, and the capability of selective removal of metals according to the following hierarchy [*3*].

$$\text{Na} \ll \text{Ca—Mg—Mn} < \text{Zn} < \text{Ni} < \text{Cu—Pb—Hg}$$

Unfortunately, xanthate precipitation also produces significant quantities of sludge, which also must be handled in accordance with RCRA requirements.

Shallow land burial is the current accepted disposal method for solid heavy metal wastes. Due to the nature of this disposal technique, waste disposed through shallow land burial has a high probability of coming in contact with water, which provides a medium for leaching and transporting contaminants into the surrounding environment. As a result, additional treatment to prevent contaminant leaching from the sludge may be required prior to disposal.

Chemical stabilization/solidification (S/S) is the current method used to enhance the heavy metal immobilization characteristics of a fluid or semifluid waste. S/S involves the mixing of waste with a binder material to enhance the physical properties and to chemically bind the free liquid. In fact, current regulations mandate that toxic wastes containing free liquid may no longer be landfilled in a shallow land burial facility. Such wastes must exhibit properties of a solid (that is, passing the paint filter tests—Method 9095 [*4*] as outlined in Fig. 1 [*5*]).

Xanthates

Xanthates are a family of organic compounds formed by reacting an organic hydroxyl-containing substrate with carbon disulfide as shown below:

$$\underset{\text{Organic Substrate}}{\text{R—OH}} + \underset{\text{Carbon Disulfide}}{\text{S=C=S}} + \underset{\text{Caustic}}{\text{NaOH}} \text{ —— } \underset{\text{Xanthate}}{\text{R—O}\overset{}{\underset{\underset{\text{S}}{\|}}{\text{C}}}\text{SNA}} + \underset{\text{Water}}{H_2O}$$

Compounds known to be effective substrates include starch, cellulose, and alcohols [*6*].

Xanthates remove metals from solution utilizing a simple substitution reaction that is much like hydroxide precipitation:

$$X(OH)_2 + M^{++} — M(OH)_2 + X^{++} \quad \text{(Hydroxide reaction)}$$

M = metal ion and X = cation

$$R{-}O\underset{\underset{\displaystyle S}{\|}}{C}SNa + M^{++} — R{-}O\underset{\underset{\displaystyle S}{\|}}{C}SM$$

(Xanthate reactions)

$$2(R{-}O\underset{\underset{\displaystyle S}{\|}}{C}Na) + M^{++} — R{-}O\underset{\underset{\displaystyle S}{\|}}{C}S{-}M{-}S\underset{\underset{\displaystyle S}{\|}}{C}O{-}R$$

This phenomenon allows easy adaptation of waste treatment facilities currently using the more classical hydroxide precipitation methods to xanthate precipitation methods.

Purpose and Scope

The purpose of this research was to investigate heavy metal immobilization by cellulose and starch xanthate precipitation and to evaluate the effects of a typical S/S technique on the sludges produced by xanthate precipitation. These results are compared with sludges prepared using typical hydroxide precipitation followed by S/S. Only a portion of the methods and results for this study could be presented in this paper due to space limitations. Full details of this study are presented in a comprehensive report entitled "Investigation and Evaluation of the Performance of Solidified Cellulose and Starch Xanthate Heavy Metal Sludges" [7].

Material and Methods

The study reported on herein was conducted in four phases: (1) preparation of the metal sludges; (2) solidification/stabilization of the various sludges; (3) preparation of test specimens; and (4) physical and chemical evaluation of the test specimens. Each phase is discussed as follows.

Preparation of the Metal Sludges

A synthetic metal plating waste was prepared by dissolving four metal nitrate salts, cadmium nitrate [$Cd(NO_3)^2 \cdot 4H_2O$], chromium nitrate [$Cr(NO_3)^3 \cdot 9H_2O$], nickelous nitrate [$Ni(NO_3)^2 \cdot 6H_2O$], and mercury nitrate [$Hg(NO_3)_2$] in reagent grade water. Cellulose xanthate, starch xanthate, and calcium hydroxide were used to remove the metals from the synthetic wastewater. This produced three synthetic metal waste sludges having average metal concentrations (Table 1).

Solidification/Stabilization of the Sludges

Each sludge was divided into two subsamples. One subsample of each sludge was solidified using portland Type I cement binder at a binder to sludge ratio of approximately 0.3:1 by

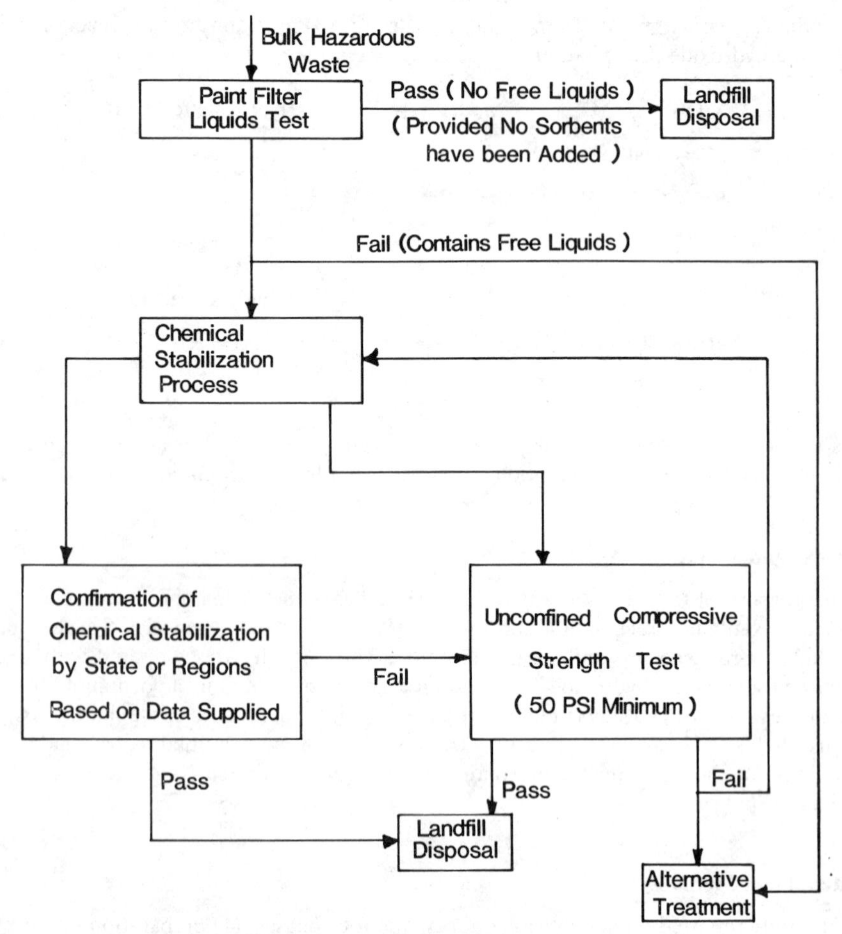

FIG. 1—*EPA testing scheme for solidified/stabilized materials.*

weight, this cement to binder ratio has been used in other S/S studies [*8*]. The other sludge subsample was not solidified.

Preparation of the Test Specimens

Solidified and unsolidified sludges were placed in 50 mm (2 in.) brass cube molds and standard proctor cylindrical molds 114.3 mm (4.5 in.) in height and 101.6 mm (4.0 in.) in

TABLE 1—*Raw sludge metal concentrations.*

Sludge	Contaminant, mg/g Dry Weight			
	Chromium	Cadmium	Mercury	Nickel
Cellulose	19.67	19.88	0.6367	19.85
Starch	17.57	33.27	1.1487	13.12
Hydroxide	74.86	16.39	0.9748	80.00

diameter. All specimens were vibrated to remove air pockets that may have developed during the molding process. The specimens were cured in the molds at 23°C and 98% relative humidity for a minimum of 24 h. They were removed from the molds when they developed sufficient strength to be free standing. After removal from the molds, the specimens were cured under the same conditions until they were utilized in the testing procedures.

Physical and Chemical Evaluation of the Test Specimens

A total of six types of sludges (unsolidified and solidified—starch xanthate, cellulose xanthate, and hydroxide sludges) were carried through the physical and chemical testing protocols described below.

Cone Penetration Test—The method described in U.S. Army Method TM 5-530 [*9*] was used to determine the cone penetration value for each specimen during the curing period of 28 days. This test measures the resistance of a material to the penetration of a 30 deg right-circular cone. The cone penetration test (CPT) is reported as force per unit area of the cone base required to push the cone through a test material at a rate of 1.83 m (72 in.) per minute. Traditionally, the CPT has been used to determine trafficability of soil or soil-like material [*10*].

Unconfined Compressive Strength Test—The unconfined compressive strength (UCS) of sludges was determined by using the ASTM Method for Compressive Strength of Hydraulic Cement Mortars (Using 2-in. or 50-mm Cube Specimens) (C 109-86) [*10*]. The UCS of the specimens was determined at 7, 14, 21, and 28 days after curing.

The USEPA Extraction Procedure (EP) Test—After the 28-day curing period, a composite sample of each sludge was prepared. The composited material was ground to pass a No. 16 sieve, the resulting solid was divided into four samples, and EP extractions were performed on each sample. Each sludge (solidified and unsolidified) was subjected to the EP test, USEPA Method 1310 [*4*]. This extraction consists of contacting approximately 100 g of solid waste with dilute acetic acid, using a liquid to solids ratio of 20 to 1. The duration of the test varies from 24 to 28 h, depending on the waste alkalinity. After extraction, the liquid was separated from the solid, and the EP extractant was analyzed according to the following USEPA methods: Cd-7131, Cr-7191, Hg-7470, and Ni-7521 [*4*]. Four replicate extractions were performed for each material. A total of 24 extractions were performed.

Serial Graded Batch Extraction Procedure (SGBEP)—The EP (now being replaced by the Toxic Characteristic Leaching Procedure [TCLP]) [*11*] was developed primarily for regulatory purposes. The data generated by the EP test are difficult to extrapolate to field situations. Several batch leaching procedures have been developed which generate test results more easily extrapolated to simulate field leaching conditions. Traditionally, sequential batch leaching procedures (SBLP) have been proposed by several authors [*12–14*]. These SBLP procedures consist of contacting a solid with a liquid until steady-state (equilibrium) conditions are achieved, removing the solid phase from the liquid, and recontacting the extracted solid with fresh liquid. This procedure is continued until the majority of the leachable contaminates are removed.

The serial graded batch extraction procedure (SGBEP) is similar to the SBLP; however, the SGBEP involves contacting the solid with a leaching fluid at varying ratios [*15,16*]. This, like the SBLP, serves to simulate varying pore volumes of water contacting the solid.

In order for the SGBEP to be applied correctly, several assumptions must be made:

1. water is the transport media,
2. only contaminants on contact with moving pore water are available for leaching,
3. contaminants that are not solubility limited are released by ion-exchange and desorption, and
4. the liquid-to-solids ratio does not affect the chemistry of the leaching process.

Since the above assumptions appear to apply to the waste system being investigated by this study and since fewer difficulties are encountered when conducting the SGBEP, this test procedure was used to investigate contaminant release.

The SGBEP used to investigate contaminant release for this study is outlined as follows; waste material used in this extraction was taken from the homogenized sludge used to conduct the EP (as previously described). The SGBEP consisted of contacting each waste (solidified and unsolidified) with distilled water on a mechanical shaker for 24 h at approximate liquid-solid ratios of 2 mL:1 g, 5 mL:1 g, 10 mL:1 g, and 50 mL:1 g. Extractions were run in triplicate in 250-mL polyethylene bottles laid in a horizontal position. After shaking, the sludge-water mixture was filtered and the extract was analyzed for cadmium, chromium, mercury, and nickel using the same methods used to analyze the EP leachate.

The concentrations of the metals in the homogenized sludges were determined by digesting of each sludges according to EPA's digestion method 3050 [*4*]. The digestate was analyzed for cadmium, chromium, mercury, and nickel using the same analytical methods used to analyze the EP leachate.

Using the preleached solid-phase concentration and the post-leach aqueous-phase concentration, the total post-leach contaminant concentration in the solid phase was determined using the formula

$$q = qo - C(V/M) \tag{1}$$

where

q = total contaminate concentration in the solid phase after leaching (mg/kg-dry weight),
qo = initial contaminate concentration of the sludge (mg/kg-dry weight),
C = contaminant concentration in the aqueous phase (mg/L),
V = volume of the aqueous phase (leachate) (L), and
M = mass of solidified waste (g-dry weight).

Desorption isotherms for each waste were developed by plotting q/qo verses C.

Results

Physical Test Results

Results of the CPT and the UCS test are presented in Figs. 2 and 3, respectively. As expected, the solidified sludges developed higher strengths than the unsolidified sludges.

Cone Penetration Test—Unsolidified starch and cellulose xanthate materials indicated only a slight resistance to penetration. The unsolidified hydroxide sludge did not develop any resistance to penetration. These CPT results for the unsolidified sludges indicate to Fig. 2 that unsolidified materials will never produce sludges with significant UCS.

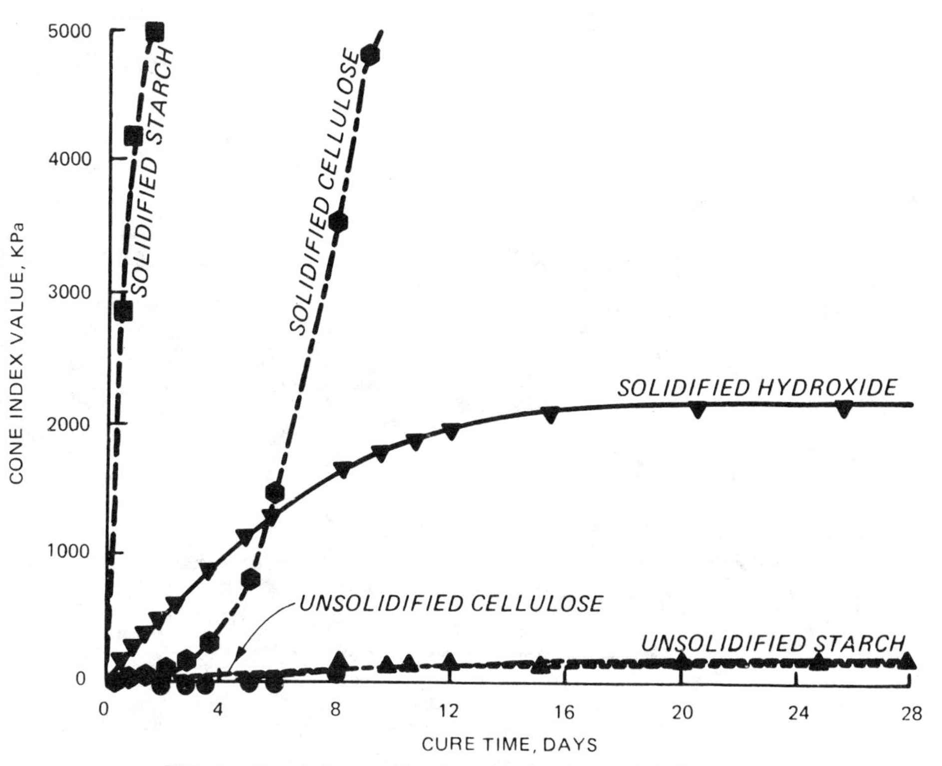

FIG. 2—*Cone index as a function of curing time and sludge type.*

Figure 2 illustrates that solidified starch and cellulose xanthate sludges developed CPT readings above 5170 kPa (750 psi) (the maximum CPT reading) in less than 10 days of curing. The solidified hydroxide sludge produced a material with a cone reading of less than 2070 kPa (300 psi) after 28 days of curing. These results indicate that the solidified starch and cellulose sludges produce sludge materials that develop significant UCS in less than 10 days, while the solidified hydroxide sludges are expected to produce sludge materials that develop much lower UCS.

Unconfined Compressive Strength Test—The results of the UCS testing are presented graphically in Fig. 3. This figure presents the UCS versus curing time for the solidified and unsolidified sludges. No curves are presented for the unsolidified hydroxide or unsolidified starch sludges because these materials did not develop a measurable UCS.

In 28 days, solidified starch and cellulose xanthate sludges developed UCSs of 2320 and 1060 kPa (337 and 154 psi respectively, while the solidified hydroxide and all unsolidified sludges developed less than 345 kPa (50 psi) in the 28-day curing period. According to the EPA's Office of Solid Waste and Emergency Response (OSWER), the "UCS test is proposed as an indirect method for determining the stability of treated waste products" and "the minimum strength recommended (to measure adequate chemical bonding) is 345 kPa (50 psi)" [*5*]. Thus, the unsolidified sludges and the solidified hydroxide sludges do not meet this recommended criteria. Based on previous research, it is assumed that the solidified hydroxide could meet the criteria if a larger binder-to-sludge ratio had been used [*8*].

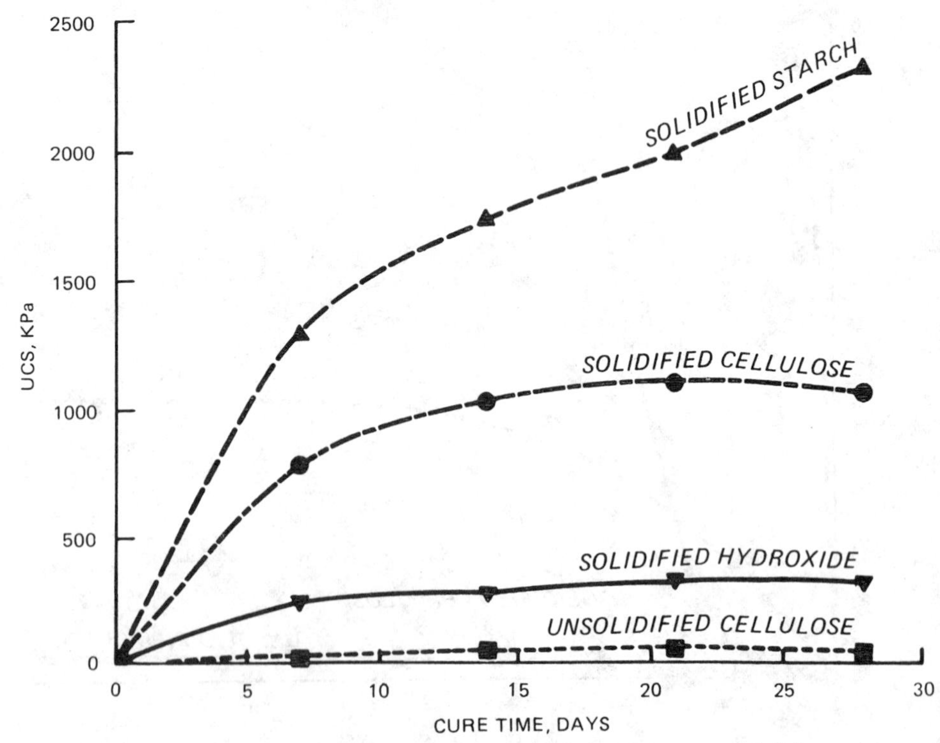

FIG. 3—*Unconfined compressive strength as a function of curing time and sludge type.*

Contaminate Release Results

EP Test Results—The results of the EP test are presented in Tables 2 and 3. Data in Table 2 indicate that all solidified sludges, except mercury for the solidified hydroxide sludges, passed the EP test. The unsolidified cellulose and starch xanthate sludges, on the other hand, failed the EP for nickel and cadmium, and the unsolidified hydroxide sludges failed the EP for each of the metals tested (cadmium, chromium, mercury, and nickel).

The data presented in Table 3 indicate that solidification is effective in reducing the leachability of the heavy metals for the xanthate and hydroxide sludges. The leachate generated from the solidified xanthate and hydroxide sludges was two to three orders of magnitudes lower than those of the leachate from the unsolidified sludges. Only the mercury data for the hydroxide sludges indicated solidification was not an effective treatment method.

Data in Table 3 also indicate that the EP leachate from the unsolidified xanthate sludges has lower metal concentrations than those observed in the unsolidified hydroxide leachate. This same trend is also reflected in the data from the solidified materials.

Serial Graded Batch Extraction Procedure—The results of the SGBEP were evaluated using desorption isotherms. A detailed discussion on the generation of isotherms is presented in an earlier report [*15*]. A desorption isotherm was generated for each treatment process and for each of the metal contaminates investigated, yielding a total of 24 desorption isotherms. In order to compare the effectiveness of the various treatment processes in immobilizing the metals, all SGBEP data were normalized by dividing q by qo. The normalized

TABLE 2—*Pass-fail results for EP test performed on solidified and unsolidified materials.*

Sludge	Type	Sample ID	Waste Contaminate Concentration Below the EP Limit[a]			
			Cd	Cr	Hg	Ni
Starch	solidified	SS4	Y	Y	Y	Y
		SS3	Y	Y	Y	Y
		SS2	Y	Y	Y	Y
		SS1	Y	Y	Y	Y
	unsolidified	US4	Y	Y	Y	N
		US3	Y	Y	Y	N
		US2	Y	N	Y	N
		US1	Y	N	Y	N
Cellulose	solidified	SC4	Y	Y	Y	Y
		SC3	Y	Y	Y	Y
		SC2	Y	Y	Y	Y
		SC1	Y	Y	Y	Y
	unsolidified	UC4	N	Y	Y	N
		UC3	N	Y	Y	N
		UC2	N	Y	Y	N
		UC1	N	Y	Y	N
Hydroxide	solidified	SH4	Y	Y	N	Y
		SH3	Y	Y	N	Y
		SH2	Y	Y	N	Y
		SH1	Y	Y	N	Y
	unsolidified	UH5	N	N	N	N
		UH3	N	N	N	N
		UH2	N	N	N	N
		UH1	N	N	N	N

[a] EP Limit Cd = 1.0 mg/L.
Cr = 5.0 mg/L.
Hg = 0.2 mg/L.
Ni = 5.0 mg/L.
The nickel EP limit is based on the "California List" for metals [*17*].

experimental data were fit to desorption models using linear regression. Three models were used that approximated the experimental data, the linear isotherm model, the langmuir isotherm model, and the Freundlich isotherm model. A fourth model, the no-release isotherm model, was also necessary to characterize the case in which the majority of the contaminant was immobilized and only small amounts of the contaminant were detected in the leachate. Figure 4 illustrates examples of each type of desorption isotherms.

The model which resulted in the largest regression coefficient (r^2) value was selected as the model that most closely fit the experimental data. Generally, the isotherm model of best fit was dependent on the metal of interest, as listed in Table 4. Examples of several desorption isotherms and their model fit are presented in Figs. 5 through 7.

The no-release isotherm models indicate that the metals were immobilized and thus the no-release isotherm is indicative of an effective treatment process. Langmuir and Freundlich isotherm models indicate that a portion of the metals are not immobilized and that there is a measurable quantity of contaminates in the leachate. These isotherm models are generally indicative of low release rates. The linear isotherm model is more appropriate where higher contaminant release rates were observed.

TABLE 3—*EP leachate concentrations for solidified and unsolidified materials.*

Sludge	Type	Sample ID	Contaminant Concentration (mg/l) Cd	Cr	Hg	Ni
Starch	solidified	1	<0.0001	0.064	<0.0008	0.13
		2	<0.0001	0.062	<0.0008	0.104
		3	<0.0001	0.064	<0.0008	0.13
		4	<0.0001	0.067	<0.0008	0.103
average			<0.0001	0.0643	<0.0008	0.1168
standard deviation			...	0.0021	...	0.0153
Starch	unsolidified	5	0.182	4.18	<0.0008	32.9
		6	0.171	4.25	<0.0008	32.8
		7	0.386	17.4	<0.0008	62.9
		8	0.473	24.8	<0.0008	75.2
average			0.3030	12.66	<0.0008	50.95
standard deviation			0.1504	10.21	...	21.49
Cellulose	solidified	10	<0.0001	0.075	0.002	0.003
		11	<0.0001	0.074	0.002	0.003
		12	<0.0001	0.071	<0.002	0.004
		13	<0.0001	0.08	<0.002	0.005
average			<0.0001	0.0750	0.002	0.0038
standard deviation			...	0.0037	...	0.0010
Cellulose	unsolidified	14	25.3	3.07	0.0105	244
		15	26.8	3.11	0.0141	254
		16	24.8	2.29	0.0135	224
		17	27.3	5.07	0.015	271
average			26.05	3.385	0.0133	248.3
standard deviation			1.190	1.185	0.0020	19.64
Hydroxide	solidified	19	<0.0001	0.011	0.843	0.002
		20	<0.0001	0.011	0.69	0.004
		21	<0.0001	0.016	0.349	0.003
		22	<0.0001	0.017	0.378	<0.001
average			<0.0001	0.0138	0.565	0.0025
standard deviation			...	0.0032	0.2412	0.0013
Hydroxide	unsolidified	23	25.9	106	0.493	50.8
		24	48.4	198	0.558	70.4
		25	90.8	383	1.54	355
		26	66.6	281	0.766	119
average			57.93	242.0	0.8392	148.8
standard deviation			27.52	118.1	0.4814	140.4

Desorption data modeled by the Langmuir, Freundlich, and linear models can be compared and evaluated graphically. As shown in Fig. 5 (the nickel data for the solidified and unsolidified starch xanthate), at a q/qo of 0.99975, the concentration of nickel in the leachate is 0.06 mg/L for the solidified sludge and 0.18 mg/L for the unsolidified sludge; thus for a q/qo of 0.99975, the solidified starch xanthate sludge is three times more effective in immobilizing the nickel than the unsolidified starch xanthate.

Using this technique, the metal immobilization potential for each treatment process was evaluated. The results of the desorption data are discussed according to the metal leached as follows:

Cadmium: As indicated in Table 4, all but the unsolidified cellulose xanthate were modeled

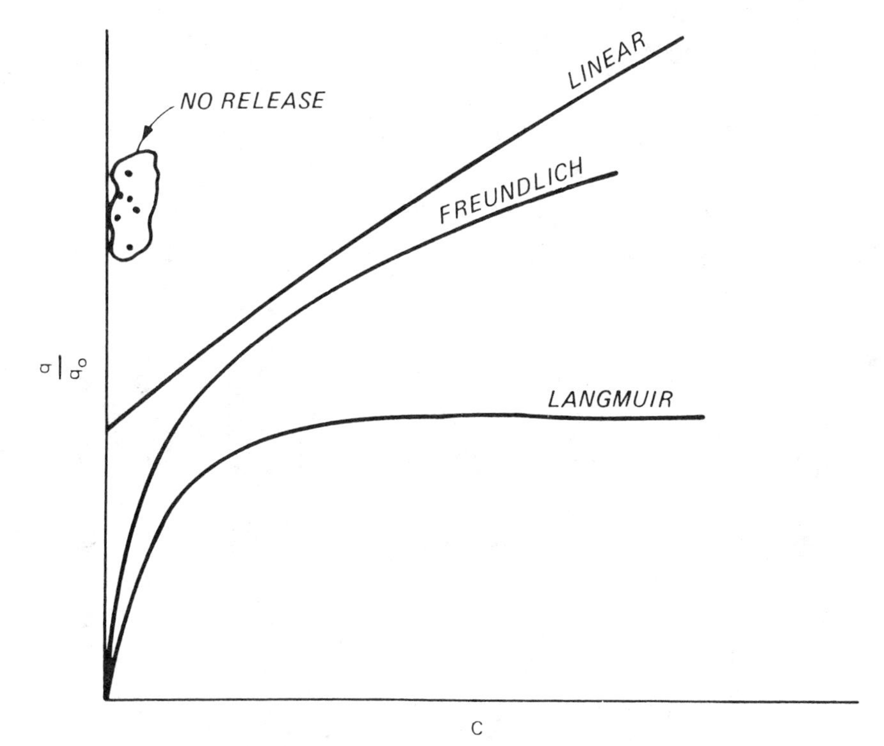

FIG. 4—*Graphical representation of desorption isotherms.*

by the no-release isotherm. Thus, all the treatment processes evaluated, except for the unsolidified cellulose xanthate, were effective in the immobilization of cadmium.

Chromium: Small amounts of chromium were measured in the leachate for each treatment process evaluated. These data were modeled using the langmuir desorption isotherm. As indicated by Fig. 6, the ability to immobilize the chromium for solidified hydroxide was greater than solidified starch xanthate which was greater than solidified cellulose xanthate. The unsolidified sludge materials had leachate concentration of chromium slightly less than the solidified sludge materials, indicating that for chromium, solidification offers no metal immobilization advantages.

Mercury: All but the hydroxide sludge materials were modeled by the no release desorption isotherm. This indicates that xanthate sludges were effective in immobilizing the mercury. Figure 7 indicates that unsolidified hydroxide sludge is more effective in immobilizing the mercury than the solidified hydroxide sludge.

Nickel: The hydroxide and solidified cellulose xanthate materials were effective in immobilizing the nickel and therefore modeled by the no-release desorption isotherm model. Unsolidified starch xanthate, solidified starch xanthate, and unsolidified cellulose xanthate were modeled by the Freundlich isotherm. As previously discussed, solidified starch is more effective in the immobilization of nickel than unsolidified starch. Unsolidified cellulose xanthate had nickel leachate concentration as large as 3.90 mg/L, indicating that it did no[t] immobilize nickel effectively.

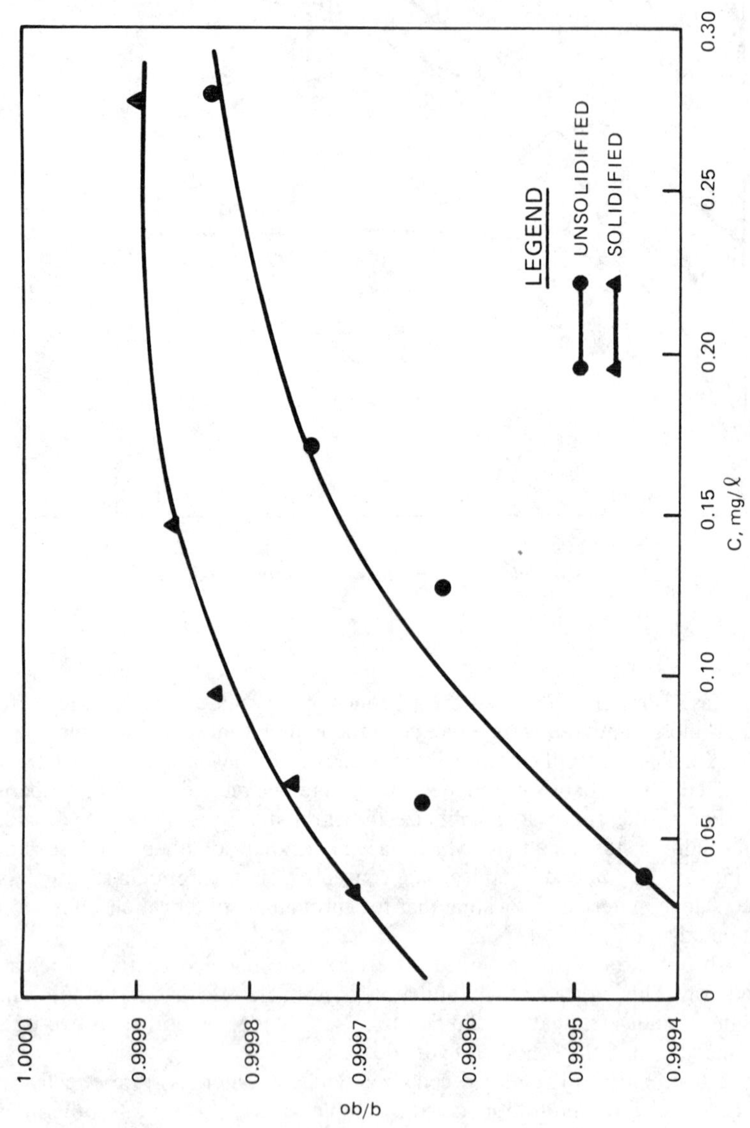

FIG. 5—*Desorption isotherm for solidified and unsolidified starch sludge, with nickel the contaminant of interest.*

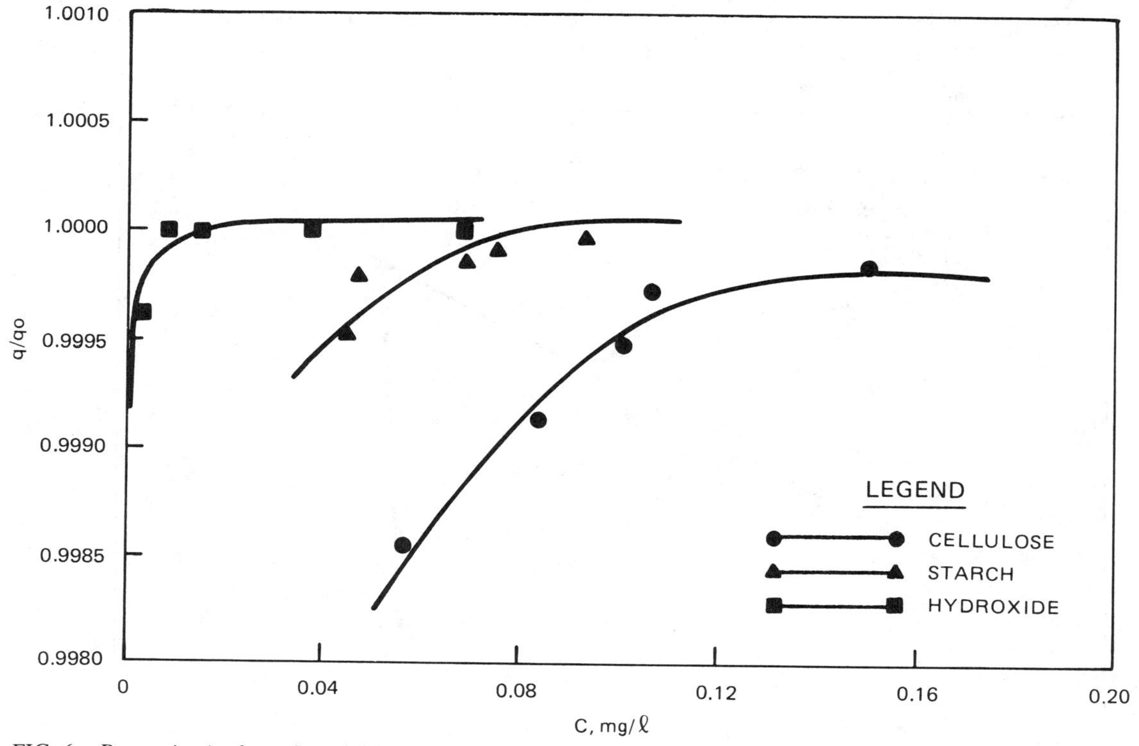

FIG. 6—*Desorption isotherm for solidified starch, hydroxide, and cellulose sludge, with chrome the contaminant of interest.*

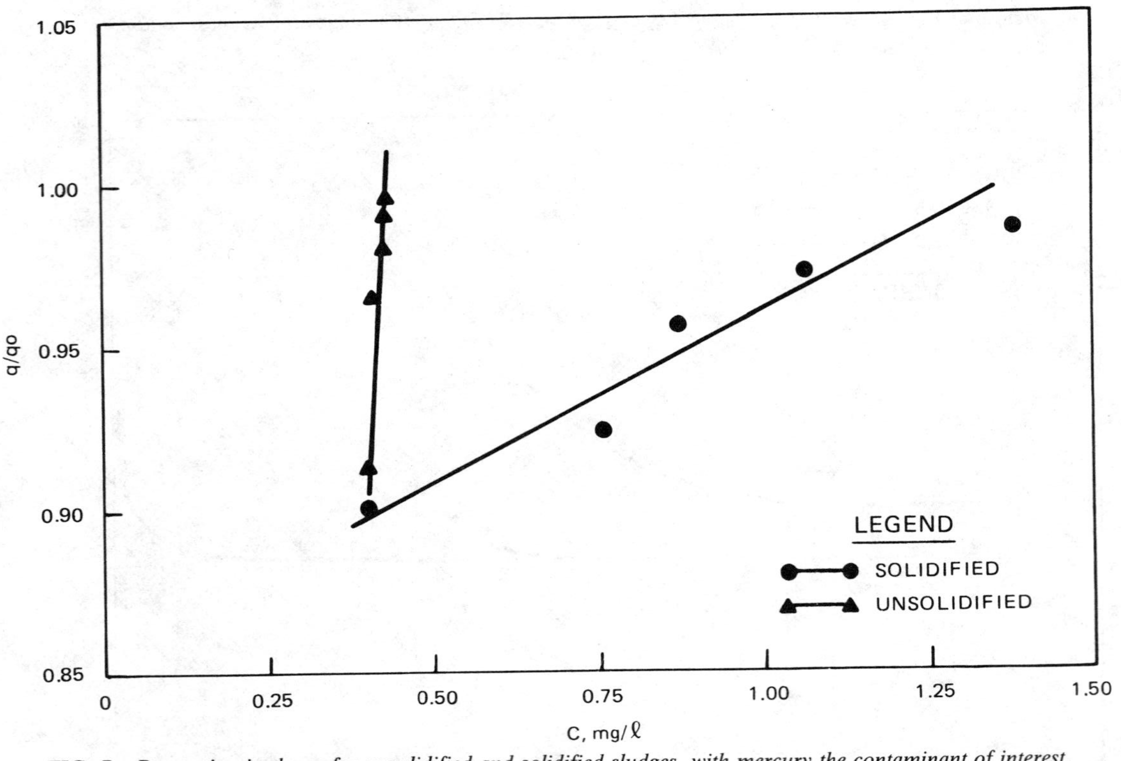

FIG. 7—*Desorption isotherm for unsolidified and solidified sludges, with mercury the contaminant of interest.*

TABLE 4—*Desorption isotherm model for unsolidified and solidified materials.*

Metal Leached	Sludge Type	Binder	r^2 Value	Isotherm Model
Cadmium	cellulose xanthate	solidified	0.909	Langmuir
	cellulose xanthate	unsolidified	...	no release
	starch xanthate	solidified	...	no release
	starch xanthate	unsolidified	...	no release
	hydroxide	solidified	...	no release
	hydroxide	unsolidified	...	no release
Chromium	cellulose xanthate	solidified	0.788	Langmuir
	cellulose xanthate	unsolidified	0.960	Langmuir
	starch xanthate	solidified	0.706	Langmuir
	starch xanthate	unsolidified	0.754	Langmuir
	hydroxide	solidified	0.753	Langmuir
	hydroxide	unsolidified	0.901	Langmuir
Mercury	cellulose xanthate	solidified	...	no release
	cellulose xanthate	unsolidified	...	no release
	starch xanthate	solidified	...	no release
	starch xanthate	unsolidified	...	no release
	hydroxide	solidified	0.869	linear
	hydroxide	unsolidified	0.926	linear
Nickel	cellulose xanthate	solidified	0.878	Freundlich
	cellulose xanthate	unsolidified	...	no release
	starch xanthate	solidified	0.839	Freundlich
	starch xanthate	unsolidified	0.920	Freundlich
	hydroxide	solidified	...	no release
	hydroxide	unsolidified	...	no release

Conclusions

1. Xanthate-precipitated sludges appear to be an effective method to immobilize the heavy metals studied. Only cadmium and nickel for the unsolidified cellulose show a tendency to leach.
2. Hydroxide-precipitated sludges do not immobilize mercury even if the hydroxide sludge is solidified.
3. Solidification in most cases appears to reduce the leachability of the four heavy metals studied.
4. Unlike the EP, the SGBEP is useful in determining the leaching characteristics of heavy metals from sludges.

Acknowledgment

The tests described and the resulting data presented herein, unless otherwise noted, were obtained from research conducted under the In-House Laboratory Independent Research Program of the U.S. Army Corps of Engineers by the U.S. Army Engineers Waterways Experiment Station. Permission was granted by the Chief of Engineers to publish this information.

References

[*1*] Wing, R. E., Doane, W. N., and Russell, C. R., "Insoluble Starch Xanthate: Use in Heavy Metal Removal," *Journal of Applied Polymer Science,* Vol. 19, 1975.

[*2*] Wing, R. E., NaVickis, L. L., Jusberg, B. K., and Rayford, W. E., "Removal of Heavy Metals from Industrial Wastewaters using Insoluble Starch Xanthates," EPA-60042-78-085, U.S. EPA-Industrial Environmental Research Laboratory Office of Research and Development, Cincinnati, OH, May 1978.

[*3*] Flynn, C. M., Jr., Carnahan, T. G., and Lindstrom, R. E., "Adsorption of Heavy Metal Ions by Xanthated Saw Dust," Report of Investigations-8427, U.S. Bureau of Mines, Reno, NV 1980.

[*4*] *Test Methods for Evaluating Solid Waste SW-846,* 2nd ed., U.S. Environmental Protection Agency-Office of Solid Waste and Emergency Response, Washington, DC, July 1982.

[*5*] "Prohibition on the Placement of Bulk Liquid Hazardous Waste in Landfills," EPA 530 SW-86-016, OSWER Policy Directive 9487, Statutory Interpretive Guidance, U.S. Environmental Protection Agency, Office of Solid Waste and Emergency Response, Washington, DC, 11 June 1986.

[*6*] Holland, M. E., "Use of Xanthates for the Removal of Metals from Waste Streams," Goodyear Atomic Corporation, Process Technology Department, Technical Division, Piketon, OH, Oct. 1975.

[*7*] Bricka, R. M., "Investigation and Evaluation of the Performance of Solidified Cellulose and Starch Xanthate Heavy Metal Sludges," Technical Report EL-88-5, USAE Waterways Experiment Station, Vicksburg, MS, Feb. 1988.

[*8*] Cullinane, J. M., Bricka, R. M., and Francingues, N. R., "An Assessment of Materials that Interfere With Stabilization/Solidification Processes," *Proceedings of the Thirteenth Annual Research Symposium,* EPA-600/9-87/015, U.S. Environmental Protection Agency Hazardous Waste Engineering Research Laboratory, Cincinnati, OH, July 1987.

[*9*] "Materials Testing," *Army Technical Manual No. 5-530,* Section 15, Department of the Army, Washington, DC, Feb. 1971.

[*10*] Myers, T. E., "A Simple Procedure for Acceptance Testing of Freshly Prepared Solidified Waste," *Hazardous and Industrial Solid Waste Testing: Fourth Symposium, ASTM STP 886,* J. K. Petros, Jr., W. J. Lacy, and R. A. Conway, Eds., American Society for Testing and Materials, Philadelphia, 1986.

[*11*] Environmental Protection Agency Part 11, Vol. 51, No. 216, Appendix I to Part 268, Federal Register, 7 Nov. 1986.

[*12*] Roy, W. R., Hassett, J. J., and Griffin, R. A., "Competitive Interactions of Phosphate and Molybdale on Arsenate Adsorption," *Soil Science,* Vol. 14L, No. 4, Oct. 1986.

[*13*] Lowenbach, W., *Compilation and Evaluation of Leaching Test Methods,* EPA-600/2-78-095, U.S. Environmental Protection Agency, Cincinnati, OH, 1978.

[*14*] Perket, C. L. and Webster, W. C., "Literature Review of Batch Laboratory Leaching and Extraction Procedures," *Hazardous and Industrial Solid Waste Testing: Second Symposium, ASTM STP 805,* R. A. Conway and W. P. Gulledge, Eds., American Society for Testing and Materials, Philadelphia, PA, 1981.

[*15*] Bettcker, J. M., "A Laboratory Study of Solidification/Stabilization Technology for Contaminated Dredged Material," thesis submitted to Virginia Polytechnic Institute and State University, Blacksburg, VA, March 1986.

[*16*] Houle, M. J. and Long, D. E., "Interpreting Results from Serial Batch Extraction Tests of Waste and Soils," *Proceedings of the Sixth Annual Research Symposium,* EPA-600/9-80-010, U.S. Environmental Protection Agency Municipal Environmental Research Laboratory, Cincinnati, OH, March 1980.

[*17*] *Hazardous Waste News,* Business Publisher, Silver Springs, MD, 27 July 1987.

Tippu S. Sheriff,[1] *Christopher J. Sollars,*[1] *Diana Montgomery,*[1] *and Roger Perry*[1]

The Use of Activated Charcoal and Tetra-Alkylammonium-Substituted Clays in Cement-Based Stabilization/Solidification of Phenols and Chlorinated Phenols

REFERENCE: Sheriff, T. S., Sollars, C. J., Montgomery, D., and Perry, R., **"The Use of Activated Charcoal and Tetra-Alkylammonium-Substituted Clays in Cement-Based Stabilization/Solidification of Phenols and Chlorinated Phenols,"** *Environmental Aspects of Stabilization and Solidification of Hazardous and Radioactive Wastes, ASTM STP 1033,* P. L. Côté and T. M. Gilliam, Eds., American Society for Testing and Materials, Philadelphia, 1989, pp. 273–286.

ABSTRACT: A number of cement-based fixation processes are available for cost-effective stabilization of inorganic wastes, but these processes are not generally so effective for treating organic wastes or organic/inorganic mixtures. With a view to possible extension of such methods to organic wastes, the use of charcoal and tetra-alkylammonium-substituted clays as pre-stabilization adsorbents has been investigated. Charcoal is a well-known adsorbent, while the use of the substituted clays exploits the hydrophobic properties of the alkyl groups to fix organic materials within the clay matrix.

Wyoming Bentonite substituted with hexadecyltrimethyl ammonium bromide and benzyldimethyltetradecylammonium chloride were found to be very effective in adsorbing chlorinated phenols with adsorption capacities of 150 mg of chlorinated phenol per gram of clay. There appears to be a clear trend between the chain length of the alkylammonium ion in the exchanged clay and the ability of the clay to adsorb a particular phenolic compound. Activated charcoal was found to effectively adsorb 180 mg of phenol or chlorinated phenols per gram of charcoal.

The exchanged clays and charcoal together with the respective adsorbed phenols were incorporated into a number of cementitious matrices. Following hydration, leachability and physical strength measurements were carried out and the results were used to make a preliminary assessment of the longer-term potential of clays and charcoal in stabilization/solidification applications.

KEY WORDS: charcoal, substituted clays, stabilization, solidification, wastes, phenols, chlorophenols

Several cement-based stabilization/solidification processes suitable for inorganic wastes have been commercially available for a number of years, but such processes are not generally so effective for treating organic wastes or inorganic/organic mixtures. Basic problems are that organic compounds are generally unreactive with many of the reagents used in stabilization/solidification processes and interfere with the physical and chemical processes which take place during solidification. Among the factors which may impair the integrity of the

[1] Post-doctoral research assistant, assistant director, post-doctoral research assistant, and director, respectively, Imperial College Centre for Toxic Waste Management, Department of Civil Engineering, Imperial College, London SW7 2BU, United Kingdom.

stabilized product are the hydrophobic nature of many organic waste materials, surface tension effects, and the presence of microorganisms [*1*]. A recent fundamental study of organic-cement interactions has highlighted the importance of the effects of different organic compounds on the rate of cement hydration [*2,3*].

A possible route to overcoming some of these problems is the use of naturally occurring clays as presolidification absorbents. Occasional reports have appeared in recent years detailing the use of clays for stabilizing organic and inorganic sludges [*4,5*], but little or no work appears to have been directed to a scientific evaluation of such methods. More recently, however, a range of clays has attracted some research interest as possible adsorbents for organic compounds in water and organic solvents. In particular, Mortland et al. [*6*] and Wolfe et al. [*7*] examined the effects of modifying montmorillonite clays by exchange with a range of alkylammonium cations to demonstrate the effect of this on the adsorptive capacities for organic molecules in water and hexane. The effect of modifying clays in this way is usually to increase the interlamellar spacing and thereby open up the adsorptive surfaces of the clays to enhance adsorption of organic molecules. Both groups of workers found that such modified clays showed some promise as adsorbents.

We have extended this concept to examine the use of such clays as presolidification adsorbents for a range of phenolic compounds which are typical of those found in some industrial waste sludges and solvents and which can pose disposal problems, particularly with regard to their free-flowing liquid content. The study reported here first examined a range of modified clays for their adsorptive capacities for phenol and chlorinated phenols. The most effective were used to adsorb known amounts of different phenolic compounds and were then incorporated into a cementitious mix. Following solidification, a range of physical and chemical tests were applied to assess the usefulness of such an approach to successful solidification of this type of compound.

A similar assessment of solidified products was carried out using activated charcoal as the adsorbent. Although activated carbon is widely used as an organic adsorbent in water and wastewater treatment, its use in solidification appears to have received very little attention [*1*]. Examination of both clays and charcoal in this work has enabled informative comparison of the effectiveness of both materials in stabilization.

Materials and Methods

Wyoming Bentonite (WB) obtained from Laporte Industries, Widnes, United Kingdom (designation RL03172), was coarsely ground using a ball mill and then finely ground in a household blender. The clay was then sieved through a 500-mesh screen and the fines used in subsequent tests. Powdered activated charcoal was used without further treatment. Reagent grade tetramethyl ammonium chloride (TMAC, BDH, London), trimethylphenylammonium chloride (TMPAC), hexadecyltrimethylammonium bromide (HDTMAB), and benzyl dimethyltetradecylammonium chloride (BDTDAC) were used throughout. Structural formulae of these salts are shown in Fig. 1.

Phenol, 3-chlorophenol, and 2,3-dichlorophenol solutions (100, 500, 1000, and 2000 mg/L) were made up by dissolving the appropriate weight of each of the solids in 1 L of deionized water (~1:10 methanol (MeOH)/water in the case of 2,3-dichlorophenol).

Exchange of Clays with Ammonium Salts

A powdered sample of WB was dried (105°C) for 20 h and then cooled in a desiccator. Four smaller specimens were then converted to different quaternary ammonium derivatives following a published method [*8*] as follows.

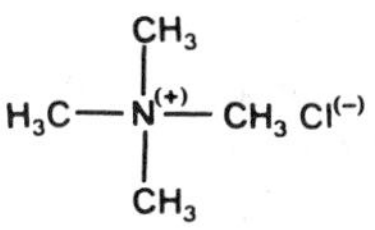

Tetramethylammonium chloride(TMAC)

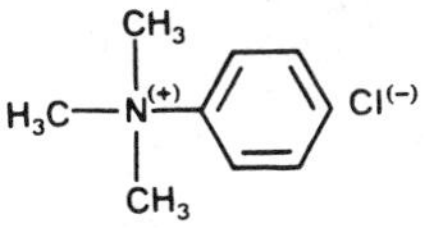

Trimethylphenylammonium chloride(TMPAC)

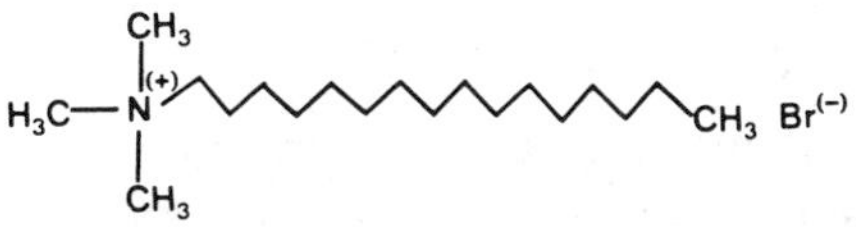

Hexadecyltrimethylammonium bromide(HDTMAB)

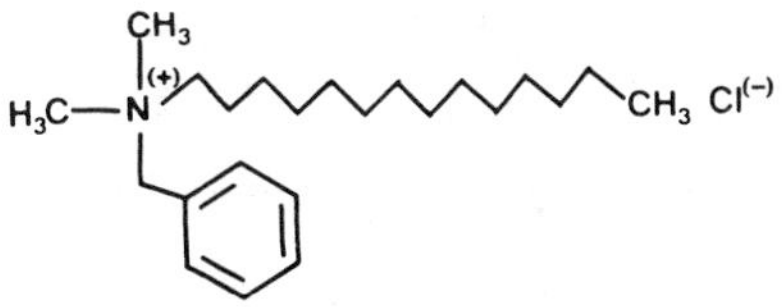

Benzyldimethyltetradecylammonium chloride(BDTDAC)

FIG. 1—*Structural formulae of quaternary ammonium salts used in exchange experiments.*

The dried clay (20 g) was placed in a plastic screw-top container and rotated end-over-end with aqueous sodium chloride (1 M; 100 mL) for a period of 20 h. The supernatant salt solution was removed by centrifugation and the clay was subjected to a further 20-h treatment with a fresh molar solution of sodium chloride.

Four portions (4 $\times$ 10 g) of the undried sodium-form were exchanged by the following alkyammonium salts for two periods of 20 and 48 h, respectively, using a fresh portion of the ammonium salt after the first period:

(1) TMAC (11 g in 100 mL water)
(2) TMPAC (17 g in 100 mL water)
(3) HDTMAB (7 g in 100 mL methanol)
(4) BDTDAC (10 mL of 50% aqueous solution in 100 mL water)

After the exchange procedure was finished, the clays were washed well (3 $\times$ 30 mL) with water or methanol (in the case of HDTMAB), dried, and then repowdered. The extent of exchange was then determined following the method of Fripiat et al. [*8*].

Measurement of Degree of Adsorption

Phenol, 3-chlorophenol, and 2,3-dichlorophenol solutions (100 mg/L, 100 mL) were placed in separate plastic screw-topped containers and 1-g quantities of either unexchanged WB, exchanged WB, or charcoal added. The containers then were continuously rotated end-over-end. At regular time intervals, rotation was halted, the containers' contents centrifuged, and a small (1-mL) aliquot of supernantant removed for phenol analysis using a standard method [9]. Supernatant and solids were then recombined before continuing mixing. In this way, a time-versus-adsorption profile was produced. This experimental protocol then was repeated with a range of higher initial concentrations of the phenolic compounds with adsorbents where the initial results indicated that the maximum adsorption capacity had not been utilized. Details of the highest initial concentrations used are given in Table 1.

Solidification with Cement

A range of solidified products was made up and subjected to a number of chemical and physical tests.

Two exchanged clays, HDTMA/WB and BDTDA/WB, were chosen for clay/cement solidification studies. Specimens (2 g) of both clays were exposed to phenolic solutions (500 mg/L, 100 mL) for 3 h, filtered, and combined with ordinary portland cement (OPC, Type 1) and deionized water. Full details are given in Table 2. The mixes were shaken well by hand for 1 min and then poured into a plastic beaker and allowed to cure for 28 days. Blanks with exchanged clay (2 g) but no phenolic compound and controls with phenolic compounds and no clay were also made up in the same way.

In the case of charcoal, two different procedures were adopted to assess the effect of mixing sequence on the leaching characteristics of the phenolic compounds:

Process T = mixing of cement (40 g) and charcoal (1 g), followed by addition of phenol solution (40 mL, 1000 mg/L)

Process W = suspension of charcoal (1 g) in phenolic solution (40 mL, 1000 mg/L) for 4 h, followed by addition of cement (40 g).

Blanks with charcoal (2 g) but no phenolics were made up using Process T, but separate control samples (phenolics but no charcoal) were not made since the controls made for the clay experiments were equally applicable.

Leach Tests

After 28 days, the cement blocks were broken into small lumps using a mortar and pestle, then ground to a powder (<1-mm mesh) using a household blender. A leach test procedure which was based upon previous work in this laboratory by Poon et al. [10] was then followed.

TABLE 1—*Highest concentration of phenol, 3-chlorophenol, and 2,3-dichlorophenol used to achieve maximum adsorption by a range of clays and charcoal.*

Adsorbent, 1 g	Phenol, mg/L	3-chlorophenol, mg/L	2,3-dichlorophenol, mg/L
Unexchanged WB	100	100	100
TMA/WB	100	100	100
TMPA/WB	500	500	500
HDTMA/WB	500	2000	2000
BDTDA/WB	500	2000	2000
Charcoal	2000	2000[a]	2000[a]

[a] Saturation not reached.

TABLE 2—*Composition of exchanged clay/cement mixes made up for leach-tests. Cement (40 g) and water[a] (40 mL) used in all cases.*

Mix No.	Phenolic Compounds	Clay Type, 2 g	Weight of Phenolic Compounds in Clay or Cement, mg
1 (control)	phenol	none	40
2	phenol	HDTMA/WB	26
3	phenol	BDTDA/WB	24
4 (control)	3-chlorophenol	none	40
5	3-chlorophenol	HDTMA/WB	45
6	3-chlorophenol	BDTDA/WB	50
7 (control)	2,3-dichlorophenol	none	40
8	2,3-dichlorophenol	HDTMA/WB	50
9	2,3-dichlorophenol	BDTDA/WB	50
10 (blank)	none	HDTMA/WB	...
11 (blank)	none	BDTDA/WB	...

[a] For controls (no clay), 40 mL of 1000 mg/L phenolic solution used.

Specimens (40 g) were placed in screw-topped plastic containers with water (100 mL), and the mixes were then rotated continuously end-over-end. The amount of phenolic compound leached out over a nine-day period was determined by filtering the cement slurries at 24-h intervals through 0.45-μm cellulose membrane filters and analyzing the filtrates. The filter cake was resuspended in a fresh sample of water (100 mL) after each filtration. Results for each of the three phenols were expressed as percentage extraction efficiencies, where

$$\text{Extraction Efficiency (\%)} = \frac{\text{weight of phenol leached out of specimen}}{\text{weight of phenol originally present}} \times 100$$

Penetrometer and Compressive Strength Measurements

Penetrometer measurements on a range of mixes (Tables 5 and 8) using a Wykeham Farrance (United Kingdom) pocket penetrometer were taken at intervals of 4, 7, 24, and 30 h for a 1.28-cm (½ in.) penetration, using a method previously described [*11*]. Compressive strength measurements were made (Tables 6 and 9) after 28 days setting at 100% relative humidity, using 51-mm (2 in.) cubes, following as far as possible the ASTM Test for Compressive Strength of Hydraulic Cement Mortars (Using 2-in. or 50-mm Cube Specimens) (C 109-86). Because of the limited amounts of exchanged clay available and the necessity of avoiding phenol leaching during curing, the main deviation from this standard was the use of hand-mixing (as for leach test specimens) and curing at 100% humidity, but not in limewater.

Results

Adsorption Studies with Exchanged Clays

The degree of exchange of the original WB with the four alkylammonium salts used was greater than 95% in all cases.

Initial adsorption studies with unexchanged and exchanged WB specimens utilized 100 mg/L solutions of phenol, 3-chlorphenol, and 2,3-dichlorophenol, respectively, in order to assess, on an application-orientated weight (of phenol) rather than molar basis, the comparative adsorptive abilities of unexchanged WB and the exchanged derivatives. Figure shows adsorption data for 2,3-dichlorophenol with the five clays and demonstrates adsorption patterns common to all three phenols with these materials. Data for all the adsorption studies using 100 mg/L phenolic solutions are summarized in Fig. 3. A major feature is the inability of unexchanged WB to adsorb any of the phenolic compounds. Furthermore exchange with the small TMA ion results in only the slight adsorption of phenol and the chlorinated phenols. On a weight basis, TMPA/WB adsorbs moderate quantities of all three phenolic compounds with a distinct preference for phenol, whereas HDTMA/WB and BDTDA/WB, containing long chained ammonium salts, adsorb the chlorinated phenol more strongly but show a reduced preference for phenol.

These basic patterns were maintained with identical experiments repeated at higher phenolic concentrations, and Table 1 indicates clearly that the phenolic concentrations needed to saturate the adsorption capacity of the clays were far higher for the exchanged WB derivatives than for WB alone. The maximum adsorption capacities (based on 4 h adsorption for the most effective adsorbents, that is, HDTMA/WB, BDTDA/WB and TMPA/WB are summarized in Table 3. After longer periods (up to 24 h) of exposure to phenolic solutions, slow desorption of the phenolic compounds was observed with both HDTMA WB and BDTDA/WB. The formation of foams (indicative of the surfactant properties of the quaternary ammonium ions used) towards the end of the 24-h periods suggested that the slow removal of the quaternary ammonium cations from the clays was taking place, with consequent leaching of the phenolic compounds into the supernatant solution, but further investigations are needed to confirm this.

Leach Tests on Solidified Clay/Cement Mixes

Because of their high relative adsorption capacity, two exchanged clays, HDTMA/WB and BDTDA/WB, were chosen for clay/cement solidification and leaching studies, with

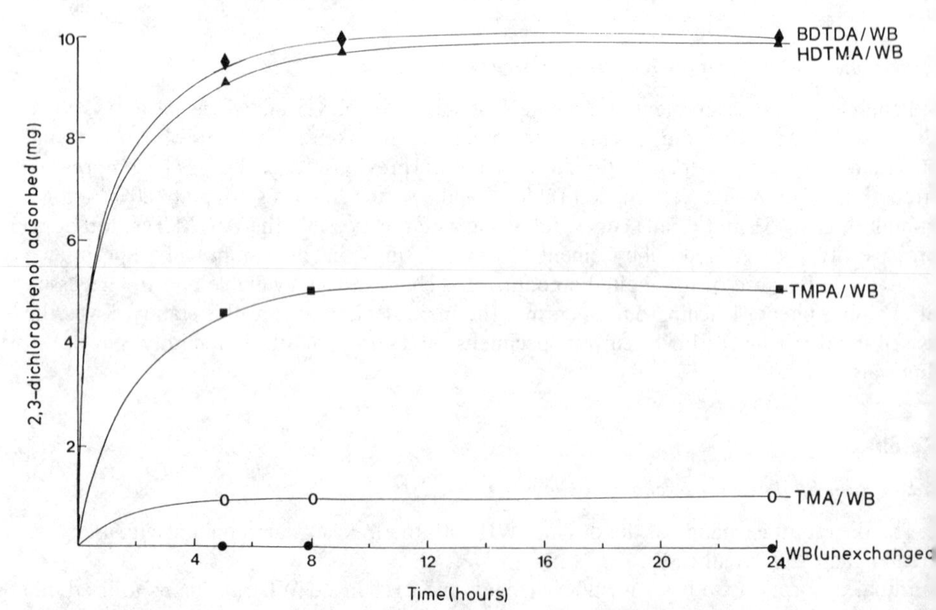

FIG. 2—*The adsorption of 2,3-dichlorophenol (100 mg/L) by WB and quaternary ammonium exchanged WB's.*

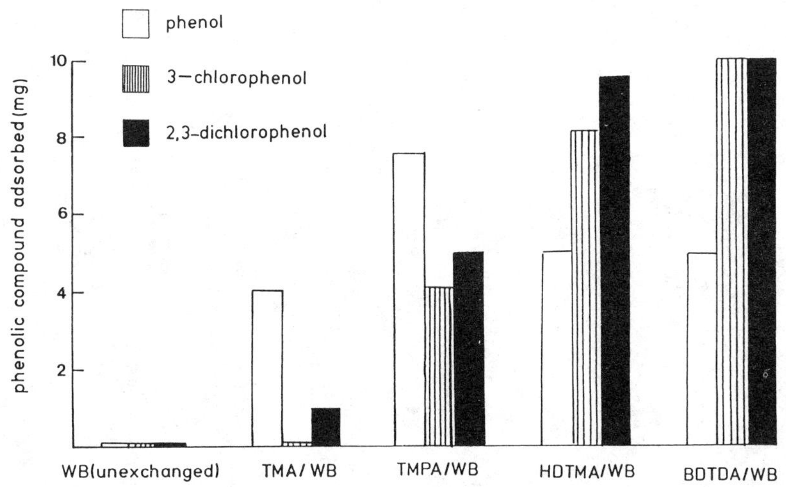

FIG. 3—*The adsorption of phenol, 3-chlorophenol, and 2,3-dichlorophenol (100 mg/L) by WB and quaternary ammonium exchanged WB's.*

cement-only (phenols but no clay) controls being used for comparison. Figures 4 and 5 give the detailed results of the leach tests over a nine-day period on these three types of sample in terms of percentage extraction efficiencies. It should be noted that in all cases, concentrations for all three phenols recorded in tests on blank samples were less than the detection limit of the method used (0.5 mg/L), and thus no corrections were made to sample concentrations. A summary of leach test data is also given in Table 4.

It is clear from Table 4 that in the absence of prior adsorption by exchanged clays, the leaching of the phenolic compounds from the solidified control (no clay) product is high and on a weight basis follows the order of increasing solubility of the phenolic compound in water, that is, phenol > 3-chlorophenol > 2,3-dichlorophenol. The stabilization factor indicates that the stabilization due to solidified HDTMA/WB/OPC and BDTDA/WB/OPC mixes for phenol and 3-chlorophenol are roughly comparable, with the former being stabilized slightly better by HDTMA/WB/OPC and the latter by BDTDA/WB/OPC. However, with 2,3-dichlorophenol the stabilization by BDTDA/WB/OPC is much greater in comparison than that achieved with both the control and HDTMA/WB/OPC.

Physical Testing of Solidified Clay/Cement Mixes

Table 5 gives the results of penetrometer experiments for various cement and clay mixes. Both the 0.6 and 1.0 ratio mixes harden to the maximum reading within 24 h, but the 0.6

TABLE 3—*Maximum adsorption capacity of various WB exchanged clays for phenol, 3-chlorophenol, and 2,3-dichlorophenol (after 4 h adsorption).*

Exchanged Clay	Adsorption Ability per Gram of Clay		
	Phenol, mg	3-chlorophenol, mg	2,3-dichlorophenol, mg
HDTMA/WB	17	150	150
BDTDA/WB	17	150	150
TMPA/WB	8	1	1

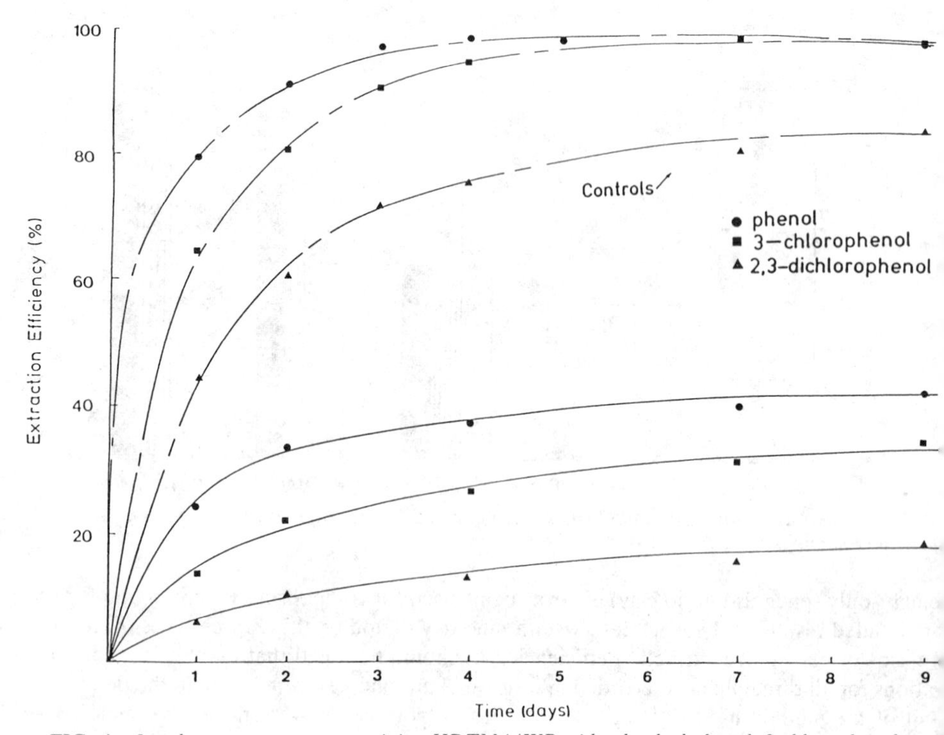

FIG. 4—*Leach tests on cement containing HDTMA/WB with adsorbed phenol, 3-chlorophenol, and 2,3-dichlorophenol.*

mixes show a predictably faster rate of hardening. The addition of clay and clay with adsorbed organic appears to have no effect on the rate of hydration of the cement, and Mix B with unexchanged WB actually sets at a faster rate.

Table 6 gives the results of compressive strength measurements on the same mixes and shows that the values vary only slightly from one mix to another. As with the penetrometer results above, the unexchanged clay appears to aid strength development. The exchange of the clay with alkylammonium salts and their subsequent adsorption of phenolic compounds results in a reduction of compressive strength of only between 5 to 15%.

Phenolic Adsorption by Charcoal

The phenol adsorption studies showed that after 3 h adsorption, the maximum capacity of 1 g of charcoal for phenol was 183 mg. For 3-chlorophenol and 2,3-dichlorophenol, the results indicated that further adsorption capacity above 183 mg was available, but this capacity was considered adequate for the present work.

Physical and Chemical Testing of Solidified Charcoal/Cement Mixes

Figure 6 shows the cumulative dynamic leach test results for a solidified sample obtained by loading 40 mg of each phenolic compound onto 1 g of charcoal followed by cement addition using Process W. Results for both Processes are given in Table 7. The amounts of phenolic compounds in leachate from the controls (containing no charcoal) are high and follow the order of the solubility of the phenolic compounds in the aqueous leachate medium.

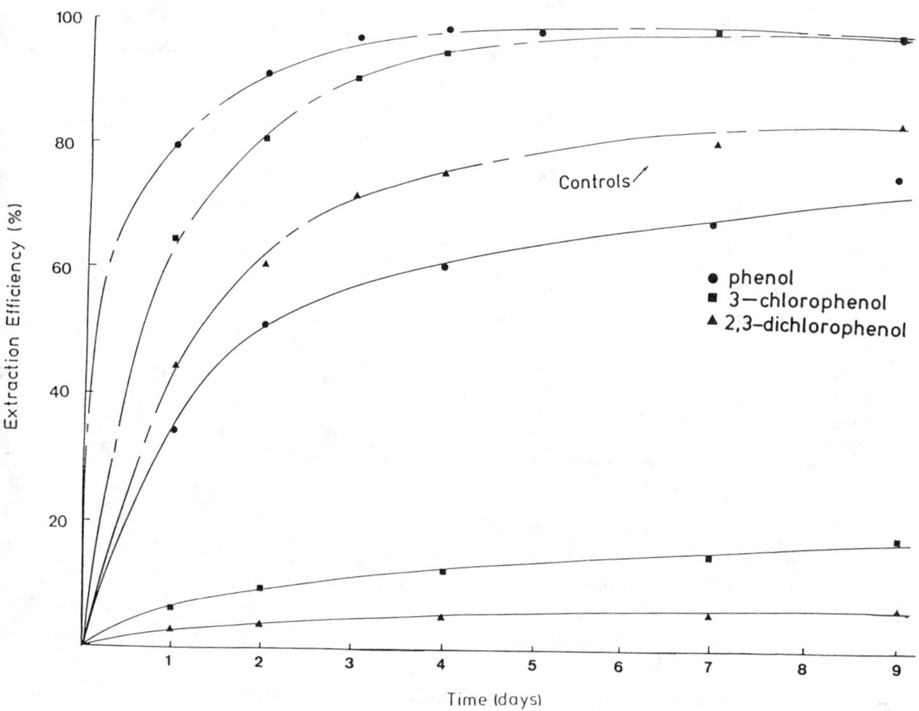

FIG. 5—*Leach tests on cement containing BDTA/WB with adsorbed phenol, 3-chlorophenol, and 2,3-dichlorophenol.*

By contrast, the mixes stabilized by charcoal are effective in keeping the cumulative amount of phenolic compound leached to below 5 mg. The induction period of 4 h left for adsorption of the phenolic compounds (W) appears to have no advantage over mixes (T) which were given no induction period. Clearly, under these conditions there is very rapid adsorption of the phenolics by charcoal. The stabilization factors are high for phenol and 2,3-dichlorophenol, but curiously somewhat lower for 3-chlorophenol.

TABLE 4—*Dynamic leach test results for cement/clay specimens (40 g) containing phenol, 3-chlorophenol, and 2,3-dichlorophenol.*

Phenolic Compound	Clay	Amount of Phenol in Specimen, mg	Amount Leached from Specimen, mg	Extraction Efficiency, %	Stabilization Factor[a]
phenol	none	28.6	28.5	99.7	...
	HDTMA/WB	18.5	7.9	42.7	2.3
	BDTDA/WB	17.1	13.0	76.0	1.3
3-chlorophenol	none	28.6	28.6	100.0	...
	HDTMA/WB	32.1	11.3	35.2	2.8
	BDTDA/WB	35.7	6.5	18.2	5.5
2,3-dichlorophenol		28.6	24.1	84.3	...
	HDTMA/WB	35.7	6.9	19.3	4.4
	BDTDA/WB	35.7	2.4	6.7	12.6

[a] Stabilization Factor = $\frac{\text{Extraction Efficiency of Control}}{\text{Extraction Efficiency of Non-Control}}$

TABLE 5—*Penetrometer readings for cement/clay mixes (10^6 N/m^2).*

	Ratio Liquid/Cement					
	0.6, h			1.0, h		
Mix[a]	4	7	24	4	7	24
A Cement + water	0.3	3.0	4.8	0.0	0.3	4.8
B Cement + 2 g WB + water	0.7	4.8	4.8	0.0	0.7	4.8
C Cement + 2 g HDTMA/WB + water	0.1	2.7	4.8	0.0	0.3	4.8
D Cement + 2 g BDTDA/WB + water	0.1	1.7	4.8	0.0	0.3	4.8
E Cement + 2 g HDTMA/WB + phenol (500 mg/L)	0.1	2.1	4.8	0.0	0.3	4.8
F Cement + 2 g BDTDA/WB + phenol (500 mg/L)	0.1	2.1	4.8	0.0	0.3	4.8

[a] In all cases, 40 g cement used.

Table 8 gives the results of penetrometer measurements for various cement and charcoal mixes. The 0.6 and 1.0 ratio mixes attained (with one exception) maximum readings within 24 and 30 h, respectively. The exception (Mix F), containing 2,3-dichlorophenol, hardened at a much slower rate at both ratios, and this is almost certainly due to the fact that the stock solution contained 10% (by volume) methanol.

An interesting trend is revealed by comparing Mixes A, B, and C containing 0, 1, and 2 g of charcoal, respectively (Table 8). The rate of hardening of the cement increases with the increasing addition of charcoal. The compressive strength data, given in Table 9, shows that the final compressive strength of the cement block (of Mixes A′, B′, and C′) also follows this general trend. Comparing Mix B′ with Mixes D′ to F′, it is clear that, with the exception of Mix F′ (containing methanol), there is no significant change in compressive strength due to the presence of the adsorbed phenolic compounds.

Discussion

Exchanged WB and charcoal have both exhibited an ability to adsorb phenol and chlorinated phenols from aqueous solution. Unexchanged clays show no affinity at all for any o

TABLE 6—*Compressive strength measurements on clay/cement mixes (0.6 water/cement ratio) after 28 days.*

Mix[a]	Compressive Strengths 10^6 N/m^2
A′ Cement + water	2.9
B′ Cement + 12 g WB + water	3.0
C′ Cement + 12 g HDTMA/WB + water	2.6
D′ Cement + 12 g BDTDA/WB + water	2.8
E′ Cement + 12 g HDTMA/WB + phenol (500 mg/L)	2.4
F′ Cement + 12 g BDTDA/WB + phenol (500 mg/L)	2.6

[a] In all cases, 150 g cement used.

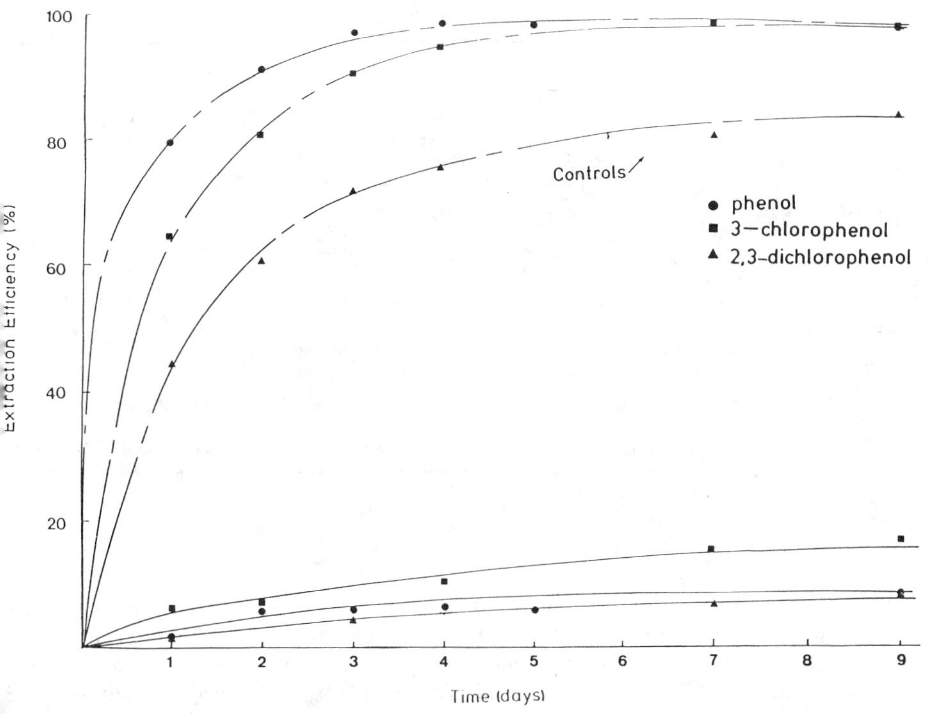

FIG. 6—*Leach tests on cement containing charcoal with adsorbed phenol, 3-chlorophenol, and 2,3-dichlorophenol.*

the phenols used. The probable reason for this is that the low polarity phenolic molecules have no tendency to penetrate either between the tightly, electrostatically held layers of silicate nor the highly polar environment of solvated cations on the surface of the clay. Clays exchanged with alkylammonium cations, particularly long chained ones, however, can displace the metallic cations from between the layers, resulting in the expansion of the interlamellar distance by intercalation of the longer organic molecule [*12*] and creation of a more

TABLE 7—*Dynamic leach test results for cement/charcoal specimens (40 g) containing phenol, 3-chlorophenol, and 2,3-dichlorophenol.*

Phenolic Compound	Mix	Amount of Phenol in Specimen, mg	Amount Leached from Specimen, mg	Extraction Efficiency, %	Stabilization Factor[a]
Phenol	control	28.6	28.5	99.7	...
	T	28.6	3.1	10.8	9.2
	W	28.6	2.4	8.4	11.9
3-chlorophenol	control	28.6	28.6	100.0	...
	T	28.6	4.8	16.8	6.0
	W	28.6	4.9	17.1	5.8
2,3-dichlorophenol	control	28.6	24.1	84.3	...
	T	28.6	2.1	7.3	11.5
	W	28.6	2.3	8.0	10.5

[a] Stabilization Factor $= \dfrac{\text{Extraction Efficiency of Control}}{\text{Extraction Efficiency of Non-Control}}$

TABLE 8—*Penetrometer readings for cement and charcoal mixes (10^6 N/m²).*

	Ratio Liquid/Cement						
	0.6, h			1.0, h			
Mix[a]	4	7	24	4	7	24	30
A Cement + water	0.0	3.7	4.8	0.0	0.0	4.1	4.8
B Cement + 1 g charcoal + water	0.3	0.7	4.8	0.0	0.0	4.5	4.8
C Cement + 2 g charcoal + water	0.3	1.7	4.8	0.0	0.0	4.8	4.8
D Cement + 1 g charcoal + phenol (1000 mg/L)	0.3	1.0	4.8	0.0	0.0	4.1	4.8
E Cement + 1 g charcoal + 3-Cl phenol (500 mg/L)	0.3	1.4	4.8	0.0	0.0	4.5	4.8
F Cement + 1 g charcoal + 2,3 diCl phenol (1000 mg/L)	0.0	0.1	4.8	0.0	0.0	1.0	3.1[b]

[a] In all cases, 40 g cement used.
[b] Went to 4.8×10^6 N/m² after 48 h.

hydrophobic environment. This allows the penetration of the organics and provides sites for their adsorption. This is clearly evidenced by the rapid adsorption of phenolic compounds into alkylammonium exchanged clays demonstrated in this study.

The mechanisms of adsorption of organic compounds by exchanged clay surfaces are complex, and detailed investigation of this was not within the scope of this work. In the case of phenol, however, hydrogen bonding between the silicate oxygens and the phenolic hydrogen may be an important factor [*12*], and further work is being undertaken by the authors to clarify this.

Although the mechanisms of phenolic adsorption are complex, the reverse in the trend of the adsorption of the phenolic compounds by (a) HDTMA/WB and BDTDA/WB and (b) TMPA/WB and TMA/WB (Fig. 3) suggests some correlation between chain length of the alkylammonium cation and the organic character of the adsorbed molecule. More specifically, clays with long chained alkylammonium cations which are very hydrophobic and organic in character strongly adsorb very hydrophobic organic molecules. Likewise, on a weight basis, shorter chain alkylammonium WB derivatives TMPA/WB and TMA/WB not

TABLE 9—*Compressive strength measurements on cement/charcoal mixes (0.6 water/cement ratio) after 28 days.*

Mix[a]	Compressive Strength, 10^6 N/m²
A′ Cement + water	2.9
B′ Cement + 6 g charcoal + water	3.6
C′ Cement + 12 g charcoal + water	4.5
D′ Cement + 6 g charcoal + phenol (1000 mg/L)	3.7
E′ Cement + 6 g charcoal + 3-Cl phenol (1000 mg/L)	3.5
F′ Cement + 6 g charcoal + 2,3 diCl phenol (1000 mg/L)	2.2

[a] In all cases, 150 g cement used.

only exhibit a lower affinity for organic molecules in general, but also a preference for less hydrophobic organic molecules. The overall conclusion is therefore that the exchanged clays used showed differential adsorptive preferences for various phenols depending on physicochemical relationships between the specific phenol and the structure of the nitrogen substituents in the quaternary ammonium cation on the clay. In broader terms, this important property suggests the useful possibility that certain clays may be "tailored" to specific wastes by exchange or other treatments to produce very cost-effective prestabilization adsorbents. This aspect of clays used in stabilization/solidification requires a rigorous approach to the analysis of all the physicochemical interactions taking place between a specific exchanged clay and a specific adsorbed organic or organics, and studies of this nature are continuing in this laboratory.

Solidification of exchanged clay or charcoal mixes with adsorbed organics offers a number of advantages in the waste disposal process. In the case of cement, it would be expected that the inorganic silicate structure of the basic clay and the surfactant characteristics of the quaternary ammonium cation would lead to at least physical compatability with the cement matrix, thus offsetting any detrimental effects on the cement hydration process due to adsorbed organics. The results of physical strength measurements on various cement/clay mixes clearly shows this to be the case at the 1:20 clay/cement and charcoal/cement ratios used, with an only 5 to 15% reduction in the compressive strength of cement due to organics. Charcoal was also found to aid cement hydration and increase compressive strengths.

Dynamic leach tests on powdered samples of the cement/clay or charcoal mixtures provided an extremely rigorous test of the ability of the phenolic compounds to be stabilized. Table 10 compares the stabilization factors obtained for WB derivatives and charcoal. BDTDA/WB is as effective as charcoal in the stabilization of 2,3-dichlorophenol, but charcoal exhibits the better overall stabilization of the three phenolic compounds. Nonetheless, at present, exchanged clays and activated charcoal retail at similar prices in the United Kingdom for industrial use, but since untreated clays are available at a tenth of these costs, the possibilities for developing cheaper clay "tailoring" treatments for specific wastes are considered to be substantial. For a truly economically viable technique for organic solidification, however, much lower clay/cement or charcoal/cement ratios (for example, 1:2) would be clearly necessary, with possibly more deleterious effects on physical strength and leaching characteristics. These aspects are also being further investigated by the authors.

General Conclusions

There is a need to extend the application of the valuable and economic hazardous waste treatment technique of cement-based stabilization/solidification to organic or mixed inorganic/organic waste sludges or liquids. Organophilic exchanged clays and charcoal have been shown to be effective presolidification adsorbents of three typical phenols found in industrial waste streams. Solidification with cement of phenols following prior adsorption by these adsorbents has demonstrated substantial reductions in leach test extraction efficiencies compared with solidified mixes made without adsorbents. The work has demonstrated the clear possibility of tailoring specific exchanged clays to specific waste organic compounds for

TABLE 10—*Comparison of the stabilization factors for exchanged clays and charcoal.*

Phenolic Compound	HDTMA/WB	BDTDA/WB	Charcoal
Phenol	2.3	1.3	10.6
[illegible]-chlorophenol	2.8	5.5	5.9
[illegible],3-dichlorophenol	4.4	12.6	11.0

maximum presolidification adsorption effectiveness and consequent solidified product integrity.

Acknowledgments

The authors gratefully acknowledge financial support for this work from the U.K. Department of the Environment. One of the authors (DM) wishes to thank Laporte Industries (U.K.) for the provision of a studentship.

References

[*1*] Tittlebaum, M. E., Seals, R. K., Cartledge, F. K., and Engels, S., *Critical Reviews in Environmental Control,* (Chemical Rubber Co.), Boca Raton, FL, Vol. 15, No. 2, 1985, pp. 179–211.
[2] Walsh, M. B., Eaton, H. C., Tittlebaum, M. E., Cartledge, F. K., and Chalasani, D., *Hazardous Waste and Hazardous Materials,* Vol. 3, No. 1, 1986, pp. 111–123.
[*3*] Chalasani, D., Cartledge, F. K., Eaton, H. C., Tittlebaum, M. E., and Walsh, M. B., *Hazardous Waste and Hazardous Materials,* Vol. 3, No. 2, 1986, pp. 167–173.
[*4*] Poon, C. S., Peters, C. J., and Perry, R., *Effluent and Water Treatment Journal,* 1983, pp. 451–459.
[*5*] Kupeic, A. R. and Escher, E. D., British Patent Application No. 2040277, Aug. 1980.
[*6*] Mortland, M. M., Shaobai, S., and Boyd, S. A., *Clay and Clay Minerals,* Vol. 34, No. 5, 1986, pp. 581–585.
[*7*] Wolfe, T. A., Demirel, T., and Baumann, E. R., *Journal Water Pollution Control Federation,* Vol. 58, No. 1, 1986, pp. 68–76.
[*8*] Fripiat, J. J., Pennequin, M., Poncelet, G., and Cloos, P. in *Clay Minerals,* Vol. 8, 1969, pp. 119–134.
[*9*] *Standard Methods for the Examination of Water and Wastewater,* 15th ed., American Public Health Association-American Water Works Association-Water Pollution Control Federation, Washington, DC, 1980, pp. 513–514.
[*10*] Poon, C. S., Peters, C. J., and Perry, R., *Science of the Total Environment,* Vol. 41, 1985, pp. 55–71.
[*11*] Poon, C. S., Clark, A. I., and Perry, R., "Investigation of the Physical Properties of Cement-Based Fixation Processes for the Disposal of Toxic Wastes," *Public Health Engineer,* Institute of Public Health Engineers, London, Vol. 13, 1985, pp. 108–111.
[*12*] Raussell-Colom, J. A. and Senotosa, J. M. in *Chemistry of Clays and Clay Minerals,* A. C. D. Newman, Ed., Mineralogical Society, London, 1987, Chapter 8, pp. 371–422.

Peter Taylor,[1] *Vincent J. Lopata,*[1] *Donald D. Wood,*[1] *and Harold Yacyshyn*[1]

Solubility and Stability of Inorganic Iodides: Candidate Waste Forms for Iodine-129*

REFERENCE: Taylor, P., Lopata, V. J., Wood, D. D., and Yacyshyn, H., **"Solubility and Stability of Inorganic Iodides: Candidate Waste Forms for Iodine-129,"** *Environmental Aspects of Stabilization and Solidification of Hazardous and Radioactive Wastes, ASTM STP 1033,* P. L. Côté and T. M. Gilliam, Eds., American Society for Testing and Materials, Philadelphia, 1989, pp. 287–301.

ABSTRACT: Iodine-129 and carbon-14 are of lesser importance than many other radionuclides in nuclear waste management because of their relatively low radiotoxicity. Their safe immobilization and disposal, however, is a significant challenge because their chemical properties are quite different from most other components of nuclear fuel waste. The authors have sought thermodynamically defensible inorganic waste forms for both of these radioisotopes.

This paper discusses the stability and solubility of inorganic iodides as they relate to the selection of a waste form for iodine-129. The best candidates appear to be silver iodide (AgI), either alone or in combination with silver chloride (AgCl), and a combination of Bi_5O_7I and Bi_2O_3. Neither is completely satisfactory. The principal drawback of AgI is its relative ease of reduction. High chloride concentrations also increase the solubility of AgI, but it is insensitive to carbonate. The main disadvantage of bismuth oxyiodide is its susceptibility to displacement reactions with dissolved chloride and carbonate; also, it is unstable under strongly reducing conditions.

KEY WORDS: bismuth oxyiodide, immobilization, inorganic iodides, iodine-129, silver iodide, solubility, stability, waste forms

Iodine-129 and carbon-14 are both long-lived beta-emitting isotopes, with respective half-lives of 17 million and 5730 years. They differ from most nuclear waste elements by forming anions, I^- and HCO_3^-/CO_3^{2-}, which are stable and soluble over a wide range of geochemical conditions, do not sorb strongly on silicate surfaces, and are incompatible with most glass and ceramic waste forms. Thus, although they are not severe radiological hazards, their disposal requires specific attention. Here, we review the solubility and stability of a number of inorganic iodides to aid the selection of appropriate waste forms for iodine-129 that would arise from nuclear fuel recycling. A similar approach to carbon-14 waste-form selection has been described elsewhere [*1*], and more detailed reports of the underlying chemistry of some iodides and carbonates are also available [*2–7*].

As well as considering the solubility or stability of a single compound, we introduce the concept of a two-phase waste form, in which one phase suppresses the solubility of the second, containing the toxic or radioactive component, essentially by a common-ion effect. The effects of temperature changes, complexation, and various alteration reactions (hydrolysis, ion exchange, oxidation, or reduction) are taken into account where information

[1] Research officer, research technologist, research technologist, and student, respectively, Research Chemistry Branch, Atomic Energy of Canada Limited, Whiteshell Nuclear Research Establishment, Pinawa, MB, R0E 1L0, Canada.

* Issued as AECL-9471.

is available. The work may thus be described as a search for thermodynamically defensible waste forms for iodine-129.[2]

This work is a contribution to the Canadian Nuclear Fuel Waste Management Program. This research and development program is assessing the possible disposal of used CANDU™ nuclear fuel, or recycle wastes, within plutonic rock formations in the Canadian Shield.

Background on Iodine-129 Waste Forms

Vance et al. [*8*] and Burger et al. [*9*] have reviewed candidate waste forms for iodine-129. Most attention has been directed at iodide-containing sodalite and other zeolites, and to silver iodide and various iodates, either alone or incorporated in portland cement. More recently, Dunn et al. [*10*] have reported further details of the synthesis and leaching behavior of iodosodalite. Processes for removing iodine from off-gases and liquid waste-streams have been reviewed by Holladay [*11*].

Based on considerations of solubility and sensitivity to hydrolysis, Burger et al. [*9*] found that silver iodide emerged as a clear choice among inorganic iodides. They dismissed PbI_2 and BiI_3 as being too sensitive to hydrolysis. However, they did not consider the possibility that the hydrolysis products themselves, that is, basic metal iodides, may be appropriate waste forms in their own right. Vance et al. [*8*] examined the thermal stability of some basic lead iodides, and dismissed them as too unstable for further consideration. Mel'nikov et al. [*12*] and Kharbanda et al. [*13*] have investigated "bismuth (III) hydroxide" as a reagent for removing iodine-131 from aqueous solutions, and lead (II) oxide is known to react with dissolved iodide and other anions to form insoluble solids [*14*]. We have directed most of our attention to basic bismuth iodides and compared their stability with silver iodide and other insoluble iodides mentioned by Burger et al. [*9*].

Aqueous Equilibria Involving Basic Iodides

General Principles

Many metal halides, MX_y, undergo sequential hydrolysis to the oxide or hydroxide, through intermediate basic halides with the general formula $a\mathrm{MO}_{y/2} \cdot b\mathrm{MX}_y \cdot c\mathrm{H_2O}$

$$\text{[Halide-rich phase]} + \mathrm{OH^-(aq)} \rightleftharpoons \text{[Oxide-rich phase]} + \mathrm{X^-\ (aq)} \pm d\mathrm{H_2O(l)} \qquad (1)$$

When one stage of the reaction is written in this way, and unit activities are assumed for the solids and water, the equilibrium constant is a simple quotient of anion activities

$$K_1 = \{X^-\}/\{OH^-\} \qquad (2)$$

Equation 3 depicts the full sequence of hydrolysis reactions

$$\mathrm{MX_y(s)} \underset{X^-}{\overset{OH^-}{\rightleftharpoons}} \mathrm{A(s)} \underset{X^-}{\overset{OH^-}{\rightleftharpoons}} \mathrm{B(s)} \underset{X^-}{\overset{OH^-}{\rightleftharpoons}} \mathrm{C(s)} \underset{X^-}{\overset{OH^-}{\rightleftharpoons}} \ldots \mathrm{MO_y} \qquad (3)$$
$$\quad K_A \qquad\qquad K_B \qquad\qquad K_C$$

If the equilibrium constants, $\{X^-\}/\{OH^-\}$, follow the order $K_A > K_B > K_C \ldots$, then each solid has a stability field in the system $\mathrm{MO}_{y/2} - \mathrm{MX}_y - \mathrm{H_2O}$ – [pH], bounded by the equilibrium constants below which it is hydrolyzed or above which it is halogenated to the

[2] Similar work on inorganic carbonates as waste forms for carbon-14 was discussed briefly in the oral presentation of this paper and has been published elsewhere [*1*].

next phase in Sequence 3. If any equilibrium constant disrupts this sequence, then one or more phases are not stable with respect to disproportionation. For example, B is stable within the range of solution conditions $K_B \geq \{I^-\}/\{OH^-\} \geq K_C$. Thus, if $K_C > K_B$, B is not stable with respect to an assemblage of A, C, and solution.

It follows that, at any given pH, the equilibrium halide activity, $\{X^-\}$, will be lowest for an assemblage of the oxide and the *most X-deficient* basic salt. In the context of iodine-129 immobilization, basic iodides are likely to be more stable waste forms than the corresponding binary iodides. From a thermodynamic standpoint, the most suitable waste form in a given system will be a composite of the metal oxide (or hydroxide) and the *most iodine-deficient* basic iodide. A full assessment, however, needs consideration of other reactions, some of which are addressed below, along with some practical considerations.

Aqueous Equilibria Between Bismuth Oxyiodides at 25°C

The following equilibrium relationships have been determined in the system Bi_2O_3-BiI_3-H_2O-[pH] at 25°C [2]

$$\alpha\text{-}Bi_5O_7I(s) + OH^-(aq) \rightleftharpoons 5/2\ \alpha\text{-}Bi_2O_3(s) + I^-(aq) + 1/2\ H_2O(l) \qquad (4)$$

$$K_4 = \{I^-\}/\{OH^-\} = 10^{-3.39\pm0.20}$$

$$5/4\ BiOI(s) + OH^-(aq) \rightleftharpoons 1/4\ \alpha\text{-}Bi_5O_7I(s) + I^-(aq) + 1/2\ H_2O(l) \qquad (5)$$

$$K_5 = \{I^-\}/\{OH^-\} = 10^{3.49\pm0.18}$$

$$1/2\ BiI_3(s) + OH^-(aq) \rightleftharpoons 1/2\ BiOI(s) + I^-(aq) + 1/2\ H_2O(l) \qquad (6)$$

$$K_6 = \{I^-\}/\{OH^-\} = 10^{10.0\pm1.0}$$

The comparative stabilities of other bismuth oxyiodides, such as $Bi_7O_9I_3$ [*15*] and β-Bi_5O_7I [*16*], are not known, but they are unlikely to be more resistant to hydrolysis than α-Bi_5O_7I. Clearly, α-Bi_5O_7I is much more resistant to hydrolysis than BiI_3 or BiOI (K_4 is 13 orders of magnitude lower than K_6, and seven orders of magnitude lower than K_5). Based on Equilibrium 4, a mixture of α-Bi_2O_3 and α-Bi_5O_7I can buffer dissolved iodide at a given pH

$$\log \{I^-\} = pH - 17.4 \pm 0.2 \text{ (at 25°C)} \qquad (7)$$

It is important to stress the role of Bi_2O_3 in this equilibrium, and hence in an iodine waste form; it is a source of dissolved bismuth, whereby the solubility of the Bi_5O_7I is suppressed by a "common-ion" effect. The solubility of α-Bi_2O_3 is about $10^{-5.4}$ mol · dm^{-3} at pH 7 to 13 and 25°C [*17*]; it rises rapidly at lower pH, so Equilibrium 4 is probably not sustainable in acidic solutions. The equilibrium iodide activities associated with [Bi_2O_3 + Bi_5O_7I] in neutral or slightly alkaline solutions are comparable to the solubility of AgI, which is 10^{-8} mol · dm^{-3}, and almost independent of pH. Other factors that may influence the release of iodine from either AgI or a bismuth oxyiodide waste form are discussed below.

Aqueous Equilibria Between Basic Lead Iodides at 25°C

A recent phase-equilibrium study revealed four lead oxyiodides in the system PbO-PbI_2 [*18*]. We have observed the sequential formation of three compounds by the hydrolysis of

PbI_2 and iodination of PbO in aqueous solutions at 25°C. Of these, only Pb(OH)I has been reported previously; the others were provisionally identified as $7PbO \cdot PbI_2 \cdot 2H_2O$ and $3PbO \cdot PbI_2 \cdot H_2O$. At least one additional phase occurs at somewhat higher temperatures (50 to 100°C). The equilibrium constants, $\{I^-\}/\{OH^-\}$ (see Eqs 4 to 6), for each stage in the sequence, $PbO \rightleftharpoons 7PbO \cdot PbI_2 \cdot 2H_2O \rightleftharpoons 3PbO \cdot PbI_2 \cdot H_2O \rightleftharpoons Pb(OH)I \rightleftharpoons PbI_2$, were estimated to be $10^{-2.7\pm0.3}$, $10^{0.0\pm0.3}$, $10^{1.4\pm0.5}$, and $10^{7.4\pm0.3}$, respectively. These estimates are based on the experimental methods described for the bismuth oxyiodide system [2].

Thus, paralleling the bismuth system, the basic lead iodides, and especially $7PbO \cdot PbI_2 \cdot 2H_2O$, are more resistant to hydrolysis than PbI_2. The phase assemblage [$7PbO \cdot PbI_2 \cdot 2H_2O$ + PbO] buffers iodide activities at levels almost as low as [Bi_5O_7I + Bi_2O_3]. However, PbO is much more soluble than Bi_2O_3 (minimum $10^{-4.5}$ $mol \cdot dm^{-3}$ at pH 10 [17]). Lead is also much more toxic than bismuth. We therefore have not considered basic lead iodides further as a waste form option.

Other Factors Affecting Iodide Solubility

The behavior of an iodine waste form in a ground water may be modified in several ways. Here, we discuss redox reactions, thermal stability, anion-displacement, and complexation reactions of both bismuth oxyiodide and silver iodide.

Iodine Redox Equilibria

Lemire et al. [19] have reviewed the aqueous redox chemistry of iodine. Iodide predominates except under strongly oxidizing conditions. At redox potentials corresponding to oxygen partial pressures above about 0.2 mPa, iodate is favored

$$I^-(aq) + 3/2\ O_2(g) \rightleftharpoons IO_3^-(aq) \qquad (8)$$

$$K_8 = \{IO_3^-\}/\{I^-\}[p(O_2)^{3/2}] = 10^{13.0\pm0.8}\ Pa^{-3/2}$$

Oxidation of dilute, nonacidic iodide solutions is extremely slow, but there is some evidence that Equilibrium 8 is attained in deep seawater [20,21]. Thus, prolonged contact of any iodide waste form with oxygenated waters would be expected to increase total dissolved iodine concentrations above the simple solubility levels. Conversely, reducing ground waters are expected to increase release of iodine from iodate waste forms by reduction to iodide. A preliminary study of bismuth iodate chemistry [22] has shown that basic bismuth iodates are all much more susceptible to hydrolysis than Bi_5O_7I and so show little promise as waste forms.

Other oxidized iodine species, such as I_2, I_3^-, HIO, and H_2IO^+, only contribute significantly to equilibrium iodine speciation within a narrow range of acidic, oxidizing conditions [20], so they are unimportant in the present context. Periodate species are not thermodynamically stable in water, although they can exist metastably. Aqueous iodine (III) species, such as IO_2^-, are unstable both thermodynamically and kinetically.

In summary, strongly oxidizing ground waters should be avoided in the disposal of an iodide waste form, but selected in the case of an iodate. This presumes that dissolution is sufficiently slow that the iodine itself does not dominate the redox chemistry of the water.

Bismuth and Silver Redox Equilibria

Bismuth is present solely as Bi^{3+} in both Bi_2O_3 and Bi_5O_7I, so Equilibrium 4 is not a redox reaction. The +5 oxidation state of bismuth is a powerful oxidant [23], and is not expected to occur in ground waters, except perhaps in strong radiation fields. Reduction of Bi^{3+} compounds to elemental bismuth, however, can occur under natural reducing conditions; native bismuth is found in a variety of ore deposits. Our measurements of Equilibrium 4 allowed us to calculate the Gibbs energy of formation of Bi_5O_7I [*3*], and hence its stability toward reductive alteration

$$Bi_5O_7I(s) + OH^-(aq) + 15/2\ H_2(g) \rightleftharpoons 5Bi(s) + 8H_2O(l) + I^-(aq) \qquad (9)$$

$$K_9 = \{I^-\}/\{OH^-\}[p(H_2)^{15/2}]$$

$$\log K_9 = 54.6 \pm 1.4\ [p(H_2)\ \text{in Pa}]$$

The reductive alteration of AgI is represented by the reaction

$$AgI(s) + OH^-(aq) + ½\ H_2(g) \rightleftharpoons Ag(s) + H_2O(l) + I^-(aq) \qquad (10)$$

$$K_{10} = \{I^-\}/\{OH^-\}[p(H_2)]^{1/2}$$

$$\log K_{10} = 9.0 \pm 0.5\ [p(H_2)\ \text{in Pa}]$$

The regions of stability of AgI and Bi_5O_7I, as functions of pH, $p(H_2)$, and $\log\{I^-\}$, are depicted in Figs. 1*a* and 1*b*. These figures indicate that Bi_5O_7I is more resistant to reduction than AgI, but is still unstable under redox conditions controlled by the Fe^{2+}/Fe^{3+} buffer, which is commonly believed to influence the redox chemistry of deep, undisturbed granitic ground waters. In sulfur-containing ground waters, alteration of AgI or Bi_5O_7I to sulfides could also occur under strongly reducing conditions.

Most other sparingly soluble iodides (for example, CuI, Hg_2I_2, HgI_2, and PdI_2) are much less stable than AgI or Bi_5O_7I towards alteration by oxidation or reduction or both, as shown in Fig. 2. The much lower stability of CuI than AgI is of some interest, since copper and silver are congeners in the periodic table. The cupric ion, Cu^{2+}, is much more stable than Ag^{2+}, and Cu_2O is much less soluble than Ag_2O. Thus, CuI is less stable than AgI with respect to both oxidation and hydrolysis, as noted by Burger et al. [*9*].

It is possible that some other inorganic iodides have superior properties to either AgI or Bi_5O_7I. Oxyiodides of the lanthanides and some of the *d*-block transition elements may merit investigation, but information on such compounds is very limited.

Thermal Stability

Stability at temperatures above 25°C is an important consideration in evaluating any waste form, especially for iodine, since volatile I_2 is usually one of the decomposition products. A metal iodide may decompose to the elements, or the iodine may be displaced by oxygen, as represented by Reactions 11 and 12, respectively

$$2AgI(s) \rightleftharpoons 2Ag(s) + I_2(g) \qquad (11)$$

$$K_{11} = p(I_2)$$

$$2Bi_5O_7I(s) + ½\ O_2(g) \rightleftharpoons 5Bi_2O_3(s) + I_2(g) \qquad (12)$$

$$K_{12} = p(I_2)/[p(O_2)]^{1/2}$$

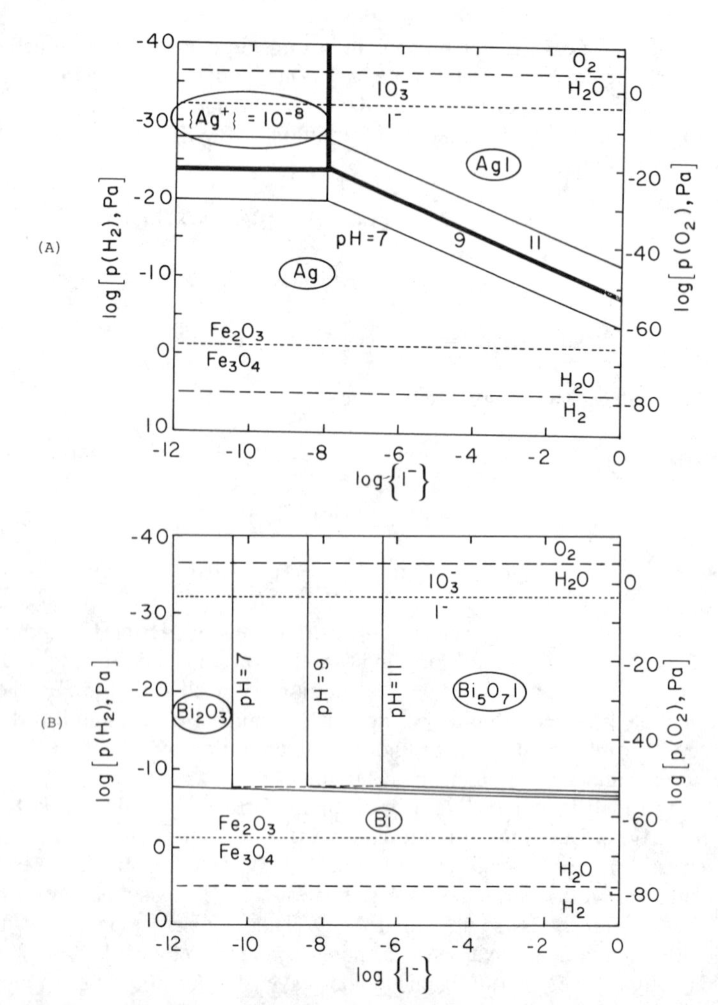

FIG. 1—*Calculated stability limits of* (a) *AgI and* (b) *Bi_5O_7I as functions of pH, iodide activity, and redox conditions at 25°C. Dashed lines represent stability limits of water; dotted lines represent I^-/IO_3^- and Fe_2O_3/Fe_3O_4 redox couples. Diagonal lines in* (a) *and near-horizontal lines in* (b) *correspond to control of dissolved iodide activity by Equilibria 10 and 9, respectively. Vertical lines in* (a) *and* (b) *correspond to control by the solubility product of AgI, and Equilibrium 4, respectively.*

Note that K_{12} is a function of oxygen partial pressure, but K_{11} is a function of temperature alone. More complex reactions may be important under some circumstances, if a volatile iodide is evolved or the metal changes oxidation state.

Decomposition reactions are often closely related to corresponding aqueous equilibria. For example, Reactions 4 and 12 involve the same pair of solids, Bi_2O_3 and Bi_5O_7I. They are related by the iodine redox reaction, 13 = 12 − (2 × 4)

$$2I^-(aq) + H_2O(l) + \tfrac{1}{2}O_2(g) \rightleftharpoons I_2(g) + 2OH^-(aq) \qquad (13)$$

Aqueous equilibrium measurements therefore can be used to calculate stability with respect to oxidative decomposition, just as they were used for calculations on reductive dissolution above.

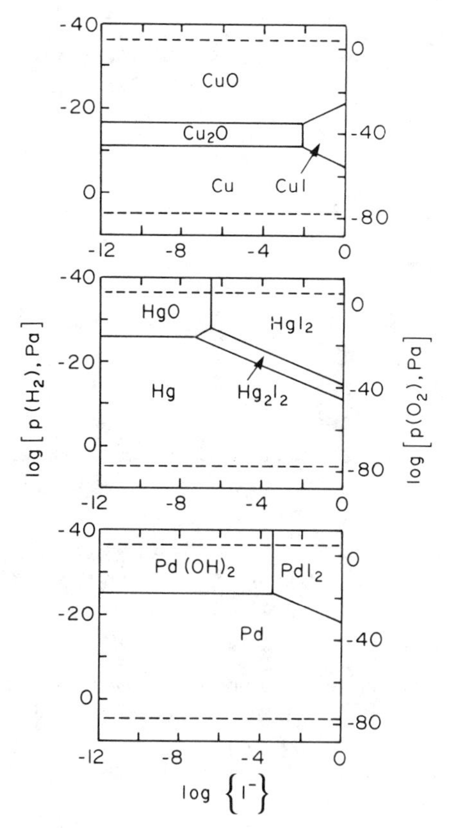

FIG. 2—*Calculated stability limits of solids in the systems M-O-I-H₂O-(pH) at pH 9 and 25°C:* (a) *M = Cu,* (b) *M = Hg,* (c) *M = Pd.*

Using this approach, we have calculated K_{12} as a function of temperature, from measurements of K_4 at 10 to 60°C, and also measured it directly by a gas-phase spectrophotometric technique at 550 to 800°C [*3*]. These two independent determinations are consistent, as shown in Fig. 3. They also concur with thermogravimetric work on Bi_5O_7I, which revealed the onset of decomposition in air at about 500°C [*2,24,25*].[3] The thermal stability of Bi_5O_7I appears to be better than for most insoluble iodides, but significantly inferior to AgI (Fig. 3). In the absence of air, Bi_5O_7I is reported to decompose by Reaction 14, at about 480°C [*24*]

$$3Bi_5O_7I(s) \longrightarrow 7Bi_2O_3(s) + BiI_3(g) \tag{14}$$

Temperature Dependence of Solubility

The following empirical relationship has been obtained from measurements of K_4 at 10 to 60°C [*3*]

$$\log K_4 = 5.24 - \frac{2573}{T} \pm 0.24$$

[3] Vance, E. R., Atomic Energy of Canada Limited, 1986, personal communication.

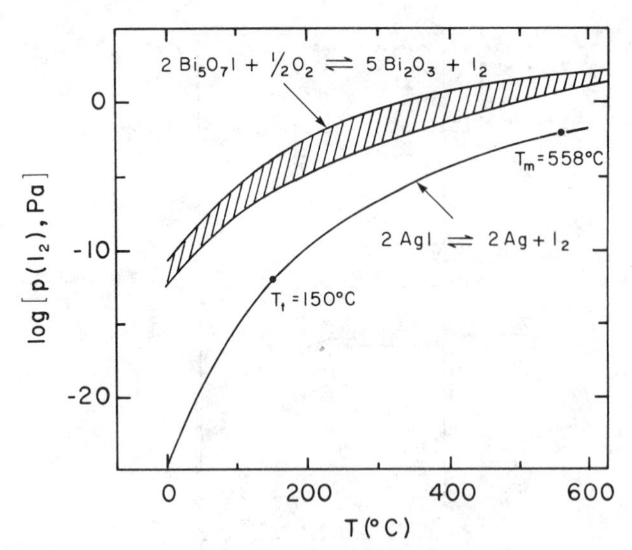

FIG. 3—*Calculated equilibrium iodine partial pressures over AgI + Ag (Eq 11, data from Ref* 43) *and* $Bi_5O_7I + Bi_2O_3$ *(in air, Eq 12, data from Ref* 4)*; boundaries of shaded area represent uncertainty limits.*

We thus estimate that K_4, and hence the equilibrium $\{I^-\}$ at a given value of $\{OH^-\}$, increases by almost two orders of magnitude between 25 and 100°C. R. J. Lemire[4] has estimated a similar increase in the solubility of AgI, from 9.2×10^{-9} mol · dm^{-3} at 25°C to 1.1×10^{-6} mol · dm^{-3} at 100°C. The two candidate waste forms are thus comparably sensitive to moderately elevated solution temperatures.

Anion Displacement Reactions

So far, we have only discussed equilibria between solid phases and aqueous iodide and hydroxide ions. We need to understand how other dissolved species can interact with silver iodide or bismuth oxyiodide, both by anion displacement and metal complexation.

Displacement of Iodide from AgI—Silver (I) salts have a simpler aqueous chemistry than bismuth (III) compounds. Their solubilities, in the absence of complexation, are represented by the general expression

$$1/n\ Ag_nX(s) \rightleftharpoons Ag^+(aq) + 1/n\ X^{n-}(aq) \quad (16)$$

$$K_{16,X} = \{Ag^+\}\{X^{n-}\}^{1/n}$$

For silver iodide, X = I and n = 1, and log $K_{16,I} = -16.1$ at 25°C, that is, AgI has a solubility of $10^{-8.0}$ mol · dm^{-3} in pure water.

If another anion, Y^{m-}, is present in sufficient quantity to exceed the solubility product of Ag_mY ($K_{16,Y}$), then Ag_mY can precipitate, and AgI will dissolve, until either the AgI is

[4] Lemire, R. J., Atomic Energy of Canada Limited, 1986, personal communication.

consumed or solution saturation is reestablished with respect to both solids

$$AgI(s) + 1/m\ Y^{m-}(aq) \rightleftharpoons 1/m\ Ag_mY(s) + I^-(aq) \tag{17}$$

$$K_{17} = \{I^-\}/\{Y^{m-}\}^{1/m}$$

$$= (\{Ag^+\}\{I^-\})/(\{Ag^+\}\{Y^{m-}\}^{1/m})$$

$$= K_{16,I}/K_{16,Y}$$

The threshold activity of Y^{m-} above which AgI may be converted to Ag_mY is then given by

$$\{Y^{m-}\} = \left[\frac{K_{16,Y} \cdot \{I^-\}}{K_{16,I}}\right]^m \tag{18}$$

Since AgI is one of the least soluble silver salts, few common anions are capable of displacing iodide in this way. These include chloride (at activities $>10^{-1.7}$ mol · dm^{-3}), bromide ($>10^{-4.2}$ mol · dm^{-3}), and sulfide ($>10^{-33.0}$ mol · dm^{-3}). These limits are derived from Gibbs energy data in Ref *26*. The extremely low threshold for sulfide is offset by its instability in nonreducing groundwaters; it does not encroach further on the stability limits of AgI defined in Fig. 1. The highest chloride and bromide activities reported in deep ground waters in the Canadian Shield are about 6 and 0.02 mol · dm^{-3}, respectively. Displacement reactions thus could raise equilibrium iodide activities in contact with AgI to 10^{-6} or even 10^{-5} mol · dm^{-3} at 25°C. For disposal in a dilute ground water environment, there may be some merit in a mixed silver halide waste form, [AgCl + AgI], since the chloride could suppress the solubility of the iodide by a silver common-ion effect, as shown in Fig. 4. The

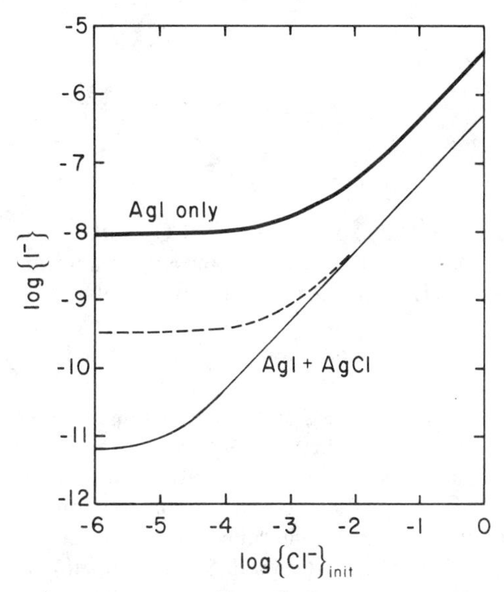

FIG. 4—*Calculated iodide activities in equilibrium with AgI and AgI + AgCl, as a function of chloride activity, based on chloride complexation (data from Ref* 26). *Dashed line indicates total dissolved iodine, including $AgI^0(aq)$.*

impact of silver chloride complexes could also be mitigated by this approach, as discussed next.

One major advantage of AgI over bismuth oxyiodide is its insensitivity to carbonate, because Ag_2CO_3 is relatively soluble. Bismuth oxyiodide is sensitive to displacement of iodide by carbonate, as described in the following section.

Displacement of Iodide from Bismuth Oxyiodide—Bismuth forms a plethora of basic salts with inorganic anions [*2,4,27,28*]. It is important to understand their stability relationships and hence their possible occurrence as alteration products of bismuth oxyiodides. Some of these salts occur as minerals, for example, bismutite $(BiO)_2CO_3$, bismoclite BiOCl, and daubréeite BiO(OH,Cl) [*29–31*]. Bismuth oxyiodides are not known in nature, whereas AgI occurs as the mineral iodargyrite (sometimes in association with bismoclite).

We have determined phase relationships in the systems Bi_2O_3 – BiI_3 – H_2O – [pH] [*2*], Bi_2O_3 – CO_2 – H_2O – [pH] [*4*], and Bi_2O_3 – $BiCl_3$ – H_2O – [pH] [*6*], and hence calculated relationships in the mixed iodide/chloride and iodide/carbonate systems at 25°C. The results are illustrated in the stability diagrams in Figs. 5 and 6. These indicate that Bi_5O_7I is susceptible to alteration by both carbonate and chloride at concentrations typically encountered in granitic ground waters (10^{-4} to 10^{-2} mol · dm^{-3} carbonate, and 10^{-4} to 1 mol · dm^{-3} or even higher chloride [*32,33*]). Shallow ground waters tend to be dominated by carbonate, and deeper waters by chloride [*32,33*].

Experiments have confirmed that Bi_5O_7I is readily converted to $(BiO)_2CO_3$ and $(BiO)_4(OH)_2CO_3$ by carbonate solutions, and to a phase resembling daubréeite by concentrated chloride solutions. The reactions are accompanied by release of iodide and a substantial increase in pH, in accordance with Eqs 19 to 21

$$2Bi_5O_7I(s) + 5CO_3^{2-}(aq) + 4H_2O(l) \longrightarrow 5(BiO)_2CO_3(s) + 2I^-(aq) + 8OH^-(aq) \quad (19)$$

$$4Bi_5O_7I(s) + 5CO_3^{2-}(aq) + 8H_2O(l) \longrightarrow 5(BiO)_4(OH)_2CO_3(s) + 4I^-(aq) + 6OH^-(aq) \quad (20)$$

$$2Bi_5O_7I(s) + 5Cl^-(aq) + 4H_2O(l) \longrightarrow 5(BiO)_2(OH)Cl(s) + 2I^-(aq) + 3OH^-(aq) \quad (21)$$

Reactions 19 and 20 can also be expressed with HCO_3^- rather than CO_3^{2-} as the reactive species. Reaction 21 has been simplified by assuming a fixed stoichiometry for BiO(OH,Cl). Note that, although BiO(OH,Cl) is much more easily hydrolyzed than Bi_5O_7I, chloride can still displace iodide from [Bi_2O_3 + Bi_5O_7I], because its relatively low reactivity is offset by its high concentration in many groundwaters. A more detailed account of these reactions is in preparation.

Experiments presently are in progress on the interaction of [Bi_2O_3 + Bi_5O_7I] with more complex salt solutions, including synthetic groundwaters. Results to date indicate that the reactions with chloride and carbonate are dominant; sulfate and dissolved silica may also play a role in iodide release. With carbonate-rich solutions, we commonly see a large transient release, with subsequent recovery of Equilibrium 4, at the increased pH, as dissolved iodide reacts with Bi_2O_3. Interactive effects between different competing anions remain unclear.

It is evident that anion displacement reactions are critical to the performance of a bismuth oxyiodide waste form. The most favorable waters would be dilute in dissolved solids, and also not strongly reducing. Effects of ionic strength on iodide solubilities are likely to be secondary to the chemical effects of dissolved anions.

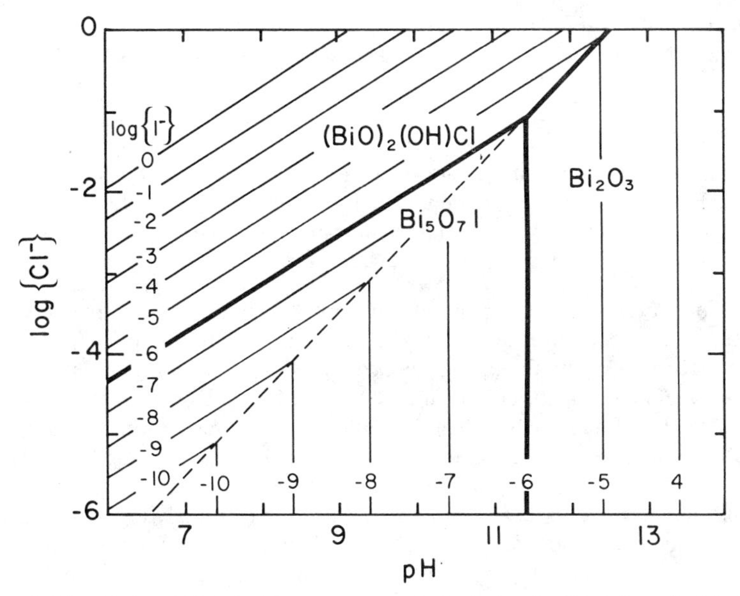

FIG. 5—*Stability limits of Bi_5O_7I with respect to synthetic daubréeite (ideal formula $(BiO)_2(OH)Cl$) with Bi_2O_3 as a function of pH and dissolved chloride and iodide activities.*

In preparing a bismuth oxyiodide waste form, it appears desirable to maximize the surface area ratio, Bi_2O_3:Bi_5O_7I, to insure the critical role of Bi_2O_3 in the equilibration. Even so, passivation of the oxide by a surface layer of carbonate could well compromise its effectiveness. Incorporation of [Bi_2O_3 + Bi_5O_7I] in cement may have a chemical advantage over and above whatever mechanical integrity it may confer. A typical portland cement would

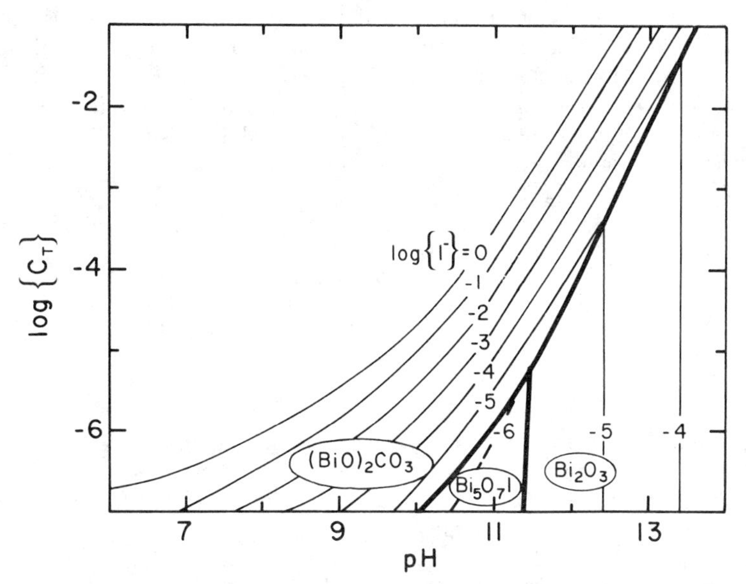

FIG. 6—*Stability limits of Bi_5O_7I with respect to $(BiO)_2CO_3$ and Bi_2O_3 as a function of pH and dissolved carbonate and iodide activities.*

increase the local pH to about 11. This would increase the inherent iodide solubility to about 10^{-6} mol · dm^{-3}, but should also reduce its sensitivity to carbonate or chloride.

Complexation

The formation of metal-ligand complexes can affect the solubility of an iodide waste form in two ways. Formation of iodine-containing complexes increases the total mass of dissolved iodine directly. Any other complexes that significantly alter the metal speciation may also increase the solubility of the metal iodide.

Bismuth Complexes—Since bismuth iodide complexes appear to be stable only in acidic, concentrated iodide solutions, they are unlikely to contribute to the solubility of a bismuth oxyiodide waste form [*34*]. Although bismuth complexation can alter the solubility of Bi_5O_7I alone, the iodide activity associated with [Bi_2O_3 + Bi_5O_7I] is independent of bismuth speciation, according to Equilibrium 4. The bismuth oxyiodide waste form is thus intrinsically insensitive to bismuth complexation, except for secondary effects arising from pH change, and possible accelerated depletion of the bismuth oxide component. Most of the published work on bismuth complexes deals with the development of analytical reagents [*35–37*]. Common ground-water anions do not appear to form strong complexes with bismuth in neutral-to-alkaline solution, although this area has been incompletely explored.

Silver Complexes—The most important silver complexes in the present context appear to be those with halide anions. Data summarized by Wagman et al. [*26*] indicate that chloride complexation could increase the solubility of silver iodide to about 10^{-5} mol · dm^{-3} in concentrated brines

$$AgI(s) + 2\,Cl^-(aq) \rightleftharpoons AgCl_2^-(aq) + I^-(aq) \quad (22)$$

$$K_{28} = \{AgCl_2^-\}\{I^-\}/\{Cl^-\}^2 = 10^{-10.8}$$

The effect of this reaction on the calculated solubility of AgI in chloride solutions is illustrated in Fig. 4. This is likely the dominant complexation reaction in ground waters, although other halide complexes, including mixed halide species, may also be important. Note that the two-phase waste form, [AgCl + AgI], can alleviate the impact of complexation, because AgCl provides a second source of silver, and the equilibrium iodide activity will be controlled by Eq 23, which is a variant of Eq 17

$$AgI(s) + Cl^-(aq) \rightleftharpoons AgCl(s) + I^-(aq) \quad (23)$$

$$K_{23} = \{I^-\}/\{Cl^-\} = 10^{-6.3}$$

Some Practical Considerations

Availability of Bismuth and Silver

Both bismuth and silver are relatively scarce, with current annual world production of about 2500 and 14 000 tonnes (t), respectively. Prices are erratic, but bismuth is considerably less expensive than silver (Bi_2O_3, U.S. $33/kg; silver bullion, U.S. $172/kg, December 1986). Bismuth also appears to have less strategic value than silver.

Current projections indicate a fission-product iodine inventory, including stable iodine-127, of about 15 t from Canadian power reactor operations up to the year 2050 [*38,39*]. If this fuel were recycled, conversion of its iodine to a bismuth oxyiodide waste form would require about 250 t of bismuth (based on equal amounts of bismuth present as Bi_2O_3 and Bi_5O_7I). Conversion to AgI would require about 13 t of silver, and a composite AgCl/AgI waste form would need about 25 t. These quantities do not appear to pose a major strain on silver or bismuth production, but broader application (that is, outside the Canadian nuclear program) likely would.

Ease of Preparation and Fabrication

Both AgI and Bi_5O_7I are easily prepared in the laboratory. The technology of silver halide production is highly developed in the photographic industry. Reliable conversion of Bi_2O_3 to Bi_5O_7I requires careful pretreatment of the oxide to eliminate traces of carbonate and initiate precipitation of the iodide [7]. It is also important to select Bi_2O_3 with a high specific surface area. After the pretreatment, contact times of the order of a few minutes are sufficient to remove over 99% of iodide from solutions with initial concentrations near 10^{-5} mol · dm^{-3}.

It may be desirable to convert a particulate product, formed by an aqueous process, to a monolithic form for disposal; this clearly requires further development. Since AgI is relatively stable at its melting point (Fig. 3), it should be amenable to sintering. This will be more difficult with bismuth oxyiodide. Bismuth oxide sinters to a rather soft ceramic at about 700°C [*40*], but Bi_5O_7I decomposes in air above 550°C [*3,24,25*]. Although it might be possible to sinter [Bi_2O_3 + Bi_5O_7I] in a sealed vessel, incorporation in a cement matrix may be more feasible.

Radiation Sensitivity

Since iodine-129 is extremely long-lived, the associated β-radiation fields are weak, so direct radiation damage is not a major concern. Accumulation of the transmutation product, xenon-129, would eventually cause structural rearrangement, but probably not for some millions of years. It is highly unlikely that any waste form would survive so long; our principal concern is that it should not alter sufficiently fast to release unacceptable concentrations of iodine.

One possible concern with a silver halide waste form is radiation-catalyzed reduction. Internal radiation may sensitize a silver halide waste form, and catalyze its reductive alteration in a manner analogous to the photographic process. Although bismuth oxyiodides are somewhat light-sensitive, they are less likely to be problematic in this regard. E. R. Vance[5] has found that aqueous dissolution of bismuth oxyiodide is insensitive to γ-radiation doses up to 10^6 Gray.

Summary

Although both silver iodide and bismuth oxyiodide have very low solubilities under some conditions, both have limited stability in geochemical environments, which must be taken into account in evaluating them as radioiodine waste forms, and selecting appropriate disposal conditions. Reductive dissolution is the most serious limitation of AgI, and anion displacement by chloride and carbonate is most important with bismuth oxyiodide. Other candidate materials, such as barium iodate and iodide-sodalite, cannot be defended on thermodynamic grounds, but may prove to have adequate kinetic stability to justify their

[5] Vance, E. R., Atomic Energy of Canada Limited, 1986, personal communication.

selection. Leaching tests on iodide-sodalite are promising, but this material has the disadvantage of requiring hot-pressing or hydrothermal techniques for efficient synthesis [*41,42*].

References

[*1*] Taylor, P., "Solubility and Stability of Inorganic Carbonates: an Approach to the Selection of a Waste Form for Carbon-14," Report AECL-9073, Atomic Energy of Canada Ltd., Pinawa, MB, Canada, 1987.
[*2*] Taylor, P. and Lopata, V. J., *Canadian Journal of Chemistry,* Vol. 64, 1986, p. 290.
[*3*] Taylor, P. and Lopata, V. J., *Canadian Journal of Chemistry,* Vol. 66, 1988, p. 2664.
[*4*] Taylor, P., Sunder, S., and Lopata, V. J., *Canadian Journal of Chemistry,* Vol. 62, 1984, p. 2863.
[*5*] Taylor, P. and Lopata, V. J., *Canadian Journal of Chemistry,* Vol. 62, 1984, p. 395.
[*6*] Taylor, P. and Lopata, V. J., *Canadian Journal of Chemistry,* Vol. 65, 1987, p. 2824.
[*7*] Taylor, P., Wood, D. D., and Lopata, V. J., "Solidification of Dissolved Aqueous Iodide by Reaction with α-Bi_2O_3 to form α-Bi_5O_7I." Report AECL-9554, Atomic Energy of Canada Ltd., Pinawa, MB, Canada, 1988.
[*8*] Vance, E. R., Agrawal, D. K., Scheetz, B. E., Pepin, J. G., Atkinson, S. D., and White, W. B., "Ceramic Phases for Immobilization of ^{129}I," Technical Report DOE/ET/41900-9, Rockwell International, Canoga Park, CA, 1981.
[*9*] Burger, L. L., Scheele, R. D., and Wiemers, K. D., "Selection of a Form for Fixation of Iodine-129," Technical Report PNL-4045, Battelle Pacific Northwest Laboratory, Richland, WA, 1981.
[*10*] Dunn, T., Montgomery, K., and Scarfe, C. M., "Synthesis and Leaching Behavior of Iodosodalite: A Potential Radioiodine Host," Technical Record TR-350, Atomic Energy of Canada Ltd., Pinawa, MB, Canada, 1985, p. 366.
[*11*] Holladay, D. W., "A Literature Survey: Methods for the Removal of Iodine Species from Off-Gases and Liquid Waste Streams of Nuclear Power and Nuclear Fuel Reprocessing Plants, with Emphasis on Solid Sorbents," Technical Report ORNL/TM-7350, Oak Ridge National Laboratory, Oak Ridge, TN, 1979.
[*12*] Mel'nikov, V. A., Moskvin, L. N., and Chetverikov, V. V., *Soviet Radiochemistry,* Vol. 25, 1983, p. 636 (*Radiokhimiya,* Vol. 25, 1983, p. 675).
[*13*] Kharbanda, J. L., Singh, I. J., and Amalraj, R. V., *Indian Journal of Chemistry,* Vol. 14A, 1976, p. 340.
[*14*] Bird, G. W. and Lopata, V. J., *Scientific Basis for Nuclear Waste Management,* Materials Research Society, Vol. 2, 1979, p. 419.
[*15*] Klimakov, A. M., Popovkin, B. A., and Novoselova, A. V., *Russian Journal of Inorganic Chemistry,* Vol. 19, 1974, p. 1394 (*Zhurnal Neorganicheskoi Khimii,* Vol. 19, 1974, p. 2553).
[*16*] Ketterer, J., Keller, E., and Krämer, V., *Zeitschrift für Kristallographie,* Vol. 172, 1985, p. 172.
[*17*] Baes, C. F., Jr., and Mesmer, R. E., *The Hydrolysis of Cations,* Wiley, New York, 1976, pp. 358–365, 375–383.
[*18*] Rolls, W., Secco, E. A., and Varadaraju, U. V., *Materials Science and Engineering,* Vol. 65, 1984, p. L5.
[*19*] Lemire, R. J., Paquette, J., Torgerson, D. F., Wren, D. J., and Fletcher, J. W., "Assessment of Iodine Behaviour in Reactor Containment Buildings from a Chemical Perspective," Report AECL-6812, Atomic Energy of Canada Ltd., Pinawa, MB, Canada, 1981.
[*20*] Liss, P. S., Herring, J. R., and Goldberg, E. D., *Nature, Physical Sciences,* Vol. 242, 1973, p. 108.
[*21*] Butler, E. C. V. and Gershey, R. M., *Analytica Chimica Acta,* Vol. 164, 1984, p. 153.
[*22*] Davidson, H., Taylor, P., and Lopata, V. J., unpublished observations, 1985.
[*23*] Smith, J. D. in *Comprehensive Inorganic Chemistry,* J. C. Bailar, Jr., H. J. Emeléus, R. Nyholm, and A. F. Trotman-Dickinson, Eds., Pergamon, Oxford, U.K., Vol. 2, 1973, p. 657.
[*24*] Schulte-Kellinghaus, M. and Krämer, V., *Angewandte Chemische Thermodynamik und Thermoanalytik Experientia Supplementum,* Vol. 37, 1979, p. 29.
[*25*] Rulmont, A., *Revue de Chimie Minérale,* Vol. 14, 1977, p. 277.
[*26*] Wagman, D. D., Evans, W. H., Parker, V. B., Schumm, R. H., Halow, I., Bailey, S. M., Churney, K. L., and Nuttall, R. L., *Journal of Physical and Chemical Reference Data,* Vol. 11, Supplement 2, 1982.
[*27*] Nurgaliev, B. Z., Vasekina, T. F., Baron, A. E., Popovkin, B. A., and Novoselova, A. V., *Russian Journal of Inorganic Chemistry,* Vol. 28, 1983, p. 415 (*Zhurnal Neorganicheskoi Khimii,* Vol. 28, 1983, p. 735).

[28] Margulis, E. V., Grishankina, N. S., Novoselova, V. N., and Maletina, E. D., *Russian Journal of Inorganic Chemistry,* Vol. 12, 1967, p. 1072 (*Zhurnal Neorganicheskoi Khimii,* Vol. 12, 1967, p. 2036).
[29] Bannister, F. A. and Hey, M. H., *Mineralogical Magazine,* Vol. 24, 1935, p. 49.
[30] Frondel, C., *American Mineralogist,* Vol. 28, 1943, p. 521.
[31] Sahama, Th. G. and Lehtinen, M., *Bulletin of the Geological Society of Finland,* Vol. 40, 1968, p. 145.
[32] Fritz, P. and Frape, S. K., *Chemical Geology,* Vol. 36, 1982, p. 179.
[33] Frape, S. K., Fritz, P., and McNutt, R. H., *Geochimica et Cosmochimica Acta,* Vol. 48, 1984, p. 1617.
[34] Eve, A. J. and Hume, D. N., *Inorganic Chemistry,* Vol. 3, 1964, p. 276 and Vol. 6, 1967, p. 331.
[35] Zhou, Nan, Yu, Ren-Qing, Yao, Xu-Zhang, and Lu, Zhi-Ren, *Talanta,* Vol. 32, 1985, p. 1125.
[36] Bidleman, T. F., *Analytica Chimica Acta,* Vol, 56, 1971, p. 221.
[37] Szczepaniak, W. and Reu, M., *Talanta,* Vol. 31, 1984, p. 212.
[38] Gillespie, P. A. et al., "Second Interim Assessment of the Canadian Concept for Nuclear Fuel Waste Disposal, Vol. 2: Background," Report AECL-8373, Atomic Energy of Canada Ltd., Pinawa, MB, Canada, 1985.
[39] Wuschke, D. M. et al., "Second Interim Assessment of the Canadian Concept for Nuclear Fuel Waste Disposal, Vol. 4: Post-Closure Assessment," Report AECL-8373, Atomic Energy of Canada Ltd., Pinawa, MB, Canada, 1985.
[40] Keski, J. R., *Journal of the American Ceramic Society,* Vol. 51, 1972, p. 527.
[41] Strachan, D. M. and Babad, H., "Iodate and Iodide Sodalites for the Long-Term Storage of ^{129}I," Technical Report RHO-SA-83, Rockwell International, Richland, WA, 1979.
[42] Tomisaka, T. and Engster, H. P., *Mineralogical Journal,* Vol. 5, 1968, p. 249.
[43] Robie, R. A., Hemingway, B. S., and Fisher, J. R., "Thermodynamic Properties of Minerals and Related Substances at 298.15 K and 1 Bar (10^5 Pascals) Pressure and at Higher Temperatures," Bulletin 1452, U.S. Geological Survey, Washington, DC, 1978 (reprinted with corrections, 1979), p. 296.

Mark Fuhrmann[1] *and Peter Colombo*[1]

Leaching-Induced Concentration Profiles in the Solid Phase of Cement

REFERENCE: Fuhrmann, M. and Colombo, P., **"Leaching-Induced Concentration Profiles in the Solid Phase of Cement,"** *Environmental Aspects of Stabilization and Solidification of Hazardous and Radioactive Wastes, ASTM STP 1033,* P. L. Côté and T. M. Gilliam, Eds., American Society for Testing and Materials, Philadelphia, 1989, pp. 302–314.

ABSTRACT: Analysis of the solid phase of portland cement specimens by energy-dispersive X-ray spectrometry before and after leaching provided elemental profiles within the cement. Releases of potassium were calculated from the solid-phase profiles and were compared with releases determined from leachate analyses of potassium and cesium-137. The fraction of potassium released in the leachate was found to correlate closely to that of cesium-137 under varying time and temperature conditions, despite the different manner in which each was originally contained in the cement. Agreement was obtained among potassium releases as determined from the solid, potassium in the leachate, and cesium-137 in the leachate. These correlations allowed the use of potassium as an analog for cesium-137 in cement.

Profiles of potassium in the solid showed varying degrees of depletion. A specimen, sectioned immediately after leaching for 471 days, contained no observable potassium to 9 mm depth from the specimen surface. From 9 mm to the center of the specimen, an apparently linear increase in concentration was observed. Specimens that had been air-dried prior to sectioning had profiles that were produced by evaporative transport of dissolved species toward the surface. Carbonation of the surface appears to have retarded migration of the dissolved material. This prevented it from reaching the outer edge and resulted in increased potassium concentrations several millimetres inside the surface.

KEY WORDS: leaching, portland cement, solid phase analysis, potassium, cesium-137

Portland cement is commonly used as a solidification agent for aqueous, low-level radioactive wastes. Hydraulic cement will also be used extensively for engineered disposal structures for these wastes. Therefore, its response to exposure to water (leaching) is important when considering both short- and long-term releases of contaminants.

A full description of the leaching behavior of cement requires consideration of the effects of leaching on the solid phase. This includes analysis of solid specimens to determine internal concentration profiles. Ideally, such profiles may be described by mass-transport equations, but concentration profiles in reactive, multiphase materials are subject to many influences. For example, alteration of the solid may occur as incongruous dissolution, causing increases in porosity and changes in elemental composition. Alternatively, authigenic mineral formation may close off porosity. Consequently, changes in both structure and chemistry must be investigated.

Unfortunately, sectioning and solid-phase analysis of radioactive specimens is often prohibitively difficult. A nonradioactive tracer (that behaves similarly to the radioactive element of interest) is difficult to find for solid-phase studies because most microprobes are not sufficiently sensitive to detect tracers at appropriate concentrations.

[1] Geochemistry associate and chemist, respectively, Nuclear Waste Research Group, Radiological Science Division, Building 703, Brookhaven National Laboratory, Associated Universitites, Inc., Upton, NY 11973.

In order for nonradioactive specimens to be used for solid-phase studies of cement waste forms, it was necessary to find an analog for cesium-137. Potassium is chemically similar to cesium and is present in detectable concentrations in cement. However, several potential problems could preclude the use of potassium for this purpose. First, the molar concentration of cesium-137 in radioactive waste is very small compared with concentrations of potassium in cement. Major differences in concentration may alter leaching behavior. There is a second related problem: at low concentrations, sorbtion of tracers by the solid may be important, whereas at high concentration it may be negligible. This is a particular concern in cement, where cesium may be incorporated into developing hydration products [*1*]. Third, cesium-137 is added to cement as an aqueous waste or tracer. In contrast, potassium is contained in the cement powder and is subjected to high temperatures during firing of the clinker [*2*]. Depending on the compounds formed, releases of this element may be limited by dissolution kinetics, causing differences in potassium and cesium-137 leaching behavior.

This paper establishes the relationship of potassium and cesium releases from cement under a variety of leaching conditions. Solid-phase analyses by scanning electron microscopy and energy dispersive X-ray spectroscopy were then used to observe changes in structure and potassium profiles within the cement. These profiles and the behavior of potassium in the solid during and after leaching can be taken as parallels to the leaching behavior of cesium-137.

Procedure

All specimens used in this study were made with portland Type I cement with a water-to-cement ratio of 0.43. They were cast in polyethylene containers and were right cylinders measuring 4.8 cm in diameter and 6.4 cm in length. Those specimens containing cesium-137 tracer were made with 10 μCi of the radionuclide included with the water used to hydrate the cement. Control samples, containing no radionuclides, were made under the same conditions.

Leach Test

The American Nuclear Society (ANS) 16.1 Leach Test [*3*] was modified for these experiments by the addition of several sampling intervals to better define the resulting curve shapes. This is a semi-dynamic test where the leachant is replaced periodically with fresh distilled water after intervals of static leaching. The volume of leachant was 1300 mL, ten times the geometric surface area of the sample. For the 20°C, long-term leach test, specimens containing radionuclides were tested in triplicate for 483 days. A similar specimen, without radionuclides, was tested for 471 days and the leachate was used for elemental analysis. Other specimens were leached for 18 days at 30 and 70°C.

Results were calculated according to Eq 1

$$\mathrm{CFR} = \frac{\Sigma A_n}{A_0} \tag{1}$$

where

CFR = cumulative fraction released,
A_n = amount of a specie released at any one leaching interval, and
A_0 = original quantity of a specie in the specimen.

Percent release is calculated by multiplying CFR by 100.

Leachate Analysis

Radionuclide gamma counting was performed with an automated sodium iodide (NaI) well-type detector and a multichannel analyzer. Analyses for calcium, silicon, aluminum, magnesium, sodium, potassium, and strontium were done by atomic absorption spectrophotometry (AAS). Acid digests of the cement powder used to make the specimens were also analyzed by AAS to determine source term concentrations of acid-soluble elements.

Solid-Phase Analysis

Leached and unleached cement samples were analyzed with a scanning electron microscope (SEM) and an energy-dispersive X-ray spectrometer (EDS). Samples were sectioned into 1-mm-thick slices measuring approximately 5 mm on a side. They were analyzed for the elements magnesium, aluminum, silicon, sulfur, potassium, and calcium using a spot size of approximately 1 mm^2. Relative elemental concentrations were based on X-ray peak intensities. Magnesium was not observed in the cement leachate (at detection limits of 0.5 μg/mL), nor did the EDS peak intensity change significantly throughout the interior of the specimen. Based on this evidence it was concluded that magnesium is immobile in cement. Consequently, the magnesium provides an internal reference to correct for any geometry and detector efficiency changes over time. The potassium data, therefore, are presented as the potassium/magnesium ratio. In addition, one of the cement samples was reanalyzed periodically to check for consistency and reproducibility of results.

Results

Solid-Phase Analysis

Solid-phase analysis by SEM/EDS was initiated to observe physical and chemical changes in specimens that may have developed as leaching progressed. Elemental profiles within the solid specimen were of particular interest. In the original survey, in order for the feasibility of detecting significant changes in elemental profiles to be determined, an EDS spectrum was obtained from the center of a leached sample. This spectrum is shown in Fig. 1. Magnesium, aluminum, silicon, sulfur, potassium, and calcium were detected. Another spectrum was obtained for the surface of the same leached specimen. The greatest change between the two was the absence of potassium from the specimen surface. This observation prompted a more detailed study of potassium profiles in leached cement specimens.

EDS analysis of sectioned, unleached portland cement specimens provided a baseline against which any changes that occurred during leaching could be compared. The average value for the potassium/magnesium ratio was 1:3. There was no observed change in this baseline value with distance into the cylinder for the unleached specimen.

Results of the solid-phase analysis for a specimen leached for 471 days are shown in Fig. 2. The original potassium/magnesium concentration ratio of unleached cement is shown as a line at 1:3. After leaching for 471 days, no potassium was observed in the specimen to a depth of 9 mm. From 9 mm to the center of the specimen, the potassium/magnesium ratio increased to a value approximately one-half the original concentration ratio. This increase appeared to be linear with distance. Comparing the area under the original concentration ratio line and the area under the specimen profile indicates that 18% of the potassium originally present in the cylinder remained, and 82% had leached out.

Examination of the leached surface by SEM showed that it was significantly altered when compared with the surface of an unleached specimen. Figure 3 is a micrograph of the surface after leaching for 471 days. When compared with Fig. 4, which is a micrograph of the surface

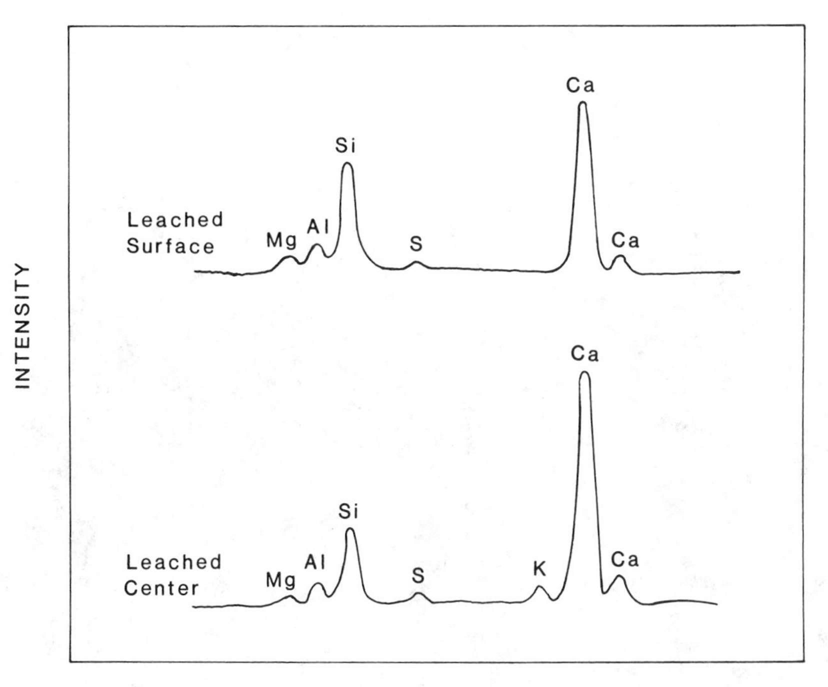

FIG. 1—*Energy dispersive X-ray spectra of specimens from center of a leached cement sample and from surface of the same leached sample. Note absence of potassium (K) peak from the surface sample.*

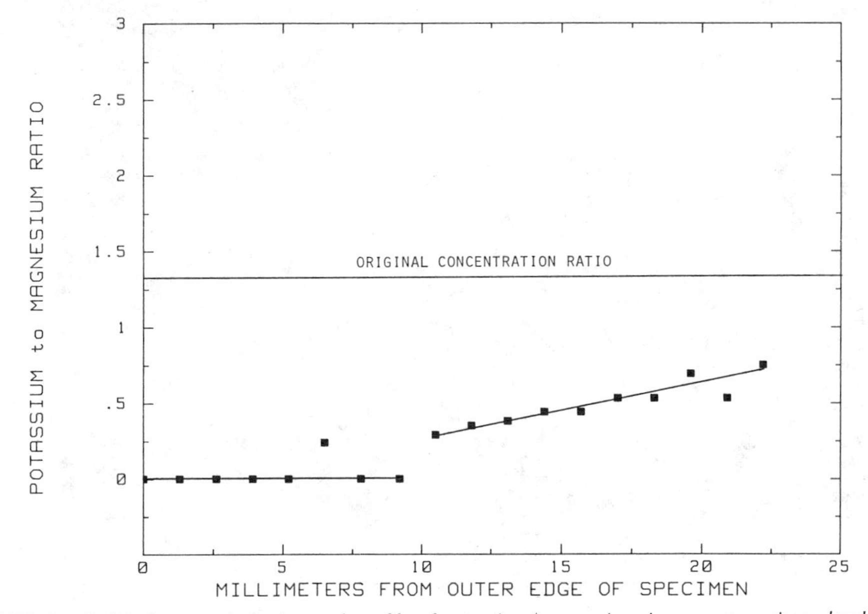

FIG. 2—*Solid-phase analysis. Internal profile of potassium/magnesium in cement specimen leached for 471 days at 20°C in distilled water. From this figure it is estimated that 82% of the potassium had leached out.*

FIG. 3—*Surface of unleached portland cement at ×1900 magnification.*

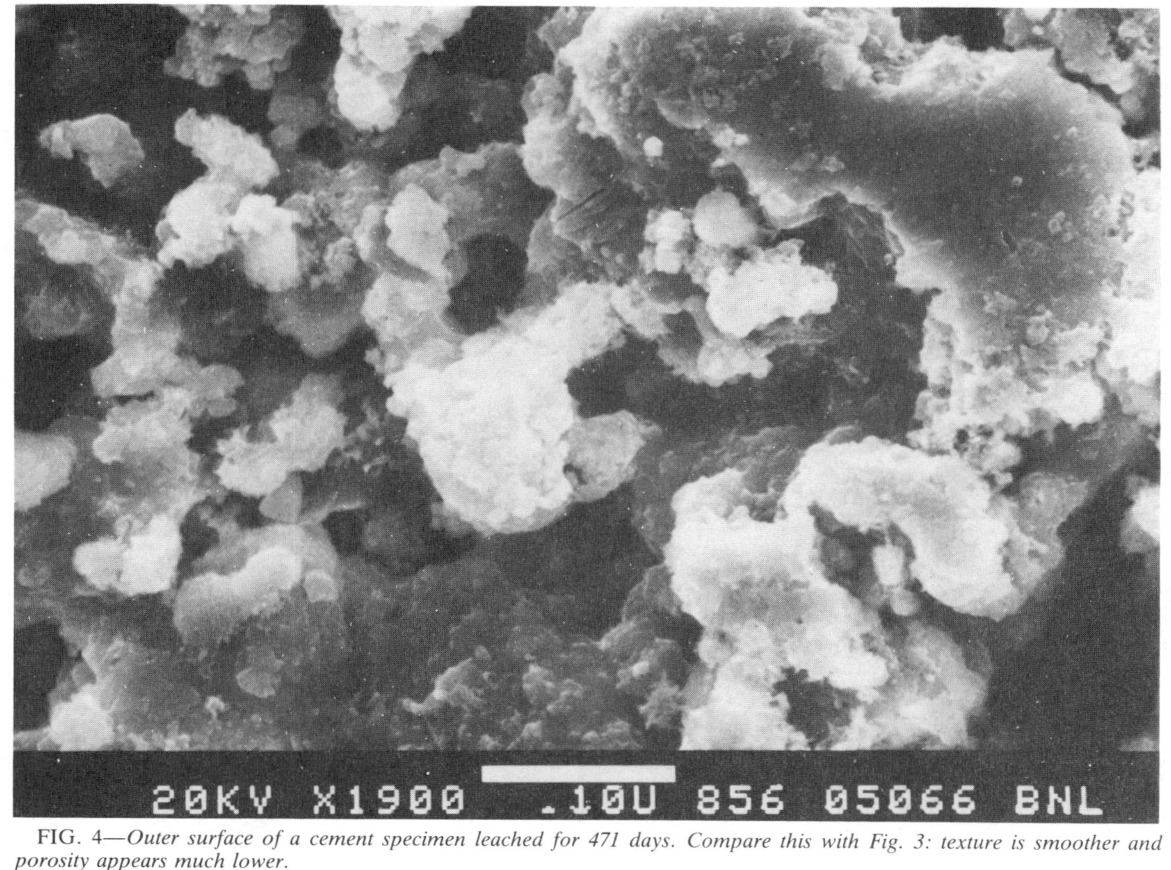

FIG. 4—*Outer surface of a cement specimen leached for 471 days. Compare this with Fig. 3: texture is smoother and porosity appears much lower.*

of an unleached specimen, the altered surface is much less porous than the original surface. Visual inspection of the sectioned specimen showed that this altered layer extended approximately 1 mm into the specimen. A specimen of the new surficial material was analyzed by X-ray diffraction to determine its mineralogical composition. The resulting diffractogram is shown in Fig. 5. The leached surface consisted of calcite ($CaCO_3$) and vaterite (μ-$CaCO_3$). Also shown in Fig. 5, to facilitate comparison, is a diffractogram of cement taken from the center of the same specimen. Distinct peaks for portlandite [$Ca(OH)_2$] and the broader peaks of the cement hydration products were present. No peaks for vaterite were observed in the center of the specimen.

Other cement cylinders were leached at 30, 40, 50, and 70°C for 18 days to determine effects of temperature on leaching. Results of these experiments relative to accelerated leaching are discussed elsewhere [*4,5*], but for this study, the 30 and 70°C cylinders were sectioned for solid-phase analysis. Unlike the long-term test specimen leached at 20°C, these specimens were allowed to air-dry prior to sectioning.

Figure 6 shows the interior potassium profile for the specimen leached at 30°C. The potassium/magnesium ratio increased dramatically (to double the original concentration ratio) to a depth of 7 mm from the surface of the specimen. Total potassium depletion was observed between 7 and 15 mm. From 15 mm to the center of the specimen, the potassium/magnesium ratio increased to a value slightly below the original concentration ratio. The percentage of potassium remaining in this specimen, as calculated from areas under the profile in Fig. 6, was 69%; the remainder, 31%, had leached out.

Figure 7 shows the potassium/magnesium ratio profile from the solid-phase analysis for a sample leached at 70°C. Most of the potassium has been leached, and no potassium was

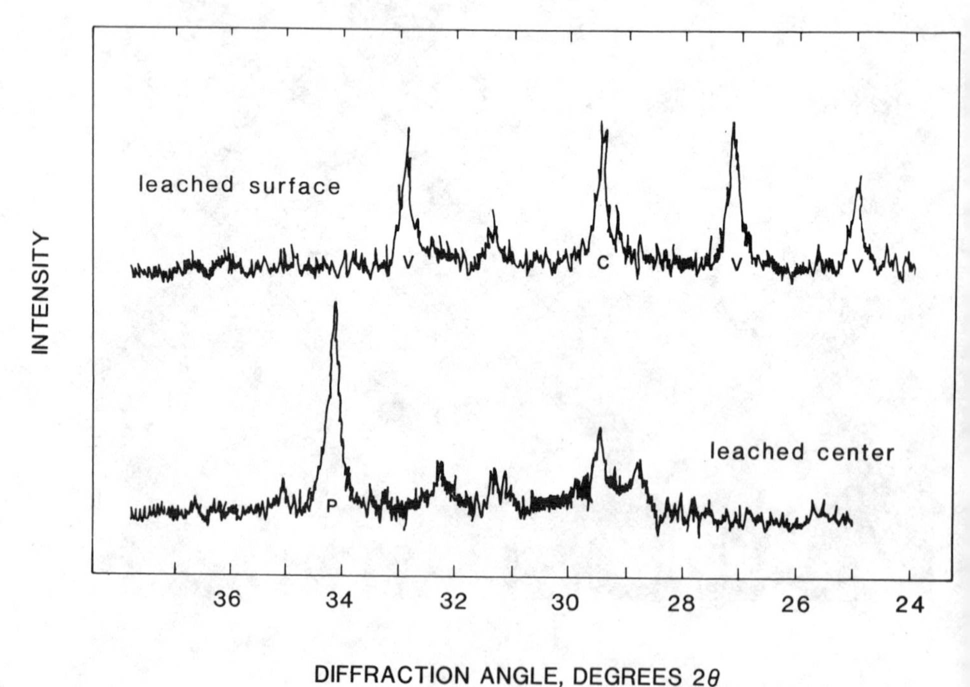

FIG. 5—*X-ray diffractogram for surface and center of a cement specimen leached for 471 days at 20°C in distilled water. Peaks labeled with V are vaterite, C are calcite, and P are portlandite. Most other peaks in the diffractogram from the center of the specimen are cement hydration products.*

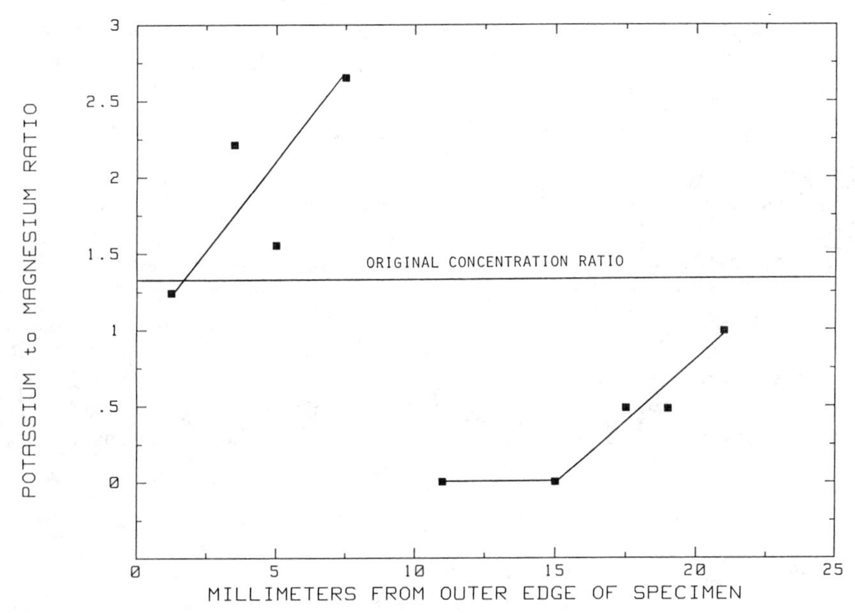

FIG. 6—*Internal potassium-magnesium profile in a cement specimen leached for 18 days at 30°C. As calculated from this figure, 31% of the potassium had leached out.*

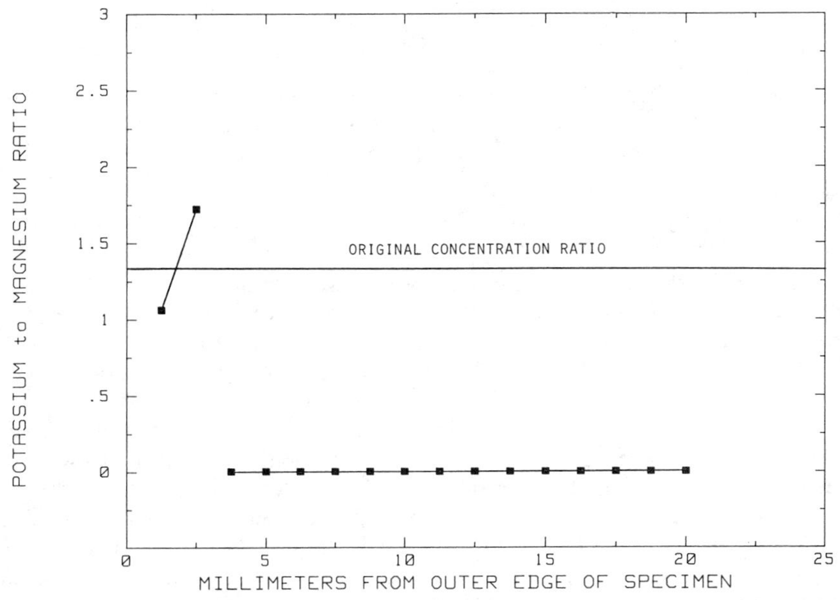

FIG. 7—*Internal potassium/magnesium profile in a cement specimen leached for 18 days at 70°C. As calculated from this figure, 85% of the potassium had leached out.*

detected at the center of the specimen. However, near the surface, potassium was observed to a depth of 3 mm at potassium/magnesium ratios similar to the original concentration ratio. The potassium remaining in the solid was calculated from the solid-phase analysis to be approximately 15% of the original value, 85% having leached out.

All specimens leached at elevated temperatures were observed to have carbonate coatings on their surfaces. These formed much faster than the coatings on specimens leached at 20°C, and carbonation was particularly rapid on first exposure to air after removal from the leachate.

An experiment was performed on triplicate specimens of the carbonate coatings from the leached sample and on specimens from the center of the same sample to determine their ability to adsorb cesium-137 (and, by correlation, potassium). Pulverized aliquots weighing 0.25 g were placed in 20 mL of distilled water, to which 1 mL of cesium-137 tracer (1000 dpm) was added. Blanks without the carbonate were also prepared. After one week, specimens of the supernate were filtered through 0.22-μm syringe filters and counted on a Geli detector. The distribution coefficient (K_D) was calculated according to the equation

$$K_D = \frac{A_s}{W_s} \cdot \frac{V_L}{A_L} \tag{2}$$

where

A_s = activity on the solid,
W_s = weight of the solid,
V_L = volume of the liquid, and
A_L = activity observed in the liquid.

The carbonate coating had an average K_D for cesium-137 of 2443 ± 2.9%, while the cement from the center of the same specimen had an average K_D of 10.5 ± 3.5%. The ability of the carbonate coating to sorb cesium-137 was 230 times greater than that of cement.

Leach Tests

Leachates from all tests were analyzed for calcium, aluminum, silicon, strontium, sodium, potassium, alkalinity, and cesium-137. Results are tabulated in Refs *4* and *5*.

Discussion

Potassium/Cesium Comparison

Comparisons of releases of the various elements were accomplished using correlation matrices such as the one shown in Table 1 for specimens leached at 70°C. This table gives the correlation coefficient (R) that approaches 1 as the correlation between any two species improves. The correlation coefficient of 0.99 for cesium-137/potassium was particularly high. A similar result was obtained for specimens leached at 30°C.

This approach was not useful for the long-term leach tests because of differences in sampling intervals between the radioactive and nonradioactive specimens. However, cumulative fraction release data for potassium from the long-term, 20°C leach test are shown on Fig. 8. The final cumulative release was 76% of the potassium originally present. Figure 8 also shows a CFR curve for cesium-137 from a replicate specimen leached under identical conditions. Although, for clarity, only one set of cesium-137 data is plotted, this specimen was one of a set of triplicates leached during this test. The final cesium-137 CFRs were

TABLE 1—*Correlation matrix for portland cement leached at 70°C.*

	Aluminum	Silicon	Strontium	Sodium	Potassium	Cesium-137	Alkalinity
Calcium	0.77	...	0.94	0.89	0.80	0.81	0.52
Aluminum		...	0.60	0.76	0.77	0.77	0.49
Silicon			...	...	...	...	...
Strontium				0.91	0.80	0.78	0.58
Sodium					0.98	0.97	0.61
Potassium						0.99	0.60
Cesium-137							0.59
Strontium-85							0.64

79.3%, 81.4%, and 81.6% (mean = 80.8 ± 2.1 at the 95% confidence interval) after 483 days.

Potassium and cesium-137 releases from the 30 and 70°C leaching experiments are compared in Fig. 9. In the 30°C test, 31% of the potassium was released, while a replicate specimen released 32% of its cesium-137. At 70°C, the final cumulative release for potassium was 92%, while 86% of the cesium-137 was leached from a replicate.

As shown in Figs. 8 and 9, the leaching behavior of potassium and cesium-137 from cement are similar for both long-term, room-temperature experiments and for short-term tests at elevated temperatures. Plotting the potassium CFR against that of the cesium-137, as in Fig. 10, for the two elevated-temperature experiments, illustrates the close correlation. Based on these data, releases of potassium from cement can be used as an analog for cesium-137 leaching releases under a variety of conditions. By extension, the behavior of potassium in the solid phase is also useful as a model for the transport behavior of cesium-137 in the solid.

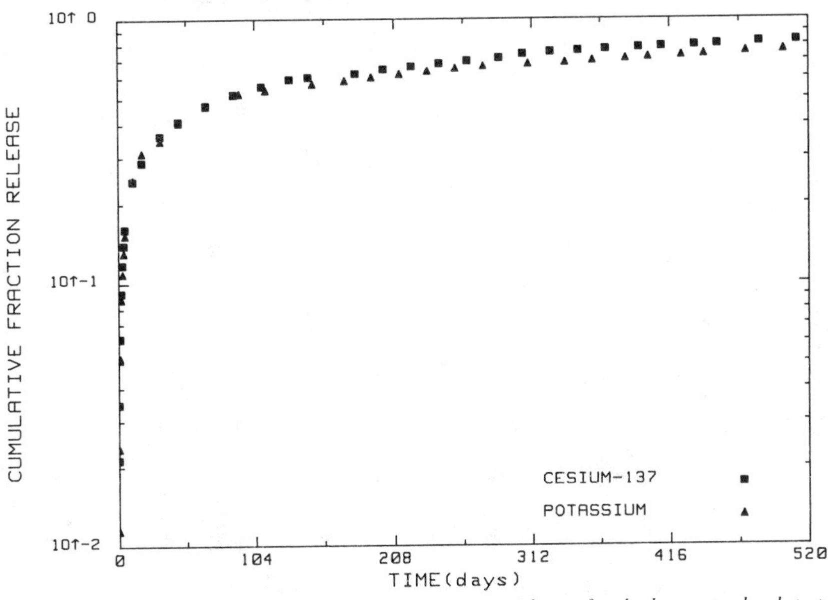

FIG. 8—*Potassium and cesium-137 cumulative fraction releases for the long-term leach test.*

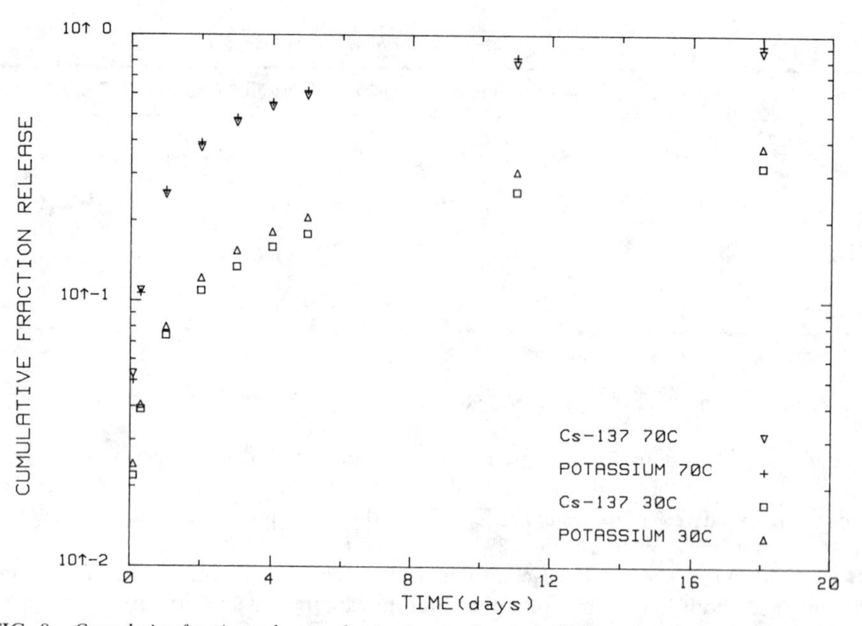

FIG. 9—*Cumulative fraction releases of potassium and cesium-137 from specimens leached at 30 and 70°C.*

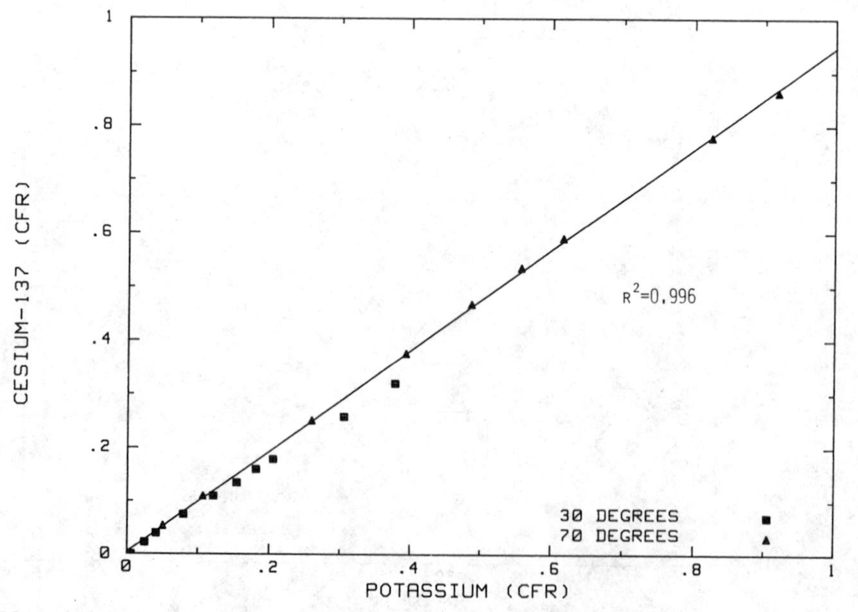

FIG. 10—*Cumulative fraction release of potassium plotted against CFR of cesium-137 for specimens leached at 30 and 70°C.*

Table 2 summarizes the releases of potassium and cesium-137 during the leach tests as well as potassium releases calculated from the solid-phase profiles. Agreement among these data is good, especially considering the different techniques used to obtain the data.

Solid-Phase Analysis

Each of the figures showing a profile of the potassium/magnesium ratio represents a snapshot of the potassium concentration gradient within the specimen at the time it was sectioned. The profile shown in Fig. 2 represented a time immediately after leaching was stopped, showing the potassium/magnesium ratio after leaching for 471 days at 20°C. Two distinct regions were present. The outer region was characterized by the absence of any detectable potassium; the inner zone was still undergoing leaching in a way that resulted in a linear increase in potassium to the center of the specimen. If the specimen had not been so depleted, a third region would presumably have been present near the center of the cylinder, with a constant ratio characteristic of the original unleached specimen. However, in the specimen analyzed, leaching had proceeded to such an extent that the unleached region was no longer present.

The outer edge of this specimen was significantly altered by both calcium depletion and carbonation. X-ray diffraction analysis of surficial material showed only peaks for vaterite (μ-$CaCO_3$) and calcite ($CaCO_3$). Another specimen taken from the center of the same sample showed no vaterite peaks, but did show strong portlandite [$Ca(OH)_2$] peaks as well as broad peaks attributed to cement hydration products. The change observed in the specimen surface is due to loss of portlandite and reaction of calcium from the cement with carbon dioxide from the atmosphere to form calcite. This surface carbonation effect is common in cement and has been reported to cause a decrease in permeability [*6,7*]. The micrographs showing the surfaces of leached and unleached specimens confirm that the surface porosity appears to be significantly reduced.

Potassium profiles for specimens leached at 30 and 70°C, shown in Figs. 6 and 7, represent potassium gradients for specimens that had been air-dried prior to sectioning. The elevated potassium/magnesium ratios observed several millimetres in from the surface of the specimens were an artifact of drying. As water evaporated from the surface of the specimen, the pore water was drawn outward, carrying dissolved elements with it. This evaporative transport has been postulated as having significant effects on leaching during wet/dry cycles in disposal environments [*8*]. Where this efflorescence process became inhibited by the surficial carbonation, which plugged the porosity of the cement, there was a buildup of dissolved species. Potassium will coprecipitate with calcite, particularly in the presence of

TABLE 2—*Final releases of cesium-137 and potassium as determined by leachate and solid-phase analysis.*

Specimen	Leachate Analysis: Cesium-137 Release, %	Leachate Analysis: Potassium Release, %	Solid-Phase Analysis: Potassium Release, %
20°C (471 days)	79, 81, 82	76	82
30°C (18 days)	32	38	31
70°C (18 days)	86	92	85

magnesium ions [9]. However, the high K_D of the carbonates indicates that adsorption onto these minerals is the cause of the profiles that were observed in the dried specimens.

Conclusions and Applications

Potassium has been confirmed as an analog for cesium-137 in leaching studies of cement. Their behavior is similar in both long-term leach tests and short-term tests at elevated temperatures. Elemental concentration profiles of potassium can be observed by EDS in the solid phase of leached cement, and these profiles appear to reflect a variety of mass-transport phenomena (leaching via diffusion and evaporative transport). There is a quantitative similarity between potassium releases calculated from the solid-phase analysis and releases of potassium and cesium-137 observed in the leachates.

Additional work can now be done to determine effects of surficial alteration and long-term aging of cementitious materials on cesium-137 releases. Application of these techniques to other radionuclides will clarify the more complex leaching behavior of other species of concern, including strontium-90 and soluble complexes of cobalt-60.

The approach taken in this paper may have important implications for efforts to determine the long-term durability of concrete barriers for radioactive waste disposal. Potassium has been shown to be an analog for cesium-137, and its behavior in the solid phase readily reflects various transport phenomena that operate in cement. Therefore, it will be useful to look for solid-phase profiles of potassium in old concrete structures. The degree of potassium depletion will be a direct indicator of the mobility of cesium through these materials. By observing this effect in a variety of cementitious materials and in a variety of environments, it will be possible to determine what factors influence concrete durability with regard to radionuclide mobility. Improved barriers for radioactive waste can be constructed with this information.

Acknowledgments

We wish to thank John Warren for his expertise and generosity with time on the electron microscope. We also thank Robert Doty for his help in the laboratory, James Calcanes for his work with electron microscopy, and David Dougherty and Richard Pietrzak for their comments. This work was performed as a contract with the Department of Energy, National Low-Level Radioactive Waste Management Program.

References

[1] Komarnei, S. and Roy, D. M., *Science*, Vol. 221, 1983, p. 647.
[2] Lea, F. M., *The Chemistry of Cement and Concrete*, 3rd ed., Chemical Publishing Co., Inc., New York, 1970.
[3] "Measurement of the Leachability of Solidified Low-Level Radioactive Wastes," ANS 16.1 Standards Committee, *Final Draft of a Standard*, American Nuclear Society, Feb. 1984.
[4] Dougherty, D., Fuhrmann, M., and Colombo, P., "Accelerated Leach Test(s) Program Annual Report," BNL-51955, Brookhaven National Laboratory, Upton, NY, Sept. 1985.
[5] Dougherty, D., Fuhrmann, M., and Colombo, P., "Accelerated Leach Test(s) Program Annual Report," BNL-52042, Brookhaven National Laboratory, Upton, NY, Sept. 1986.
[6] Soroka, I., *Portland Cement Paste and Concrete*, Chemical Publishing Co., New York, 1980.
[7] Swenson, E. G. and Sereda, P. J., "Mechanism of the Carbonation Shrinkage of Lime and Hydrated Cement," *Journal of Applied Chemistry*, Vol. 18, No. 4, 1968, pp. 111–117.
[8] Dayal, R., Schweitzer, D. G., and Davis, R. E., "Wet/Dry Cycle Leaching Aspects of Release in the Unsaturated Zone," BNL-33580, Brookhaven National Laboratory, Upton, NY, 1983.
[9] Okumura, M. and Kitano, Y., "Coprecipitation of Alkali Metal Ions with Calcium Carbonate," *Geochemica et Cosochimica Acta*, Vol. 50, No. 1, Jan. 1986, pp. 49–58.

Large-Scale Evaluation or Demonstration

ohn P. Englert,[1] Carlyle J. Roberts,[1] and Eugene E. Smeltzer[2]

Strategy for Management of Mixed Wastes at the West Valley Demonstration Project

REFERENCE: Englert, J. P., Roberts, C. J., and Smeltzer, E. E., **"Strategy for Management of Mixed Wastes at the West Valley Demonstration Project,"** *Environmental Aspects of Stabilization and Solidification of Hazardous and Radioactive Wastes, ASTM STP 1033,* P. L. Côté and T. M. Gilliam, Eds., American Society for Testing and Materials, Philadelphia, 1989, pp. 317–329.

ABSTRACT: The evolution of regulations covering the disposal of low-level radioactive wastes (10 CFR 61) and hazardous wastes (40 CFR 260-265) has created a dilemma for generators of wastes that are both radioactive and hazardous, that is, mixed wastes. Because these regulations were developed independently, compliance with one set of regulations may create conflicts with the requirements of the other regulations. The challenge confronting mixed-waste generators is to develop waste management programs that eliminate these conflicts.

The West Valley Demonstration Project is employing cement solidification on several characteristic mixed-waste streams to remove the final waste form from the hazardous waste universe and simplify the regulatory requirements for subsequent storage and disposal. This paper discusses the regulatory basis for specific waste management practices and illustrates their application to a corrosive mixed-waste stream and a radioactive-waste stream that is hazardous based on the extraction procedure (EP) toxicity characteristic.

KEY WORDS: hazardous waste, radioactive waste, mixed waste, waste treatment, waste management, cementation, neutralization, solidification, waste stabilization

The West Valley Demonstration Project (WVDP) is a radioactive-waste management roject managed by the U.S. Department of Energy (DOE) at the Western New York Juclear Services Center near West Valley, New York (Fig. 1). This is the only commercial uclear fuel reprocessing facility to have operated in the United States. The wastes to be ıanaged include all three of the major radioactive waste categories: high-level waste (HLW, ıe waste generated by the reprocessing of spent nuclear fuel); transuranic waste (TRU, aste containing greater than 100 nCi/g of alpha-emitting radionuclides with half-lives reater than five years); and low-level waste (LLW, radioactive waste that is neither HLW or TRU).

For each category of waste, hazards are associated with the radioactive nature of the ıaterial and other hazards are associated with the chemical nature of individual waste treams. Clearly, the hazards of the intense-radiation field associated with HLW or the long-ved radiotoxicity of TRU waste exceed the potential chemical hazards of typical HLW or RU waste streams. However, this statement cannot be applied consistently to LLW, some

[1] Senior environmental scientist and safety and environmental assistant manager, respectively, Dames Moore, c/o U.S. Department of Energy, West Valley Demonstration Project, P.O. Box 191, West alley, NY 14171.

[2] Senior engineer, Westinghouse Electric Corporation, Research and Development Center, Pittsırgh, PA.

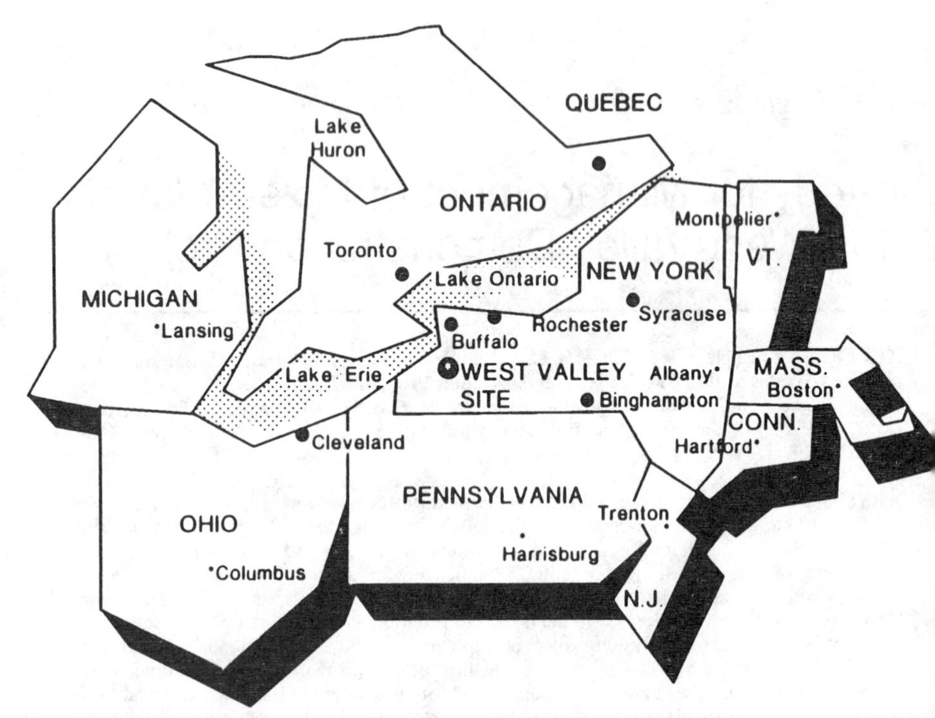

FIG. 1—*Location of the West Valley demonstration project, West Valley, New York.*

of which possesses minor amounts of radioactivity in a chemical matrix which may be more hazardous. The waste-management practices developed for HLW and TRU waste mitigate the radioactive hazards and provide ample protection against the potential chemical hazard of these wastes. In contrast, LLW-management practices must consider both radioactive hazards and potential chemical hazards because the latter may require mitigative measure in addition to what would otherwise be provided.

Low-level radioactive-waste management requirements for U.S. Nuclear Regulatory Commission (NRC) regulated facilities are identified in Title 10 of the Code of Federal Regulations, Part 20-Standards for Protection Against Radiation [*1*], and Part 61-Licensing Requirements for Land Disposal of Radioactive Waste [*2*] (10 CFR 20 and 10 CFR 61). DOE Orders 5820.2-Radioactive Waste Management [*3*], and 5480.1A-Environmental Radiation, Safety, and Health Protection Programs for DOE Operations [*4*] contain the requirements which are applicable to DOE facilities. DOE has also issued Order 5480.2 Hazardous and Radioactive Mixed Waste Management [*5*]. The WVDP, as a DOE operation complies with the DOE Orders, but also utilizes the NRC regulations for additional guidance. This is particularly important because the DOE tenure at West Valley is limited by the WVDP Act (PL 96-368), and control of the site will revert to New York State under a NRC license at the conclusion of the demonstration project. To assure that LLW generated by the WVDP will meet NRC requirements, they are classified in accordance with the technical requirements of 10 CFR 61 [*2*].

Chemically hazardous wastes are identified by the U.S. Environmental Protection Agency (EPA) in the hazardous waste regulations at Title 40 of the Code of Federal Regulations Part 261 (40 CFR 261) [*6*]. Wastes which are identified in this part, either by virtue c characteristic (ignitability, reactivity, corrosivity, or extraction procedure [EP] toxicity), or by being on one of several hazardous waste lists contained in this part, are subject to EPA regulation from the point of generation through storage, transport, treatment, and disposal in accordance with the remaining hazardous waste regulations found primarily at 40 CFR 260-265 [*7–11*].

Specifically excluded from the regulatory definition of solid wastes, and hence hazardous wastes, are materials regulated under the Atomic Energy Act of 1954 as source, special nuclear, or byproduct material (40 CFR 261.4) [*6*]. Source material, as defined at 10 CFR 40 [*12*], is uranium or thorium, or ores containing greater than 0.05% uranium or thorium or any combination thereof. Special nuclear material, defined in the same part, is plutonium, uranium 233 or uranium enriched in the isotope 233 or 235. Byproduct material, as defined at 10 CFR 30 [*13*], is any radioactive material except special nuclear material made radioactive in the production or utilization of special nuclear material. This exclusion from regulation by EPA was initially interpreted to apply to the entire waste stream containing these materials regardless of the chemical nature and content of the waste; however, subsequent clarification [*14*] indicates that low-level radioactive waste containing hazardous waste or possessing hazardous characteristic(s) (that is, mixed waste) is subject to dual regulation—the hazardous fraction by EPA and the radioactive fraction by either DOE or NRC. DOE does not distinguish between the different categories of radioactive waste (that is, LLW, HLW, and TRU) in its interpretation of this exclusion, thus any category of radioactive waste may include mixed-waste streams.

Mixed wastes represent a small fraction of either the hazardous waste universe or the radioactive waste universe. The annual mass of low-level radioactive waste generated in commercial operations is estimated to be approximately 0.04% of the hazardous waste mass generated annually in the United States [*15*], and a survey across the spectrum of LLW generators indicates that the largest fraction of potentially mixed waste represents only about 2.3% of the LLW identified in the survey [*16*].

The data presented in recent studies [*15–17*] and earlier data used to develop the technical basis for 10 CFR 61 [*18*] indicate that mixed wastes at nuclear-fuel cycle facilities are comprised to a large extent of wastes that are hazardous because they exhibit at least one of the hazardous waste characteristics identified previously. If these wastes can be treated so that they do not demonstrate the hazardous characteristic(s), they are no longer hazardous and hence are no longer mixed wastes.

The EPA regulations provide several mechanisms for generators to perform such treatment without obtaining a hazardous-waste-treatment permit. Treatment methods exempted from the permitting process include totally enclosed treatment, elementary neutralization, and treatment in accumulation tanks or containers. The WVDP has or will employ these options for managing characteristic mixed-waste streams until such time that it is determined that a hazardous-waste-treatment permit is required.

WVDP Waste Streams and Treatment Methods

High-Level Waste

One of the main objectives of the WVDP is to solidify the approximately 2.1 million litres of liquid high-level waste remaining from the nuclear-fuel reprocessing activities during the period 1966 to 1972. There are two types of HLW stored in underground tanks at the site. The majority of the waste (approximately 98% by volume) was formed by adding excess sodium hydroxide to a nitric-acid-based waste stream generated by the PUREX process. This waste stream contains the bulk of the fission products which were separated from the uranium and plutonium in the spent fuel. The neutralization operation resulted in the formation of a two-phase waste in the storage tank (Tank 8D-2): a sludge layer approximately 0.5 m thick and a liquid supernatant layer. The chemical composition of the sludge and supernatant are provided in Tables 1 and 2.

The remaining HLW, approximately 45 000 L, resulted from a single reprocessing campaign of thorium-enriched fuel using the THOREX process. This waste is stored as an unneutralized nitric-acid solution in a stainless-steel tank (Tank 8D-4). The chemical composition of this waste is shown in Table 3. These data, and additional detailed information on HLW characterization, are provided in Ref *19*.

Processing the HLW into a borosilicate glass involves several steps. First, the PUREX supernatant, which contains nearly all of the cesium 137 in this waste, will be processed through a series of ion-exchange columns installed inside the spare HLW storage tank (Tank 8D-1). The columns are loaded with a zeolite (IONSIV IE-96), which is very selective for cesium removal. Bench scale tests indicate that four columns in series will provide a cesium decontamination factor greater than 1000 [*19*]. The cesium-loaded resin will be discharged to the floor of Tank 8D-1, and fresh zeolite will be charged to the columns as needed. After all the supernatant has been removed from Tank 8D-2, the sludge will be washed several times with deionized water to remove the remaining water-soluble activity. The sludge (containing the bulk of the strontium 90, actinides, and rare earth nuclides), the cesium-loaded zeolite, and the THOREX waste will be transferred to the vitrification facility. Here it will be mixed with glass formers and fed as a slurry to a glass melter where it will be

TABLE 1—*Purex high-level sludge solids chemical composition.*

Component	Reference, kg
$Fe(OH)_3$	66 040
$FePO_4$	6 351
$Al(OH)_3$	5 852
AlF_3	536
MnO_2	4 581
$CaCO_3$	3 208
$UO_2(OH)_2$	3 087
$Ni(OH)_2$	1 088
SiO_2	1 263
$Zr(OH)_4$	159[a]
$MgCO_3$	826
$Cu(OH)_2$	376
$Zn(OH)_2$	128
$Cr(OH)_3$	65
$Hg(OH)_2$	23
FISSION PRODUCTS	
fission product hydroxides	1 485
rare earth hydroxides	1 484
fission product sulfates	520
TRANSURANICS	
NpO_2	42
PuO_2	37
AmO_2	27
CmO_2	0.3
TOTAL	97 178

[a] Excludes fission product zirconium.

TABLE 2—*Purex high-level supernatant chemical composition.*[a]

Compound	Total kg in Supernatant
$NaNo_3$	602 659
$NaNO_2$	311 326
Na_2SO_4	76 261
$NaHCO_3$	42 557
KNO_3	36 274
Na_2CO_3	25 249
NaOH	17 537
K_2CrO_4	5 113
NaCl	4 684
Na_3PO_4	3 799
Na_2MoO_4	691
Na_3BO_3	597
$CsNO_3$	534
NaF	503
$Sn(NO_3)_4$	245
$Na_2U_2O_7$	231
$Si(NO_3)_4$	230
$NaTcO_4$	177
$RbNO_3$	119
Na_2TeO_4	82
AlF_3	77
$Fe(NO_3)_3$	43
Na_2SeO_4	15
$LiNO_3$	14
H_2CO_3	9
$Cu(NO_3)_2$	6
$Sr(NO_3)_2$	4
$Mg(NO_3)_2$	2
TOTAL	1 129 038
H_2O (by difference)	1 727 164

[a] pH = 10.0.

heated to nearly 1200°C to form borosilicate glass. The high-level waste processing flow sheet is shown in Fig. 2.

Low-Level Waste

The treated supernatant and sludge wash solutions will be LLW because the bulk of the radioactivity will have been removed by the zeolite. This extracted salt solution will be piped from the supernatant treatment system to the Radioactive Waste Treatment System (RTS), which is composed of the Liquid Waste Treatment Subsystem (LWTS) and the Cement Solidification Subsystem (CSS). In the LWTS, extracted salt solution will be concentrated to approximately 40 weight percent solids prior to transfer to the CSS.

The CSS is a fully automated, remotely operated system equipped with two high-shear mixers. Waste solutions are routed from the LWTS to a waste-dispensing vessel in the CSS process cell, where it is metered into the high-shear mixers. Dry Type I Portland Cement is gravimetrically metered into the mixers, which operate in a batch mode, and is mixed with the waste solution for a predetermined period of time. The waste-cement mixture is then pumped into a 269-L (70 gal) square steel drum, three mixer loads per drum. The

TABLE 3—*Thorex waste chemical composition.*

Compound	Mass, kg	Compound	Mass, kg
		SOLUTION	
$Th(NO_3)_4$	11 633	$NaTcO_4$	12
$Fe(NO_3)_3$	8 462	$Sm(NO_3)_3$	14
$Al(NO_3)_3$	4 175	$Zr(NO_3)_4$	12
HNO_3	2 129	$Y(NO_3)_3$	11
$Cr(NO_3)_3$	1 918	$Rh(NO_3)_4$	11
$Ni(NO_3)_2$	791	$Zn(NO_3)_2$	10
H_3BO_3	480	$Pd(NO_3)_4$	8
$NaNO_3$	227	$UO_2(NO_3)_2$	6
Na_2SO_4	180	$RbNO_3$	6
KNO_3	128	$NaTeO_4$	5
Na_2SiO_3	126	$Co(NO_3)_2$	3
K_2MnO_4	122	Na_2SeO_4	1
$Mg(NO_3)_3$	57	NaF	1
Na_2MoO_4	54	$Eu(NO_3)_3$	1
NaCl	50	$Sn(NO_3)_3$	0.9
$Nd(NO_3)_3$	46	$Cu(NO_3)_2$	0.8
$Ce(NO_3)_4$	43	$Pu(NO_3)_3$	0.7
$Ru(NO_3)_4$	42	$Gd(NO_3)_3$	0.3
$Ca(NO_3)_2$	30	$X^a(NO_3)_4$	0.3
$CsNO_3$	28	$Cd(NO_3)_2$	0.3
$Ba(NO_3)_2$	27	$Sb(NO_3)_3$	0.1
$La(NO_3)_3$	22	$AgNO_3$	0.1
$Pr(NO_3)_3$	21	$In(NO_3)_3$	0.04
$Sr(NO_3)_2$	16	$Pm(NO_3)_2$	0.02
Na_3PO_4	12	TOTAL	43 587
		H_2O (by difference)	12 663
		SOLIDS	
$Th(NO_3)_4$	19 421	Insolubles	35

[a] Np, Am, and Cm.

drum is then capped, sealed, and decontaminated. The filled and sealed drums are then removed from the process cell and transferred by truck to a storage facility. This storage facility is designed for eventual conversion to an above-grade disposal tumulus. The flow sheet for the RTS is shown in Fig. 3.

The cement-stabilized extracted salt solution will be Class C waste per the waste classification criteria of 10 CFR 61 [*2*]. In order for it to be assured that the cemented waste meets the requirements for Class C waste forms, various cement recipes were developed and the selected waste recipe was tested against the six stability criteria identified by the NRC in the Branch Technical Position Paper on waste form (that is, compressive strength, leachability, immersion stability, radiation stability, thermal stability, and biological stability). Recipe development and cement testing were performed by the Westinghouse Research and Development Center; results are reported in Refs *20* to *27* and indicate that the cement-stabilized, extracted salt solution will meet the NRC waste form criteria.

Focusing on the hazardous nature of the low-level, extracted salt solution, the data in Table 2 indicate that the HLW supernatant contains appreciable amounts of hexavalent chromium and a lesser amount of selenium. These metals are of interest because they are identified in the EPA extraction procedure toxicity method for determining characteristic hazardous waste. Because the IE-96 resin used in supernatant processing is so specific for cesium, the extracted salt solution is expected to contain the same mass of all other con-

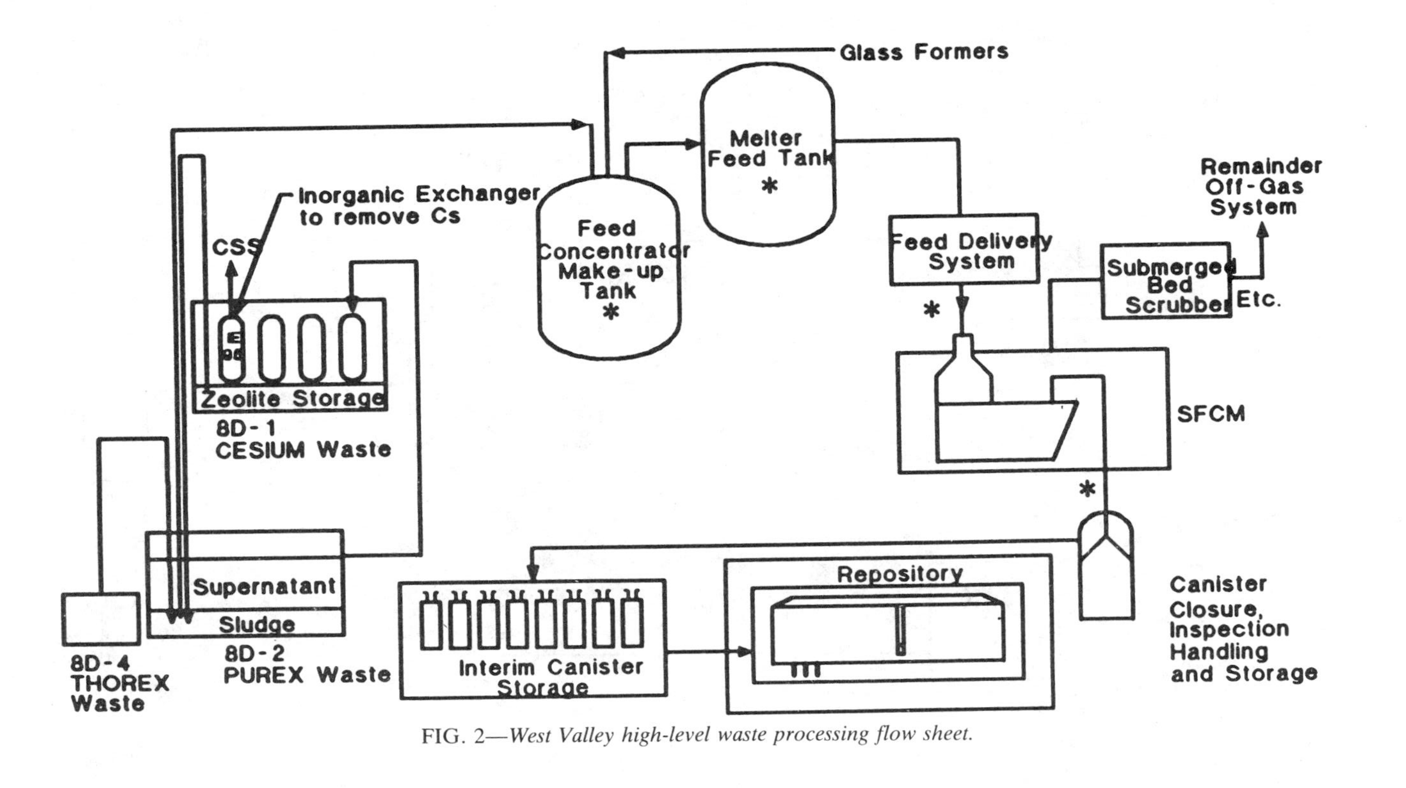

FIG. 2—*West Valley high-level waste processing flow sheet.*

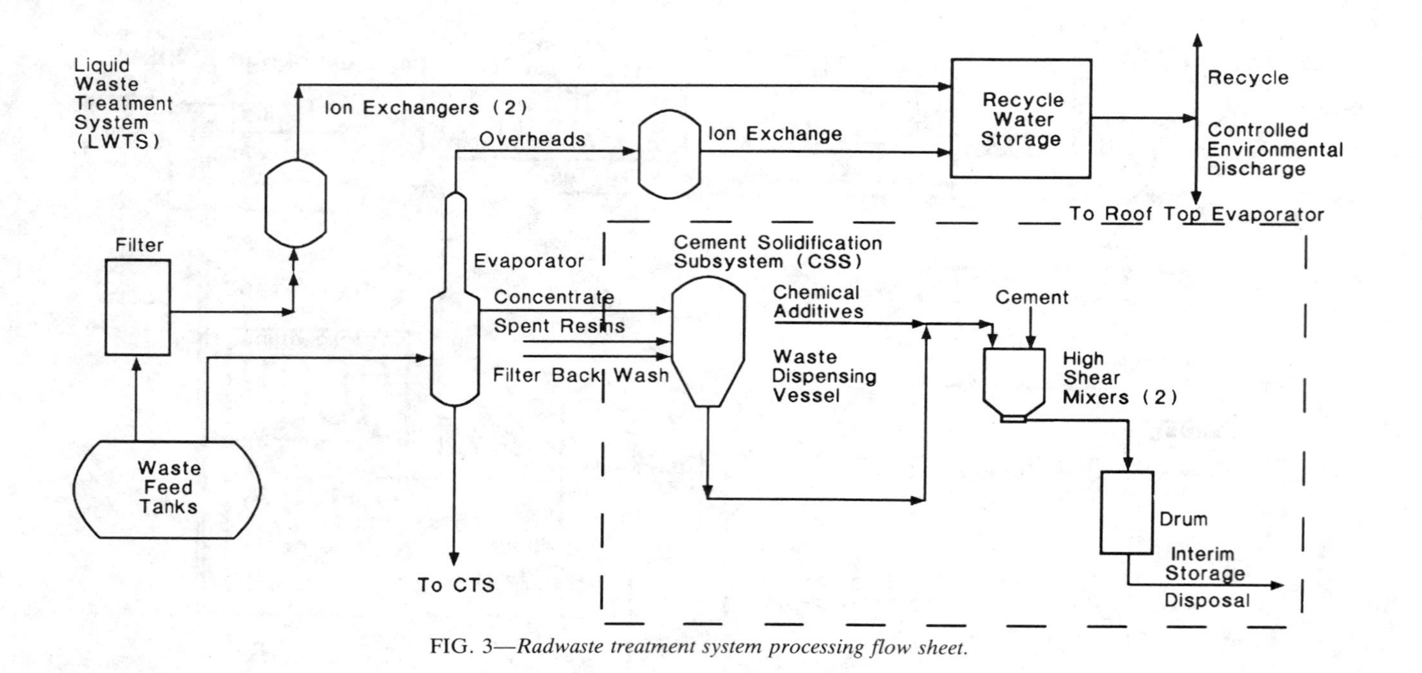

FIG. 3—*Radwaste treatment system processing flow sheet.*

stituents, including toxic metals as the supernatant. While the selenium is not present in sufficient quantity to cause the extracted salt solution to be classified as a hazardous waste, the chromium is. Thus, the HLW supernatant and the extracted salt solution are mixed wastes.

Stabilization of this solution in a cement matrix immobilizes the chromium and other metals in addition to providing structural stability. Cemented waste produced from a simulated (nonradioactive) extracted salt solution loaded with hexavalent chromium was subjected to the EPA reference extraction procedure (EP toxicity) [*28*] and the proposed toxicity characteristic leaching procedure (TCLP) (51 FR 114). The simulated solutions were prepared at 30, 42, and 50 weight percent salts with the composition shown in Table 4. These solutions were then mixed with Type I Portland Cement to cover a water-to-cement ratio range of 0.6 to 0.7. A 3.2-cm-diameter by 8.3-cm-long mold size was used to provide at least 100-g samples for TCLP testing. Several samples were rapid cured at 49°C for 24 h and then cured at room temperature for an additional five days before being subjected to the TCLP. Other samples were cured at room temperature for 85 days prior to TCLP testing and the EP Toxicity Test. The data indicate that the long-cured waste form at the proposed waste loading will not leach chromium in sufficient quantity to cause the cement to be classified as hazardous waste. The results of these tests are shown in Fig. 4.

Throughout the entire process, from supernatant treatment to cement drum filling, the waste is contained within the closed process system. This was a design requirement of the system in order to provide containment for the radioactive materials being processed. The closed, contained system prevents the release of radioactive and hazardous materials to the environment, providing the same level of protection as the "totally enclosed treatment" defined by EPA at 40 CFR 260.10 [*7*]. EPA has excluded such treatment facilities from the requirements imposed on other treatment, storage, and disposal facilities (TSDFs) by the regulations at 40 CFR 264 [*10*] and 265 [*11*]. It is the intention of the WVDP to demonstrate that this system meets the requirements for this exclusion.

Other Wastes

Obviously, the management of large volumes of extremely hazardous or radioactive waste requires a great deal of planning and control to provide safety to the workers and protection of the environment. In such cases, enclosed treatment systems are a necessity, at least to the point at which the hazards have been sufficiently mitigated to allow a lesser degree of containment. However, in the case of many waste streams, neither the volume nor the hazards are of such magnitude to warrant elaborate measures to effect treatment. Generators of mixed wastes in this category generally use more conventional management methods and must address regulatory compliance in a different manner. If the generator has a listed mixed waste, he may attempt to have it delisted. This process involves a demonstration that the waste no longer meets any of the criteria for which it was originally listed and that it is not hazardous with regard to any other factors. If the delisting petition is accepted, the waste can be managed as LLW, but other conditions may be imposed as a part of the delisting. If the waste is not delisted, the generator must store the waste until it can be sent to a licensed mixed-waste TSDF. Waste storage must be in accordance with the EPA regulations and may require a hazardous-waste storage permit.

If the mixed waste is hazardous by characteristic, the generator has several other options to render the waste nonhazardous and avoid the dilemma of dual regulation. Corrosive wastes may be neutralized or solidified. Neutralization is allowed without the need for a treatment permit per 40 CFR 264.1 [*10*]. Solidification renders the waste nonhazardous because the corrosive characteristic requires that the waste be liquid, and this may be

TABLE 4—*Simulated supernatant solution composition at three different concentrations.*

Compound	Weight Fraction	Weight of Salt Added, g, for 30 weight %	42 weight %	50 weight %
$NaNO_3$	0.21100	561.38	785.9	935.7
$NaNO_2$	0.10900	290.0	406.0	483.4
Na_2SO_4	0.02670	71.04	99.45	118.4
$NaHCO_3$	0.01490	39.64	55.50	66.07
KNO_3	0.01270	33.79	47.30	56.32
Na_2CO_3	0.00884	23.52	32.93	39.20
NaOH	0.00614	16.34	22.87	27.23
K_2CrO_4	0.00179	4.76	6.67	7.94
NaCl	0.00164	4.36	6.11	7.27
Na_3PO_4	0.00133	3.54	4.95	5.90
Na_2MoO_4	0.00024	0.64	0.90	1.07
NaF	0.00016	0.43	0.60	0.71
$Na_2B_4O_7$	0.00009	0.22	0.31	0.36
H_2O	0.60547	2454.97	2036.95	1758.29
Cr weight fraction	4.79×10^{-4}	3.64×10^{-4}	5.09×10^{-4}	6.06×10^{-4}

performed without a permit under certain conditions. Solidification or other treatment methods may also be employed to eliminate the other hazardous characteristics of ignitability, reactivity, and extraction procedure toxicity. In each instance, treatment may be performed without a hazardous-waste treatment permit under the following conditions.

EPA has identified three conditions which must be met in order to treat hazardous waste without obtaining a permit. First, the treatment must be done in the generator's accumulation tanks or containers. Second, the tanks and containers must be managed in strict compliance with the applicable standards; and third, the hazardous waste can not be stored longer than the time period prescribed for generators. The regulations codified at 40 CFR 262, 264, and

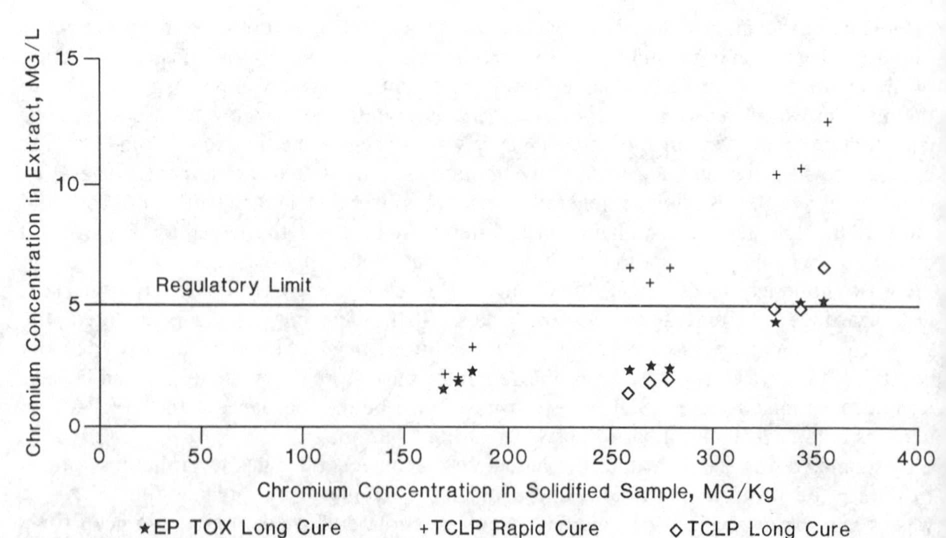

FIG. 4—*Chromium leachability of cemented extracted salt solution by EP toxicity and TCLP methods.*

265 [*8,10,11*] do not explicitly state these conditions, but EPA, in the response to comments section of the final regulations for generators of 100 to 1000 kg of hazardous wastes per month, describes these provisions and discusses the conditions which apply to various size generators (51 FR 56, page 10168). Although these conditions limit the utility of this management strategy for hazardous wastes (often it is easier and more economical to ship the waste to a licensed TSDF), they provide potential relief to the mixed-waste generator.

The treatment of a small quantity of zinc bromide waste at the WVDP provides an example of how this strategy may be employed. Zinc bromide solution is used at the WVDP in shield windows and is a corrosive liquid at the concentration used for this application [*29*]. Approximately 400 L of this solution leaked out of a shield window in the fuel reprocessing plant, and about half of the solution ran into an area containing radioactive surface contamination. The spilled material was collected in polyethylene-lined drums, with the mixed waste segregated from the uncontaminated hazardous waste. The containers were labelled and stored in accordance with the container-management requirements of 40 CFR 264, Subpart I [*10*], until they could be solidified in the collection drums with Portland Type I Cement. Once solidified, the mixed waste no longer displayed the hazardous characteristic and is presently being stored as low-level (Class A) radioactive waste. The uncontaminated waste was solidified and is being stored pending approval from the New York State Department of Environmental Conservation for disposal as nonhazardous industrial waste.

In this case, the state agency responsible for hazardous and solid waste regulation became involved with the nonradioactive portion of the waste. This is because New York, like many other states, has been authorized by EPA to implement and enforce a hazardous-waste program. The program must be at least equivalent to the Federal regulations for the state to receive authorization. The state requirements may be more restrictive than the federal, and requirements may vary among the states. Few states are authorized to regulate mixed waste, and regulatory authority for these wastes remains with EPA in unauthorized states. Thus, the mixed waste management at West Valley is presently subject to the EPA regulations.

Conclusion

Mixed-waste management involves compliance with dual regulations promulgated by the EPA and either NRC or DOE. Identification of the regulatory requirements and resolution of potential inconsistencies between the regulations are challenges facing mixed-waste generators and TSDF's as they develop compliance strategies. At the WVDP, several exclusion options are being exercised for characteristic mixed-waste streams. A totally enclosed treatment exemption is being pursued for cemented extracted salt solutions and a combination of neutralization and treatment in accumulation tanks and containers is used for smaller, less radioactive mixed-waste streams. The objective of this strategy is to eliminate the hazardous characteristic and allow the waste to be stored and disposed of solely as radioactive waste. If listed mixed wastes are encountered, the Project will have to revise this strategy to provide for mixed-waste treatment, storage, and disposal in accordance with both the radioactive and hazardous waste regulations.

References

[*1*] United States Code of Federal Regulations, Title 10—Energy—Part 20, "Standards for Protection Against Radiation."

[*2*] United States Code of Federal Regulations, Title 10—Energy—Part 61, "Licensing Requirements for Land Disposal of Radioactive Waste."

[3] United States Department of Energy Order 5820.2, "Radioactive Waste Management."

[4] United States Department of Energy Order 5480.1A, "Environmental Radiation, Safety, and Health Protection Programs for DOE Operations."

[5] United States Department of Energy Order 5480.2, "Hazardous and Radioactive Mixed Waste Management."

[6] United States Code of Federal Regulations, Title 40—Environment—Part 261, "Identification and Listing of Hazardous Waste."

[7] United States Code of Federal Regulations, Title 40—Environment—Part 260, "Hazardous Waste Management System: General."

[8] United States Code of Federal Regulations, Title 40—Environment—Part 262, "Standards Applicable to Generators of Hazardous Waste."

[9] United States Code of Federal Regulations, Title 40—Environment—Part 263, "Standards Applicable to Transporters of Hazardous Waste."

[10] United States Code of Federal Regulations, Title 40—Environment—Part 264, "Standards for Owners and Operators of Hazardous Waste Treatment, Storage, and Disposal Facilities."

[11] United States Code of Federal Regulations, Title 40—Environment—Part 265, "Interim Status Standards for Owners and Operators of Hazardous Waste Treatment, Storage, and Disposal Facilities."

[12] United States Code of Federal Regulations, Title 10—Energy—Part 40, "Domestic Licensing of Source Material."

[13] United States Code of Federal Regulations, Title 10—Energy—Part 30, "Rules of General Applicability to Domestic Licensing of Byproduct Material."

[14] U.S. Environmental Protection Agency, "Guidance on the Definition and Identification of Radioactive Mixed Waste Office of Solid Waste and Emergency Response Directive 9440.00-1."

[15] Kempf, C. R., MacKenzie, D. R., and Bowerman, B. S., "Management of Radioactive Mixed Wastes in Commercial Low-Level Wastes," NUREG/CR-4450, Brookhaven National Laboratory, Brookhaven, NY, Jan. 1986.

[16] Bowerman, B. S., Kempf, C. R., MacKenzie, D. R., Siskind, B., and Piciulo, P. L., "An Analysis of Low-Level Waste: Review of Hazardous Waste Regulations and Identification of Radioactive Mixed Wastes," NUREG/CR-4406, Brookhaven National Laboratory, Brookhaven, NY, Dec. 1985.

[17] Bowerman, B. S., Davis, R. E., and Siskind, B., "Document Review Regarding Hazardous Chemical Characteristics of Low-Level Waste," NUREG/CR-4433, Brookhaven National Laboratory, Brookhaven, NY, March 1986.

[18] Wild, R. E., Oztunali, O. I., Clancy, J. J., Pitt, C. J., and Picazo, E. D., "Data Base for Radioactive Waste Management—Waste Source Options Report," NUREG/CR-1759, Dames & Moore, Pearl River, NY, Nov. 1981.

[19] Rykken, L. E., "High-Level Waste Characterization at West Valley—Progress Report for the Period 1982–1985," DOE/NE/44139-14, Department of Energy, June 1986.

[20] Grant, D. C., Smeltzer, E. E., and Skriba, M. C., "Low-Level Waste Cement Encapsulation for West Valley—Recipe Development," 83-8E4-EASTV-R5, Westinghouse R&D Report, Westinghouse Research and Development Center, Pittsburgh, PA, 2 Dec. 1983.

[21] Smeltzer, E. E., Grant, D. C., and Skriba, M. C., "Low-Level Waste Cement Encapsulation for West Valley—Compressive Strength," 83-8E4-EASTV-R6, Westinghouse R&D Report, Westinghouse Research and Development Center, Pittsburgh, PA, 9 Dec. 1983.

[22] Smeltzer, E. E., Grant, D. C., and Skriba, M. C., "Low-Level Waste Cement Encapsulation for West Valley—Radiation Stability," 84-8B3-EASTV-R1, Westinghouse R&D Report, Westinghouse Research and Development Center, Pittsburgh, PA, 27 Jan. 1984.

[23] Grant, D. C., Smeltzer, E. E., and Skriba, M. C., "Leachability of Cement Encapsulated West Valley Waste Streams," 84-8B3-EASTV-R2, Westinghouse R&D Report, Westinghouse Research and Development Center, Pittsburgh, PA, 15 Feb. 1984.

[24] Smeltzer, E. E., Grant, D. C., and Skriba, M. C., "Low-Level Waste Cement Encapsulation for West Valley—Thermal Cycling Stability," 84-8B3-EASTV-R5, Westinghouse R&D Report, Westinghouse Research and Development Center, Pittsburgh, PA, 27 June 1984.

[25] Smeltzer, E. E., Grant, D. C., and Skriba, M. C., "Low-Level Waste Cement Encapsulation for West Valley—Biological Stability," 84-9B3-EASTV-R4, Westinghouse R&D Report, Westinghouse Research and Development Center, Pittsburgh, PA, 8 June 1984.

[26] Smeltzer, E. E., Grant, D. C., and Skriba, M. C., "Low-Level Waste Cement Encapsulation for West Valley—Immersion Stability," 84-8B3-EASTV-R3, Westinghouse R&D Report, Westinghouse Research and Development Center, Pittsburgh, PA, 29 June 1984.

[27] Grant, D. C., Smeltzer, E. E., and Skriba, M. C., "Cement Encapsulation and Waste Qualification Testing of West Valley Low-Level Waste Streams," 85-8B3-EASTV-R1, Westinghouse R&D Report, Westinghouse Research and Development Center, Pittsburgh, PA, Aug. 1985.
[28] U.S. Environmental Protection Agency, "Test Methods for Evaluating Solid Waste," SW846, Office of Solid Waste and Emergency Response, 1982.
[29] Smokowski, R. T., "Analytical Cell Decontamination and Shielding Window Refurbishment," DOE/NE/44139-1, Department of Energy, 1 Dec. 1985.

Leo P. Buckley,[1] Nancy B. Tosello,[1] and Brent L. Woods[1]

Leaching Low-Level Radioactive Waste in Simulated Disposal Conditions

REFERENCE: Buckley, L. P., Tosello, N. B., and Woods, B. L., **"Leaching Low-Level Radioactive Waste in Simulated Disposal Conditions,"** *Environmental Aspects of Stabilization and Solidification of Hazardous and Radioactive Wastes, ASTM STP 1033,* P. L. Côté and T. M. Gilliam, Eds., American Society for Testing and Materials, Philadelphia, 1989, pp. 330–342.

ABSTRACT: Leaching tests formally recognized by the International Atomic Energy Agency (IAEA) or by the American Nuclear Society (ANS) are used to evaluate the release of radioactivity from a waste form under idealized conditions. These conditions do not represent the reality of disposal environments, where engineered barriers are in place to limit access of water and to retard the migration of radionuclides into the biosphere. A test program underway at Chalk River, Ontario is attempting to provide an understanding of the release and transport of radionuclides under more realistic conditions. Preliminary results indicate that releases are suppressed by at least an order of magnitude when the effects of barriers in a repository failure are considered. However, migration mechanisms cannot be fully explained by classical concepts involving saturated flow. Research effort is now focused on measurements in an unsaturated, reducing environment, the condition which will be expected to be found in a well-designed repository.

KEY WORDS: leaching, lysimeters, low-level radioactive waste, disposal, engineered barriers, backfill, ground-water flow, ground-water chemistry, waste form, diffusion, retardation, saturated, unsaturated

Low-level radioactive wastes generated at nuclear reactors, from laboratory research, and from radioisotope users are currently stored in Canada. At Chalk River Nuclear Laboratories plans are underway to dispose of rather than carry on the practice of storage [*1,2*]. In an effort to define the source term to be used to predict the movement of radionuclides from a disposal facility, comparisons of the release of radionuclides have been performed using a number of leaching techniques.

Most of the radioactivity to be placed in the disposal facility will be concentrated chemical wastes immobilized in bitumen. The sources of the chemicals, mainly sodium phosphate, are decontamination facilities located at the Chalk River site. A second major waste form will be baled wastes, which contain substantially less radioactivity, but will comprise most of the volume of waste to be placed in the disposal facility. The compacted solid waste arises from daily activities taking place in the laboratories and research reactors. The waste is a mixture of plastics, cloth, metal, glass, paper, and rubber which cannot be incinerated.

Laboratory-scale samples of these waste forms were examined to determine the release rates. Initial tests were performed in the usual manner of soaking the specimens in water,

[1] Section leader, research technologist, and research technologist, respectively, Waste Management Technology Division, Atomic Energy of Canada Limited, Chalk River Nuclear Laboratories, Chalk River, ON, Canada K0J 1J0.

replenishing the water at prescribed intervals, and measuring the transfer of radioactivity from the waste form to the leachant solution. The present leaching methods such as the modified International Atomic Energy Agency (IAEA) or the American Nuclear Society (ANS) procedures (Measurement of the Leachability of Solidified Low-Level Radioactive Wastes by a Short-term Test Procedure, ANSI/ANS-16.1-1986) [*3*] appear to overestimate the amounts of radioactivity released from the waste forms. More realistic attempts were made to establish the release rates under possible disposal conditions.

Experiments in which the leaching procedures were modified to include nonadsorbing backfill materials indicated that releases from the waste form were reduced. These tests have caused concern with the use of present leaching procedures to generate data for the source term in predictive models. The program was extended to the use of bench-scale lysimeters to mimic the conditions of a humid disposal site. One of the reasons these tests were conducted was to determine whether or not the conclusions reached in the earlier leaching program could be supported. The radionuclide releases measured from the lysimeter program indicate that a more realistic release rate can be established and used as the source term for predictive models. A second vital reason for the lysimeter program was to obtain information on the movement of radionuclides during failure conditions, that is, when infiltration or flooding takes place in the disposal facility.

Background

Earlier experimental work at Chalk River and elsewhere has focused mainly on how different matrix materials retained radionuclides when the waste forms were leached in water. The procedures used were based on a modification to the IAEA leaching procedure [*3*]. Samples were submersed in water, exposing the entire surface area rather than only a fraction. Tests performed in which only a single surface was exposed, that is, the top of a right cylinder, generated misleading data, since water eventually contacted areas that were not initially exposed. Otherwise, the leachant replacement procedure was followed. The leachant was changed daily for two weeks, then weekly for eight weeks, and then monthly.

The samples used for these particular leach tests were bituminized sodium phosphate. The salt content was 40 weight percent, and trace amounts of cobalt-60 and cesium-137 were added to the salt before it was immobilized. Specimen sizes were right cylinders approximately 2.5 cm in diameter and 6 cm in height. The average volume of each cylinder was 25 cm^3, with a surface area of 50 cm^2.

The leaching results are presented as a plot of the cumulative fraction released, f, from the specimen as a function of the total time of leaching, Σt_n

$$f = \Sigma A_n / A_0 (V/S) \quad (1)$$

where

A_n = radioactivity leached during leachant renewal period (Bq),
A_0 = radioactivity initially present in specimen (Bq),
S = exposed surface area of specimen (cm^2),
V = volume of specimen (cm^3), and
t_n = duration of leachant renewal period (s).

By plotting the data in this manner, they can be readily used to estimate the fraction of radioactivity which might be released from a full-scale waste form, assuming the waste forms approximate a semi-infinite medium.

The results for the release of cobalt-60 and cesium-137 are given in Fig. 1. The radionuclide release from the waste form can be approximated by Fick's law of diffusion such that the cumulative fraction released is dependent on the rate of diffusion. Expressed mathematically the release is described by

$$f = 2\sqrt{Dt/\pi} + \alpha \tag{2}$$

where D is the effective diffusion coefficient (cm^2/s) whose value can be calculated from the slope of the line upon plotting the cumulative fraction released against $\sqrt{t}$. The term α(cm) is the value of the intercept of the line with the ordinate at $t = 0$. This value is a representation of the subsurface or near surface contaminants washed off the sample during the initial leaching period. The terms of the above expression were determined using a multiple regression curve fitting routine. For the results plotted in Fig. 1, the diffusion coefficient was found to be 1.54×10^{-10} cm^2/s for cobalt-60 and 1.37×10^{-10} cm^2/s for cesium-137.

The test was repeated but with the container in which the specimen and water were contacted filled with inert silica beads. The silica did not absorb activity released from the waste sample. This was established by acid washing of the silica at the end of the leaching period and analyzing the acid rinse for cobalt-60 and cesium-137. The same volume of water was used, and the same leachant replenish schedule was followed. The results are presented in Fig. 2 for cobalt-60 released from the specimen. The results obtained previously for cobalt-60 are also reproduced on the second figure for comparison. The reduction in the cumulative fraction released is nearly one order of magnitude less in the presence of the silica packing. But what is of greater significance is the change in the shape of the curve

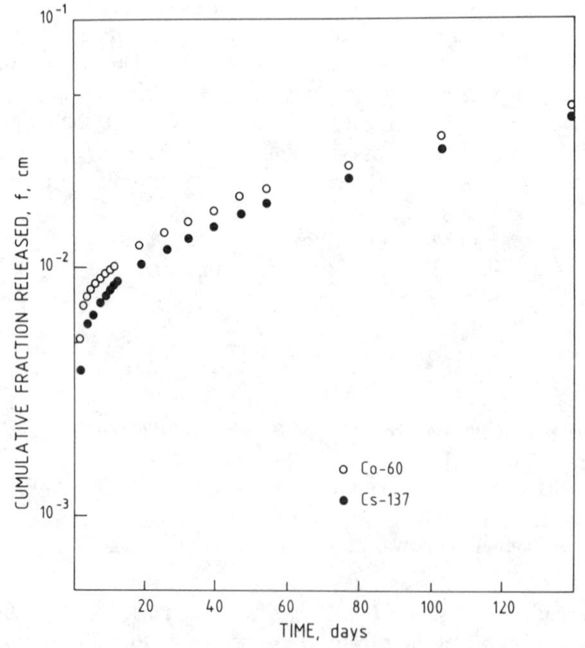

FIG. 1—*Release of radioactivity from bituminized sodium phosphate.*

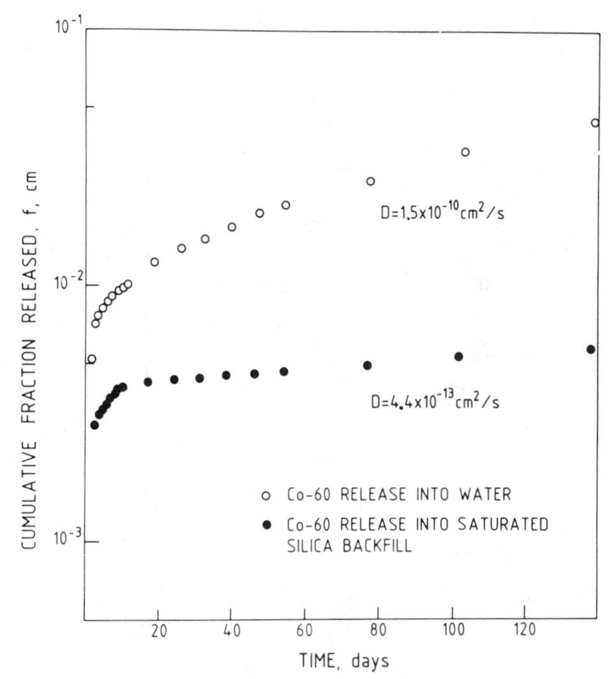

FIG. 2—*Comparison of cobalt-60 releases under different leaching techniques.*

which translates into a reduction in the diffusion coefficient. The diffusion value obtained was 4.4×10^{-13} cm^2/s, a drop of 350 times the value determined while leaching the specimen in a container holding only water.

The implications of the findings suggested that, under disposal conditions, the release rates were going to be lower than for the quite conservative values measured by leaching the waste form in water alone. In a repository designed to take advantage of additional engineered barriers, the use of diffusion data determined by leaching in water would lead to predicted releases that were conservatively too large. In turn, this would cause more or thicker engineered barriers to be placed in a repository, thus creating additional unnecessary costs for the disposal of low-level radioactive waste.

Experimental Setup

In order for the concerns generated by the above leaching experiments to be examined, a program was established where the waste form and engineered barriers could be evaluated under repository failure conditions, namely, flooding or infiltration. Bench-scale lysimeters which are extensively used in agricultural studies to measure water movement in soils [*4*] and which have been used for radionuclide migration studies [*5*] were assembled to perform the work.

The lysimeters were designed to simulate the effect of engineered barriers and used artificial precipitation cycles to accelerate the effect of time. The lysimeters are approximately 30 cm high and 30 cm in diameter (Fig. 3). Each lysimeter was equipped with a water-spray nozzle, a ventilation fan, and a heat lamp. The water was introduced over short time intervals

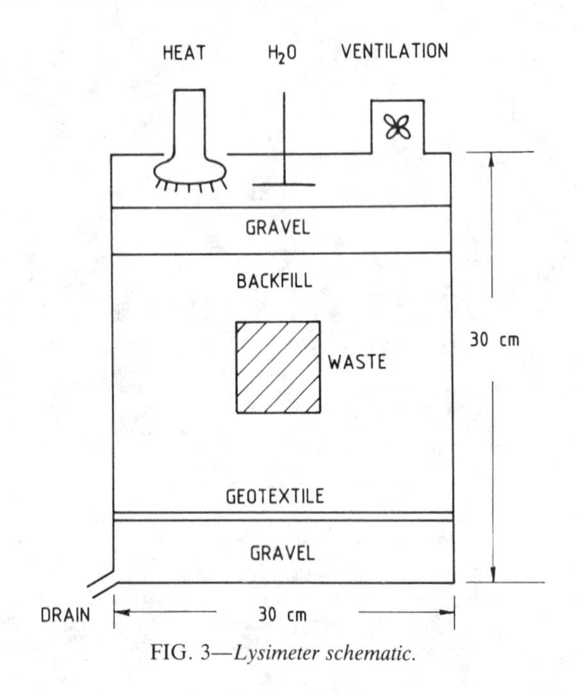

FIG. 3—*Lysimeter schematic.*

every eight hours. The heat lamp and fan were cycled on and off every hour. The purpose of the heat lamp and fan was to mimic the evaporation-transportation cycle, while the short burst of water simulated rainfall. On average, the amount of water passing through the lysimeter was about 30% to 40% of the amount added to the top of the lysimeter. This was comparable to the natural movement of water through the unsaturated zone to the water table in the environs of Chalk River.

There were two different waste forms prepared for the test program; compacted solid waste and bituminized phosphate. The compacted form was prepared from shredded paper, polyethylene, polyvinyl chloride, rubber, cloth, and wood. The shredded material was soaked in an aqueous solution containing cobalt-60, strontium-85, and cesium-137. The material was then dried and compacted to reduce the volume by a factor of 5. The average waste form size was 6.5 cm in height and 4.7 cm in diameter. The shape of the compacted waste form was retained by enclosing the waste form in a stainless-steel screen. The bituminized waste forms were prepared with the phosphate salt traced with the three isotopes, dried to include the removal of the hydration water, and then immobilized in an oxidized bitumen. The average waste form was 4.9 cm and 5.0 cm in diameter.

The bottom 4 cm of each lysimeter was filled with crushed, washed gravel. A geotextile, a fine-woven synthetic material made from fiberglass, polyethylene, or polypropylene (for example, Dupont Typar) was placed over the gravel to prevent the fine-grained backfill from being washed into the voids in the gravel bed. Backfill was compacted in place to a depth of 15 cm, and then the waste form was centered in the lysimeter in a bored hole with a depth of 6.5 cm to the gravel interface. The space above the specimen was then filled in and compacted.

The lysimeters were kept either saturated, where the waste form remains fully wetted, or unsaturated, where the water drained past the waste form. The water chemistry was adjusted in some lysimeters by first equilibrating the water with cement before passing it through the lysimeter. A cement barrier was placed around one waste form, and either compacted waste or bituminized salt waste forms have been placed in each lysimeter. The final variable examined was the backfill composition placed around the waste form in the lysimeter. The operating conditions for each lysimeter ,are outlined in Table 1.

Results

The rather ambitious program to examine the migration of radionuclides while simulating many variables has provided interesting data. The results are presented by discussing in turn the effects of each variable, that is, waste form, backfill, water chemistry, and degree of saturation, and by placing an additional barrier (package) around the waste form. The total amounts released from each of the lysimeters are provided in Table 2, where the total fractional releases after 170 days of operation are given for cobalt, strontium, and cesium isotopes.

Waste Form

Two groupings of the lysimeters (1 and 2; 3 and 4) were possible for the comparison of the radionuclide differences expected based on the waste form. Lysimeters 1 and 2 were identical except for the different waste forms. As one would expect, the release of cesium-137 was greater from the compacted waste form. However, the release of more cobalt-60 and strontium-85 from the bituminized waste (Lysimeter 1) was unexpected. Parallel studies on the measurement of adsorption coefficients indicated that there was a synergistic effect from the chemical ions co-leached from the bituminized waste form. When Lysimeters 3 and 4 were compared, again, the release of cesium-137 was lower for the bituminized waste form and also for the strontium-85, but not for cobalt-60. Here the dominating variable for retention of the strontium-85 is the presence of clay in the backfill. The higher release of cobalt-60 from the lysimeters containing the bituminized waste form appears to be as a consequence of the water chemistry and not due to the differences in the diffusion rates from the two different waste forms.

Backfill

Comparisons were made on two groups of lysimeters (1 and 3; 2 and 4). The lysimeters in each grouping had the same waste form, groundwater and were unsaturated. The sand available locally has an iron oxyhydroxide coating which has some capability to adsorb

TABLE 1—*Lysimeter identification.*

Lysimeter	Waste Form	Backfill	Groundwater	Operation	Added Barrier
1	bituminized salt	sand	demineralized	unsaturated	no
2	compacted waste	sand	demineralized	unsaturated	no
3	bituminized salt	sand/5% clay	demineralized	unsaturated	no
4	compacted waste	sand/5% clay	demineralized	unsaturated	no
5	bituminized salt	sand/5% clay	cement equilibrated	unsaturated	no
6	bituminized salt	sand/5% clay	cement equilibrated	saturated	no
7	compacted waste	sand	demineralized	saturated	no
8	compacted waste	sand	demineralized	unsaturated	cement

TABLE 2—*Radionuclide releases.*

Lysimeter	Total Fraction Released, $\times 10^6$, After 170 Days		
	Co-60	Cs-137	Sr-85
1	17050	110	620[a]
2	20580	620	490
3	4880	40	...
4	7810	200	10
5	6380	20	...
6	5090	20	...
7	5690	140	420
8	...	...	...

[a] 620×10^{-6}.

radionuclides, particularly cobalt-60, strontium-90, and cesium-137 [*6*]. It was expected that some holdup of the radioisotopes would take place even with the lysimeters containing just sand. The addition of illite clay had the effect of reducing by more than half the losses of cobalt-60 and cesium-137. However, the release of any cesium-137 in the lysimeters containing clay was entirely unexpected. The lysimeters were designed so that the presence of clay which had a high adsorbing capacity for cesium would have prevented any release from the lysimeter. Subsequently to these initial tests, it has been discovered that the high velocities of water indeed have led to particle transport, with particles containing adsorbed radionuclides. The combination of high flow rates and a porous geotextile lead to a loss of clay from the four lysimeters and perhaps an underestimate of the radioactivity released.

For the lysimeters containing the bituminized waste form, the presence of clay prevented any release of strontium-85, and clay had the effect of cutting the strontium-85 losses from the compacted wastes by an order of magnitude. The presence of illite clay appears to reduce both the quantity and the rate at which the radionuclides are released from the lysimeters. Not only does the effect of the addition of clay to the sand provide adsorption sites for the released radionuclides, but the presence of clay alters the physical properties, reducing the porosity and thus increasing the contact time between the liquid and solid phases. However, given the flow rate, the contact time may not have been sufficient to use equilibrium distribution coefficients (see mathematical modeling section).

Additional Barrier

An additional barrier of cement, 1 cm in thickness, was molded around a compacted waste form before it was placed in the lysimeter. The analysis of water leaving Lysimeter 8 was compared with the effluent from Lysimeter 2. All other variables were identical except for the presence of the cement barrier around the compacted waste form. There were no measurable releases of any of the radionuclides from the additionally protected waste form. The barrier served two purposes: first, because the cement is not very permeable, the amount of water in contact with the waste form is reduced; and, second, the cement acts as a diffusion barrier to radionuclide movement. Only if a much longer period of time were taken would there have been a noticeable measurement of radionuclides. Over a short time period, the additional barrier served its purpose of preventing release.

Ground-Water Chemistry

To evaluate the effect of ground-water chemistry, Lysimeters 3 and 5 were compared. Again, the remaining variables were the same: backfill, waste form, and unsaturated flow.

The ground-water variable appeared to be radionuclide specific. The presence of the alkaline ground-water apparently increased the cobalt-60 and decreased the cesium-137 mobility. Adsorption of the cobalt-60 to the backfill material in the presence of alkaline ground-water appears to be reduced. We speculate that cobalt-60 is forming as colloidal cobalt hydroxide, which is carried out of the lysimeter during subsequent artificial precipitation cycles. The cesium-137 nuclide is able to compete effectively for the adsorption sites in the backfill with the calcium and magnesium ions released from the cement. No strontium-85 was observed in the effluent from either lysimeter. The conjecture at this time is that co-leached phosphate ions from the bituminized waste form are reacting with the strontium-85 radionuclides and precipitating the product on the surfaces of the backfill particles.

Saturated/Unsaturated Effects

Two groups of lysimeters could be compared to examine the effect of operating the lysimeters in a flooded or drained condition. The two groups were Lysimeters 2 and 7 and Lysimeters 5 and 6, containing compacted waste and bituminized salts, respectively. Both cobalt-60 and cesium-137 releases from the unsaturated lysimeters were greater than from the flooded lysimeters, while the strontium-85 release was higher from the flooded lysimeter containing the compacted waste form. No releases of strontium-85 were observed from the bituminized waste from lysimeters. The sharply lower releases from the flooded, compact waste form lysimeters are due to the longer residence time of water contact with the backfill media which permits higher adsorption of cobalt-60 and cesium-137. The differences observed with the lysimeters containing the bituminized salt waste specimens are much less, which may be due to the presence of sodium phosphate co-leached from the waste forms. The sodium ions are known to be adsorbed preferentially to clay.

In summary, the highest releases were observed for compacted waste forms placed in sand backfill. Substantial drops in releases were observed when clay was mixed with the sand, an expected result based on the increased adsorption capacity of the clay. However, some cesium-137 was released and not expected in the effluent, based on the adsorption capacity of the clay. It is suggested at this point that other processes have an influence such as particle transport or microbial action to keep radionuclides soluble. Changes in ground-water chemistry increased the releases of cobalt-60 and cesium-137. With the co-leaching of sodium phosphate from the bituminized waste forms, cobalt-60 releases were greater than expected, but, based on comparable leaching results for the two waste forms, higher releases were observed for the compacted waste. Perhaps most dramatic was the complete retention of strontium-85 in the lysimeters containing both clay and phosphate. Finally, based on previous work with diffusion processes in different matrices, the complete retention of radionuclides in the lysimeter containing a compacted waste form protected by a coating of cement was an expected result. In conclusion, effects were observed based on flow, on interferences from co-leached chemicals, on the different adsorption characteristics of the backfill materials, and on the type of waste form.

Modeling Studies

The results discussed above also were examined by comparing the data with mathematical models developed to predict the release of radioactivity from a combination of waste forms and barriers. The difficulty in doing the modeling is that there is no simple method to account for unsaturated flow. As noted in the previous sections, only with the compacted waste form lysimeters was there any large difference in retention. So, one of the first uncertainties was the way to handle the flow through the lysimeters.

The movement of solutes through porous media has been developed by several authors [*6–8*]. Radionuclide movement can be assessed with the aid of a general mass transport model. The model incorporates the effects of dispersion, ground-water flow, retardation, and radioactive decay. For saturated one-dimensional flow in a granular medium, the expression is

$$\partial C/\partial t = (D/R)\, \partial^2 C/\partial^2 x - (V/R)\, \partial C/\partial x - \lambda C \tag{3}$$

where

C = ratio of radionuclide concentration in water to source concentration at some time (t),
D = dispersion coefficient of radionuclide in porous medium (cm^2/s),
R = retardation factor of radionuclide in porous medium,
V = ground-water velocity (cm/s),
λ = decay constant (L/s), and
x = thickness of granular medium (cm).

The dispersion coefficient for granular medium can be expressed as [9]

$$D = D_0 \tau / R \tag{4}$$

where

D_0 = diffusion of ion in water (cm^2/s),
τ = tortuosity factor which describes the tortuous path length, and
R = retardation factor.

The granular medium (backfill) is assumed to adsorb radionuclides in an amount linearly dependent on the concentration of the radionuclides. Thus the retardation factor can be expressed by

$$R = 1 + \rho K_d / \eta \tag{5}$$

where

ρ = dry bulk density of backfill (kg/L),
η = porosity of backfill, and
K_d = distribution coefficient (L/kg).

The releases of cobalt-60 from the lysimeters containing the bituminized waste form are presented in Fig. 4. Superimposed on the same figure are the data from leaching the waste form in water using the IAEA procedure. The graphical comparison points out the large differences in the choice of technique taken to determine release rates.

The partial differential equation describing the movement of radionuclides given above was solved using a computer code which uses finite differences to approximate the differential equation. Different values of the ground-water velocity and distribution coefficient were screened to obtain a fit to the lysimeter data. The results of fitting the expression to one of the release curves are given in Fig. 5, where different ground-water velocities were fitted with a constant distribution coefficient. A similar family of curves is presented in Fig. 6, where the velocity is held constant while the distribution coefficient is altered. Examination of these figures would indicate that ground-water velocity is the dominant factor. In order for the best fit to be provided, the estimated ground-water velocities are extremely high, far greater than one could expect in a normal repository. However, these velocities were obtained in the lysimeters. The water was not added over long time periods, but in short

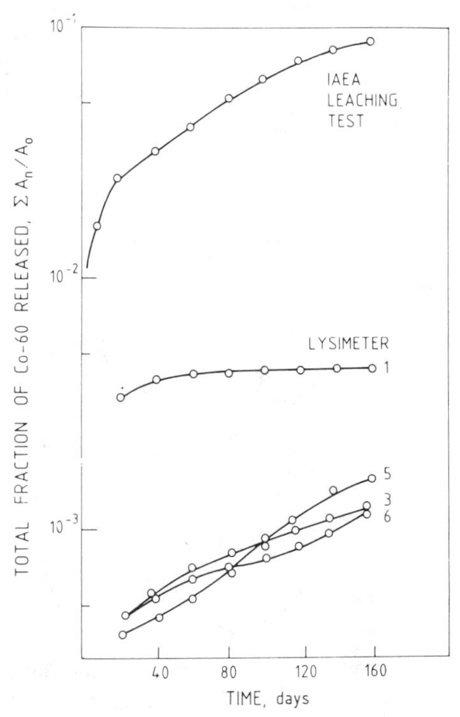

FIG. 4—*Behavior of cobalt-60 releases from bituminized salt waste forms in a simulated disposal environment.*

bursts. Estimates of the flow rates through the lysimeters range from instantaneous rates of 400 cm/day to normalized rates of 1.7 cm/day.

The higher rates are greater than might be considered during spring thaws, when snow melts and flows downward through the granular medium to the water table. The lower rate is simply the time averaged rate of water movement through the backfill. The attempt to accelerate the precipitation cycle should not have been made because of the effect observed.

With the assumed high ground-water flows, the distribution coefficients are no longer appropriate. They are based on equilibrium conditions between the granular medium and water. If water is in contact for short time intervals, then the K_d values must be affected. Data from equilibrium studies [*10*] are compared in Table 3 with the fitted values for each of the lysimeters for cobalt-60, cesium-137, and, where appropriate, strontium-85. The disparities are significant. The distribution coefficients derived from cobalt-60 are one to two orders of magnitude below the values obtained under equilibrium conditions. The values for cesium-137 are lower by less than an order of magnitude. The strontium-85 values extracted from the lysimeter experiments seem to be similar, except with the compacted wastes where the retention is exceedingly high, without any apparent reason.

There have been several contradictions to the expected normal operation of the lysimeters. The movement of cesium-137 through the lysimeters was surprising, based on earlier determinations of equilibrium distribution coefficients. The presence of clay fines in the effluent water may have contributed to the cesium release, with the adsorbed radionuclides carried through the system by the high flow velocities. On the other hand, strontium-85 releases were lower and not similar to cesium-137 had the release been solely due to particle transport.

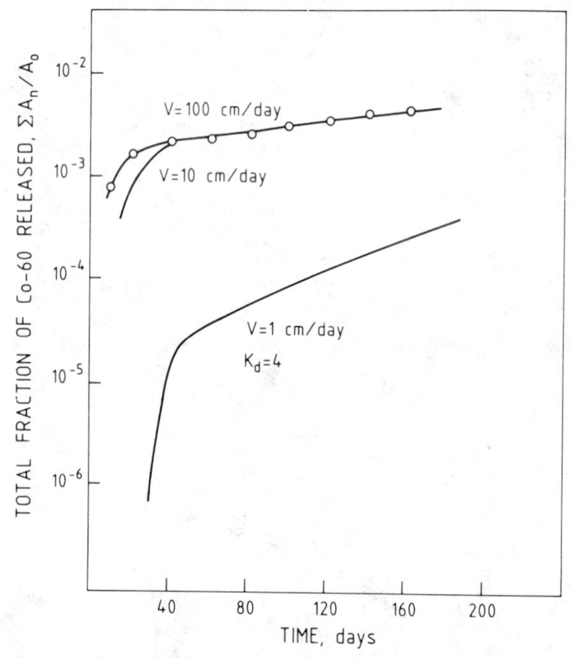

FIG. 5—*Fitting cobalt-60 release data from lysimeter 3, assuming constant distribution coefficients.*

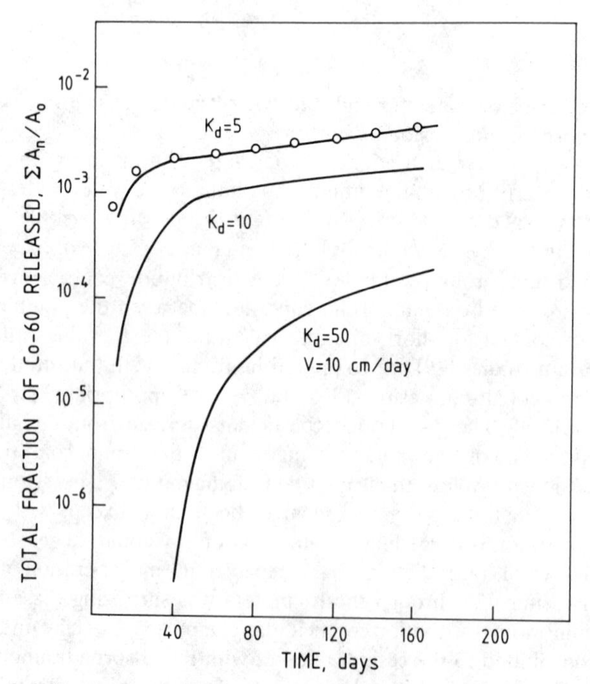

FIG. 6—*Fitting cobalt-60 release data from Lysimeter 3, assuming constant ground-water velocity.*

TABLE 3—*Estimates of distribution coefficients.*

	Distribution Coefficients, L/kg		
	Co-60	Cs-137	Sr-85
Demineralized Water	900	7300	14
Lysimeter 2	11	980	1250
Lysimeter 4	26	1120	$>10^5$
Lysimeter 7	40	1600	530
Demineralized Water with phosphate	64	520	$>10^5$
Lysimeter 1	1	180	32
Lysimeter 3	4	490	$>10^5$
Demineralized Water with phosphate and cement-equilibrated	32	3500	$>10^5$
Lysimeter 5	3	990	$>10^5$
Lysimeter 6	4	990	$>10^5$

Indeed, the ion-exchange column may have been able to extract cesium and cobalt but not strontium from the particles passing through the column. Yet again, the presence of microorganisms may have affected the transport by selectively keeping some of the radionuclides soluble and easily transportable.

The experimental attempt to mimic failure conditions in a low-level radioactive waste repository have not been completely successful. The program has pointed out the difficulties in trying to accelerate the effects of time, unknowns associated with subjecting the waste forms to unsaturated flow conditions, and the extent to which retardation mechanisms may be affected by nonequilibrium conditions. Indeed, serious questions have been raised about the effects of water movement. For instance, is there a movement of fine clay particles through the backfill, and does the presence of backfill enhance the release from the waste form?

On the other hand, the experience gained with these first series of lysimeters has generated a more realistic approach to assessing the environment within a disposal vault. The movement of radionuclides in unsaturated conditions, with and without flow, the interaction of degradation products and the backfill, additional backfill materials, and the chemistry of the vault water must be studied in more detail to predict the movement of water and radionuclides under expected and failure conditions of a disposal facility.

Summary

From these experiments, several points have emerged. The lower releases of radionuclides observed with changing the leaching protocol were also observed with lysimeter experiments which were designed to mimic repository failures. In more detail, the presence of clay in the backfill significantly retarded the movement of radionuclides. The unsaturated lysimeters appear to release more radioactivity than similar lysimeters which had remained fully saturated. Fitting the data to an advective, dispersive transport equation to model the release has shown that the release from the waste forms can be approximated by leaching the waste forms packed in nonreactive media. Simply leaching waste forms by present procedures overestimates the radionuclide release. Other results from the program indicate that strontium can be retarded in the presence of phosphate, which in this case was the chemical co-leached with radioactivity from the waste form. The use of cement-equilibrated water also increased the retention of cesium. The study indicated that there is a deficiency of knowledge in unsaturated systems and that the chemistry and flow rate of the groundwater, perhaps coupled to particle transport, are important variables in understanding the migration of radioactivity in a disposal environment.

While it was difficult to separate completely all the effects in studying the waste form releases under simulated disposal conditions, the releases are more indicative of what can be expected in reality. Better source terms can be derived through the use of proper experimental conditions. These in turn can provide more realism to the predictive modeling needed to perform the safety assessments on future sites chosen for disposal of radioactive wastes. Indeed, our current research effort is now focused on determining release rates in an unsaturated, reducing environment typical of the expected conditions in a well-designed repository.

References

[1] Dixon, D. F., "Strategy for the Disposal of Low- and Intermediate-Level Radwastes in Canada," AECL-7439, Atomic Energy of Canada Limited Report, Chalk River, ON, Canada, June 1981.

[2] Hardy, D. G. and Dixon, D. F., "The Transition from Storage to Permanent Disposal of Low and Intermediate-Level Wastes at the Chalk River Nuclear Laboratories," AECL-8728, Atomic Energy of Canada Limited Report, Chalk River, ON, Canada, March 1985.

[3] Hespe, E. D., "Leach Testing of Immobilized Radioactive Waste Solids. A Proposal for Standard Methods," *Atomic Energy Review,* Vol. 9, No. 1, 1971, pp. 195–215.

[4] Fodor-Csanyi, P. et al., "Penetration and Distribution of Tritiated Water in Soils of a Lysimeter," *Nordic Hydrology,* Vol. 11, 1980, pp. 169–186.

[5] Hooker, R. L. and Root, R. W. Jr., "Lysimeter Tests of SRP Waste Forms," Savannah River Laboratory, Report DP-1591, E. I. duPont de Nemours & Co., Aiken, SC, 1981.

[6] Grisak, G. E. and Jackson, R. E., "An Appraisal of the Hydrogeological Processes Involved in Shallow Subsurface Radioactive Waste Management in Canadian Terrain," Scientific Series No 84, Inland Waters Directorate, Water Resources Branch, Ottawa, Canada, 1978.

[7] Bear, J., *Dynamics of Fluids in Porous Media,* Elsevier Publishing, New York, 1972.

[8] Ogota, A., "Theory of Dispersion in a Granular Medium," Geological Survey Professional Paper 411-I, U.S. Government Printing Office, Washington, DC, 1970.

[9] Buckley, L. P., Arbique, G. M., Tosello, N. B., and Woods, B. L., "Evaluation of Backfill Materials for a Shallow-Depth Repository," AECL-9337, Atomic Energy of Canada Limited Report, Chalk River, ON, Canada, Nov. 1986.

[10] Buckley, L. P., "Waste Packages and Engineered Barriers for the Chalk River Nuclear Laboratories' Disposal Program," AECL-8853, Atomic Energy of Canada Limited Report, Chalk River, ON, Canada, July 1985.

Ann M. Boehmer,[1] Robert L. Gillins,[1] and Milo M. Larsen[1]

Stabilization of Mixed Waste at the Idaho National Engineering Laboratory

REFERENCE: Boehmer, A. M., Gillins, R. L., and Larsen, M. M., **"Stabilization of Mixed Waste at the Idaho National Engineering Laboratory,"** *Environmental Aspects of Stabilization and Solidification of Hazardous and Radioactive Wastes, ASTM STP 1033,* P. L. Côté and T. M. Gilliam, Eds., American Society for Testing and Materials, Philadelphia, 1989, pp. 343–357.

ABSTRACT: EG&G Idaho, Inc. has initiated a program to develop safe, efficient, cost-effective treatment methods for the stabilization of some of the hazardous and mixed wastes generated at the Idaho National Engineering Laboratory. Laboratory-scale testing has shown that Extraction Procedure toxic wastes can be successfully stabilized by solidification, using various binders to produce nontoxic, stable waste forms for safe, long-term disposal as either landfill waste or low-level radioactive waste, depending upon the radioactivity content.

This paper presents the results of drum-scale solidification testing conducted on hazardous, low-level incinerator fly ash generated at the Waste Experimental Reduction Facility. The drum-scale test program was conducted to verify that laboratory-scale results could be successfully adapted into a production operation.

KEY WORDS: stabilization, fly ash, solidification, Toxicity Characteristic Leaching Procedure (TCLP) test, metals leachability

Radioactive waste which is also hazardous [as defined by the Resource Conservation and Recovery Act (RCRA) in 40 CFR 261] is considered a radioactive mixed waste (RMW). The Department of Energy-Idaho Operations (DOE-ID) Office has decided that the Radioactive Waste Management Complex (RWMC), which is the Idaho National Engineering Laboratory's (INEL) low-level waste disposal facility, shall not accept RMW. Proper disposition of a hazardous waste (HW) requires that the HW be sent to an Environmental Protection Agency (EPA)-permitted disposal facility. Existing EPA-permitted disposal facilities will not accept radioactively contaminated hazardous waste since they do not have a Nuclear Regulatory Commission (NRC) license and are not designed to handle radioactivity.

There are two other options for dealing with RMW: to treat it so that it is no longer hazardous or radioactive, or to store it until it can be treated or legally disposed of. Treatment and storage of a RMW also require an RCRA permit by EPA or a state.

The INEL has applied to the EPA for a RCRA Part B permit. The permit application includes a storage facility for HW, a storage facility for RMW, incineration of HW and RMW, and stabilization of HW and RMW. This paper describes the stabilization development activities conducted at the INEL by EG&G Idaho, Inc. for the DOE.

There are two primary purposes for stabilizing a RMW. One is to enhance the waste form

[1] Engineer, engineer, and manager, Waste Engineering Development Unit, respectively, Idaho National Engineering Laboratory, EG&G Idaho, Inc., P.O. Box 1625, Idaho Falls, ID 83415.

for optimum storage conditions; the second is to treat the waste so that hazardous characteristics are eliminated and the waste can then be disposed of as a low-level waste.

The stabilization development plan consists of four primary activities: (1) Characterize the HW and RMW at the INEL to determine volumes, levels of radioactivity, and which wastes can be stabilized; (2) conduct laboratory-scale stabilization tests to evaluate stabilization binders and processes and to determine the optimum binder-to-waste ratios; (3) conduct drum-scale tests to ensure that 208-L (55 gal) drums of waste can be successfully stabilized and to obtain the necessary test data to support the RCRA permit application and (4) provide a production stabilization capability at the Waste Experimental Reduction Facility (WERF).

The first activity, waste characterization, has been completed for existing identified HW and RMW at the INEL. However, this will be a continuing activity as new wastes are identified or as changes to the regulations add chemicals or materials to the list of regulated wastes, or both. Also included in this activity was a determination of which wastes might be candidates for stabilization. Detailed waste characterization information is contained in Ref *1*.

Wastes which are hazardous due to toxicity, as determined by the Toxicity Characteristic Leaching Procedure (TCLP), are prime candidates for stabilization. Proper stabilization of these wastes, so that they will pass the TCLP test, will allow them to be disposed of as LLRW. Stabilization development activities at the INEL have concentrated on these wastes.

The second activity, laboratory-scale development, was conducted on fly ash from incineration of low-level waste, photochemical wastes, aqueous potassium chromate solutions, and lead refining dross. Binders that were economical and easy to use were chosen for testing. These included cement, combinations of cement and sodium silicate, and ENVIROSTONE. Recipes were developed for binder-to-waste ratios which resulted in successful stabilization of each waste, as demonstrated by success in passing the TCLP test. Reference *2* provides detailed information on the laboratory-scale development.

The third activity in stabilization development is the drum-scale testing required to ensure that 208-L (55 gal) quantities can be successfully stabilized and pass the TCLP test, and to obtain the data required to support the RCRA Part B permit application. The only waste to be tested to date in drum-scale quantities is the fly ash generated from incineration of LLRW at WERF. The WERF fly ash is considered a RMW due to leachable lead and cadmium levels detected by the TCLP test. This fly ash is collected in 208-L drums.

To support drum-scale development, a stabilization development facility was installed in the Waste Engineering Development Facility (WEDF, the deactivated SPERT II Reactor building). The facility consists of a drum-tumbler for mixing the waste and binders; a HEPA-filtered air-exhaust system; a scale for weighing the waste, water, and binder additions; a water supply system; and miscellaneous support equipment.

Drum-scale testing evolved through four phases before an acceptable process was settled on to yield consistently successful results. The fly ash was sampled before and after solidification, and samples were submitted to two independent laboratories for analysis, using the TCLP to verify binding of the hazardous constituents. Eight drums of fly ash were solidified. Two of the first five drums did not pass the leach tests on all samples submitted; the final three drums passed the leach tests on all samples submitted.

Statistical evaluation of the data obtained from the final three drums showed that the number of samples analyzed from these drums was sufficiently large to determine that cadmium and lead are not leachable from the solidified fly ash in toxic concentrations.

Following the successful completion of the drum-scale testing on WERF fly ash, conducted at the WEDF, a production solidification system was installed at WERF. The equipment and procedures to be used for the production solidification system for fly ash were evaluated for operability and completeness during the system checkout test. Fly ash Drum 9 was solidified during the checkout test. All the samples of solidified fly ash from this drum passed the leach test (these results are included in this paper).

Based on the data presented in this paper, the production solidification system which has been installed at WERF will produce a stabilized waste monolith that contains no free liquids and no leachable metal levels in excess of EPA limits and can therefore be disposed of as a low-level radioactive waste.

Description of Operation

The existing drums of fly ash contain solid liners with lids which provide no access for the addition of binder materials. Consequently, all of the existing liner lids were replaced with modified lids. The modified lids for the first two drums contained three bungs to allow for drum ventilation and the addition of cement and water separately. The lid configuration was changed to two bungs for the last six drums, one for ventilation and the other for material addition. Changeout of the first two liner lids was conducted at the WERF incinerator ash-collection system glovebox. The liner lids on the remaining six drums were changed out in the WEDF solidification room after it was determined that the operation, properly conducted, did not result in fly ash dispersion.

The fly ash was sampled during the lid changeout operation to provide baseline data for leach tests. Samples were taken at various depths in each drum, using a grain-probe (thief) sampler, and combined to provide a composite sample for each drum.

Solidification mixture sampling was accomplished for all drums using a core sampler inserted to the bottom of the drum at two radial locations. This yielded a good cross section of samples to give an indication of homogeneity.

The actual development of the stabilization process was conducted in four phases (described in the following paragraphs), using the same ingredients, (with the exception of the addition of 3.8 L [1 gal] of liquid sodium silicate to Drum 8), but with variations in recipe, sequence, method of addition, and method of mixing. These phases and their variations are detailed in Table 1.

Drum 1 (Phase A, see Table 1) used the initial recipe developed in the lab-scale program, with an additional amount of water intended to facilitate thorough mixing, but the mixture proved to be too wet. After 24 h, the mixture remained in slurry form under a 15.24-cm (6 in.) water layer. Additional dry cement was then added and the drum retumbled. A slurry mixture under a 10.2-cm (4 in.) water layer still remained. At this point, additional laboratory-scale testing was performed on the mixture to determine the amount of dry cement needed to set up the slurry. The water layer on top of the mixture was removed, and, using additional dry cement, the slurry was set up into a monolith with no free liquid. All samples of this monolith passed the leach test.

The water layer that was removed was sampled and the samples sent for analysis. The results of this analysis showed that the water contained a cadmium level of 108 mg/L, which is in excess of the EPA limit of 1.0 mg/L. The other seven toxic metals levels were below the EPA limits (from 40 CFR 261). The water was placed into a 113.6-L (30 gal) drum and solidified with cement. Analysis of a sample of the solidified water/cement mixture showed a leachable cadmium content of <0.01 mg/L, which is below the limit. This drum therefore will be disposed of as a low-level radioactive waste.

The recipe, based on additional laboratory-scale tests, was adjusted for Drum 2 (Phase B). However, the fly ash in this drum had been compacted during drumming so that it contained 50% more fly ash by weight than Drum 1, even though the fill level was only slightly higher. As a consequence, the drum became so full with the basic recipe ingredients

TABLE 1—*Stabilization development phase description.*

	Phase A Drum 1	Phase B Drum 2	Phase C Drums 3–5	Phase D Drums 6–8	Operations Drum 9
	Lid Configuration				
	Lid with 3 Bungs, Threaded Fittings	Lid with 3 Bungs, Threaded Fittings	Lid with 2 Bungs, Slip Fittings	Lid with 2 Bungs, Slip Fittings	Lid with 2 Bungs, Slip Fittings
	Ash Weight, lb (kg)				
	106(48.0)	150(68.0)	87(39.5), 98(44.5), 97(44.0)	88(40.0), 89(40.4), 97(44.0)	100(45.4)
Step 1	slurry ash at a water-to-ash ratio of 2.25:1[a] by tumbling with mixing bars	slurry ash at a ratio of 1.5:1[a] by tumbling with mixing bars	same as Drum 2	slurry ash at a ratio of 1.25:1[a] by tumbling with mixing bars	same as Phase D
Step 2	add dry cement at 0.72:1[a] and water at 0.75:1;[a] tumble again	add dry cement at 2:1[a] and water at 0.5:1;[a] tumble again	same as Drum 2	slurry cement in mixer. Cement-to-ash ratio of 2:1,[a] water-to-ash ratio of 1:1[a]	same as Phase D
Step 3	measure void space and fill with equal parts dry cement and water; tumble again	would have been the same as Drum 1 but was omitted due to full drum	sample mixture	pour cement slurry into drum. Mix with motorized mixing paddle	same as Phase D
Step 4	add additional dry cement to thicken mixture; tumble again	sample mixture	...	add dry cement to adjust consistency for best sampling, and mix with mixing paddle	same as Phase D
Step 5	skim off surface water	...	...	sample mixture	sample mixture
Step 6	sample mixture	...	...		
Comments	*Mixture too wet*	*Drum too full*	*Inconsistent mixing*	*No problems*	*No problems*
Leach test	passed	failed	No. 3 and 4 passed; No. 5 failed	all passed	passed

[a] All ratios shown are of the ingredient mentioned to ash, by weight.

that there was not sufficient headspace to allow proper mixing. Samples of this drum, prior to setup, showed layers of wet and dry ingredients, indicating nonhomogenity.

The quantity of ash per drum was limited to 45.4 kg (100 lb) for all subsequent operations. Drums 3, 4, and 5 (Phase C) were solidified with the same recipe as that used for Drum 2. Sampling of these drums showed inconsistent mixing from drum to drum, indicating that the tumbling process was not adequate, particularly when using dry cement.

During the processing of Drums 3, 4, and 5, the decision was made not to fill the void space remaining after mixing the basic ingredients because it was felt that this should be accomplished in a separate operation after mixture setup. It was also determined that at the time the void space was filled, another visual inspection for the presence of free liquid would be conducted.

The initial inspection for free liquid is conducted during the solidification processing. The operators inspect the surface of the monolith during the sampling operation, and, if free liquids are present, they would also be observed when the sample is extruded from the sample probe.

No free liquids have been observed in any of the monoliths solidified in the development program, and none are expected in production operations, but the visual inspections will be conducted as a verification. The void space in the drum will be filled with sand or dry cement. The void space is filled only to minimize future subsidence in the disposal facility and will not alter the leaching characteristics of the monolith in any way.

The final three drums, 6, 7, and 8 (Phase D), were stabilized by adding cement (slurried in a cement mixer) to the slurried ash and mixing with a motorized paddle inside the drum. Dry cement was then mixed in, using the paddle mixer, to obtain the desired consistency for sampling, and probe samples of the mixture were removed.

Following the completion of the production solidification facility installation at WERF, Drum 9 was stabilized using the production system equipment. The procedure used contained the final 2:2.25:1 cement-to-water-to-ash recipe, and the mixing of slurried cement and slurried ash with the paddle mixer, as was done for Drums No. 6, 7, and 8.

Results

The testing conducted during this development activity to identify the leachable levels of hazardous metals in the fly ash, solidified fly ash, and quality-control samples was completed using the TCLP. This procedure was developed by the EPA as a replacement for the Extraction Procedure (EP) toxicity leach test.

The TCLP test differs from the EP toxicity leach test in that the TCLP requires elimination of the structural integrity procedure; tumbling method of agitation only; change of leaching solution; and crushing of the sample so that it passes through a 0.95-cm (⅜ in.) sieve before leaching. These changes make the TCLP a more stringent test; it was used, therefore, as a worst-case procedure for sample leachable toxic metals evaluation. The TCLP was also used to ensure that the process will provide a waste form which will meet the requirements of the new method and will be in compliance with anticipated EPA regulations.

Results of the drum-scale testing are contained in the following sections.

Fly Ash Analytical Results

The TCLP was the first test conducted on the fly ash and was used to verify that the fly ash was hazardous, to give an indication of the homogeneity of the leachable levels of metals in the fly ash, and to provide a basis for comparison of the leachable metal levels before and after solidification.

As shown by the data presented in Table 2, leachate from the fly ash was definitely above the EPA toxicity limits for cadmium and lead, 1.0 and 5.0 mg/L, respectively, thus verifying that the fly ash was a hazardous waste.

Multiple samples of fly ash from Drum No. 2, taken from five different levels in the drum, were analyzed to determine the homogeneity of the leachable metals in the fly ash. The data from Laboratory 1 tests on Drum No. 2 indicate good homogeneity of the leachable metals throughout the drum. The data from all five levels of Drum 2, as well as that from the composite sample, are consistent. Based on the results of this analysis, it was decided that analysis of composite samples of unsolidified fly ash from the remaining test drums would provide the required data.

The data obtained from Laboratory 2 analyses are reported, but are not used in the evaluation because of incorrectly performed procedures. Laboratory 2 used mineral acid digestion of part of the solid samples, in addition to the TCLP, to obtain the results. This would not be an indication of leachable levels of metals, but rather an indication of the total amounts of metals in the sample, which is substantially higher.

The fly ash samples were also analyzed for radionuclide content. Major isotopes found consistently in the fly ash are cobalt-60, cesium-137, cesium-134, and antimony-125. Silver-110m, zinc-65, and magnanese-54 were also found (in lesser amounts) in some of the drums of fly ash.

TABLE 2—*Fly ash TCLP results.*

Sample No.	Laboratory No.	Analyte Concentration Detected, mg/L							
		Arsenic	Barium	Cadmium	Chromium	Lead	Mercury	Selenium	Silver
D1T1	1	<0.01	0.8	155	<0.01	34.6	0.0016	0.02	0.11
D1T2	1	<0.01	0.8	138	<0.01	29.8	0.0011	0.01	0.10
D1T3	1	0.06	0.5	149	0.01	13.0	<0.0004	0.03	0.06
D1T4	2	0.80	1.1	310	1.8	420.0	0.047	<0.01	0.61
D2C1	1	<0.01	0.2	283	0.19	82.1	0.0007	0.03	0.02
D2C2	2	0.76	7.1	260	8.3	1400.0	0.012	0.028	0.58
D2P1	1	<0.01	0.2	273	0.17	61.1	0.0006	0.01	0.02
D2P2	1	<0.01	0.1	287	0.15	88.5	0.0004	0.03	0.03
D2P3	1	<0.01	0.2	297	0.14	73.1	0.0005	0.05	0.02
D2P4	1	<0.01	0.2	282	0.18	75.0	0.0004	0.05	0.02
D2P5	1	<0.01	0.2	280	0.17	71.2	0.0006	0.04	0.02
D2SP1	2	0.078	0.9	260	1.4	350.0	<0.002	<0.01	0.23
D2SP3	2	<0.01	<0.5	300	<0.1	120.0	<0.002	<0.01	0.08
D2SP5	2	0.33	2.8	290	3.4	740.0	0.012	0.018	0.34
D3C1	1	0.03	0.3	136	0.3	28.0	0.0092	0.06	0.09
D4C1	1	0.02	<0.1	169	<0.1	212.0	<0.0004	0.05	0.19
D5C1	1	<0.01	<0.1	438	<0.1	130.0	<0.0004	0.05	0.14
D6C1	1	<0.01	0.4	77	0.1	33.5	0.0037	0.04	0.13
D6C2	3	0.16	0.9	69	0.71	26.0	0.0052	<0.02	0.22
D7C1	1	<0.01	<0.1	366	<0.1	125.0	<0.0004	0.03	0.14
D7C2	3	<0.001	<0.1	340	<0.01	120.0	<0.0002	<0.04	0.32
D8C1	1	<0.01	<0.1	305	0.5	168.0	<0.0004	0.02	0.13
D8C2	3	<0.001	<0.1	280	0.02	94.0	0.0004	<0.04	0.27
D9C1	1	<0.01	0.2	187	0.05	59.0	<0.0004	0.04	0.51
D9C2	1	<0.01	0.2	202	0.05	57.3	<0.0004	0.03	0.68

[a] D = drum number.
T = top sample.
C = composite sample.
P = first probe sample and level in drum, 1 = top to 5 = bottom.
SP = second probe and level in drum, 1 = top to 5 = bottom.

Analyses of fly ash samples have been conducted on a random basis since the incinerator began contaminated operations. These analyses were done to provide an indication of the composition of the ash. The results of spectrochemical analyses have been consistent in identifying the principal constituents of the fly ash. These results are summarized in Table 3.

Moisture content was also evaluated and found to be consistently low, ranging from 0.5 to 4.1%, and averaging about 2%. No correlation was found between the moisture content of the fly ash and the variation in the amount of water required to stabilize the fly ash.

From the consistency of the fly ash composition analytical results, it seems reasonable to conclude that large variations in waste feed makeup have not occurred or have not negatively affected fly ash composition. The waste feed makeup is administratively controlled and randomly checked to minimize inclusion of prohibited materials (chlorinated materials, free liquids, large metal objects, etc.), but it is impractical to attempt to quantify the specific waste feed makeup. It therefore is not possible to relate fly ash composition to waste feed makeup.

The WERF incinerator was designed to operate with a minimum combustion efficiency of 99.9% and a minimum destruction efficiency of 99.99%. Operation at these efficiencies ensures that the fly ash will be relatively inert with respect to reactivity and organic content, regardless of the waste feed mix. The absence of significant quantities of organic compounds increases the efficiency of cement as a binder in the solidification process.

Solidified Fly Ash Analytical Results

The results of the TCLP testing were used to verify that the solidified monolith did not contain leachable metal levels in excess of EPA limits and to give an indication of the homogeneity of the monolith. As shown by the data presented in Table 4, Drum 1 did not contain leachable levels of metals in excess of EPA limits, with the exception of one sample analyzed by Laboratory 2. As stated previously, the data from Laboratory 2 are considered invalid because of improper procedures.

The data obtained from the analysis of Drum 2 samples show leachable levels of metals in excess of EPA limits in some sections of the drum; the contents of this drum therefore must still be considered a mixed waste. These results were the first indication of a lack of homogeneity, resulting from poor mixing within the drum. The poor mixing was a result of

TABLE 3—*Principal ash constituents.*

Constituent	Concentration Range ~% Weight
carbon (fixed)	11–37
calcium	0.1–5
cadmium	0.1–0.6
chromium	<0.1–<5
iron	0.8–4.9
potassium	1.0–2.1
sodium	<0.1–3.1
phosphorus	0.4–3.1
lead	0.1–9.6
sulfur	1.3–2.0
zinc	4.9–>30
chlorides	1.7–21.6
sulphates	1.6–3.4
H_2O	0.5–4.1

TABLE 4—*Solidified fly ash TCLP results.*

Sample No.[a]	Laboratory No.	Analyte Concentration Detected, mg/L[a]	
		Cadmium	Lead
	Phase A		
D1C1	1	<0.001	0.3
D1C2	2	23.000	80.0
D1R1	2	0.050	0.4
D1R2T	1	<0.001	0.2
D1R2M	1	<0.010	0.3
D1R2B	1	<0.001	0.2
	Phase B		
D2C1B	2	240.000	14.0
D2C2T	1	0.400	<0.1
D2C2M	1	<0.001	2.3
D2C2B	1	<0.001	2.4
D2C3T	1	25.800	2.4
D2C3M	1	<0.001	6.0
D2C3B	1	<0.001	1.4
D2R1	2	30.000	5.9
D2R2T	2	360.000	55.0
D2R2B	1	24.900	2.8
D2R3T	1	46.200	37.5
D2R3M	1	48.300	16.1
D2R3B	1	<0.001	0.3
	Phase C		
D31T	1	<0.001	2.2
D31B	1	<0.001	2.0
D32T	1	<0.001	1.9
D32B	3	<0.005	0.5
D33T	3	<0.005	0.4
D33B	3	<0.005	0.4
D41SA	1	<0.001	2.0
D41SB	1	<0.001	1.7
D42T	3	<0.005	0.7
D42B	3	<0.005	0.7
D51T	1	66.400	12.0
D51B	3	59.000	10.0
D52T	3	56.000	8.2
D52B	3	63.000	12.0
D53T	1	68.400	9.7
D53B-C[b]	1	<0.001	1.5
	Phase D		
D61T	1	<0.001	1.7
D61M	1	<0.001	1.8
D61B	1	<0.001	2.0
D62T	3	<0.005	1.5
D62M	3	<0.005	1.2
D62B	3	<0.005	1.2
D63T	3	<0.005	1.5
D63B	1	<0.001	1.3
D71T	3	<0.005	1.0
D71M	3	<0.005	1.1
D71B	3	<0.005	1.2
D72T	1	<0.001	<0.1
D72M	1	<0.001	<0.1
D72B	1	<0.001	0.1

TABLE 4—*Continued.*

Sample No.[a]	Laboratory No.	Analyte Concentration Detected, mg/L[a] Cadmium	Lead
D82SA	3	<0.005	0.1
D82SB	1	<0.001	0.0
D83SA	3	<0.005	0.1
D83SB	1	<0.001	0.0
D84SSA	3	<0.005	0.0
D84SSB	1	<0.001	<0.1
D85SSA	3	<0.005	<0.1
D85SSB	1	<0.001	<0.1
	OPERATIONS		
D9C1T	1	0.03	1.2
D9C1B	1	0.02	1.5
D9R1T	1	0.02	1.0
D9R1B	1	0.01	1.5

[a] D = drum number.
C = probe sample taken from center bung of drum.
R = probe sample taken from a radial bung in drum.
1,2,3,4 = probe numbers in order taken from drum.
T = top of probe sample.
M = middle of probe sample.
B = bottom of probe sample.
S = split sample.
SS = split sample after silicate addition.
[b] This sample was a chunk of material, mostly cement, which came out in the bottom of the probe from the layer of material on the bottom of the drum.

insufficient headspace, caused by too much fly ash in the drum. Addition of the binder and water required for stabilization filled all of the headspace, preventing the agitation required for good mixing.

As a result of these findings, the weight of ash in the drums was limited to 45.4 kg (100 lb) per drum, and solidification of Drums 3, 4, and 5 was initiated.

The data obtained from Drums 3 and 4 sample analysis showed no leachable levels of metals in excess of EPA limits. However, the sampling operations revealed inconsistent mixing within the drums. As shown by the leach test results obtained from Drum 5, the inconsistent mixing caused a lack of homogeneity in the solidified fly ash. Therefore, Drum No. 5 did not pass the leach test and remains a mixed waste. Homogeneity of the drum contents is required to ensure that the chemical reactions are adequate to bind the metallic ions and prevent leaching. It became apparent that tumbling as a mixing method, or addition of dry cement, or both were inadequate to produce homogeneous drum mixtures. Therefore, the decision was made for future drums, to slurry the cement with water prior to adding it to the drum, and to mix the drum contents with a paddle mixer in the drum and not rely on tumbling for final mixing. Solidification of Drums 6, 7, and 8 was then initiated using these changes.

In the solidification of Drum 6, 34.1 kg (75 lb) of dry cement were added and mixed into the drum mixture following the addition and mixing of the slurried ash and slurried cement. This cement was added to provide a mixture viscous enough to permit immediate removal of probe samples. In Drum 7, a small amount 10 kg (22 lb) of dry cement was added and mixed into the drum after mixing of the cement and fly ash slurries. This amount of cement was not enough to provide a consistency sufficient to permit immediate sampling. The drum mixture was allowed to set for 20 min, at which time sampling was performed.

Drum 8 was solidified using only the slurried ash and slurried cement mixtures. Samples D82SA&B and D83SA&B were then taken from the drum. One gallon (3.8 L) of liquid sodium silicate was then mixed into the drum contents and Samples D84SSA&B and D85SSA&B were taken.

The data obtained from the analysis of the solidified samples of Drums 6, 7, and 8 show that this method of solidification (adding slurried cement to slurried ash and mixing with the paddle mixer) provides a monolith which does not contain leachable levels of toxic metals in excess of EPA limits. The results also show a very good homogeneity of the drum contents. The results of Drum 6 in particular show an excellent consistency throughout the drum and also between the two laboratories. The additional dry cement which was added to Drums 6 and 7 appears to have had little or no effect on the leachable levels of lead in the monolith. The sodium silicate which was added to Drum 8 had no effect on the consistency of the drum mixture and showed little effect on the leachable lead levels. Sodium silicate used in the laboratory-scale testing caused an almost instantaneous set of the fly ash mixture and did reduce the amounts of leachable lead. The quantities of silicate used in the laboratory-scale testing were in a greater ratio than the 3.8 L (1 gal) used in Drum 8, thus explaining why the addition of sodium-silicate reduced leachable lead levels in laboratory-scale tests but not in drum scale.

Homogeneity of the drum mixture is verified in two ways. First, the sample probe extracts a full-length core sample of the drum mixture. A visual inspection of the sample as it is extruded from the probe gives an indication of homogeneity. The sample of a nonhomogeneous drum will show layers or spots of cement intermixed with the fly ash, whereas a homogeneous drum sample is a single color and does not break apart when extruded.

These observations have been supported by the second way of verifying homogeneity, that is, analytical results. Drum 1 samples appeared to be quite homogeneous, and all passed the leach test. Some layering was observed in Drum 2 samples, and some of the samples passed, but others did not. Drums 3 and 4 showed a fairly homogeneous sample. Some very small cement pockets were observed, but these samples did pass the leach test. Drum 5 samples showed considerable layering, and the samples from this drum failed to pass the leach test. Samples from Drums 6, 7, 8, and 9 appeared homogeneous, and all of these samples passed the leach tests with consistent results.

The results from Drums 6, 7, 8, and 9 were also evaluated with respect to what effect dilution would have had. In earlier laboratory-scale testing, samples of the fly ash were solidified using only water. The water does set up the fly ash, but, as shown in Samples 1 and 2 in Table 5, it does not affect the leachable levels of either cadmium or lead. Therefore, the water has no dilution effect on the leachable metals levels.

The cement was added at a 2:1 ratio with the fly ash, thereby cutting nondiluted leachable levels in half. Therefore, doubling the final results would account for any dilution resulting from the process. As shown in Table 5, the results from the drums, even when doubled, are still below EPA limits.

Statistical evaluation of the data obtained from Drums 6, 7, and 8 was performed per EPA-SW-846 (see Ref *3*). The statistical evaluation was conducted only on the leachable lead levels because all the leachable cadmium levels were below the detectable limits.

Stratified random sampling was appropriate for analyzing the lead concentrations in the solidified fly ash from Drums 6, 7, and 8. Because the drums were solidified using slightly different methods for each drum, it was known that nonrandom chemical heterogeneity could have existed. Each drum is thus considered a stratum and samples are drawn from each drum. Stratified random sampling results in greater precision than simple random sampling when it is evident that the population can be efficiently divided into strata that maximize the variability among strata and minimize the variability within each stratum.

Drum	Stratum, k,	n_k	$\overline{X}_k$	S_k^2	$W_k = n_k/N$
6	1	8	1.525	0.085	0.364
7	2	6	0.562	0.352	0.272
8	3	8	0.039	0.001	0.364

where

n_k = number of samples from stratum,
$\overline{X}_k$ = sample mean from stratum,
$t_{0.20}$ = probable error rate,
S_k^2 = variance of sample from stratum,
W_k = fraction of population represented by stratum,
N = total number of samples,
RT = regulatory threshold (RT for lead is 5.0 mg/L), and
$S_{\bar{x}}$ = standard error.

SW-846 Box 2, Strategy for Determining if Chemical Contaminants of Solid Waste are Present at Hazardous Levels—Stratified Random Sampling of Wastes (Strategies section, page 19 of Ref *3*) outlines the method for calculating $\overline{X}$ and S^2

$$\overline{X} = \sum_{k=1}^{3} W_k \overline{X}_k = 0.722$$

$$S^2 = \sum_{k=1}^{3} W_k S_k^2 = 0.127$$

The appropriate number of samples, n, is determined by

$$n = \frac{t^2_{0.20}\, S^2}{\Delta^2}$$

where $t_{0.20}$ equals 1.323 (from Table 2 SW846 with degrees of freedom 21) and Δ^2 equals $(RT - \overline{X})^2 = (5 - 0.722)^2 = 18.301$. This yields $n = 0.1$. This shows that the samples of size 8, 6, and 8 from the three drums were greater than the appropriate number of samples, $n = 0.1$, and were sufficiently large to draw the following conclusion.

An 80% confidence interval (the level is specified in EPA-SW-846) is determined by

$$CI = \overline{X} \pm t_{0.20}\, S_{\bar{x}} = 0.722 \pm 0.101 = 0.621, 0.823$$

The upper limit of the confidence interval (0.823 mg/L) is well below the regulatory threshold of 5.0. It therefore can be concluded that lead is not present in the solidified fly ash at a hazardous concentration.

Following the successful completion of the drum-scale testing at the WEDF, a production solidification system was installed at WERF. This system duplicated the hardware/equipment used in the solidification of Drums 6, 7, and 8. The operating procedures which will be used for the production operations were used to solidify Drum 9 at WERF during the system checkout testing.

The results of the analysis of solidified samples from Drum 9 showed that the production method to be used for stabilization of the WERF fly ash produces a homogeneous monolith

TABLE 5—*Dilution effect evaluation.*

	Average Analyte Concentration Detected, mg/L				
	Before Solidification		After Solidification		Accounting for the Dilution Effect
Sample No.	Cadmium	Lead	Cadmium	Lead	Lead[a]
1 (D1A)	32.0	443.0	46.000	407.0	...[b]
2 (D2A)	37.0	332.0	29.000	399.0	...[b]
D6	73.1	29.8	<0.005	1.5	3.0[c]
D7	353.0	122.5	<0.005	0.8	1.6[c]
D8	292.5	131.0	<0.005	0.1	0.1[c]
D9	194.5	58.2	0.020	1.3	2.6[c]

[a] The results for cadmium are below the detectable limits; therefore, the dilution effect is considered negligible.
[b] These are the laboratory-scale samples solidified using water only, which has no dilution effect as demonstrated by these results.
[c] Cement was added at a ratio of 2 to 1, so the results were doubled to account for dilution.

which contains no free liquids and does not contain leachable levels of toxic metals in excess of EPA limits.

Quality Assurance/Quality Control

Quality assurance/quality control (QA/QC) of the sample gathering and analysis was considered important to ensure that the methods used would provide valid, reliable, consistent, and representative analytical results. Therefore, a formal QA/QC plan was prepared for each phase of development.

Each laboratory used for sample analysis also has its own QA/QC plan, which includes procedural controls, specifications, and standard methods for performing analysis, verifying results, performing instrument calibration, instrument maintenance procedures, and internal quality control. All analyte concentration analyses were performed using EPA-approved or laboratory standard procedures. Each of the laboratories used for analysis uses blanks periodically in analytical procedures for background data and quality control.

The procedure used for leaching the samples, to provide the extract for metals analysis, was the TCLP. EG&G personnel were familiar with the development and intent of this procedure and provided resolution to questions the laboratories had regarding this method. However, Laboratory 2 did not question EG&G personnel before using an acid digestion of solids that passed through the filter during their analyses of samples. The mineral acid digestion would have released metals from the solids which would not have leached under normal test procedure, thereby accounting for the higher lead levels reported by Laboratory 2. Because this procedure was not properly performed, the results from Laboratory 2 were considered invalid. A contract was then negotiated with Laboratory 3, so that there would still be two independent laboratories performing analysis.

The data in Table 6 are the results of the standards and blanks submitted to the laboratories for analysis. These samples were submitted blind to provide a check of laboratory accuracy and were analyzed simultaneously with the samples of fly ash and solidifed fly ash.

The sample series labeled WEDSS1 and WEDSS2 were standard solutions, prepared by EG&G Chemical Sciences personnel, and contained 1.0 mg/L cadmium and 5.0 mg/L lead. Samples WEDSS3 and WEDSS4 were a standard solution, purchased from a chemical-supply company, which contained 1000 mg/L lead. As shown by the analytical results, all laboratories performed within acceptable tolerance bands.

The sample series labeled SPC consisted of solidified blanks, prepared by WED personnel, which contained only cement and demineralized water. These samples were used to determine if any bias existed in the laboratory procedures for specific metals, or if any tramp metals were present in the cement used in the solidification tests. The data obtained from all the laboratories are consistent and indicate neither bias for any metal nor the presence of any tramp metals in the cement used.

The sample series labeled WEDFAS consisted of fly ash standards, purchased from the National Bureau of Standards. The leachable levels content of the fly ash standards was not known initially; however, with the small amounts of total metals present in the standard as compared to the total metal levels in the WERF fly ash, the leachable levels in the standards were assumed to be very low. The intent of the analysis of fly ash standards was to develop a leachable levels standard of a medium similar to that of the WERF fly ash. The analytical results of the fly ash standards are fairly consistent and will be evaluated over a period of time (as more samples are analyzed) to provide an average leachable level as a basis for future analyses.

In addition to the standards and blanks used as quality-control samples, the fly ash and solidified fly ash samples also contained quality-control samples.

The fly ash samples for all the drums were splits of a composite sample that was drawn from levels throughout the drum and tumbled together to provide a representative sample

TABLE 6—*QA/QC sample analytical results.*

Sample No.	Laboratory No.	Analyte Concentration Detected, mg/L							
		Arsenic	Barium	Cadmium	Chromium	Lead	Mercury	Selenium	Silver
WEDSS1A	1	...[a]	...[a]	1.1	...[a]	5.6	...[a]	...[a]	...[a]
WEDSS1B	2	...[a]	...[a]	0.95	...[a]	4.6	...[a]	...[a]	...[a]
WEDSS1C	3	...[a]	...[a]	0.99	...[a]	4.6	...[a]	...[a]	...[a]
WEDSS2A	2	...[a]	...[a]	1.2	...[a]	5.2	...[a]	...[a]	...[a]
WEDSS2B	1	...[a]	...[a]	1.03	...[a]	5.8	...[a]	...[a]	...[a]
WEDSS2C	1	...[a]	...[a]	1.1	...[a]	5.0	...[a]	...[a]	...[a]
WEDSS3A	1	...[b]	...[b]	...[b]	...[b]	984	...[b]	...[b]	...[b]
WEDSS3B	3	...[b]	...[b]	...[b]	...[b]	1000	...[b]	...[b]	...[b]
WEDSS3C	1	...[b]	...[b]	...[b]	...[b]	1002	...[b]	...[b]	...[b]
WEDSS4A	1	...[b]	...[b]	...[b]	...[b]	974	...[b]	...[b]	...[b]
SPC1	1	<0.01	1.7	<0.001	0.01	<0.01	<0.0004	<0.01	<0.01
SPC2	2	<0.01	1.9	<0.04	0.1	<0.01	<0.002	<0.01	0.35
SPC3	1	<0.01	1.9	<0.001	<0.01	<0.01	<0.0004	<0.01	<0.01
SPC4	2	0.065	3.2	<0.04	0.5	0.26	<0.002	<0.01	0.06
SPC5	1	<0.01	2.4	<0.001	0.01	<0.01	<0.0004	<0.01	<0.01
SPC6	2	<0.01	1.5	<0.04	<0.1	<0.1	<0.002	<0.01	0.04
SPC7	3	<0.001	1.5	<0.005	0.01	<0.03	<0.0002	<0.004	<0.01
SPC8	1	<0.01	2.4	<0.001	<0.1	<0.1	<0.0004	<0.01	<0.01
SPC9	3	<0.001	2.1	<0.005	<0.01	<0.03	<0.0002	<0.004	0.03
SPC10	1	<0.01	1.9	<0.001	<0.1	0.06	<0.0004	<0.01	<0.01
SPC11	1	<0.01	2.0	<0.03	0.3	0.01	<0.0004	0.01	<0.01
WEDFAS1	1	0.22	0.2	0.007	0.14	<0.01	<0.0004	0.01	<0.01
WEDFAS2	3	0.13	0.4	<0.005	0.23	<0.03	<0.0002	0.044	<0.01
WEDFAS3	1	0.06	0.2	0.008	<0.1	0.19	<0.0004	0.02	<0.01

[a] Standard solution of 1.0 mg/L cadmium and 5.0 mg/L lead only; therefore, the sample was not analyzed for this element.

[b] Standard solution of 1000 mg/L lead only; therefore, the sample was not analyzed for this element.

of the drum population. The results of the analysis of these samples show a good consistency within and between laboratories, with the exception of Laboratory 2 results, which are considered invalid because of improper procedure. The probe samples of fly ash analyzed from the individual levels throughout Drum 2 were collocated samples which were used to give an indication of the homogeneity of the leachable metal levels throughout the drum and to determine that a composite sample drawn from all levels throughout the drum was a representative sample of the drum population.

The solidified fly ash samples were collocated samples, drawn in the same method at the same time from the drum; these gave an indication of the homogeneity of the solidified mixture. In addition, the samples from a single probe were split into two or three separate samples to provide further indication of the homogeneity of the drum and consistency between and within the laboratories. The results of this analysis showed Drums 1, 3, 4, 6, 7, 8, and 9 to be homogeneous and also showed a good consistency of analytical results between and within the laboratories.

The data obtained for this program have been shown to be valid, reliable, consistent, and representative of the sample population by the consistency and accuracy of the analysis results both between and within the laboratories.

Conclusions

The primary goal of the drum-scale stabilization development was to verify that the successful results obtained in the laboratory-scale program could be adapted into a production-scale operation. The result of the production-scale will be a stabilized waste monolith that contains no free liquids, no leachable metal levels in excess of EPA limits, and can therefore be disposed of as a general or low-level radioactive waste. The objectives set for meeting this goal were:

1. verification of the laboratory-scale binder-to-waste ratios (recipes) at a drum-scale level,
2. development of mixing and sampling methods to ensure and verify homogeneity of the mixture in the drum,
3. verification that the monolith leachable metal levels are below EPA toxicity leaching criteria, and
4. development of procedures for production-scale solidification.

As discussed in the text, the laboratory-scale recipe was adjusted for Drum 1 processing; the resulting mixture was too wet. Additional cement was added, and samples of the resulting monolith did pass the leach test. Drum 1 contained no free liquids after solidification was completed. The recipe was adjusted and Drum 2 processed. The weight of fly ash in Drum 2 was too great to allow sufficient headspace for mixing once the recipe ingredients were added. Samples of this monolith did not pass the leach test. The weight of fly ash in the drums was limited to 45.4 kg (100 lb), and processing of Drums 3, 4, and 5 was initiated. Since processing of these drums showed mixing problems, the mixing method was changed to incorporate use of a paddle mixer. The processing of Drums 6, 7, 8, and 9 involved using the final recipe of 2:2.25:1 cement-to-water-to-fly ash, addition of cement in slurry form, and mixing with the paddle mixer. This process produced satisfactory monoliths of the WERF fly ash; however, testing will be required for each new waste stream.

The homogeneity of the drum mixture is initially determined by visual inspection of samples taken from the drum mixture after mixing, but before final setup of the monolith.

Homogeneity is verified by the analysis of the drum mixture samples. The visual inspection and analyses from the last four drums have shown the monoliths to be homogeneous.

The final drum of fly ash was solidified using the production facility and operation procedures. The samples from this drum passed the leach test and provided verification of the operating procedures and production facility equipment.

The results obtained from the final four drums processed show, through both analytical and statistical evaluations, that the monolith resulting from the solidification process does not contain leachable toxic metal levels in excess of EPA limits and can be disposed of as a low-level radioactive waste.

Acknowledgment

This work was supported by the U.S. Department of Energy under DOE Contract No. DE-AC07-76-ID01570.

References

[1] Boehmer, A. M., "Waste Characterization and Analysis Activities Conducted in Support of the Solidification Development Program at the Idaho National Engineering Laboratory," EGG-WM-7175, EG&G Idaho, Inc., Idaho Falls, ID, March 1986.
[2] Boehmer, A. M. and Larsen, M. M., "Hazardous and Mixed Waste Solidification Development Conducted at the Idaho National Engineering Laboratory," EGG-WM-7225, EG&G Idaho, Inc., Idaho Falls, ID, April 1986.
[3] *Test Methods for Evaluating Solid Waste—Physical/Chemical Methods,* EPA-SW-846, 2nd ed., U.S. Environmental Protection Agency, 1982.

Bibliography

Boehmer, A. M., Gillins, R. L., and Larsen, M. M., "Drum-Scale Fly Ash Stabilization Development," EGG-WT-7393, EG&G Idaho, Inc., Idaho Falls, ID, Nov. 1986.
Boehmer, A. M. and Larsen, M. M., "Solidification of Hazardous and Mixed Waste at the Idaho National Engineering Laboratory," *Proceedings,* Waste Management '86 Conference, Tucson, AZ, March 1986.

Tim L. Jones[1] *and Richard L. Skaggs*[1]

Influence of Hydrologic Factors on Leaching of Solidified Low-Level Waste Forms at an Arid-Site Field-Scale Lysimeter Facility

REFERENCE: Jones, T. L. and Skaggs, R. L., **"Influence of Hydrologic Factors on Leaching of Solidified Low-Level Waste Forms at an Arid-Site Field-Scale Lysimeter Facility,"** *Environmental Aspects of Stabilization and Solidification of Hazardous and Radioactive Wastes, ASTM STP 1033,* P. L. Côté and T. M. Gilliam, Eds., American Society for Testing and Materials, Philadelphia, 1989, pp. 358–380.

ABSTRACT: The release of contaminants from solidified low-level waste forms is being studied in a field lysimeter facility at the Hanford Site in southeastern Washington State. The lysimeter facility, constructed in 1984, consists of ten 3-m-deep by 1.8-m-diameter steel caissons surrounding a 4-m-deep central instrument caisson. Each lysimeter contains one field-scale waste form (210-L barrel), obtained from a commercial power reactor. Solidification agents being tested include masonry cement, Portland III cement, Dow polymer, and bitumen.

Most of the precipitation at the Hanford Site arrives as winter snow; this contributes to a strong seasonal pattern in water storage and drainage observed in the lysimeters. This seasonal pattern in storage corresponds to an annual range in the volumetric soil water content of 11% in late winter to 7% in the late summer and early fall. Annual changes in drainage rates cause pore water velocities to vary by nearly two orders of magnitude, from approximately 4 cm/week in early spring to less than 0.01 cm/week in early fall.

Measurable quantities of tritium and cobalt-60 are being collected in lysimeter drainage water. Approximately 30% of the original tritium inventory has been leached from the only two waste forms containing tritium. Cobalt-60 is contained in all waste form samples and is consistently being leached from five lysimeters. Total cobalt-60 collected from each of the five lysimeters varies, but in each case is less than 0.1% of the original cobalt inventory of the waste sample.

KEY WORDS: solidified waste, lysimeter, leaching, radionuclide, low-level waste

Federal regulations [*1*] prohibit the near-surface disposal of low-level radioactive waste in a liquid form. Consequently, generators of low-level waste, such as commercial power reactors, solidify their liquid waste streams before shipment to a waste disposal facility. A liquid waste is considered solidified if less than 0.5% by volume remains as a free liquid [*2*]. The U.S. Nuclear Regulatory Commission (NRC) has devised a three-tier waste classification system (Classes A, B, and C) to guide the solidification process. Class A waste contains the lowest concentration of radionuclides and can be solidified by adding any absorbent, such as sawdust or vermiculite, that will reduce the free liquid content. Class B and C waste contain higher concentrations of radionuclides and must meet additional requirements of stability such as minimum compressive strength, resistance to corrosion, dissolution, biodegradation, and leaching [*2*]. Therefore, Class B and C wastes are commonly

[1] Research scientist and department manager, respectively, Pacific Northwest Laboratory, P.O. Box 999, Richland, WA 99352.

solidified with materials such as cement, bitumen, or vinyl-ester styrene (for example, Dow polymer[2]).

The stability requirements are primarily designed to prevent the slumping and subsidence of surface covers that result from compaction and degradation of unstabilized buried waste. However, solidification can also affect the release rate and speciation of radionuclides entering the soil. Understanding these latter effects becomes important when predicting the performance of a buried-waste site. There are a minimum of two performance assessments required by 10 CFR 61 [*1*] in the life of a burial site. The first is in the initial licensing process and the second is at site closure. Additional assessments may be required if significant migration of radionuclides is detected during the operational phase. Each assessment includes extensive modeling of radionuclide transport from the waste site by pertinent pathways such as groundwater, air, plants, and burrowing animals.

Although these diverse pathways are different in terms of transport mechanisms and modeling strategy, they do share a common origin or source: the waste form. To enter any of the pathways, contaminants must first leave the waste form. Therefore, the initial modeling step, regardless of the pathway, is to characterize the release of contaminants from the waste form.

There are three characteristics of a waste form that determine its role as a contamination source: (1) the total amount (that is, inventory) of contaminants contained in the waste form; (2) the chemical form (that is, speciation) of the contaminants being released from the waste; and (3) the tendency of the contaminants to leave the waste form. This latter characteristic differs from the other two in that it may depend as heavily on the burial environment (that is, soil type and climate) as on properties of the waste form. The tendency of contaminants to leave a waste form may be quantified as a known flux of contaminant at the waste form surface or as a known solution concentration at some point in the soil profile. Together, inventory, species, and release rate or concentration comprise what in modeling is known as the source term. Any transport calculation, regardless of the pathway involved, must make sure assumption about or prediction of the source term associated with the burial site.

Program Summary

In 1983 a research program was initiated to evaluate radionuclide source terms associated with low-level waste being produced and solidified at commercial power reactors. The research is sponsored by the U.S. Department of Energy Low-Level Waste Management Progam and is being conducted at the Hanford Site near Richland, Washington, by the Pacific Northwest Laboratory. The objectives of the research program are fourfold: (1) to monitor the release of radionuclides from solidified waste forms under field conditions; (2) to correlate laboratory estimates of source term and transport with measured field results; (3) to determine, where possible, the chemical and physical mechanisms controlling source terms and the relative role played by waste form and environment; and (4) to make recommendations on how source terms for solidified waste should be characterized and modeled to support a burial site performance assessment.

The first objective is being met by the construction and operation of a field lysimeter facility where large-scale (210-L drum) samples of soildified waste have been buried in soil-filled caissons since the spring of 1984. The lysimeters are free of vegetation and receive only natural precipitation. The inventory of the waste forms was measured for each sample before burial. Leaching of the waste forms currently is being monitored through analysis of

[2] Product of the Dow Chemical Co., Midland, MI 48640.

drainage water collected at the bottom of the lysimeter. Future measurements of leaching will come from exhumation of the waste forms, which will permit remeasurement of radionuclide inventories, together with detailed measurement of radionuclide concentrations within the lysimeter.

The samples of solidified waste being studied in the lysimeter facility were obtained from three commercial power reactors currently operating in the United States. The samples were manufactured as part of routine waste solidification operations at the reactor and with actual waste streams. As such, they represent waste forms routinely produced by the reactors and shipped to near-surface burial sites. Table 1 lists the waste streams and solidification agents used to manufacture the waste forms. The approximate radionuclide inventory of each waste form is also provided. The lysimeter facility contains duplicate samples (individual inventories vary by about 10% to 15%) for each of the five waste forms. The waste streams were solidified with agents designed to meet the structural specifications of Class B or C waste, although the radionuclide concentrations are well below the upper limit of Class A waste set by 10 CFR 61.

In addition to the lysimeter study, the program also includes laboratory leaching studies and a transport modeling effort. The laboratory leaching studies include traditional water leaching tests such as ANS 16.1 [*3*], soil column leach tests, as well as soil column and batch sorption tests [*4*]. Historically, field performance of solid waste forms has been predicted solely on the basis of these types of laboratory tests. An important objective in this research program is to evaluate predictions based on laboratory measurements by comparison with measurements made in the lysimeter facility.

Modeling will be used extensively to help interpret experimental results and formulate recommendations. The solution chemistry of laboratory and field leachate samples will be analyzed with geochemical models such as MINTEQ [*5*], to aid in estimating the speciation component of source terms and to evaluate sorption parameters. Two-dimensional hydrologic models such as UNSAT2 [*6*] will be used to evaluate the effects of waste-form geometry on water contents and flow rates, which are difficult to measure experimentally.

The results of laboratory leaching tests (that is, estimated diffusion coefficients, adsorption surfaces, etc.) will not be directly comparable to data collected in the field (that is, leachate concentrations, soil radionuclide distributions, water drainage rates, etc). Thus hydrologic, geochemical, and dispersive transport models will be used to translate laboratory-measured parameters into field-measured results. Using the models to make predictions based on laboratory data will clarify which laboratory measurements are most descriptive and which conceptual models of leaching and transport are most effective in predicting field performance.

TABLE 1—*Description of waste streams, waste forms, and inventories of waste samples buried in the lysimeter facility.*

Waste Stream	Solid Matrix	Lysimeter Nos.	Inventory per Sample, mCi				
			^{60}Co	^{137}Cs	^{134}Cs	^{54}Mn	^{3}H
Evaporator concentrate (boric acid)	masonry cement	1,7	0.5	12	8	0.1	5
	bitumen	5,6	1	0.8	0.3	0.1	0
Evaporator concentrate (Na_2SO_4)	Portland III cement	2,8	95	16	2	6	0
Evaporator concentrate + exchange resin (Na_2SO_4)	Portland III cement	3,9	150	30	3	10	0
	Dow polymer	4,10	130	<11	<12	5	0

The remainder of this paper discusses the design and operation of the field lysimeter facility and the preliminary results obtained over the past three years. The important characteristics of the annual water balance are explained together with the relationship between lysimeter hydrology and local soil and climatic conditions. This discussion is followed by a summary of radionuclide leaching data and some preliminary observations on the relationship between the site hydrology and leaching history of the waste forms.

Description of Lysimeter Facility

The lysimeter complex (Fig. 1) was constructed in the summer of 1983 and consists of ten soil-filled caissons (183 by 305 cm) placed concentrically around a central access caisson (365 by 365 cm). Each lysimeter contains one waste form sample that has been removed from its 210-L barrel, leaving the waste form in direct contact with the soil. Each lysimeter is equipped with a gravity drain leading to the central access caisson for collection of leachate water. The central caisson also provides access to each lysimeter through horizontal sampling ports. A 15-cm-diameter well is installed at the side of each lysimeter to allow downwell gamma-ray scanning and neutron probe measurements of soil water content. Lysimeters are instrumented with thermocouples to measure the vertical soil temperature profile, fiberglass resistance blocks to monitor soil water potential, and suction candles to allow extraction of soil solution samples.

The bottom of the waste sample in each lysimeter is at 250 cm; however, the height of individual samples range from 60 to 80 cm. The waste samples were located deep in the lysimeter to minimize the transport distance to the drain (gravel layer), thereby isolating the source term as much as possible. Fiberglass moisture blocks were placed above the waste samples in an effort to detect ponding of water on top of the sample. Suction candles were placed immediately below the waste and between the waste and the drain to allow samples to be collected along the transport stream. Selected meteorological measurements are made on site and these are supplemented with data from the Hanford Meteorological Station

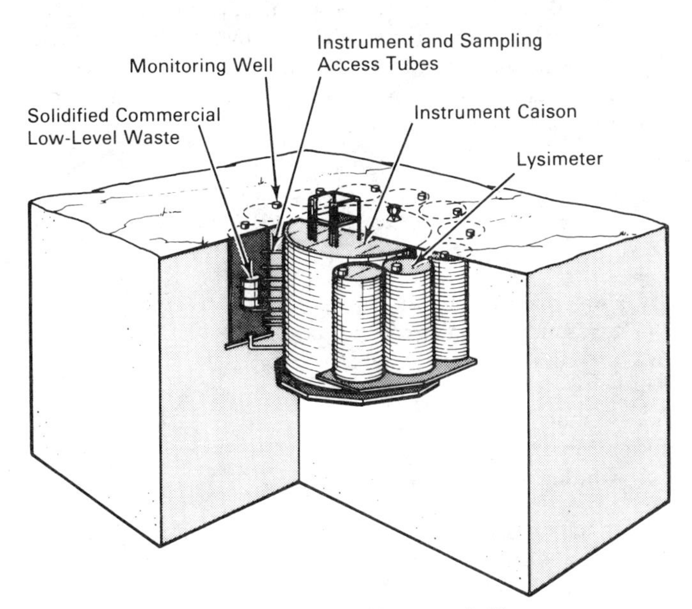

FIG. 1—*Schematic of lysimeter facility.*

located 48 km (30 miles) north of the lysimeter facility. A complete description of all instrumentation and their placement is given by Walter et al. [7].

Eight of the ten lysimeters were filled in March 1984; these included Lysimeters 1 to 4 and 7 to 10, which contain the waste samples solidified with cement and Dow polymer. The bitumen waste forms were buried in Lysimeters 5 and 6 in September 1984. Drainage measurements were made approximately once a month during 1985, and once every two to three weeks during 1986 and 1987. All drainage samples are analyzed for radionuclides, and a complete cation/anion analysis is performed quarterly. Soil water contents are measured approximately twice a month, whereas soil temperature and the soil moisture blocks are monitored continuously by a data logger system. To date, samples have not been collected with the suction candles.

Lysimeter Hydrology

The soil material used to fill the lysimeters is from flood deposits found throughout the Hanford Site. These sediments consist of coarse sands and gravels, with minimal silt and clay fractions. The material used to fill the lysimeters was screened to remove the gravel and the remaining material consists of 92% sand, 6% silt, and 2% clay [7].

The coarse nature of the sandy material is illustrated by the water retention curve and hydraulic conductivity function shown in Fig. 2. Measurements of water retention and saturated hydraulic conductivity were made on samples from each lysimeter individually to check for variability among lysimeters. These data are reported in Walter et al. [7]. No obvious trend in variability was discovered, and the data shown in Fig. 2 are average values. The unsaturated hydraulic conductivity data shown in Fig. 2 are calculated from field measurements of soil water content and drainage rates. The composite data represent at least two datapoints from each of the ten lysimeters. Four observations can be made from the data in Fig. 2 that are all consistent with the known coarse nature of the lysimeter fill material: the saturated hydraulic conductivity is relatively high; the air entry potential is approximately −1 J/kg (−10 cm of water); the water content (~8.0%) at a water potential of only − 10 J/kg (−100 cm of water) is relatively low; and the hydraulic conductivity drops four orders of magnitude in the 0 to −10 J/kg range.

The smooth curves in Fig. 2 come from an analysis described by van Genuchten [8]. The equation describing the water retention curve is of the form

$$\theta = \theta_r + (\theta_s - \theta_r)\left(\frac{1}{1 + (\alpha\psi)^n}\right)^M \quad (1)$$

where

θ = volumetric water content, m^3/m^3,
θ_s = SAT WTR = saturated water content, m^3/m^3,
θ_r = RES WTR = residual water content, m^3/m^3,
ψ = soil water potential, J/kg,
α = ALPHA = fitting parameter,
n = N = fitting parameter, and
$M = 1 - 1/n$ = fitting parameter.

The saturated water content was calculated from the bulk density of the samples, and the residual water content was taken as zero. The parameters α and n were curve fit from the data according to the procedure described by van Genuchten [8].

The hydraulic conductivity curve shown in Fig. 2 is predicted from the measured desorption

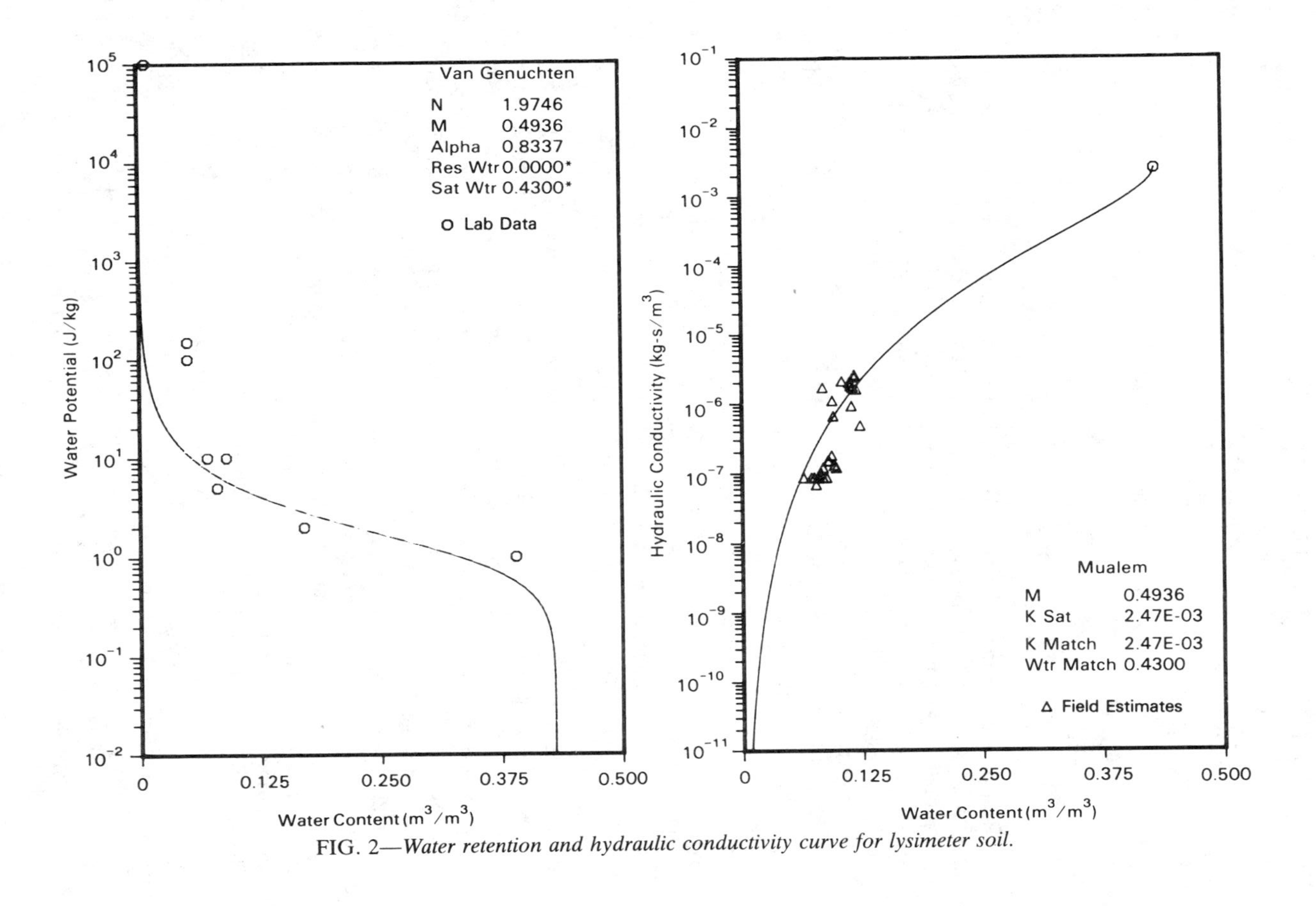

FIG. 2—*Water retention and hydraulic conductivity curve for lysimeter soil.*

curve following the theory of Mualem [9]. The theory requires one matching point. In Fig. 2, the curve was matched to the measured saturated hydraulic conductivity as reported by Walter et al. [7]. The equation is of the form

$$K(\theta) = K_s * \theta^{1/2} [1 - (1 - \theta^{1/M})^M]^2 \quad (2)$$

where

$K_s = K\ \mathrm{SAT}$ = saturated hydraulic conductivity,
$K(\theta)$ = hydraulic conductivity at θ, and
$\theta = \dfrac{(\theta - \theta_r)}{(\theta_s - \theta_r)}$.

The soil physical properties measured for the lysimeter soil (that is, particle size distribution, water retention, and saturated hydraulic conductivity) show no systematic difference in soil properties among lysimeters. Therefore, the water balance of each lysimeter should be similar. A preliminary analysis of the data has confirmed that the water balance of the lysimeters are similar. Figure 3 shows the initial water contents of Lysimeters 1, 4, 7, and 9. The water contents are quite similar both with depth and among lysimeters. A partial exception to this is the slightly dryer surface soil of Lysimeter 4. The wetter surface soil found in other lysimeters is due to rain events during and immediately after filling of the lysimeters.

The seasonal pattern of water storage in Lysimeters 1, 4, and 7 is shown in Fig. 4. The small differences in initial storage values are due to the differences in initial water contents already mentioned. These differences quickly disappear as the winter precipitation wets the soil and soil water storage increases. By late winter, the amount of water stored in all three lysimeters is essentially equal and remains so for most of the next two years.

Figure 4 also shows the distinctive seasonal pattern in water storage observed in all lysimeters. The peak soil-water storage occurs during the late winter months, which corresponds with the season of peak precipitation at the Hanford Site. In addition, most of this winter precipitation occurs as snow. Both the winter of 1984–1985 and the winter of 1985–1986 were extremely cold, preventing significant snow melt during the winter. In both years most of the winter snowpack melted rapidly in February and produced the sharp peaks in seasonal storage.

The high water storage in February is followed by a steady decline throughout the spring and summer, with minimum storage occurring in early fall just prior to the onset of the next winter's precipitation. Note again the similarity in water storage among the three lysimeters shown in Fig. 4 particularly during 1985 and the winter of 1986. Lysimeter 1 shows less storage during the summer of 1986, but not to any large degree.

The drainage pattern for Lysimeters 1, 4, and 7 is shown in Fig. 5. As expected, the patterns are somewhat the inverse of the storage patterns shown in Fig. 4. As the water storage is declining from February to October each year, the cumulative drainage is increasing. The drainage rate however decreases as the soil profile dries.

The similarity in the water balance behavior of Lysimeters 1, 4, and 7 also holds for Lysimeters 2 and 3. This similarity was expected because of the similarity in soil properties discussed above. However, there are factors other than soil type that can influence the water balance of a soil profile. There were two anomalies that made the water balance of Lysimeters 5, 6, and 8 to 10 temporarily different from the other five.

The water balance of Lysimeters 5 and 6 was significantly affected by their low initial water contents. As stated above, Lysimeters 5 and 6 were filled in September 1984, six

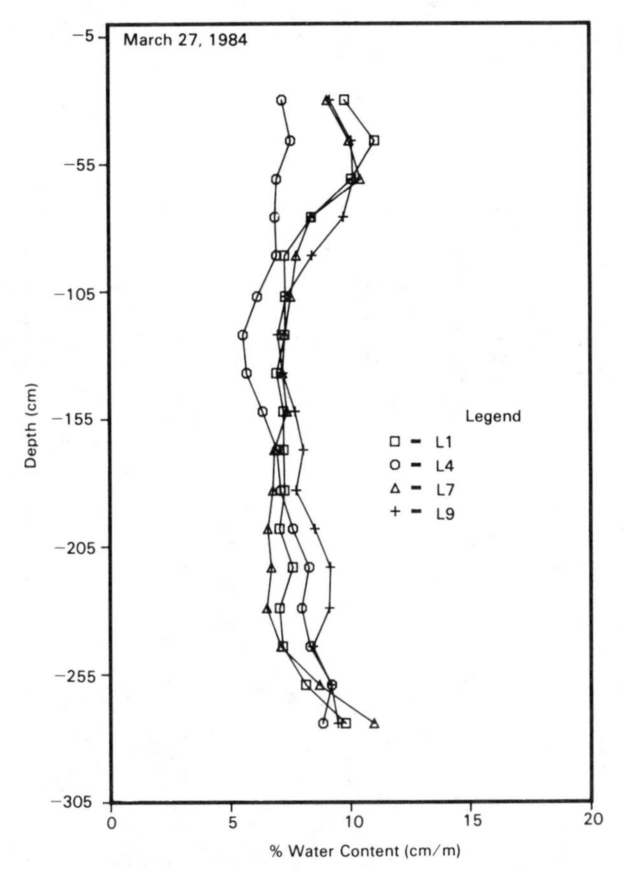

FIG. 3—*Initial water content profile for Lysimeters 1, 4, 7, and 9.*

months later than the others. As a result, the soil used to fill the top 2 m of Lysimeters 5 and 6 had been dried to a relatively low water content during the summer. Figure 6 shows that in September 1984, the volumetric water content in the top 1.5 m of Lysimeters 5 and 6 was approximately 2.5%, in contrast to the 7% water content of Lysimeters 1 and 7. The higher water content illustrated by Lysimeters 1 and 9 is also representative of the other six lysimeters that had been filled in March.

The effect of these dry initial conditions was to retard the wetting of the soil profile in Lysimeters 5 and 6 during the winter of 1984–1985. Figure 7 shows how the storage of water in Lysimeters 5 and 6 lagged behind Lysimeter 1. Similar comparisons can be made with other lysimeters, as well. By the end of the first wetting and drying cycle (September 1985), the effect of the initially dry conditions was effectively removed. Lysimeter 6 still lags behind during the second cycle (1985–1986), but not by as much as in 1984–1985. All three lysimeters seemed to be responding similarly thus far during the winter of 1986–1987.

The dry initial conditions also affected drainage from Lysimeters 5 and 6. Figure 8 shows how the initial drainage of water from Lysimeters 5 and 6 was delayed by almost one year. This delay in initiation of drainage was because the precipitation received during the winter of 1984–1985 was not enough to wet the dryer soil of Lysimeters 5 and 6 sufficiently to allow measurable drainage. However, by the second winter (1985–1986), the drainage rates

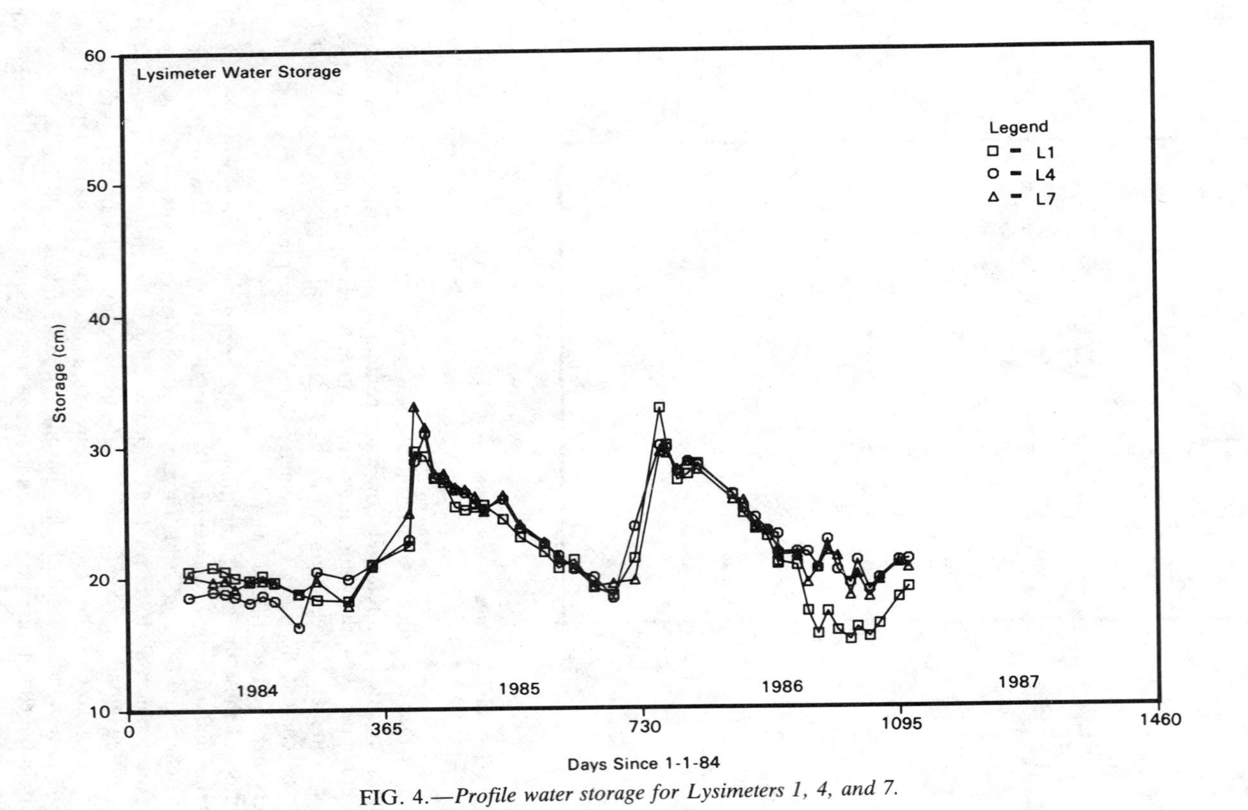

FIG. 4.—*Profile water storage for Lysimeters 1, 4, and 7.*

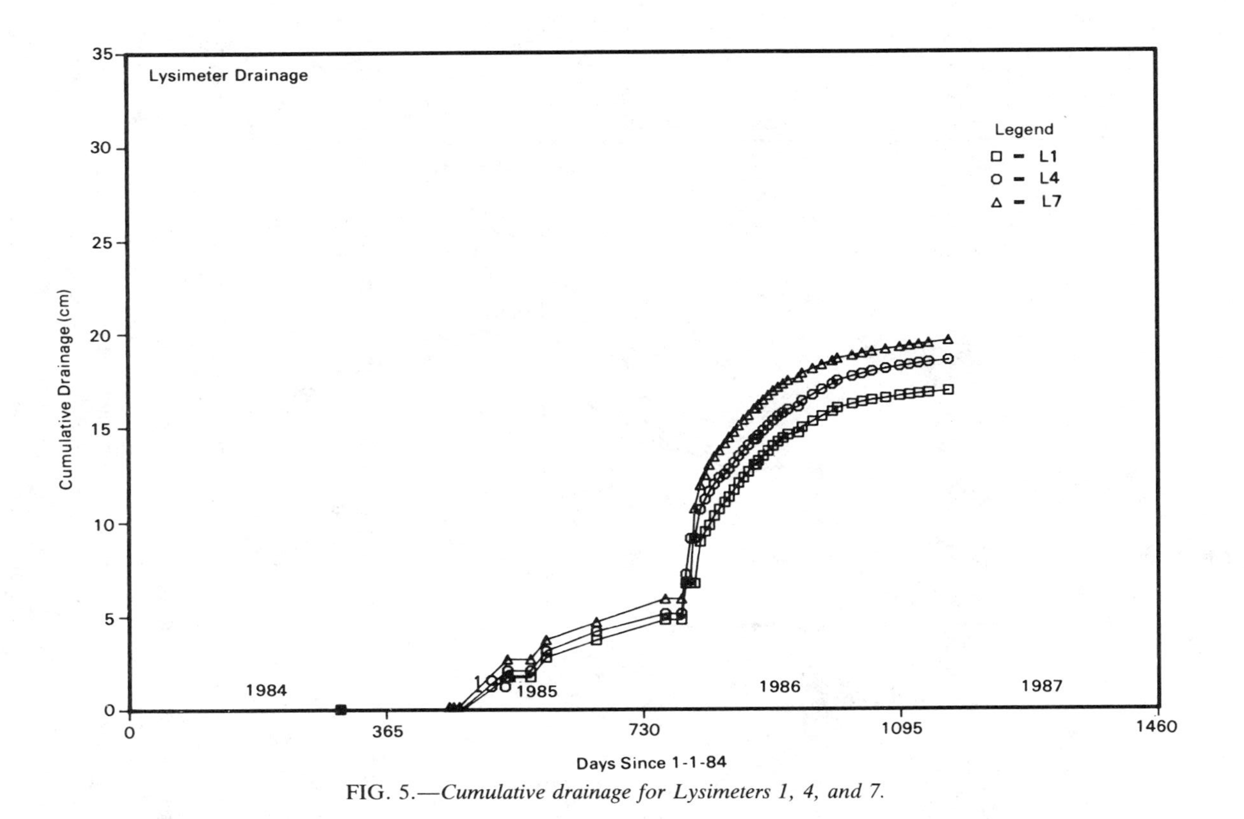

FIG. 5.—*Cumulative drainage for Lysimeters 1, 4, and 7.*

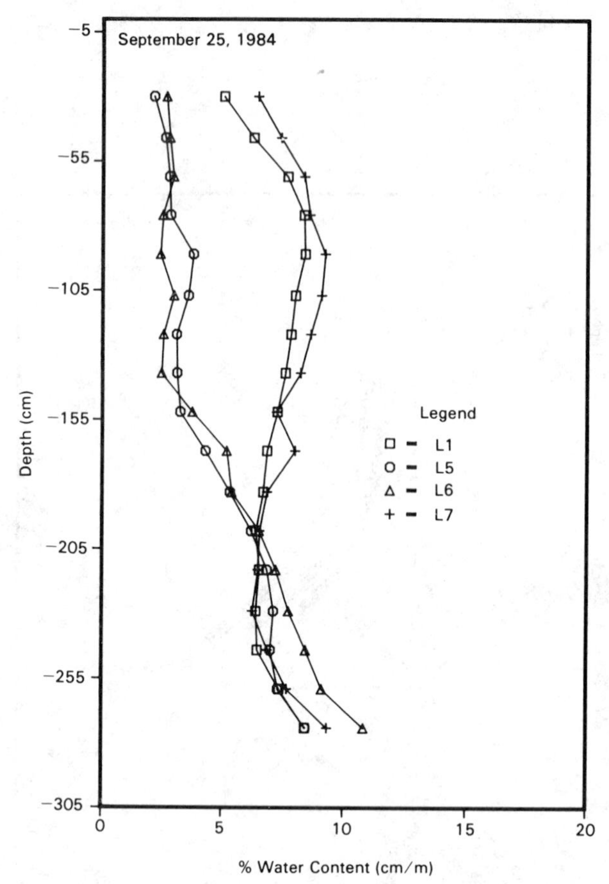

FIG. 6.—*Water content profile for Lysimeters 1, 5, 6, and 7 for 25 Sept. 1984.*

of Lysimeters 5 and 6 were comparable to the rest of the lysimeters as represented by Lysimeter 1 in Fig. 8. In fact, the cumulative drainage from Lysimeter 6 caught up to the rest of the lysimeters, even with the one-year delay in the start of drainage.

The second anomaly that affected the hydrology of the lysimeter facility occurred in February 1985. Due to rapid snow melt together with frozen soil conditions, Lysimeters 8 to 10 received from 5 to 15 cm extra water supplied by runoff from areas adjacent to the lysimeter facility. This additional water resulted in the flooding of the three lysimeters and a temporary, yet dramatic increase, in both storage and drainage.

Figure 9 compares the three distinct storage patterns that were seen during 1984–1985 at the lysimeter facility. The curves show the reduced storage of Lysimeter 5 as a result of dry initial conditions, and the temporary spike in water storage of Lysimeter 8 caused by flooding in early 1985. The different water storage patterns illustrated by these three lysimeters were not caused so much by differences in soil properties, but by the factors of initial conditions and runoff discussed previously. Figure 9 also shows how, when these factors were not present (1985–1986), the storage patterns were very similar among these and the other lysimeters. The same results are also shown for the case of drainage in Fig. 10. Lysimeter 5 shows delayed onset of drainage relative to lysimeter 1 because of dry initial conditions,

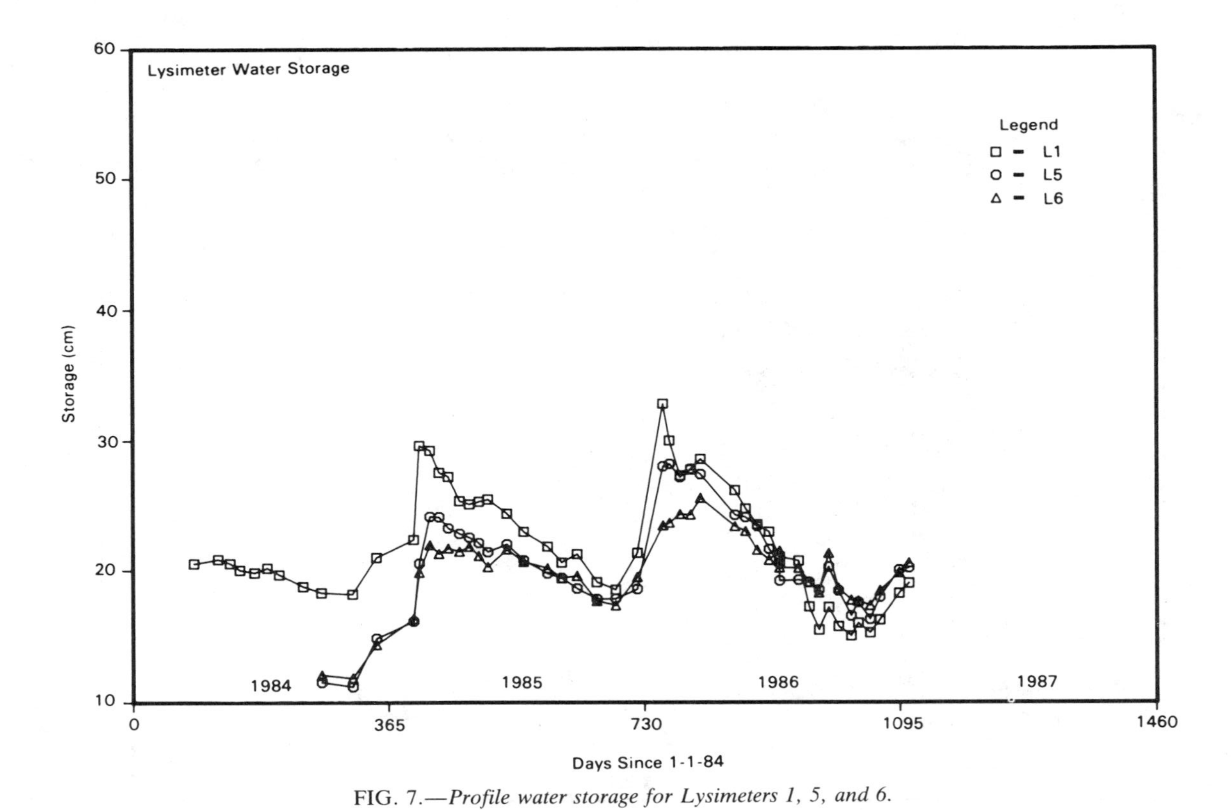

FIG. 7.—*Profile water storage for Lysimeters 1, 5, and 6.*

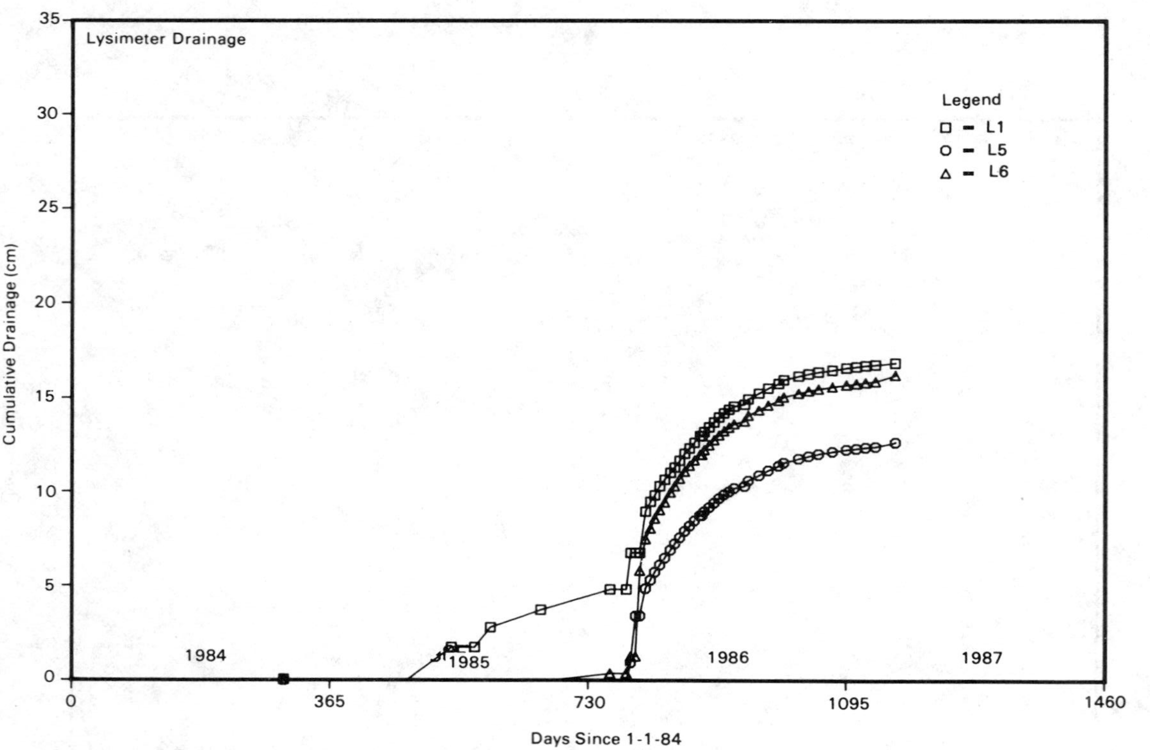

FIG. 8.—*Cumulative drainage for Lysimeters 1, 5, and 6.*

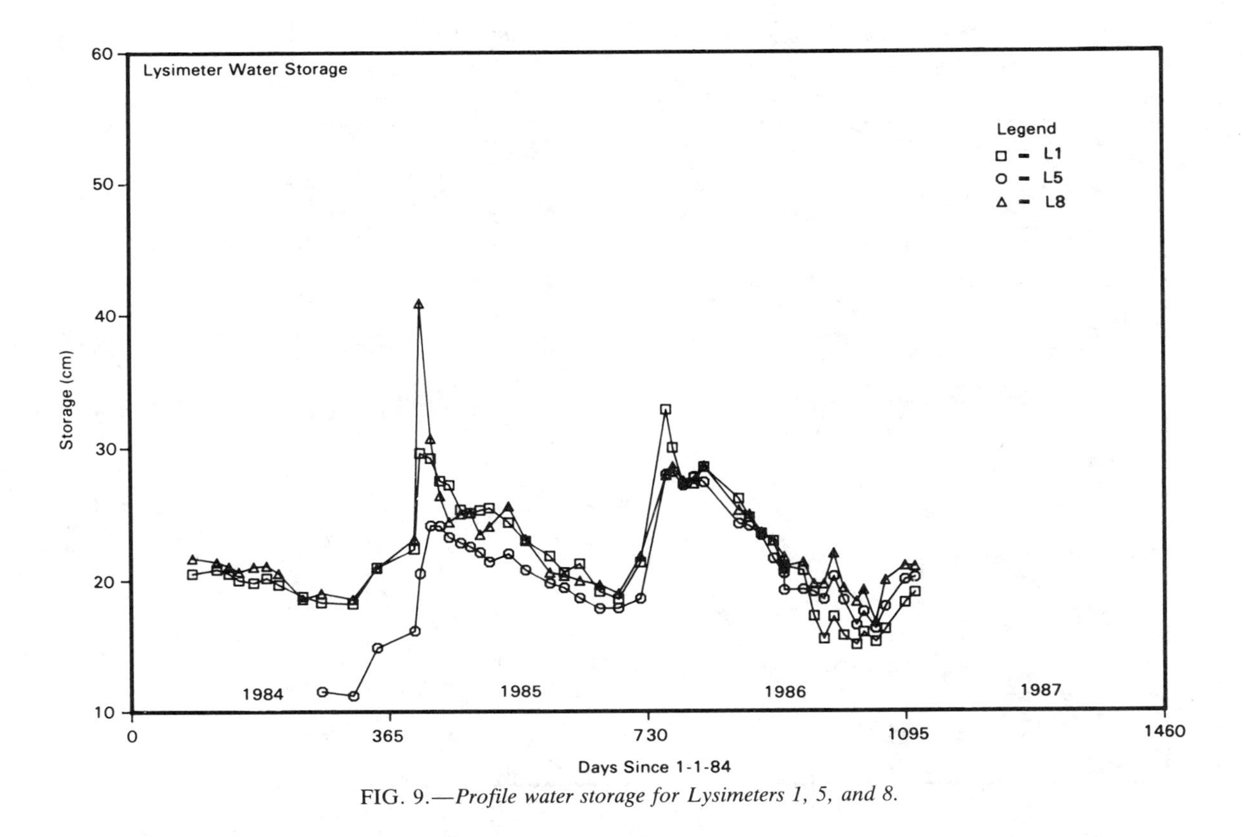

FIG. 9.—*Profile water storage for Lysimeters 1, 5, and 8.*

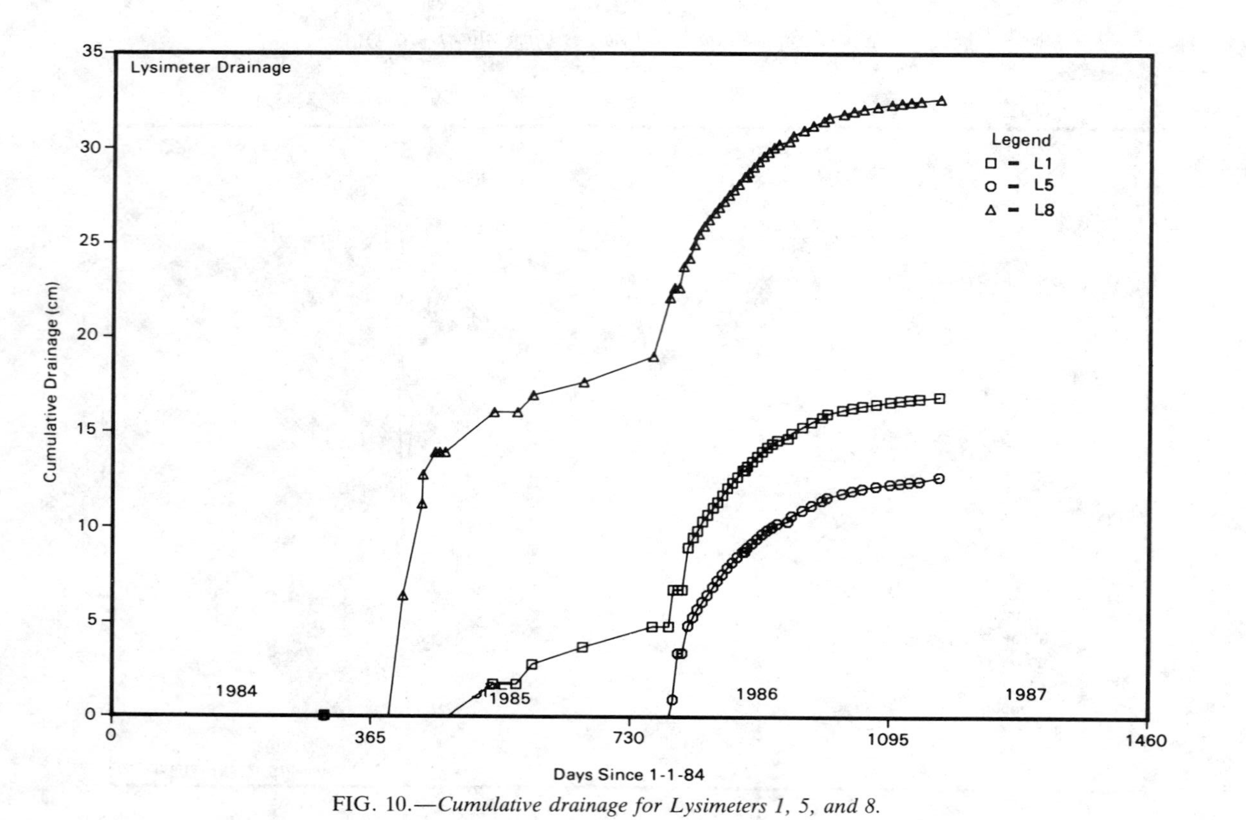

FIG. 10.—*Cumulative drainage for Lysimeters 1, 5, and 8.*

while Lysimeter 8 shows significantly more drainage caused by the flooding of 1985. Both drainage rates and amounts, however, are very similar during 1986.

The overall water balance of the lysimeters is summarized in Table 2. The calculations for the year 1984–1985 exclude Lysimeters 8 to 10 because the exact water input is not known. Lysimeters 5 and 6 are also excluded from the drainage averaging for reasons discussed previously. The average values and standard deviations listed in Table 2 represent parameters for the lysimeters, excluding temporary anomalies such as those mentioned above. The relatively small standard deviations shown for water-balance components illustrate the small variability in behavior among lysimeters due to variability in soil properties. Another interesting observation is the similarity in estimated evaporation for the two years even though 1985–1986 received 50% more precipitation. This similarity illustrates the complex relationship among precipitation and either drainage or evaporation. In this case, it is due to the extra precipitation in the form of snow, which quickly entered the soil as snow melt during late winter, when evaporation was low.

The seasonal changes in total water storage shown above correspond to changes in average soil water contents. As shown in Table 2, the volumetric water content at the depth of the waste samples ranged from 7 to 11% in both years. This is a fairly narrow range, considering the significant amount of drainage water moving through the soil profile. However, this is typical of coarse soils. Drainage rates also fluctuate during the year. The annual rates shown in Table 2 are annual averages. During 1986, the drainage rate ranged from a high of 0.4 cm/week in February to less than 0.001 cm/week in November.

There are two contaminant transport processes that are known to vary with water content and drainage rate. The first, hydrodynamic dispersion, is the process that causes a solute pulse to spread (that is, disperse) as it moves through a soil profile. It is usually characterized by a dispersion coefficient, which is commonly modeled as a linear function of pore water velocity. Pore water velocity in the case of the lysimeters is the drainage rate divided by the volumetric water content. Winter drainage rates are associated with a water content of about 11%, and summer rates with 7%. Therefore, pore water velocities are ranging from a high of 4 cm/week to less than 0.01 cm/week. This change in pore water velocity can result in an order of magnitude change in dispersion.

The other transport parameter thought to vary with water content is solute retardation. The most popular model for sorption and transport of radionuclides in soil is the linear Freundlich isotherm combined with the classical convective-dispersive equation. In this model, sorption is expressed as a retardation factor R, defined as the ratio of water velocity to solute velocity. In the usual formulation of R, retardation goes down as soil water content

TABLE 2—*Average water balance parameters for 1984–1986.*

Parameter	9/1/84[a] to 8/31/85	9/1/85 to 8/31/86
Total precipitation, cm	13.8	21.7
Snow, cm	55.6	86.6
Δ storage, cm	2.1 ± 0.5	−0.6 ± 2.2
Drainage, cm	4.2 ± 0.5	12.3 ± 1.2
Evaporation, cm	7.5 ± 0.6	10.0 ± 2.8
Δ water content, cm/m	6.8 − 11.4	6.9 ± 10.7
Δ matric potential, cm	−40 to −80	−40 to −80

[a] Calculations for 9/1/84 to 8/31/85 are only for Lysimeters 1, 2, 3, 4, and 7.

increases. A change in water content of 7 to 11% would reduce the retardation factor by 60%. Therefore, in the winter when drainage rates are at their highest, sorption and retardation may be at their lowest.

Radionuclide Migration

Drainage water collected from the bottom of each lysimeter is analyzed for radionuclide content as one measure of release and transport. Thus far, only tritium and cobalt-60 have been identified in any of the leachates. Tritium is found in only the boric acid/masonry cement waste forms and is present in the leachate of Lysimeters 1 and 7, which contain samples of this waste form. The total amount collected from the bottom of the lysimeter represents approximately 30% of the estimated tritium inventory of these waste forms.

Cobalt-60 is present in all waste streams but is a primary constituent in the sodium sulphate evaporator concentrate waste streams. Cobalt-60 has been identified in all four cement waste forms constructed with the concentrate or concentrate plus exchange resin (Lysimeters 2, 8 and 3, 9, respectively). Cobalt-60 has also been detected in one of the lysimeters containing a polymer waste form (Lysimeter 4). The total amount of cobalt-60 collected from the lysimeters is less than 0.1% of the estimated cobalt-60 inventory for each of the waste forms.

The pattern of radionuclide release from the lysimeters is shown in Figs. 11 to 14. The cumulative release is plotted versus time and also versus cumulative drainage from each of the lysimeters. Figure 11 shows the release of tritium from Lysimeters 1 and 7 versus time. The total amount and temporal pattern of tritium arrival at the lysimeter drain are nearly identical for each lysimeter. Figure 11 indicates that there is a seasonal nature to tritium arrival at the lysimeter drain. More tritium arrives at the lysimeter bottom with the spring drainage of winter precipitation and snowmelt than with the drainage water arriving in the summer and fall. This seasonal variability is greatly reduced when the cumulative amount of tritium is plotted or normalized by the cumulative drainage (Fig. 12). Figure 12 shows that there is less variability in the amount of tritium contained in a unit of drainage water than drains in a unit of time. However, Fig. 12 also shows that there is more variability among lysimeters, in how long it takes the tritium to arrive and how much tritium arrives at the lysimeter bottom when cumulative release is measured against cumulative drainage.

Using plots such as Figs. 11 and 12 to infer source-term mechanisms is speculative at best because the drainage of tritium involves the release of tritium from the waste form as well as the transport from the waste form to the lysimeter drain. Ignoring the transport process would lead to one of two conclusions regarding the release of tritium from the waste form: (1) release from the waste form is seasonal or (2) the seasonal increases in mass released is caused by the concentration being controlled outside of the waste form, perhaps by a solubility reaction causing more mass to be released with more drainage water. Obviously conclusion 2 cannot be valid for tritium. Conclusion 1 would imply that release from the waste form is somewhat dependent on soil water content.

A third hypothesis is that the release of tritium from the waste form is fairly constant throughout the year. Consequently, because of the low pore water velocities during much of the year, tritium is stored in the profile. When the spring drainage occurs, the water is moving so much faster that it flushes much of the tritium that has been released throughout the year. This explanation illustrates the difficulty in interpreting effluent data directly as source-term data without accounting for any of the transport processes. Data collected over the next two to five years together with the use of transport models should help interpret these effluent curves in a more conclusive way.

The pattern of cobalt-60 drainage is shown in Figs. 13 and 14. Drawing source-term conclusions from these curves is more speculative than for the tritium because these releases

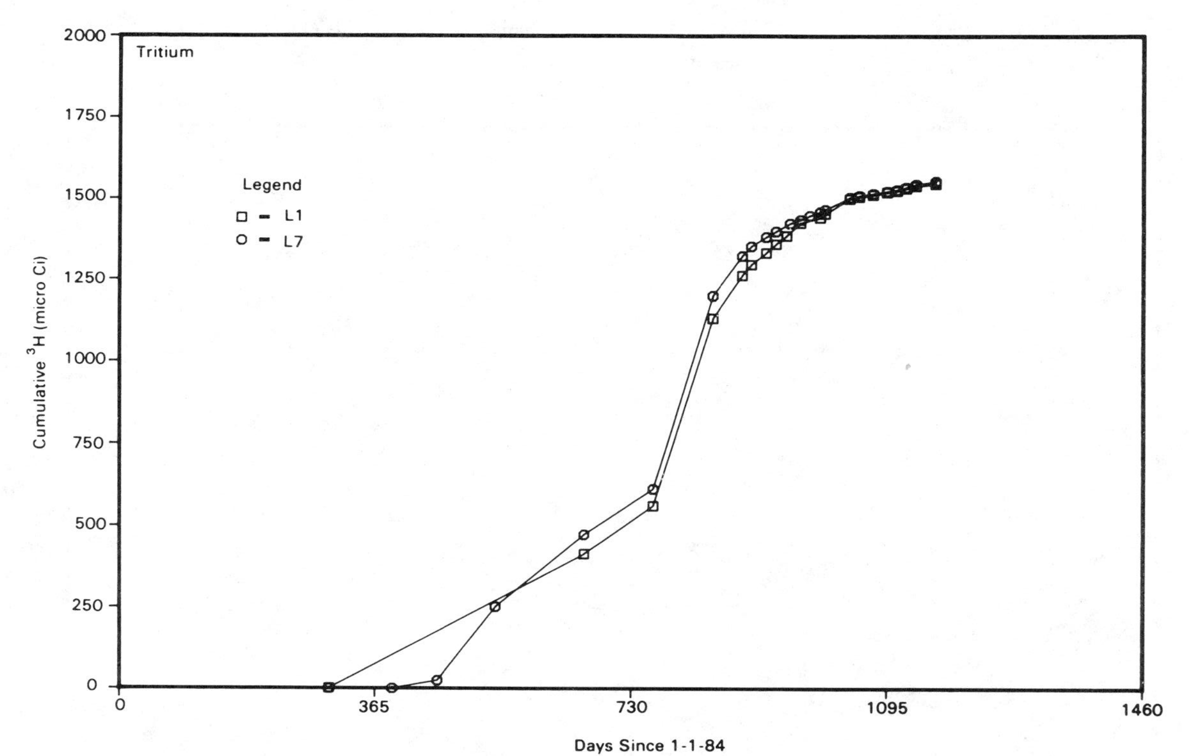

FIG. 11.—*Cumulative release of tritium from Lysimeters 1 and 7 versus time.*

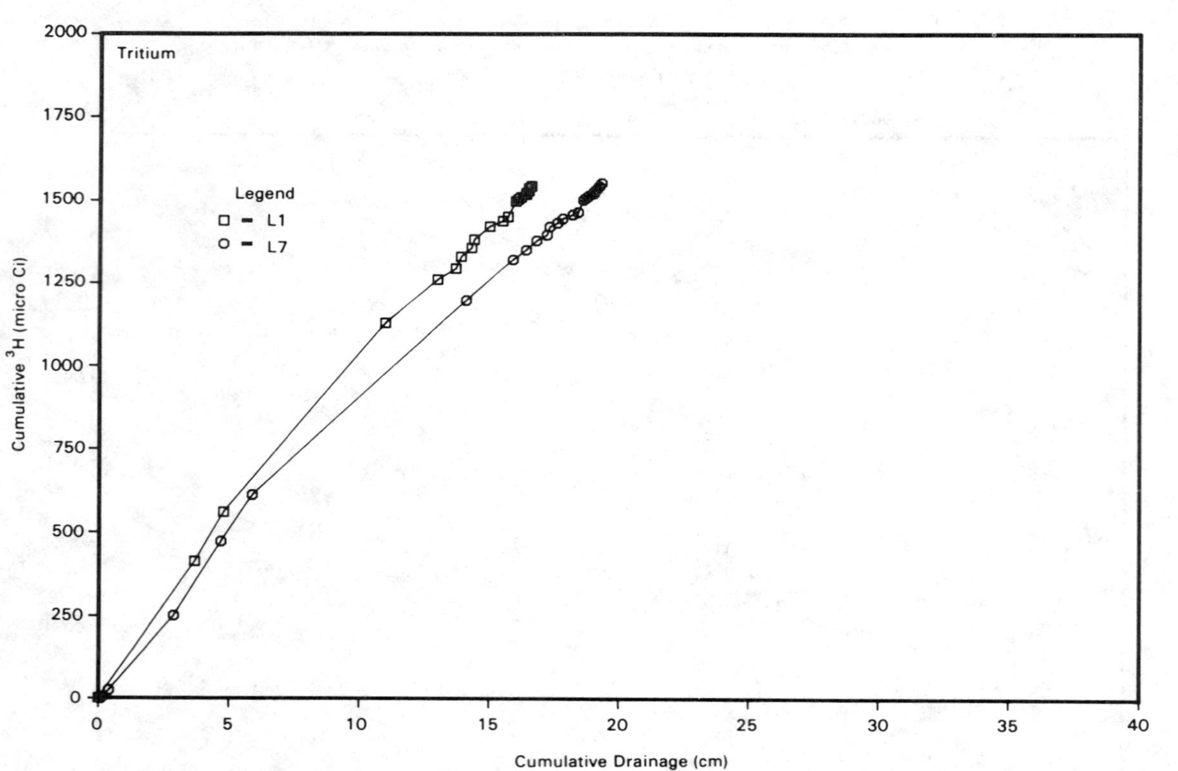

FIG. 12.—*Cumulative release of tritium from Lysimeters 1 and 7 versus cumulative drainage.*

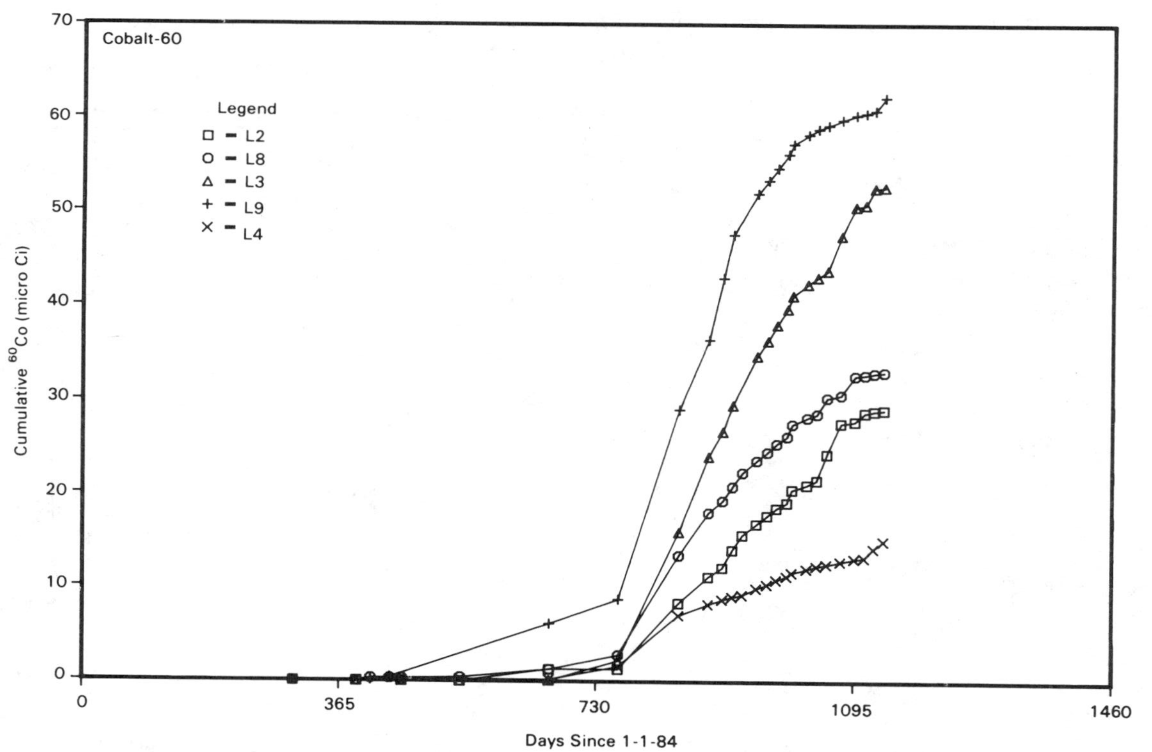

FIG. 13.—*Cumulative release of cobalt-60 from Lysimeters 2, 3, 4, 8, and 9 versus time.*

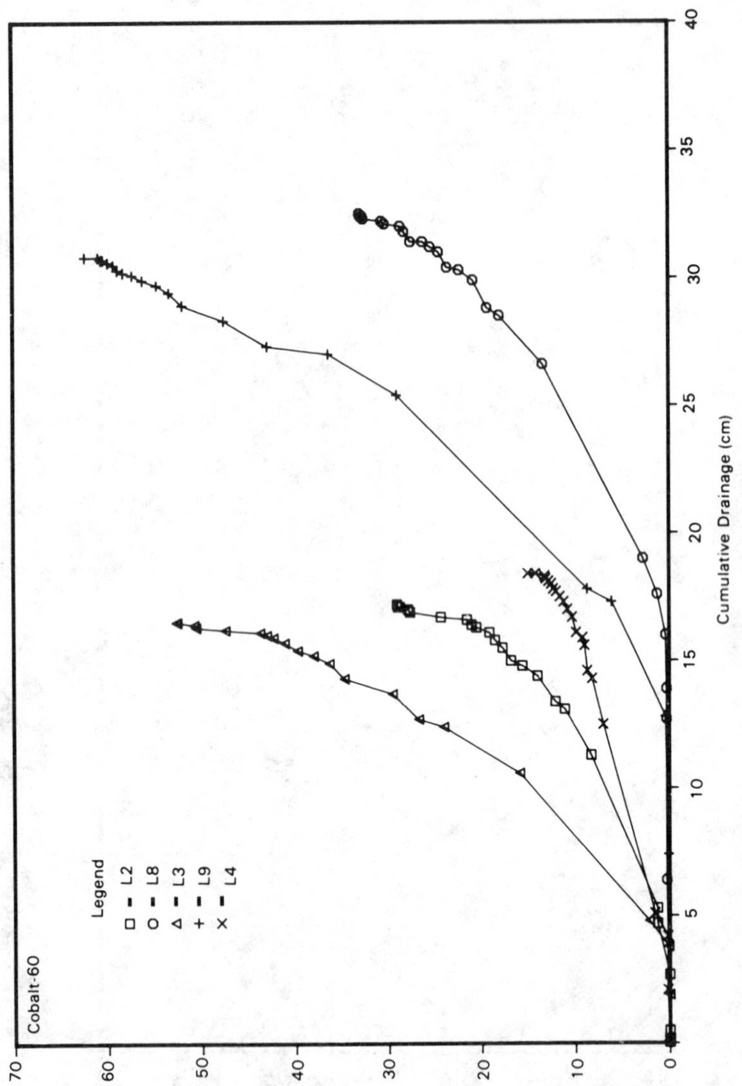

FIG. 14.—*Cumulative release of cobalt-60 from Lysimeters 2, 3, 4, 8, and 9 versus cumulative drainage.*

represent less than 0.1% of the total inventory of cobalt in the waste forms. Lysimeters 3 and 9 are replicate lysimeters that contain the same waste form; this is also true of Lysimeters 2 and 8. These replicate lysimeters seem to be releasing similar amounts of cobalt. Figures 13 and 14 show less variability in cumulative release among replicate lysimeters than among waste forms. The replicate of Lysimeter 4 is Lysimeter 10, which has not shown any levels of cobalt above limits of detection. The polymer waste forms (Lysimeters 4 and 10) are releasing less cobalt than the cement (Lysimeters 2, 8 and 3, 9). The higher amount of cobalt released from Lysimeters 3 and 9 relative to Lysimeters 2 and 8 is probably due to the higher cobalt inventory of Lysimeters 3 and 9.

Interpreting the cumulative release curves in terms of mechanisms is as difficult for cobalt as it was for tritium. Figure 13 indicates a seasonal variability in drainage as was discussed above for tritium. The possible conclusions are also the same. Mobile cobalt-60 generally is thought to result from complex formation with some chelating agent. Chelating agents are present in the waste steams, and the formation of complexes is a likely possibility. Solubility controls do not seem plausible for the mobile fraction of the cobalt. A comparison of Fig. 13 with Fig. 14 shows that the release is more strongly dependent on time than on cumulative drainage. The flooding of Lysimeters 8, 9, and 10 does not seem to have had much effect on total amount or time of release. A cumulative release that depends more on time than drainage seems more consistent with some sort of diffusion-controlled source term, although more data and extensive modeling will be necessary to confirm the most representative source-term formulation.

Additional Data Collection

Current data collection activities have produced a fairly comprehensive picture of the hydrologic conditions at the lysimeter facility. However, certain instruments have not proven as useful as originally anticipated. For example, the fiberglass moisture blocks have proven ineffective for monitoring the presence of ponding on top of the waste form. This is because the drained water potentials of -4 to -7 J/kg (40 to 70-cm suction) are too high for these instruments to respond. Future plans include installing tensiometers for this purpose. Tensiometers will also be installed laterally between the waste form and the lysimeter wall to monitor lateral gradients that may develop as a result of two-dimensional water flow around the waste form. These data will be important to validate the results of model calculations designed to evaluate the possible effects of higher water contents near the waste form on leaching and transport.

Current data-collection activities also exclude certain information. For example, there is no sampling capable of monitoring the leaching and transport of radionuclides that are sorbed onto the soil matrix. Unless the radionuclides reach the bottom of the lysimeter, they are not detected. There are two efforts planned to collect this additional information. The first is to use suction candles to sample the soil solution closer to the waste form. Candles were installed at two depths between the waste form and the gravel layer during construction, but have not been used to date. The second technique that will be used to collect information on sorbed radionuclides is destructive sampling. This will initially be done through the horizontal access ports located in every lysimeter. More detailed sampling is scheduled for 1989, when half of the lysimeters will be exhumed. Excavation will allow detailed sampling of radionuclide distributions in the lysimeter soil and will also allow the radionuclide inventory of the waste form to be remeasured. Exhumation of the second half of the facility is currently scheduled for 1991.

Future Work

A substantial database describing the lysimeter hydrology and waste leaching has been collected during the past three years. This data collection will continue; however, the emphasis in the project will shift to data interpretation and modeling. The overall goal of the modeling task is to test the predictions of several source term/transport models against the lysimeter data and document which, if any, models do an adequate job of predicting the leaching of the waste forms. Two kinds of modeling are currently being done in preparation for the transport modeling. The first is geochemical modeling of the lysimeter leachate, and the second is modeling of the lysimeter water balance. A geochemical model is being used to analyze lysimeter leachates primarily to look for any solubility control that might be limiting the concentration of any constituents. These data will be useful in developing possible leaching and sorption models that will be combined with hydrologic models to form the transport models to be tested later. The water balance models are necessary to describe the convective flow of water through the lysimeter, which then helps transport the waste form leachate to the bottom of the lysimeter. Water balance models are currently being validated against measurements made at the lysimeter facility.

Acknowledgments

The authors wish to acknowledge the support of Marcia Walter, Virginia LeGore, and Paula Heller for their help in collecting and managing the data contained in this paper.

This work was supported by the U.S. Department of Energy under Contract No. DE-AC06-76RL0 1830.

References

[*1*] "Licensing Requirements for Land Disposal of Radioactive Waste," 10 CFR Part 61, *Code of Federal Regulations,* U.S. Nuclear Regulatory Commission, Washington, DC, 1986.

[*2*] *Technical Position on Waste Form,* U.S. Nuclear Regulatory Commission, Washington, DC, 1983.

[*3*] *Measurement of the Leachability of Solidified Low-Level Radioactive Wastes,* Standards Committee, Working Group ANS 16.1, American Nuclear Society, June 1984.

[*4*] Walter, M. B., Serne, R. J., Jones, T. L., and Mclaurine, S. B., "Chemical Characterization, Leach, and Adsorption Studies of Solidified Low-Level Wastes," PNL-6047, Pacific Northwest Laboratory, Richland, Washington, Dec. 1986.

[*5*] Peterson, S. R., Hostetler, C. J., Deutsch, W. J., and Cowan, C. E., "MINTEQ User's Manual," NUREG/CR-4808, PNL-6106, Pacific Northwest Laboratory, Richland, WA, Feb. 1987.

[*6*] Davis, L. A. and Neuman, S. P., "Documentation and User's Guide: UNSAT2—Variably Saturated Flow Model," NUREG/CR-3390, prepared by Water, Waste and Land, Inc., Fort Collins, CO, for the U.S. Nuclear Regulatory Commission, Washington, DC, 1983.

[*7*] Walter, M. B., Graham, M. J., and Gee, G. W., "A Field Lysimeter Facility for Evaluating the Performance of Commercial Solidified Low-Level Waste," PNL-5253, Pacific Northwest Laboratory, Richland, WA, Nov. 1984.

[*8*] van Genuchten, R., "Calculating the Unsaturated Hydraulic Conductivity with a New Closed-Form Analytical Model," 78-WR-08, Department of Civil Engineering, Princeton University, Princeton, NJ, 1978.

[*9*] Mualem, Y., *Water Resources Research,* Vol. 12, No. 3, June 1976, pp. 513–522.

C. Nomine[1] and A. Billon[2]

xperience Acquired in the Field of ong-Term Leaching Tests on Blocks f Radioactive Waste

REFERENCE: Nomine, J. C. and Billon, A., "**Experience Acquired in the Field of Long-Term Leaching Tests on Blocks of Radioactive Waste,**" *Environmental Aspects of Stabilization and Solidification of Hazardous and Radioactive Waste, ASTM STP 1033,* P. L. Côté and T. M. Gilliam, Eds., American Society for Testing and Materials, Philadelphia, 1989, pp. 381–391.

ABSTRACT: Long-term leaching tests on full-size radioactive blocks have been undertaken since 1979 in the leaching facility of the Nuclear Research Center of Saclay. The aims of the tests are the assessment of the release of radionuclides into the biosphere, measurements of ion transfer between the blocks and the environment, and understanding of the transfer mechanisms. After a general presentation of the blocks submitted at the tests, and a description of the leaching facility, examples of results are given; they concern the balances of activities released, post-leaching appraisal, and investigations of mechanisms of leaching experiments.

KEY WORDS: block, encapsulates, ion transfer, leaching, radionuclides

omenclature

an = Activity released in Ci or Bq over a cycle measured in days t_n

Σa_n = Accumulated activity release in Ci or Bq over a cycle measured in days Σt_n

$A0$ = Initial activity of the block in Ci or Bq

V = Volume of the block in cm^3

S = Surface area of the block in cm^2 (geometric surface area)

$a_n = a_\ell + a_f + a_c + a_b$

a_ℓ = Activity measured in the leachant

a_b = Activity of sludge

a_c = Fixed activity by contamination

a_f = Fixed activity on filters

Among the characteristics of solidified radioactive waste, leaching is generally recognized one of the main criteria involved in the qualification of blocks of waste of low and medium tivity. With this in mind, the Atomic Energy Commission has been carrying out long-rm leaching tests on full-scale blocks of such waste since 1979, the aims being

[1] Research and development engineer, Atomic Energy Commission, Saclay, Nuclear Studies Center, 191 GIF Sur Yvette CEDEX, France.

[2] Atomic Energy Commission, Nuclear Studies Center, Fontenay-aux-Roses, France.

[a] Assessment of the release of radionuclides into the biosphere,
[b] Measurement of ion transfer between the blocks and the environment likely to sho their aging, and
[c] understanding and advancement of our knowledge of the transfer mechanisms.

The first two aspects, which concern essentially the preparation of a qualification fil with a view to acceptance of the blocks by the National Radioactive Management Agenc (ANDRA) at a surface storage site, is carried out in accordance with the basic safety rule published by the Ministry of Industry. The third aim is more of a research and developmer aspect, which is why we looked at problems as diverse as the study of scale-effect, th drawing up of the most exhaustive leaching balances possible, and the identification c leached chemical forms.

Experimental Procedures

Nature of the Blocks or Encapsulates

The blocks or encapulates studied are of varied origin: EDF (French Electricity Boarc Nuclear Power Plants, COGEMA Reprocessing Plants, and French or ECC Country Nuclea Research Centres. The size and weight characteristics cover a very wide range, from a fe litres to several cubic metres for weights of up to 5000 kg; the encapsulation matrices, whic are hydraulic binder, bitumen or polymer base, contain either homogeneous or heteroge neous waste, which carries α-β-γ emitters; the dose rates on contact may reach 200 rad/h Table 1 illustrates and summarizes the characteristics of a few blocks either leached or bein tested.

Test Method

The leaching tests, designed to characterize blocks of waste of low or medium activity are carried out in accordance with a very strict procedure. The test protocol followed i essentially based on national policy, such as fundamental safety rules and ANDRA spec fication, but also takes into account certain international recommendations. Generally speak ing, the test is a static test, performed using standard synthetic water (Table 2) simulatin the behavior of the ground-water table at the shallow land burial. This standard water i maintained at a temperature of 23 ± 3°C and renewed periodically (cycles of 2 × 1: 2 × 30, and 4 × 90 days). At the end of each cycle, leachates samples are taken for

(*a*) spectrometer measurements: the main radionuclides looked for are ^{239}Pu, ^{241}Am, ^{244}C for α-emitters; ^{137}Cs, ^{90}Sr, ^{60}Co, ^{3}H, etc., for β-γ-emitters, and
(*b*) chemical and physicochemical measurements, showing the ion transfers between th blocks and the water; the ions looked for depend on the nature of the matrix an the waste; thus, in the case of a cement-nitrate block (evaporator concentrate), w would be looking for the sodium, the nitrates, and the dry extract.

Note that renewal of the leachant required 1 or 2 days, which made it impossible to perfor the test sequences in rapid succession.

The results given in Table 2 are a very useful indication of the stability to water of th block examined; we shall see later that it is quite possible to achieve very good confineme of the radionuclides with a considerable amount of the waste contained being in solutio The volume of leachant/surface area-to-block ratio is between 10 and 20 cm in all cases.

TABLE 1—*Main features of waste forms studied in leaching facility.*

Specimen Composition: Matrix		Specimen Composition: Waste	Volume, L	Radionuclides Embedded (in Ci) (Total Initial Activity)	Leaching Test Duration, days
Concrete (AEC)	CPA55 sand aggregates adsorbent	evaporator concentrate ($NaNO_3$)	200	^{239}Pu: 0.16 ^{241}Am: 0.80 ^{137}Cs: 2.00 ^{90}Sr: 0.20	1500
Concrete (AEC)	CPA55 sand	evaporator concentrate (H_3BO_3, NaOH)	110	^{137}Cs: 0.007 ^{9}OSr: 0.006 ^{60}Co: 0.004	1215
Concrete (EEC-RFA)	pozzolanic cement	evaporator concentrate ($NaNO_3$...)	170	^{239}Pu: 0.005 ^{241}Am: 0.002 ^{137}Cs: 0.020 ^{90}Sr: 0.010 ^{60}Co: 0.001	446
Polymer (CEA)	epoxy resin sand	solid waste	200	^{239}Pu: 0.006 ^{241}Am: 0.018 ^{238}Pu: 0.111 ^{244}Cm: 0.030	360
Cement (AEC)	bituminous-emulsion	pre-compacted solid waste	870	^{239}Pu: 1.48 ^{241}Am: / ^{238}Pu: 0.39	1865
Cement polymer (AEC)		solid tritiated waste		^{3}H: 200	367
Bitumen (AEC)	Mexphalte 40/50	evaporate concentrate ($NaNO_3$...)	170	^{239}Pu: 0.27 ^{241}Am: 0.20 ^{90}Sr: 0.8	1785
Bitumen (AEC)	Mexphalte 40/50	Sludge	75	^{239}Pu: 0.014 ^{137}Cs: 0.83 ^{90}Sr: 0.026 ^{60}Co: 0.016	1255
Concrete and concrete container (EdF)		evaporator concentrate (borates)	2000	^{54}Mn: ^{137}Cs: traces ^{60}Co: 0.003	120

In order to achieve the most exhaustive leaching balance possible, we also make additional spectrometer measurements:

(*a*) on the "bottom of tank" sludge (inside of the leaching loop), which shows any mechanical deterioration of the block over a period of time and
(*b*) on the tank flushing solutions, used to recover all deposits from the walls of the bank (fixed contamination).

All these different measurements are used to calculate the Annual Leached Fraction (ALF), for which the equation is

$$\text{ALF} = \Sigma \frac{a_n}{A0}$$

TABLE 2—*Average chemical composition of leachant (simulation "center of the column" water).*

Component	Quantity in mg · L^{-1}
NaCl	117
$Ca(HCO_3)_2$	203
$K2\ SO_4$	87
$Mg(NO_3)_2$	39.8
$CaSO_4$	34
$MgSO_4$	13

and the corresponding leaching rate

$$v = \Sigma \frac{a_n}{Ao} \times \frac{V}{S} \times \frac{1}{\Sigma\, t_n}$$

Today we have two cells, one of which receives blocks with a dose rate on contact limited to 10 rad/h, the other with a dose rate on contact of up to 200 rad/h. The "low-activity" cell comprises twelve work loops which can accept blocks between 200 and 2000 L, and twelve stations for smaller blocks (for research and development). The "medium activity" cell contains two shielded test loops. The tanks are in fact designed on the model of certain transfer casks, thereby facilitating the transfer of blocks. The leachant is supplied by a small plant with a storage tank of a few cubic millimetres.

Leaching Test Station

It is necessary to know that the leaching facilities for full-scale at the Nuclear Research Center of Saclay have undergone considerable modifications since entry into service at the start of 1979, in order to enable more realistic *in situ* tests to be carried out. One modification was to air-condition the two cells to ensure a constant temperature throughout the experiments. Figure 1 shows a typical work station and Fig. 2 a more general view of the leaching facility.

Examples of Results and Leaching Balance

The leaching tests on full-scale blocks, which we have carried out in our facility for several years now, suggest comments which, although they are not of a general nature, nevertheless point the way to certain trends, as the examples which follow will illustrate.

Leaching of the Cement Block: Pressurized Water Reactor (PWR) Evaporator Concentrate Block—This test, which lasted 1215 days, was performed on a block with a hydraulic binder sand and lime matrix, containing evaporate concentrate (boric acid, neutralized by sodium hydroxide) bearing ^{137}Cs and ^{90}Sr. At the end of the test on this 110-L block, we drew up a balance of the activity released, which is given in Table 3.

Leaching of Cement: Spent PWR Fuel-Cladding Hulls Block—These tests, which lasted 333 days, were performed on two small blocks (2 L), composed of a cement-sand matrix for encapsulation of fragments of spent PWR fuel-cladding hulls (German Reactor Obrig-

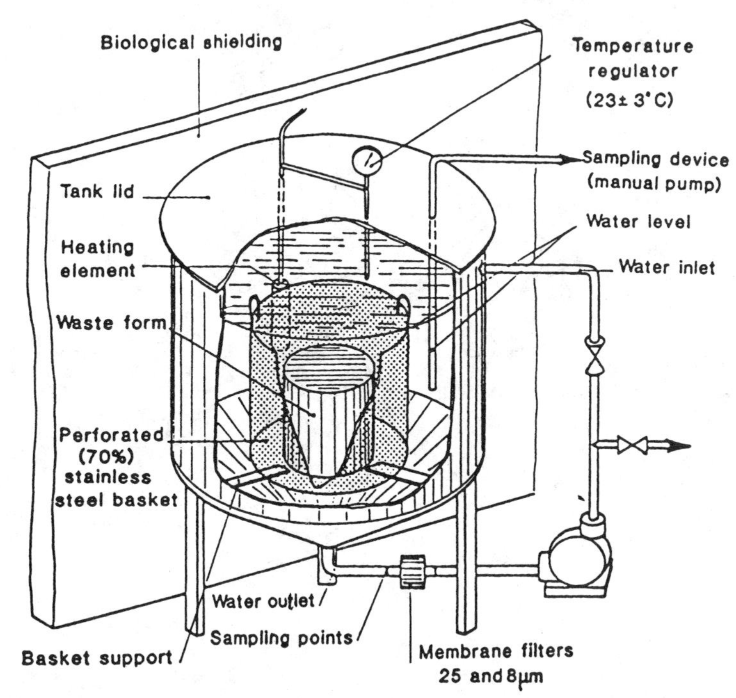

FIG. 1—*Leaching tank at the Saclay station.*

heim), which were β-γ and α-emitters, with a dose rate on contact of around 300 rad/h. After analysis and interpretation of the results, we noted the following:

(*a*) good behavior similarity between two blocks,
(*b*) leaching of β-γ is controlled by the release of ^{137}Cs; release of the ^{60}Co is not significant, mainly because of its low initial activity. Insofar as concerns the ^{137}Cs, the order of magnitude of the average leaching rate throughout the whole duration of the experiment is 10^{-4} g × cm^{-2} × d^{-1}; in the *activity balance* drawn up, the leachant's share is predominant and leaching is de facto majority governed by the initial phenomena.
(*c*) the α-leaching is very small (average rate at 333 days of 6.10^{-7} g × cm^{-2}, d^{-1}); in the *activity balance* drawn up, the activity deposited on the loop can reach more than 90% of the total of the activity released.

Leaching of a Bitumen: Evaporator Concentrate Block—This test, which lasted 1785 days, was performed on a bitumen matrix block, containing evaporator concentrate [sodium nitrate ($NaNO_3$), essentially] and bearing α-emitters, which was dry before encapsulation. The 170-L block initially contained 37% in weight of salts. At the end of the test, we attempted to draw up a *balance of the activity released,* as shown in Table 4. Examination of these values shows that the activity released is shared between the activity of the leachates and that fixed by contamination to the walls of the leaching tank, *which can even be very much higher than that of the leachates* (case of the α-emitters). On the other hand, the activity fixed to the sludge at the bottom of the tank is negligible, compared to the other activities.

It is also shown that of all events for β-γ emitters, the highest fraction of activity released

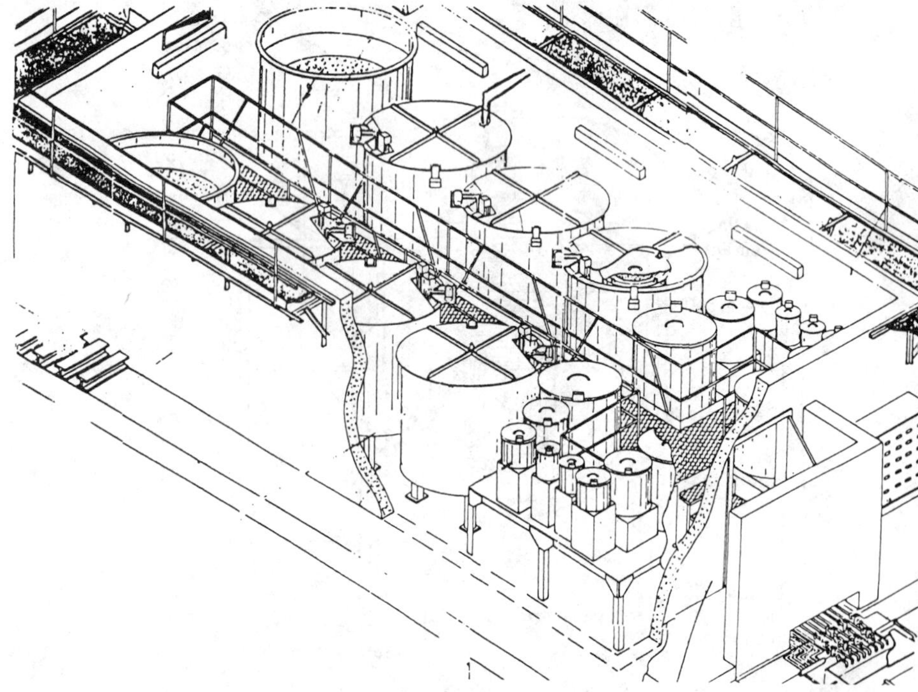

FIG. 2—*Leaching facility for full-scale, low-level-waste blocks.*

during a leaching test is due to the leachates (a_l); the fractions of activity released from sludges, deposits, or filters are almost negligible; the fraction concerning contamination of the tank (decontamination carried out using 4 N nitric solution kept in contact with the tank for 24 h) by the leachate is also negligible. This means that, in principle, the values obtained using the leachates only are the values which are significant of leaching of the block under the test conditions obtained.

Leaching of a Polymer: Solid Waste Block—This test was performed on a 200-L block with an epoxy plus sand matrix, designed to encapsulate technological waste (metal, glass, plastic). After the first 360 days of leaching, we drew up an initial balance of the radioactivity released, which is given in Table 5. The highest fraction of radioactivity released is again in the part fixed to the leaching tank by contamination.

The examples given above clearly demonstrate the importance of making the most com-

TABLE 3—*Balance of activity released from a borate-cement block.*

Radionuclide	Initial Activity, $A0$, Ci	Leachate Activity, a_l, Ci	Sludges Activity, a_s, Ci	Decontamination Activity, a_c, Ci	Filters Activity a_f, Ci
^{137}Cs	6.3×10^{-3}	3.3×10^{-4}	7.7×10^{-7}	1.5×10^{-7}	7×10^{-6}
^{90}Sr	6.0×10^{-3}	$\leq 7.5 \times 10^{-6}$	4×10^{-8}	4.6×10^{-7}	not measured

TABLE 4—*Balance of activity released from a nitrate-bitumen block.*

Radionuclide	Initial Activity, $A0$, Ci	Leachate Activity, a_l, Ci	Sludges Activity, a_s, Ci	Decontamination Activity, a_c, Ci
^{239}Pu (+^{240}Pu)	0.27	4.9×10^{-5}	1.2×10^{-6}	10^{-6}
^{241}Am (+^{238}Pu)	0.20	2.5×10^{-6}	1.2×10^{-8}	10^{-5}
^{90}Sr	5.4×10^{-3}	$\leq 1.5 \times 10^{-6}$	not detected	10^{-7}

plete analysis possible in order to obtain release balances. In particular, we note that in the case of α-emitters, although the leachate activity is often small, as is that of the sludge, the same does not apply to the activity fixed by contamination. This activity often represents more than 90% of the total activity released by the blocks. On the other hand, in the case of ^{137}Cs-type emitters, only the leachates are representative of the activity released. In this case, the leachate activity can reach more than 90% of the total activity evacuated from the block.

Post-Leaching Expert Appraisal

The leached blocks are often examined by experts after leaching as a complement to the information already obtained. These examinations, which are both destructive and non-destructive, generally start with a visual inspection of the structure. The block is then either sectioned by sawing or core sampled to show any faults such as cracks, air pockets and foreign bodies. Nondestructive tests may be made at this point, as illustrated by Fig. 3, which shows the relative variation in distribution of ^{137}Cs and ^{60}Co in a core sample taken transversely from the concrete block leached for 1215 days. The results were obtained by γ-spectrometry measurement, using a collimated germanium-lithium detector. We note that the thickness of the layer disturbed by leaching is 50 to 60 mm in places. This type of examination also shows the distribution of radionuclides in the matrix (homogeneity in the block).

A second example is given in Table 6, showing measurement of the residual quantitites of salts and α-emitters ($NaNO_3$ traced with ^{239}Pu and ^{241}Am) in a block with a bitumen matrix, leached for 1785 days. We first took a core sample from the block, using a cooling system, and then sawed up the sample obtained. The outer parts of the block (external surface) and the central parts were placed in solution with mineralization-extraction and then analyzed.

TABLE 5—*Balance of activity released from a polymer-solid waste block.*

Radionuclide	Initial Activity, $A0$, Ci	Leachate Activity, a_l, Ci	Sludges Activity, a_s, Ci	Decontamination Activity, a_c, Ci	Filters Activity, a_f, Ci
^{239}Pu (+^{240}Pu)	6.5×10^{-2}	4.3×10^{-6}	10^{-11}	4.5×10^{-5}	4.3×10^{-6}
^{241}Pu (+^{238}Pu)	1.3×10^{-1}	5.3×10^{-5}	6×10^{-10}	15×10^{-6}	7.5×10^{-5}

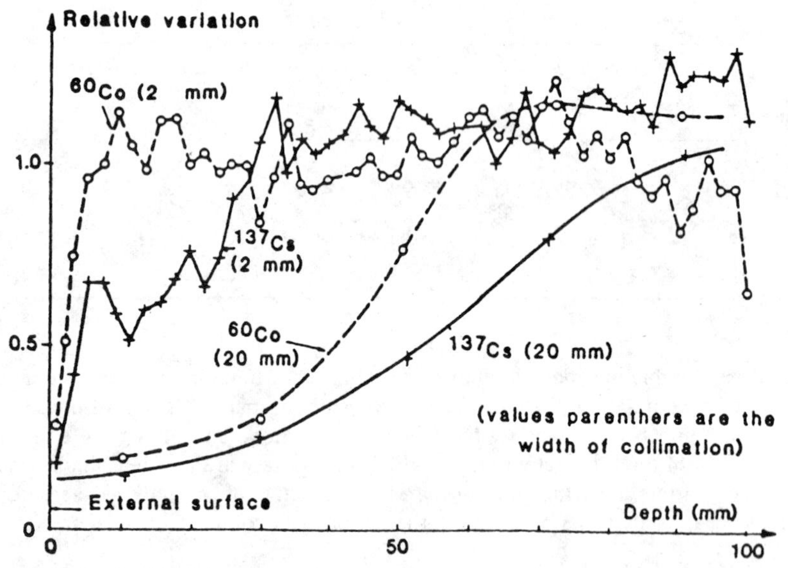

FIG. 3—*Post-leaching examination of drilled specimens by continuous γ-scanning.*

We noted that the outer parts showed depletion of α-emitters, thus confirming the results of leaching, and the same applied to the final salt content. We would note in this respect that the salt content of the central part (27.8%) was much less than the initial declared content (37%); we cannot exclude the fact that core sampling in water-splash core drilling may be partly responsible for this result.

Finally, Fig. 4 shows a cement-borate block, which was leached for more than 1200 days and which has been cored. Figure 5 shows a cement-nitrate block after leaching for 1500 days.

Example of Action in Research and Development

In order to improve the encapsulation formula and to select the most appropriate composition, we must proceed very rapidly with short-duration leaching tests. Faced with this

TABLE 6—*Post-leaching control of core drilled samples taken from a $NaNO_3$-bitumen matrix.*

Sample Position	Weight, g	Salt Content[a] Initial	Salt Content[a] Final	Radionuclides Activity[b] ^{238}Pu Initial	^{238}Pu Final	^{241}Am Initial	^{241}Am Final
External right	71.7	...	25.6	...	1.2	...	1.06
Center	41.5	37	27.8	1.6	1.3	1.18	1.28
External left	77.5	...	23.4	...	1.2	...	1.00

[a] Expressed in percent by weight of the waste form.
[b] Expressed in 10^{-3} Ci · kg^{-1} of waste form.

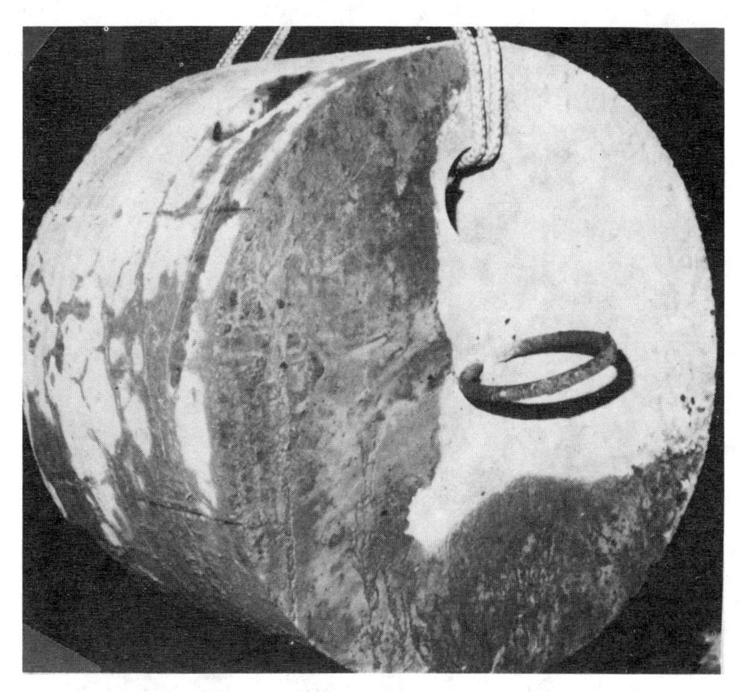

FIG. 4—*Cement-borate block of 110 L after 1215 days of leaching tests; coring operation.*

situation, it is permissible, for reasons of simplicity, cost, and available time, to carry out leaching tests on laboratory samples, which are easier to prepare than full-scale blocks, but whose representatively may need to be treated with caution. It was in a context such as this that part of our work in this field examined what is known as the "scale-up effect."

The first experiments were made using hydraulic binder-based blocks of different sizes (200, 20, 4 by 2, and 0.2 L), doped with ^{137}Cs with the same specific activity (10 mCi/L) and tested for over one year. The leaching test, carried out under strictly identical conditions for all the specimens (same curing time, ratio of the volume of leachate to the surface area of the blocks constant and equal to 15 cm), was preceded by preliminary shaping tests used in order to check for perfect homogeneity of the various samples (density and doping distribution measurements, etc.). After eight months experimentation, a number of provisional conclusions were reached. Among the more important were

1. There is an initial leaching period (non-totally diffusional process) with a duration located between 60 and 90 days and independent of the size of the specimen.
2. The four 2-L specimens, in which diameter/height ratio is not similar, showed a total similarity in their leachability; results are known with errors within a few percent.
3. It is possible to proceed at an extrapolation from the 2- and 20-L samples to the full-size blocks, using the volume/surface (V/S) ratio as a correcting factor; errors generated do not exceed the range 20% to 50%.
4. The 0.2-L specimen is out of range, showing a singular behavior; no explanation is available today, and the V/S correction is not adapted. We think (verification has not been possible) that other phenomena distinct from the "dimensional scale effect" interfere; for example, differences in the structures, that is, in the porosity spectrum,

FIG. 5—*Cement-bitrate block of 200 L after 1500 days of leaching tests.*

of the internal microcrack distribution, between the different sizes of specimens and in relation to other parameters such as thermal gradient developed during the setting period of the hydraulic binder, can influence the results.

Conclusion

The full-scale leaching tests conducted on radioactive waste blocks during several years at the leaching facility of the Nuclear Research Center of Saclay lead to the following conclusions:

1. In spite of some difficulties encountered during the experimentations, it has been possible to make a judgment on the quality of the package by establishing the more exhaustive balance of the total activity released, especially with α-emitters.
2. Appraisals of the post-leached packages are very important; they enhance the knowledge on their behavior under leaching tests with realistic conditions.
3. Research and development, like scale-up effect studies, showed that extrapolation of leaching results from small laboratory samples to full-size blocks is not always easy.

For these reasons, the long-term, full-scale leaching tests are to be continued.

Bibliography

[*1*] Nomine, J. C., Bernard, A., and Farges, L., "Long-Term Leaching Tests on Full-Scale Blocks of Radioactive Wastes," presented at the Oak Ridge National Laboratory Conference on the Leachability of Radioactive Solids, Gatlinburg, TN, 9–12 Dec. 1980.

[2] Marcaillou, J., Faure, J. C., Bernard, A., and Nomine, J. C., "Pratique en Matière de Conditionnement des Déchets Solides Radioactifs en Vue de Leur Stockage Temporaire sur le CEN-Cadarache en Attente d'Évacuation," presented at the AIEA-CCE Conference, Utrecht, the Netherlands, 1–25 June 1982.
[3] Nouguier, H., Nomine, J. C., and Vaunois, P., "Example of a Quality Control Operation Performed on a Nuclear Reactor Waste Package," Waste Management, Tucson, AZ, March 1983.
[4] Nomine, J. C. and Vejmelka, P., "Experience with Full-Scale Leaching of Low and Medium Level Waste," presented at the CEC Seminar on Testing, Evaluation and Shallow Land Burial of Low and Medium Radioactive Waste Forms, Geel, Belgium, Sept. 1983.
[5] Pottier, P. E. and Glasser, F. P., "Characterization of Low and Medium Level Radioactive Waste Forms," CEC Final Report EUR 10579, 2nd program, 1980–1984.
[6] Dayal, R., Arora, H., and Morcos, N., "Estimation of Cesium-137 Release from Waste/Amount Composites Using Data from Small-Scale Specimens," NUREG CR 3382, Nuclear Regulatory Commission, 1983.
[7] Becker, W. et al., "Correlation of ^{137}Cs Leachability from Small-Scale (Laboratory) Samples to Large-Scale Waste Forms," NUREG CR 2617, Nuclear Regulatory Commission, 1982, p. 41.62.
[8] Morcos, N. et al., "Correlation of ^{137}Cs Release from Small Scale to Large Scale Cement Waste Form," Scientific Basis for Nuclear Waste Management.

P. F. McIntyre,[1] *S. B. Oblath,*[2] *and E. L. Wilhite*[2]

Large-Scale Demonstration of Low-Level Waste Solidification in Saltstone

REFERENCE: McIntyre, P. F., Oblath, S. B., and Wilhite, E. L., **"Large-Scale Demonstration of Low-Level Waste Solidification in Saltstone,"** *Environmental Aspects of Stabilization and Solidification of Hazardous and Radioactive Wastes, ASTM STP 1033,* P. L. Côté and T. M. Gilliam, Eds., American Society for Testing and Materials, Philadelphia, 1989, pp. 392–403.

ABSTRACT: The saltstone lysimeters are a large-scale demonstration of a disposal concept for decontaminated salt solution resulting from in-tank processing of defense waste. The lysimeter experiment has provided data on the leaching behavior of large saltstone monoliths under realistic field conditions. The results also will be used to compare the effect of capping the waste form on contaminant release.

Three saltstone lysimeters were installed in December 1983 and January 1984, each containing a buried monolith formed from 9500 L of decontaminated salt solution solidified with a blended cement. The decontaminated salt solution was produced using actual Savannah River Plant waste. One lysimeter has a gravel cap and another has a clay cap placed over the buried monolith. The monolith in the third lysimeter is uncapped. Each lysimeter has provision to collect leachate from a sump below the buried monolith and soil moisture samplers for gathering samples adjacent to the waste form.

Biweekly monitoring of sump leachate from all three lysimeters has continued on a routine basis for approximately three years. The uncapped lysimeter has shown the highest levels of nitrate and ^{99}Tc release. The gravel-and-clay-capped lysimeters have shown levels equivalent to or slightly higher than background rainwater levels.

Mathematical model predictions are compared with lysimeter results and are applied to predict the impact of saltstone disposal on ground-water quality.

KEY WORDS: leaching, lysimeter, saltstone

High-level nuclear wastes are stored in large underground tanks at the Savannah River Plant (SRP). Processing of these wastes for ultimate disposal began in 1988. Nuclear wastes are processed to separate the high-level radioactive fraction from the low-level radioactive fraction. The separation is made in existing waste tanks by a process combining precipitation with tetraphenyl borate to remove cesium, adsorption onto titanate to remove plutonium and strontium, and filtration [*1*]. The high-level fraction is vitrified into borosilicate glass in the Defense Waste Processing Facility (DWPF) for disposal in a federal repository [*2–4*]. The low-level fraction (decontaminated salt solution, ~375 million litres) is mixed with a blend of cementitious blast furnace slag and fly ash [*5*]. The resulting waste form, called "saltstone," is disposed of on-site by emplacement in an engineered facility.

[1] E. I. duPont de Nemours and Company, Marshall Labs, Greys Ferry Road, Philadelphia, PA 19103.

[2] E. I. duPont de Nemours and Company, Savannah River Laboratory, Aiken, SC 29808.

Saltstone Lysimeter Design

The saltstone lysimeters used in this experiment are field lysimeters. A field lysimeter measures infiltration of water and migration rate of soluble components through soil by collection and analyses of leachate from a drainage sump.

Each buried saltstone monolith was constructed using 9500 L of actual radioactive waste consisting of decontaminated salt solution (~70 μCi/L radioactivity and pH >12.5), which was solidified with a blended cement (80 weight percent ASTM Class C fly ash, 20 weight percent API Class H cement) [*6*]. Table 1 shows the average composition of the decontaminated salt solution used in the saltstone formulation. The monoliths were formed by pouring the saltstone grout into a trench formed inside an earthen basin lined with Hypalon®. The liner in each of the three basins was 45 mil in thickness and was one continuous piece so that no seams were present. Each liner was leak-tested by the vendor before shipment and hydrostatically tested by Savannah River Laboratory (SRL) after installation. Figure 1 is a diagram of the saltstone lysimeters showing the relative position and dimensions of the buried saltstone monoliths.

Three separate lysimeters were constructed to evaluate the performance of saltstone disposal in earthen trenches. The first lysimeter, called the uncapped lysimeter, contains only a buried saltstone monolith. The second lysimeter, or clay cap lysimeter, contains a buried saltstone monolith covered with a clay cap made of 18% bentonite mixed with local soil. The third lysimeter, the gravel cap lysimeter, contains a buried saltstone monolith covered with a cap made of pea-sized gravel. The soil used to fill the lysimeter basins is a sandy-clay soil with saturated hydraulic conductivity of about 1×10^{-4} cm/s^{-1} and porosity of 42%. Laboratory studies of the clay cap material showed a porosity of 40% and saturated hydraulic conductivity of about 3×10^{-9} cm/s^{-1}. Figure 1 shows that the gravel and clay caps covered only the area above the saltstone monoliths. The caps act as an umbrella to divert infiltrating rainwater from direct contact with the monoliths while permitting collection of all infiltration in the sump.

The rainwater that percolated through the soil was collected in individual prefabricated cement sumps installed at the bottom of each lysimeter. Each sump was filled with a

TABLE 1—*Chemical and radionuclide composition of decontaminated salt solution.*

Chemical	Weight %	Radionuclide	nCi/g of Solution
H_2O	70	^{99}Tc	35
$NaNO_3$	17.7	^{3}H	12
Na_2CO_3	2.1	^{106}Rh	6.6
$NaAl(OH)_4$	2.6	^{106}Ru	6.6
NaOH	3.6	^{125}Sb	3.5
Na_2SO_4	1.7	^{137}Cs	1.7
$NaNO_2$	0.9	^{137m}Ba	1.7
$CaSO_4$	0.24	^{90}Sr	<0.3
NaCl	0.005	^{90}Y	<0.3
$NaB(C_6H_5)_4$	0.04	^{239}Pu	<0.4
NaF	0.004		
$NaSiO_3$	0.006		
Na_2CrO_4	0.006		
Na_2MoO_4	0.0004		
NaHgO(OH)	2×10^{-6}		
$NaAg(OH)_2$	$<1 \times 10^{-7}$		

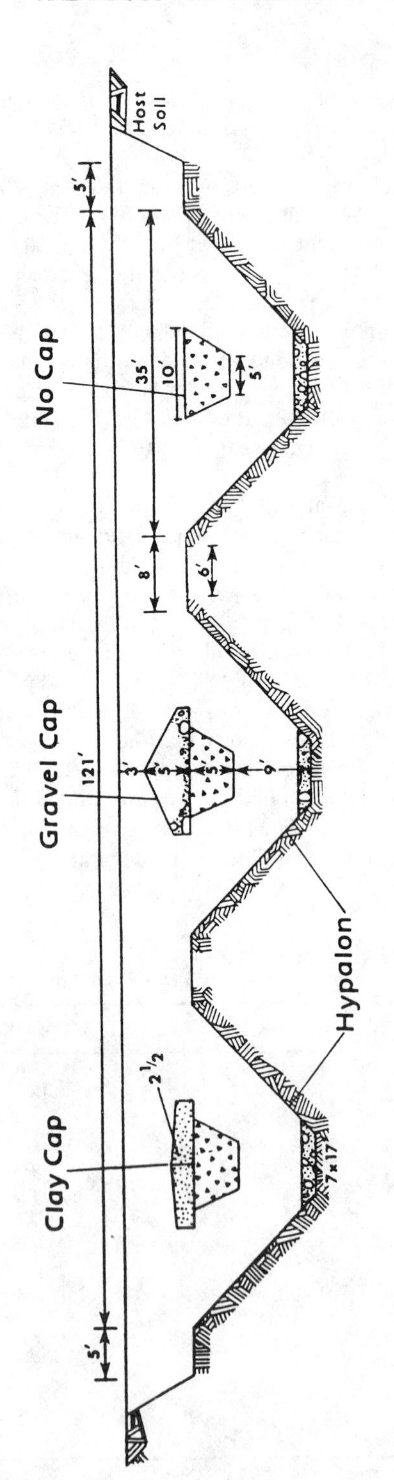

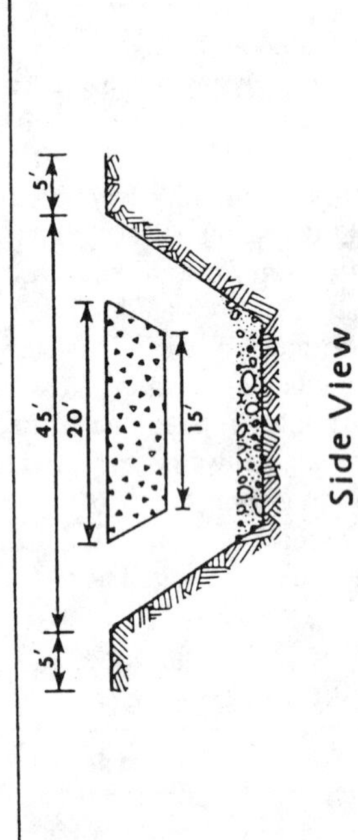

FIG. 1—*Schematic of saltstone lysimeters.*

0.3-m (1 ft) layer of smooth, pea-sized gravel to facilitate drainage to the sump. The plumbing system consisted of a 20-cm-diameter polyvinyl chloride (PVC) pipe which extended from the soil surface down into each lysimeter sump. A submersible pump was lowered down each of the pipes and into their respective sumps. The pump and its associated tubing was constructed from PVC. None of the materials used to construct the plumbing system interfered with components being analyzed for from the three waste forms. Each lysimeter has its own pump and holding tank for independent evaluation of sump leachate. Samples were collected on a biweekly basis from a sampling tee in the plumbing system before the sump effluent reached the hold tank. This eliminated possible cross-contamination with previous samples being stored in the hold tank.

Soil moisture samplers were installed in each lysimeter basin within 15 cm of the monoliths prior to the pouring of the saltstone. The soil moisture samplers detected the early release of radionuclides and other diffusing species near the wasteform.

A rain gage was installed at the lysimeter site to measure accurately the amount of rainfall. Individual rain events as low as 0.25 mm could be measured with this instrument. The bargraph in Fig. 2 shows the amount of rainfall at the lysimeter site. Each bar represents the total amount of rainfall that occurred during each biweekly sampling interval. This provided essential data for modeling rain infiltration into the lysimeter. Samples of rainwater were also submitted for chemical analysis to provide background levels for all components measured. A listing of background levels for nitrate, mercury, calcium, sodium, sulfate, gross alpha, technetium and tritium is given in Table 2. These levels represent an average of 48 biweekly rainwater samples which were collected and measured between January 1984 and February 1987.

Sampling Techniques and Analytical Procedures

Starting in August 1984, biweekly samples of leachate were collected from the sumps of each saltstone lysimeter. Chemical analysis of each sample for major anions (nitrate, nitrite, and sulfate) was performed using ion chromatography. Inductively coupled plasma-emission spectroscopy was used to determine concentrations of major cations such as calcium, sodium, and chromium present in samples acidified with ultrapure nitric acid. Analysis of each sample for mercury was performed using atomic absorption spectroscopy. The pH and conductivity of each sample was also measured.

Analysis of radiochemical species included the measurement of gross alpha and gross nonvolatile beta. This was done by concentrating 100 mL of sample to dryness followed by dissolution in 5 mL nitric acid and evaporation onto a counting planchet. These samples were analyzed using a Baird low-background alpha-beta-gamma spectrometer and counting each sample for 30 min. The tritium level in each sample was determined by distilling 30 mL of the sample and collecting the first 5 to 10 mL of distillate. Three milliliters of the distillate was added to 20 mL of scintillation cocktail and counted on a Packard Model 4000 liquid scintillation counter.

Analysis was performed for benzene as a potential contaminant because of possible degradation of residual sodium tetraphenylborate used to precipitate cesium during in-tank processing. Monthly samples for benzene analysis were submitted using precleaned glass vials with tetrafluoroethylene septum caps. The analysis was performed using gas chromatography (purge and trap) with a sensitivity of 1 ppb.

The soil moisture samples were collected by placing a vacuum on the porous cap sampler for a set time period and using capillary action and suction to collect the sample. The soil moisture samples have been collected only periodically (about once a year) and were analyzed by the same methods as the sump samples, provided sufficient sample was collected.

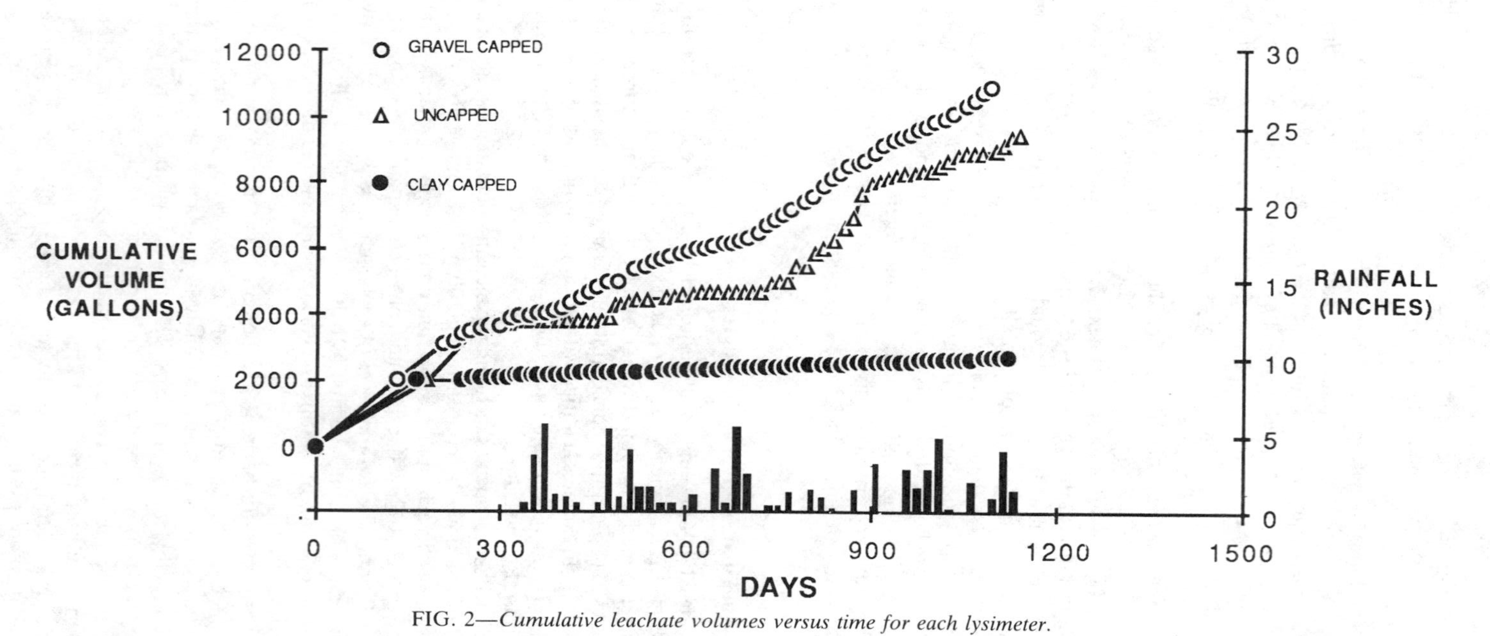

FIG. 2—*Cumulative leachate volumes versus time for each lysimeter.*

TABLE 2—*Average background levels of major components of saltstone obtained from rainwater analyses.*[a]

Component	Average Background Level
nitrate	<1.5 ppm
sulfate	2.0 ppm
calcium	0.5 ppm
sodium	1.0 ppm
mercury	<1.0 ppb
chromium	15 ppb
gross alpha	<1 pCi/L
technetium	15 pCi/L
tritium	35 nCi/L

[a] Levels listed represent average of 50 biweekly samples taken between January 1984 and February 1987.

For quality assurance, all data collected and reported were obtained using suitably calibrated instruments. The samples were analyzed with standards being submitted along with at least every other set of samples. Radionuclide analysis was performed on instruments calibrated against National Bureau of Standards (NBS) traceable standards. Sampling and analysis methods were partially in compliance with Environmental Protection Agency (EPA) guidance as found in SW-846 [7]. Sample treatment and analyses for metals (Methods 6010 and 7470) and benzene (Methods 8020 and 5030) conform to EPA guidance. Analyses for anions and radioactivity do not.

Results and Discussion

Disposing of chemical wastes in an engineered landfill is a common practice. The performance of such a facility is strongly dependent on the effectiveness of the waste form and the hydrology of the particular site. The primary goal is to maintain levels of dissolved components in the groundwater such that they do not exceed drinking water standards. The results reported in this paper reflect biweekly monitoring from August 1984 to January 1987. Continued biweekly monitoring of all three lysimeters is planned for the foreseeable future.

The uncapped lysimeter is the only one to show significant nitrate levels above background. Nitrate is a primary component of saltstone. Lab tests show that the leach rate for nitrate in saltstone is higher or as high as any other constituent. Its effective diffusion coefficient in saltstone has been determined to be 8×10^{-9} cm^2/s^{-1} [8]. The maximum nitrate concentration observed to date in the uncapped lysimeter is 209 ppm. The gravel-cap lysimeter is starting to show nitrate and technetium-99 (^{99}Tc) levels above background, as expected. The clay-cap lysimeter continues to show levels equivalent to background. Figure 3 shows a comparative plot of cumulative fraction of nitrate leached as a function of time for all three lysimeters.

Low levels of nitrite and sulfate continue to be detected in the leachate from all three lysimeters. The average nitrite level in each lysimeter is below the detection limit (<5 ppm). The average levels of sulfate detected are <5, <5, and 20 ppm for the uncapped, gravel-capped, and clay-capped lysimeters, respectively. The absence of nitrite in the uncapped lysimeter sump indicates its destruction (probably oxidation to nitrate) in the environment. Based on its concentration in the waste solution and its leaching properties, nitrite should have been observed above background levels.

Mercury concentrations have remained below the analytical detection limit of 1 ppb in

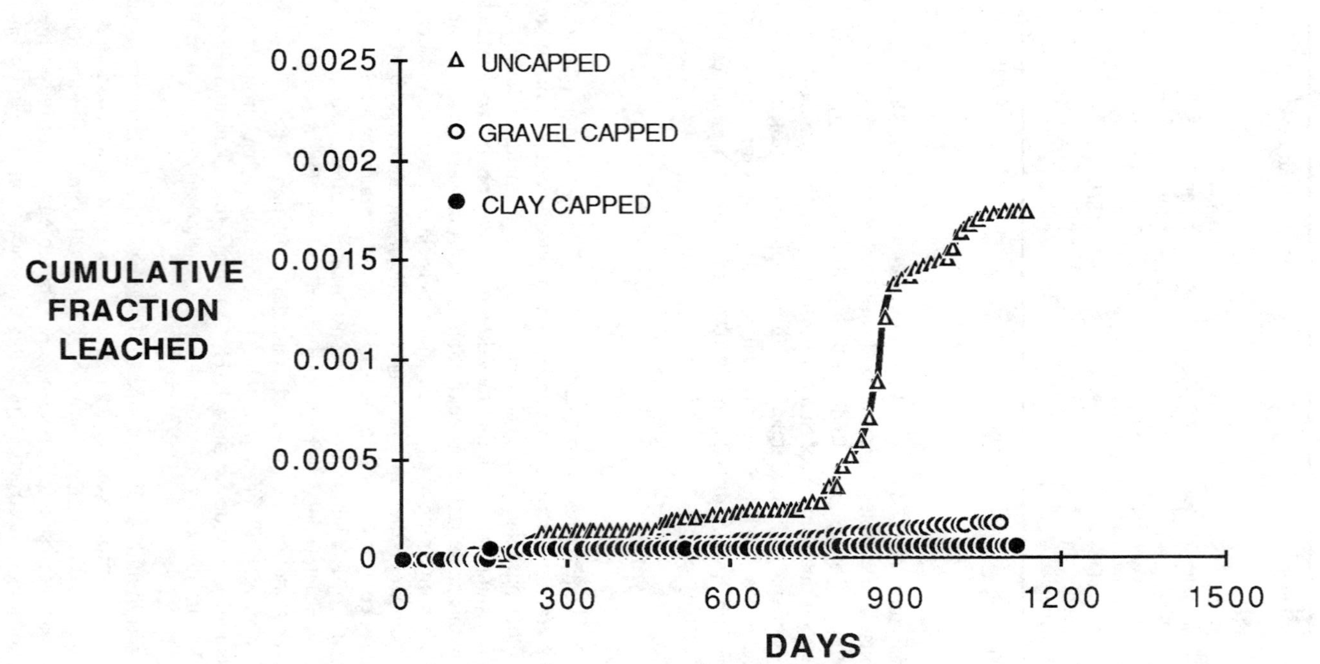

FIG. 3—*Cumulative fraction of nitrate leached versus time for all three lysimeters.*

the leachate of all three lysimeters. Average sodium concentrations measured in leachate from the uncapped, the gravel-capped, and the clay-capped lysimeters were 40, 3, and 6 ppm, respectively. The average concentration of chromium in leachate from the three lysimeters was also below the detection limit of 10 ppb. The average concentrations of calcium in leachate from the uncapped, the gravel-capped, and the clay-capped lysimeters were 5, 5, and 30 ppm, respectively.

The most abundant radionuclide detected in the sump leachate of the uncapped lysimeter is technetium-99. Radiochemical analysis performed at SRL has shown that the levels of nonvolatile beta measured are equivalent to ^{99}Tc levels. The maximum level observed in the uncapped lysimeter is 11.9 nCi/L, whereas levels in both capped lysimeters remain at background or slightly above. Figure 4 plots ^{99}Tc concentration as a function of time for all three lysimeters. A similar behavior was noted in release of nitrate and ^{99}Tc from the saltstone waste form of the uncapped lysimeter (see Figs. 4 and 5).

The second most abundant radionuclide present in saltstone is tritium, which is present in the decontaminated salt solution at a concentration of 12 nCi/g. Tritium is typically present in the infiltrating rainwater at levels of 35 nCi/L (see Table 2), which is the level observed in the sump leachate from the clay-cap and gravel-cap lysimeters. The uncapped lysimeter shows slightly higher tritium levels at 50 to 55 nCi/L. This increase above background can be attributed to hydrogen-3 diffusion from the wasteform and migration through the soil to the sump by infiltrating rainwater.

Analysis of leachate samples from each lysimeter show no detectable alpha activity (<1 pCi/L). This is consistent with the low level of alpha emitting species present in the decontaminated supernate (see Table 1).

Leachate samples from all three lysimeters have pH levels ranging from 5.8 to 6.2. The conductivity of leachate from the uncapped lysimeter is typically in the range of 140 to 170 μmho/cm [probably due to sodium nitrate ($NaNO_3$)]. The gravel-cap lysimeter has conductivity levels of 30 to 60 μmho/cm and the clay-cap lysimeter has levels of 100 to 180 μmho/cm (consistent with the higher calcium and sulfate levels).

Soil moisture samples were taken and analyzed during November 1986 from all three lysimeters (see Table 3). The results indicate significantly higher concentrations of species diffusing adjacent to the waste form versus levels measured in the sump leachate. The levels observed in the sump leachate of the capped lysimeters are much lower than the uncapped lysimeter, demonstrating the effectiveness of capping in reducing waste migration through the soil.

An unexpected result observed during this lysimeter test was a hydrologic imbalance between the clay-capped lysimeter versus the other two. It was expected that all three lysimeters would yield roughly the same volume of leachate. The uncapped and the gravel-capped lysimeters have yielded similar volumes, but volumes collected from the clay cap have been substantially lower.

Figure 2 plots cumulative volume of leachate collected versus time for all three lysimeters. Presently, there is no adequate explanation for the lower volume collected from the clay-cap lysimeter. This imbalance is being investigated.

Saltstone Release Modeling

Both numerical and analytical mathematical models have been developed to predict the releases of nitrate and other materials from various saltstone disposal designs. One model was developed at SRL [9], and the other was developed by Intera Technologies (Austin, Texas).

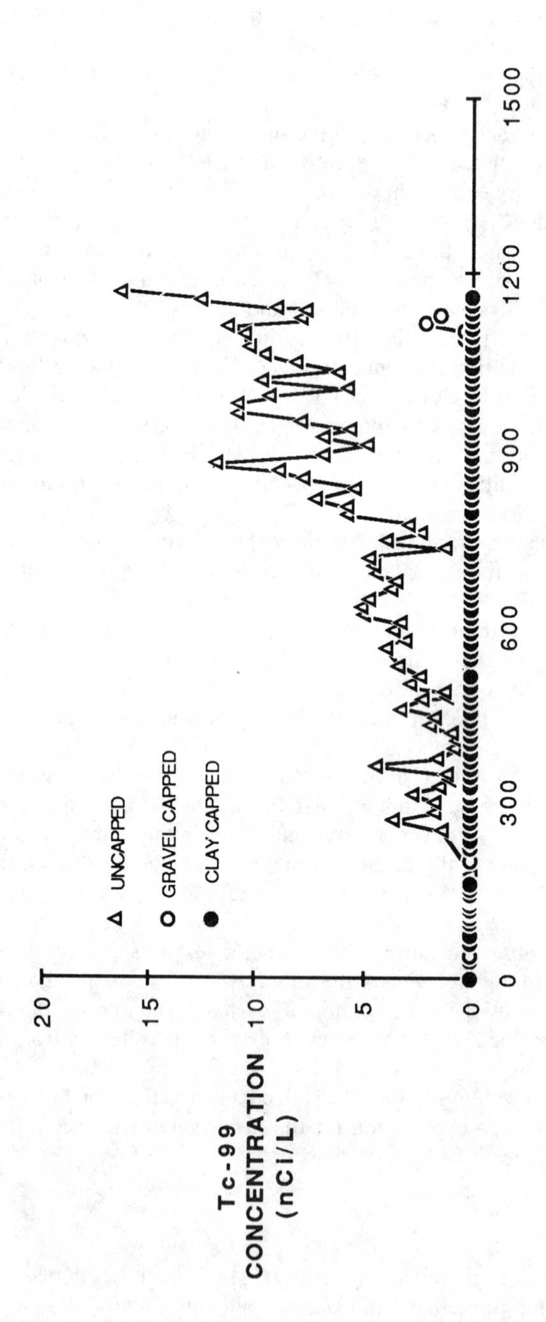

FIG. 4—^{99}*Tc concentration versus time for each lysimeter.*

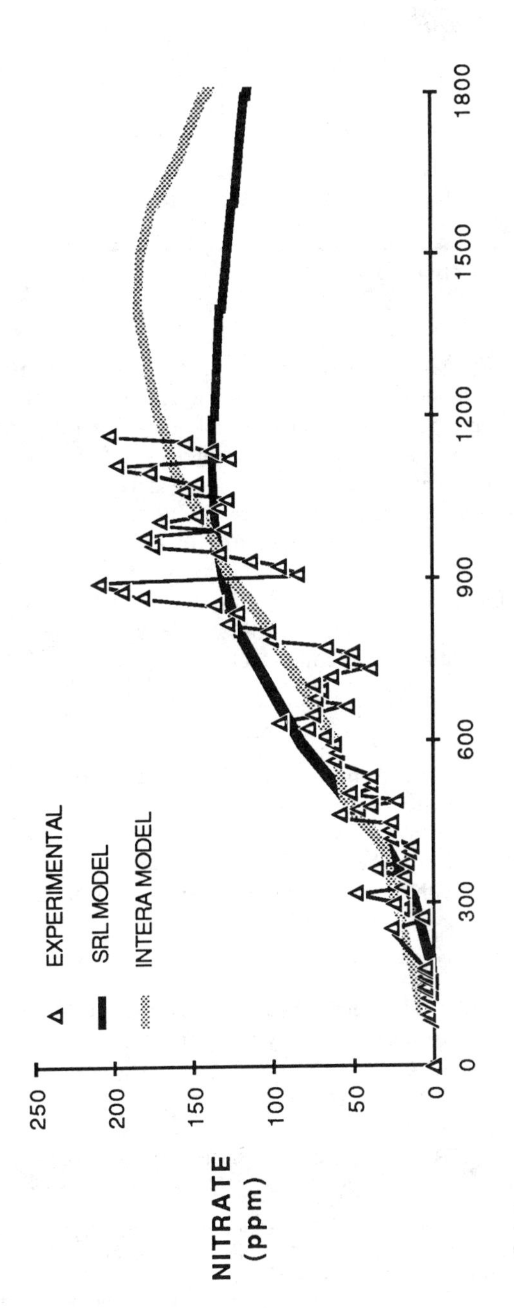

FIG. 5—*Comparison of models with nitrate release from uncapped lysimeter.*

TABLE 3—*Soil moisture sampler results for November 1986 from all three lysimeters.*[a]

	Uncapped Lysimeter	Gravel-Capped Lysimeter	Clay-Capped Lysimeter
NO_3 (ppm)	16 050	25 550	34 470
NO_2 (ppm)	4 950	1 480	<1 000
SO_4 (ppm)	2 960	3 620	2 400
Ca (ppm)	6	6	58
Na (ppm)	6 340	7 550	8 570
Cr (ppb)	<200	<200	<200
^{99}Tc mCi/L	1.1	1.2	1.5

[a] Results obtained from soil moisture samplers located in a region underneath each wasteform.

The models for transporting material released from the monolith were applied to the experimental results from the uncapped saltstone lysimeter. The SRL model assumed that the water moved through water saturated soil (potential flow). The Intera model (FAMOS) is a proprietary, three-dimensional, finite-difference model. FAMOS is a full two-phase model which simulates the movement of both air and water phases in unsaturated soil. Model inputs included measured rainfall, estimates of runoff, and evaporation, as well as measured unsaturated flow properties and diffusivities of soil, saltstone, and clay-cap materials.

A favorable comparison was achieved between the Intera model and experimental results (see Fig. 5). Predicted release levels using the SRL model are somewhat lower, but also show fairly good agreement with experimental results.

Conclusions

The experimental results identified nitrate and ^{99}Tc as the major components released from the large-scale saltstone waste forms. Capping of the waste forms significantly reduced the rate at which the released components are transported through the soil to the sump below. Modeling efforts continue to show good agreement with experiment. Monitoring of the lysimeter effluents will be continued indefinitely in support of modeling efforts.

Acknowledgment

The information contained in this article was developed under Contract No. DE-AC09-76SR00001 with the U.S. Department of Energy.

References

[1] D'Entremont, P. D. and Walker, D. D., "Tank Farm Processing of High-Level Waste for the Defense Waste Processing Facility," *Waste Management '87,* Vol. 2, R. G. Post, Ed., American Nuclear Society, Tucson, AZ, 1987, pp. 69–73.

[2] Boersma, M. D., "Process Technology for Vitrification of Defense High-Level Waste at the Savannah River Plant," *Proceedings,* American Nuclear Society Meeting—Fuel Reprocessing and Waste Management, Jackson Hole, WY, 26–29 Aug. 1984.

[3] Maher, R., Shafranek, L. F., Kelley, J. A., and Zeyfang, R. W., "Solidification of Savannah River Plant High-Level Waste," *American Nuclear Society Transactions,* Vol. 39, 1981, p. 228.

[4] Baxter, R. G., "Design and Construction of the Defense Waste Processing Facility Project at the Savannah River Plant," *Waste Management '86,* Vol. 2, R. G. Post, Ed., American Nuclear Society, Symposium on Waste Management, Tucson, AZ, 1986, pp. 449–454.

[5] Langton, C. A., Dukes, M. D., and Simmons, R. V., "Cement Based Waste Forms for Disposal of Savannah River Plant Low-Level Radioactive Salt Wastes," *Scientific Basis for Nuclear Waste Management,* Vol. 7, G. L. McVay, Ed., North Holland, NY, 1984, pp. 575–582.
[6] Wolf, H. C., *Large-Scale Demonstration of Disposal of Decontaminated Salt as Saltstone—Part 1: Construction, Loading and Capping of Lysimeters,* DPST-84-497, June 1984.
[7] "Test Methods for Evaluating Solid Waste," US EPA SW-846, U.S. Environmental Protection Agency, Washington, DC.
[8] Wilhite, E. L., "Waste Salt Disposal at the Savannah River Plant," *Proceedings,* American Nuclear Society International Meeting: Low, Intermediate, and High-Level Waste Management and Decontamination and Decommissioning, Niagara Falls, NY, 14–18 Sept. 1986.
[9] Pepper, D. W., "Transport of Nitrate from a Large, Cement-Based Waste Form," presented at the 6th International Conference of Finite Elements in Water Resources, Lisbon, Portugal, 1–5 June 1986.

Robert D. Rogers,[1] *John W. McDonnel,*[1] *Edward C. Davis,*[2] *and Melvin W. Findley*[3]

Field Testing of Waste Forms Using Lysimeters

REFERENCE: Rogers, R. D., McConnell, J. W., Davis, E. C., and Findley, M. W., **"Field Testing of Waste Forms Using Lysimeters,"** *Environmental Aspects of Stabilization and Solidification of Hazardous and Radioactive Wastes, ASTM STP 1033,* P. L. Côté and T. M. Gilliam, Eds., American Society for Testing and Materials, Philadelphia, 1989, pp. 404–417.

ABSTRACT: The Low-Level Waste Data Base Development—EPICOR-II Resin/Liner Investigation Program—funded by the U.S. Nuclear Regulatory Commission (NRC) is obtaining information on performance of radioactive waste in a disposal environment. This paper presents a description of the field testing and gives preliminary findings. Solidified ion-exchange resin materials from EPICOR-II prefilters used in the cleanup of the Three-Mile Island Nuclear Power Station are being field tested to develop a low-level waste data base and to obtain information on survivability of waste forms composed of ion exchange media loaded with radionuclides and solidified in matrices of cement and Dow polymer. Emphasis is placed on evaluating the requirements of 10 CFR 61 "Licensing Requirements for Land Disposal of Radioactive Waste" by obtaining data on performance of waste in a disposal environment using lysimeter arrays at Oak Ridge National Laboratory (ORNL) and Argonne National Laboratory in Illinois (ANL).

KEY WORDS: lysimeters, field testing, EPICOR-II, waste form, leaching, cesium, cobalt, antimony, strontium

Lysimeters are used in field testing because when properly designed they can be used to isolate and manipulate soil systems under actual environmental conditions. Lysimeters have been used for many years for field tests to determine the amount of leaching of various elements from soil due to percolation of water. In fact, the word lysimeter comes from the two Greek roots *lysi,* meaning loosening, and *meters,* to measure.

The shape and size of lysimeters are determined by the imagination and experimental requirements of the investigator. They can range from small volumes of field soil isolated from surrounding areas by impervious dividers to concrete or metal tanks. If tanks are used, they can consist of an upper and lower compartment and are placed into the field so that the open end is level with the soil surface. The upper compartment serves as a containment for soil while the lower level serves as a collection and storage compartment for water which has paste through the overlying soil. Those lysimeters without bottom compartments will have some porous material in the bottom such as gravel to separate percolating water from soil.

[1] Scientific specialist and senior engineering specialist, respectively, Biotechnology and Waste Management Programs, Idaho National Engineering Laboratory, P.O. Box 1625, Idaho Falls, ID 83415.

[2] Assistant soil scientist, Environmental Effects Research Program, Argonne National Laboratory, Argonne, IL 60439.

[3] Scientist, Environmental Sciences Division, Oak Ridge National Laboratory, Oak Ridge, TN 37831.

Lysimeters lend themselves to instrumentation. They can be placed on top of weighing evices so that water content can be determined gravimetrically and they can have various ater sampling, temperature, and moisture-sensing devices implanted as they are filled with oil. The amount of sophistication depends on experimental objectives.

Lysimeters are the obvious tool to use in the field-testing of solidified waste forms. In ne study reported here, the waste forms are composed of solidified ion-exchange resin naterials from EPICOR-II prefilters used in the cleanup of the Three-Mile Island Nuclear ower Station. They are being subjected to long term testing to develop a low-level waste atabase and to obtain information on survivability of waste forms composed of ion exchange nedia loaded with radionuclides and solidified in matrices of cement and Dow polymer. Emphasis has been placed on evaluating the requirements of 10 CFR 61 "Licensing Requirements for Land Disposal of Radioactive Waste" [1] by obtaining data on performance f waste in a disposal environment.

Methods and Materials

Waste forms used in the experiment contain a mixture of synthetic organic ion-exchange esins or the organic exchange resins mixed with an inorganic zeolite. Solidification agents

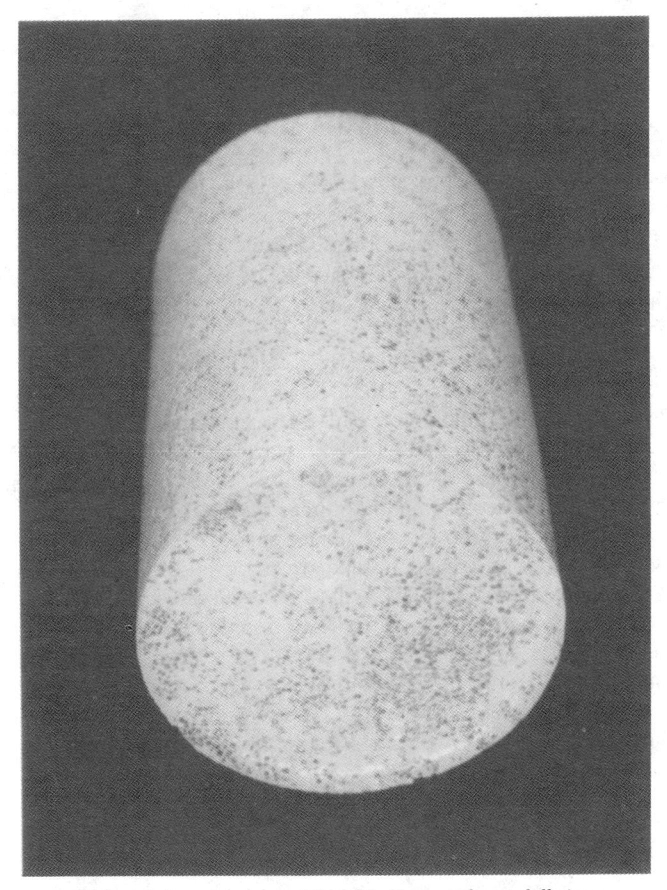

FIG. 1—*Example of an EPICOR-II waste form, full size.*

TABLE 1—*Lysimeter waste form inventory.*

Lysimeter	Number and Type of Waste Forms	Total Curie Content[a]	Lysimeter	Number and Type of Waste Forms	Total Curie Content[a]
ANL 1	2 C1A[b]		ORNL 1	2 C1	
	5 C1	3.5×10^{-1}		5 C1A	$3.5 \times 10^{-}$
ANL 2	7 C2B	15.3×10^{-1}	ONRL 2	7 C2B	$15.3 \times 10^{-}$
ANL 3	7 D1A[c]	5.2×10^{-1}	ORNL 3	7 D1A	$5.2 \times 10^{-}$
ANL 4	7 D2	20.6×10^{-1}	ORNL 4	7 D2	$20.6 \times 10^{-}$
ANL 5	4 C1A	3.5×10^{-1}	ORNL 5	7 C2B	$15.3 \times 10^{-}$
	3 C1				

[a] Waste forms have the following average curies contents:

	^{134}Cs	^{137}Cs	^{90}Sr	Total
D1 and D1A	4.38×10^{-3}	66.22×10^{-3}	3.92×10^{-3}	74.52×10^{-3}
D2 and D2A	18.22×10^{-3}	275.45×10^{-3}	0.64×10^{-3}	294.31×10^{-3}
C1 and C1A	2.95×10^{-3}	44.58×10^{-3}	2.64×10^{-3}	50.17×10^{-3}
C2A and C2B	13.53×10^{-3}	204.59×10^{-3}	0.47×10^{-3}	218.59×10^{-3}

[b] Portland Type II cement.
[c] Dow vinyl ester-syrene.

which were used to produce the 4.8-by-7.6-cm cylindrical waste forms (Fig. 1) examined ir the study were Portland Type I-II cement and Dow vinyl ester-syrene (VES) [*2*]. Sever of these waste forms were stacked end-to-end in each lysimeter [*3*] (five lysimeters a Oak Ridge National Laboratory [ORNL] and five at Argonne National Laboratory-East [ANL-E]) [*4,5*] to provide a 1-L volume. The inventory and approximate nuclide content of waste forms used in each lysimeter is found in Table 1.

Lysimeters used in this study were designed to be self-contained units which will be disposed at the termination of the 20-year study. Each is a 0.91-by-3.12-m right-circular cylinder divided into an upper compartment, which contains fill material, waste forms, and instrumentation, and a lower compartment, which collects leachate (Fig. 2). Four lysimeters at each site are filled with soil, while a fifth (used as a control) is filled with inert silica oxide sand. Table 2 provides textural data on the fill materials. Figure 3 shows the placement of the lysimeters. Instrumentation within each lysimeter includes five Teflon porous cup soil-water samplers and three soil moisture/temperature probes [*3*] (Fig. 2). The probes are connected to an on-site data acquisition and storage system (DAS), which also collects data from a field meteorological station located at each site [*3*].

Each month, data stored on a cassette tape are retrieved from the DAS and translated into an IBM PC-compatible disk file. These data are reduced using LOTUS for tabular and graphic displays. At least quarterly, water is drawn from the porous cup soil-water samplers and the lysimeter leachate collection compartment and selected samples are analyzed for beta- and gamma-producing nuclides. Soil moisture/temperature at three elevations in each lysimeter, along with a complete weather history, are recorded on a continuing basis by the DAS. Highlights of the results of testing are presented in this paper. A more detailed presentation of the data is given in Ref *6*.

Results and Discussion

Results obtained from 17 months of field testing are presented in this section. There was a period from late August 1985 until November 1985 when the ORNL DAS was inoperable, due to the necessity of returning the system to the manufacturer for repairs.

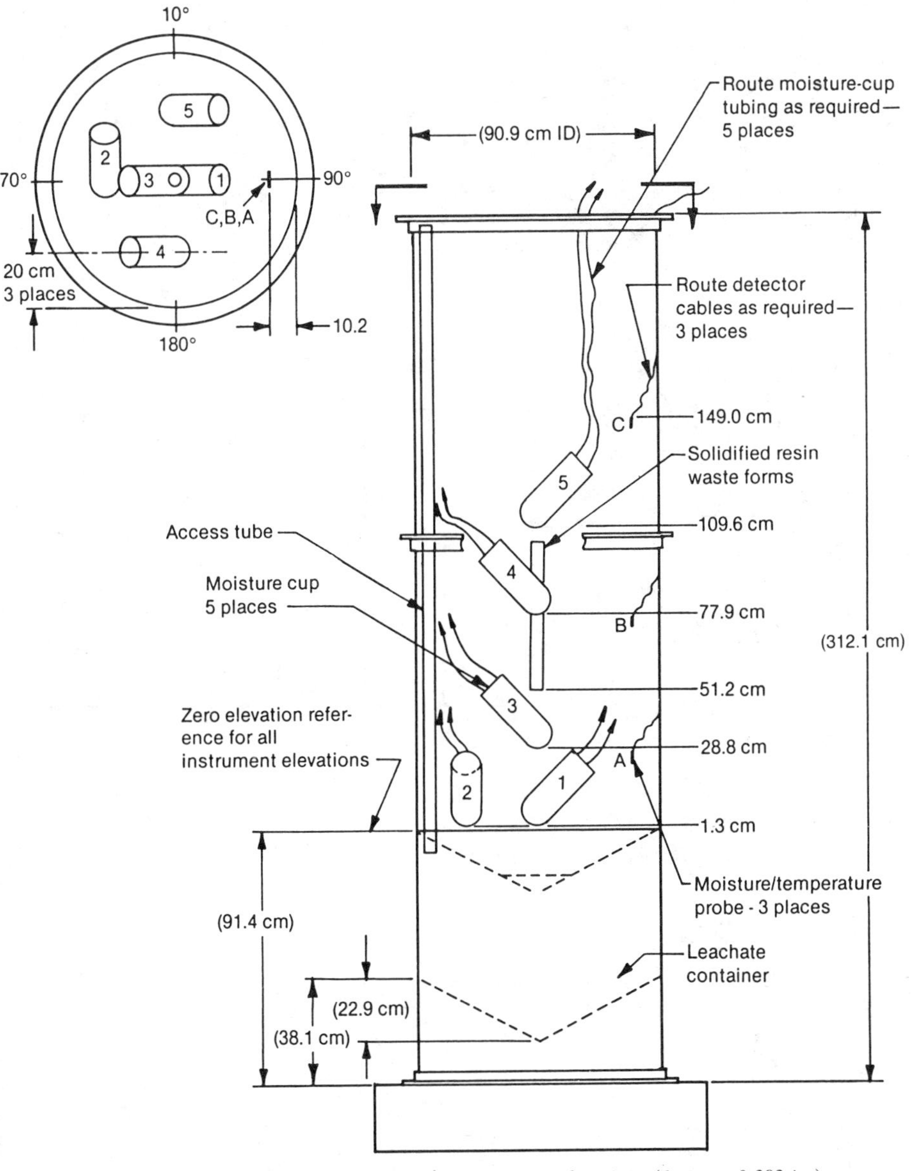

FIG. 2—*EPICOR-II lysimeter vessel with component locations (1 cm = 0.393 in.).*

Instrument Operation and Data

Wind speed, air temperature, relative humidity, and rainfall are recorded over a 12-month period by the DAS systems for the ANL and ORNL sites. Air temperature data from ANL (Fig. 4) show that there were days of freezing temperatures from mid-November 1985 until mid-March 1986, while there were very few days with air temperatures of 0°C at ORNL (Fig. 5). Rainfall data from the ORNL site appeared to be higher than normal. This trend became apparent during December 1985, and indications were than the Weather Measure tipping bucket rain gage supplied with the DAS system was not capable of accurately

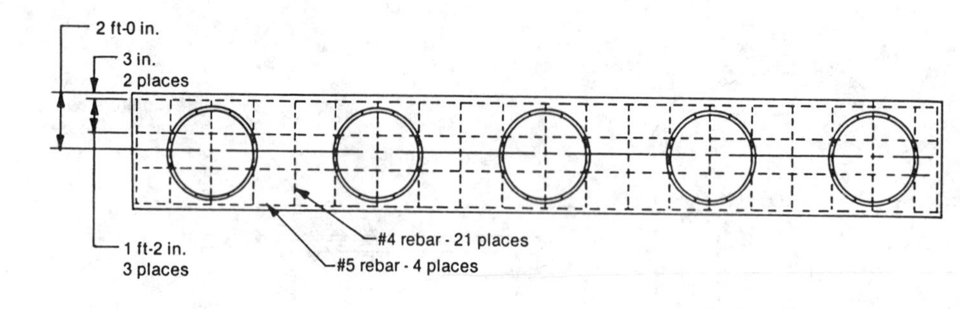

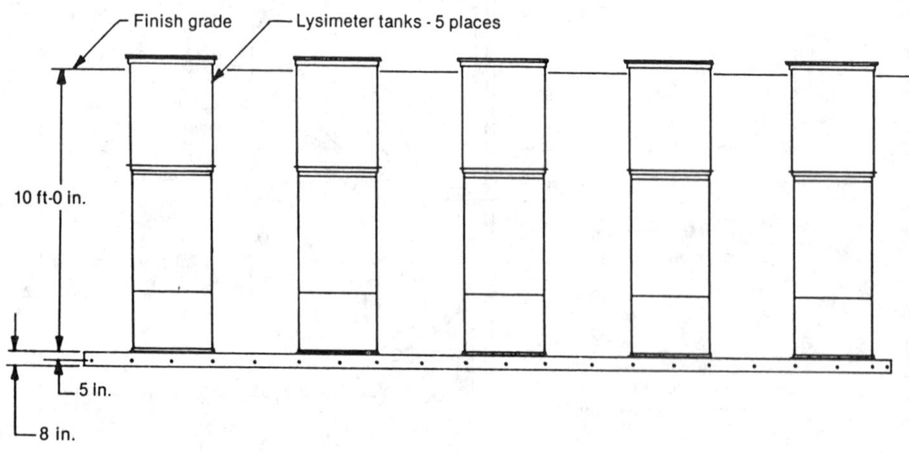

FIG. 3—*Placement of EPICOR-II lysimeters (1 ft = 0.3048 m; 1 in. = 2.54 cm).*

TABLE 2—*Physical and chemical characteristics of soils used at ANL -E and ORNL.*

Characteristic	Soil	
	ANL	ORNL
soil bulk density, g/cm^3	1.74	[a]
texture, %		
sand	29	58
silt	29	2
clay	42	39
clay mineralogy, %		
vermiculite	[a]	10
kaolinite	[a]	80
percent carbon	4.20	0.07
cation exchange capacity (meq/100 g)	8.4	4.9
pH (1:1 paste method)	8.3	6.2
percent moisture-holding capacity	40.6	44.5

[a] Not available.

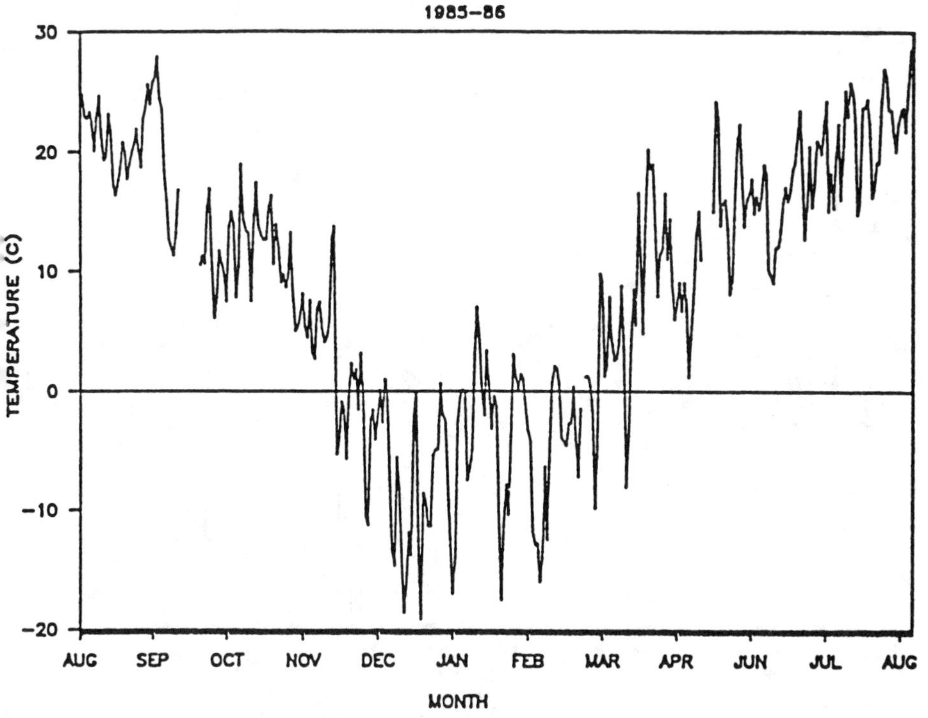

FIG. 4—*ANL weather data—air temperature.*

responding to periods of intense rainfall. In June 1986, that rain gage was replaced with a Climatronics tipping bucket gage designed for episodic high-intensity rainfall. Data from this gage appear to be accurate most of the time; however, the rainfall data recorded by the DAS contain occasional, erroneously high data points. The malfunctions have not resulted in a loss of rainfall data, since both ANL-E and ORNL have mechanical recording rain gages in close proximity to the lysimeter sites. Data from those nearby rain gages (Table 3) were used to calculate the total quantities of precipitation received by each site.

Temperature probes located in all ten lysimeters at the depth of the waste forms (77.0 cm) indicate that at no time were the waste forms exposed to freezing temperatures (Figs. 6 and 7). The soil (or sand) temperature data further show (as would be expected) that the near-surface soil temperatures (elevation 149.0 cm, 66.7 cm below the soil surface) fluctuate more than the intermediate (elevation 77.9 cm) or bottom (elevation 28.8 cm) soils. It is also noted from the data that the frost line in the soil did not move as deep as the first probe (66.7 cm below the soil surface).

Some abnormally low soil temperature readings were observed from the intermediate and bottom probes in Lysimeter ANL-3 in January 1986 and in ANL-4 by June 1986. There were no such occurrences with near-surface probes. One possible explanation for the malfunction is related to an average soil subsidence of 30 cm in all ANL soil-filled lysimeters. That subsidence was manifested as a general settling of the soils and waste forms. It is hypothesized that subsiding soil may have caused damage to the lead wires connecting the lower probes to the DAS system. These probes were replaced with new ones, and recent data from the replacements show that they are functioning normally.

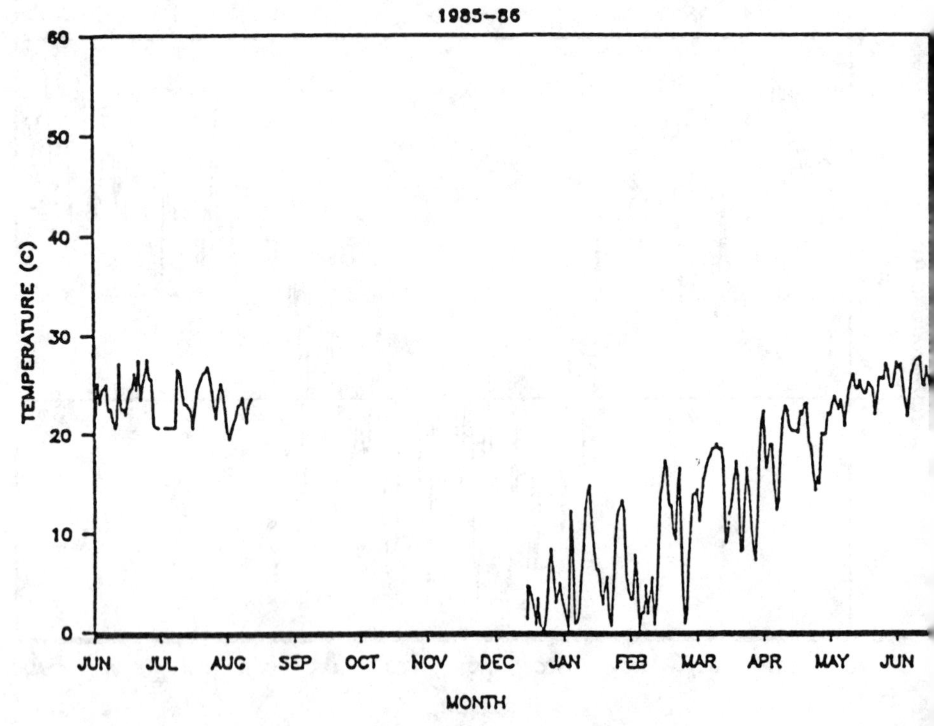

FIG. 5—*ORNL weather data—air temperature.*

TABLE 3—*Yearly precipitation at ANL and ORNL as measured by back-up instrumentation—July 1985 through July 1986.*

Month	Precipitation, cm	
	ANL	ORNL
July[a]	...	13.3
Aug.[b]	5.6	23.1
Sept.[c]	6.3	4.3
Oct.	11.6	7.6
Nov.	18.9	10.2
Dec.	1.7	5.3
Jan.	0.6	3.1
Feb.[c]	6.4	10.4
March	7.6	7.2
April[c]	4.0	5.1
May	7.7	7.7
June[c]	11.1	2.6
July	8.6	...
TOTAL	93.5	99.0

[a] ORNL lysimeter experiment initiated in July.
[b] ANL-E lysimeter experiment initiated in August.
[c] Months leachate was retrieved for analyses.

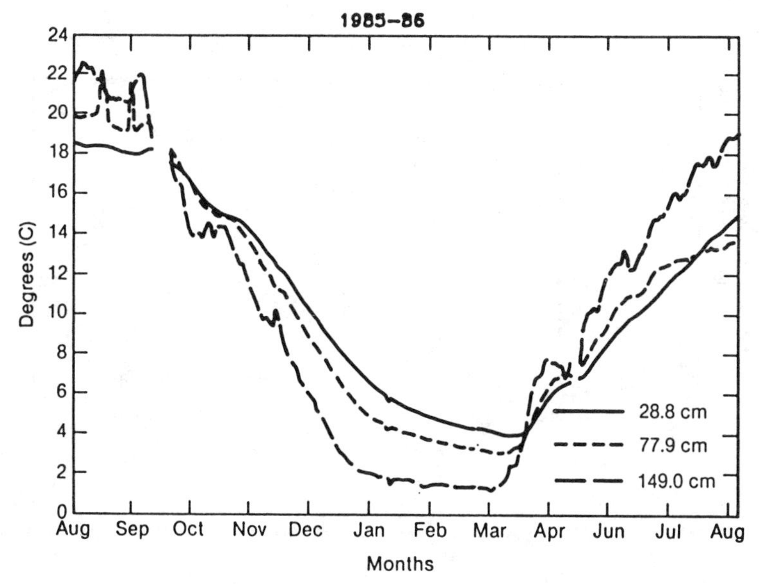

FIG. 6—*ANL Lysimeter 1 soil temperatures.*

The bottom temperature probes in ORNL-3 and -5 has consistently indicated high temperatures as compared to temperatures measured by probes in those and nearby lysimeters. Because the abnormal readings began close to the time of lysimeter installation, it is possible that probes or wiring were damaged during installation.

Moisture probes at the two sites show that two (ORNL) and three (ANL) months were required after the lysimeters were filled with soil for the soil to reach saturation (Figs. 8

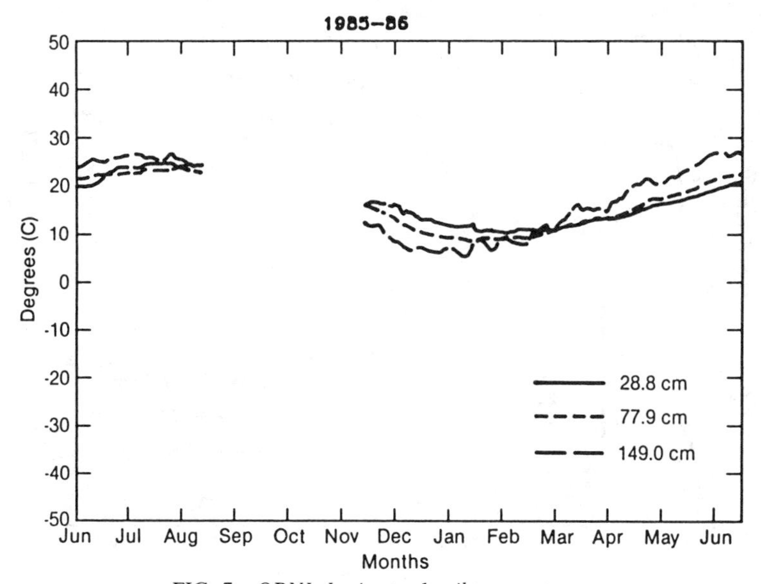

FIG. 7—*ORNL Lysimeter 1 soil temperatures.*

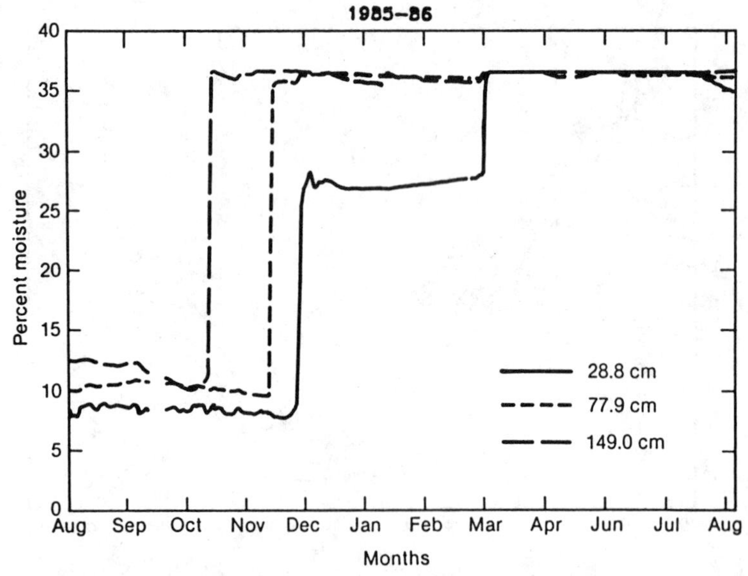

FIG. 8—*ANL Lysimeter 2 soil moistures.*

and 9). As a precaution, the accuracy of the probes in the soil-filled lysimeters was determined by comparing their data against the gravimetric water content of soil cores retrieved from all four ORNL lysimeters and one ANL lysimeter. From those comparisons it was apparent that the probes are overestimating the soil moisture content and that the soils may not have been in saturated condition as indicated by the probes. Corrective action consisted of re-

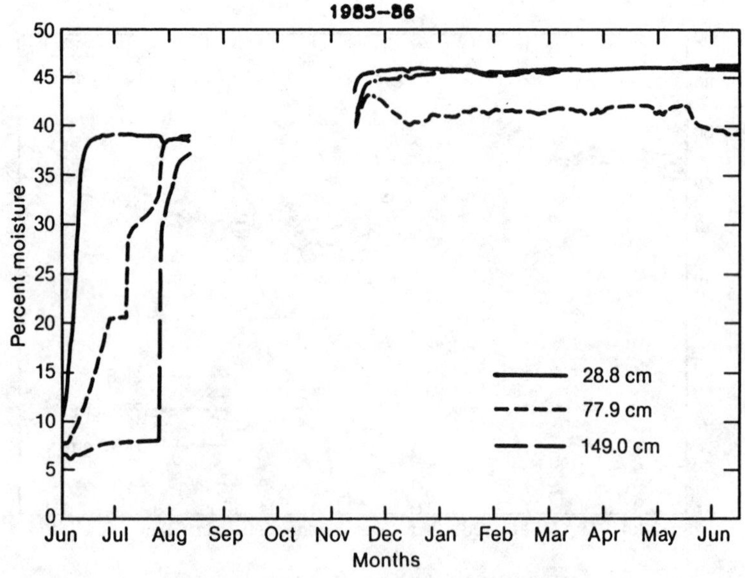

FIG. 9—*ORNL Lysimeter 2 soil moistures.*

calculation of the polynomial equation which transforms probe input into percent moisture using laboratory and field soil gravimetric measurements.

Lysimeter Water

During a 12-month period, the ANL site had 93.5 cm of precipitation while ORNL received 99.0 cm, well below the normal of 134 cm. Using these values and the area of exposed lysimeter (6489.5 cm^2), it was calculated that the ANL and ORNL lysimeters received 607 L and 643 L of water, respectively. Total water retrieved from the leachate collectors of each lysimeter during the period from July 1985 to July 1986 is shown in Table 4. On the average, the leachate collectors of the soil-filled ANL lysimeters contained 128.7 ± 22.6 L; those at ORNL, 441.7 ± 20.9 L. The leachate collectors of the sand-filled lysimeters at ANL and ORNL contained 337.9 L and 528.0 L, respectively.

The two sites received comparable volumes of precipitation; however, water moved through the lysimeters at the sites in unequal amounts. Because vegetation is removed from each lysimeter, evapotranspiration did not factor into water loss through surface evaporation could have contributed to some loss. It appears that the differences in water movement were due more to soil texture and weather conditions. Soil used at ANL is heavier (contains more fine material such as silts and swelling clay) than the soil used at ORNL [*3*]. Therefore, infiltration and percolation of water through the ANL soil would be reduced in comparing the sand-filled control lysimeters at the two sites. At ANL, 55% of the volume of precipitation passed through the lysimeter versus 82% for ORNL.

Based on the amount of water retrieved from the lysimeters, the ANL soil-filled lysimeters had 0.18 pore volumes of water pass through them, while 0.62 pore volumes passed through similar ORNL lysimeters. Pore volume for the control lysimeters (sand-filled) were 0.57 for ANL and 0.94 at ORNL. Theoretically, then, 18% of the water held in the soil pore space of the ANL lysimeters was replaced during the first 12 months, while 62% was replaced in the ORNL soil lysimeters. Similarly, 57 and 94% was replaced in the ANL and ORNL control lysimeters, respectively. Therefore, if nuclides were in the water surrounding the waste forms, the greatest opportunity for detection would be found in water from the ORNL site. (This is based on two assumptions: that the nuclide is water soluble and that the soil column does not interfere with nuclide movement.) It is noted that cobalt-60 (^{60}Co) and strontium-90 (^{90}Sr) move through soils freely, while antimony-125 (^{125}Sb) and cesium-137 (^{137}Cs) are readily retained by most soils. The inert sand used in the No. 5 lysimeters was selected because it does not interfere with the movement of the nuclides under investigation.

Radionuclide Analysis

Water samples have been collected from the leachate collectors and Moisture Cup 3 from each lysimeter on five occasions since experiment initiation. The first two times, water

TABLE 4—*Total quantities of leachate retrieved from lysimeters during a 12-month period.*

	Quantity of Water, L Lysimeter No.				
Site	1	2	3	4	5
ORNL	415	438	449	465	528
ANL-E	113	132	160	112	338

samples were analyzed only for gamma-producing nuclides. The last three water samples, taken at both sites in April, June, and October 1986, were analyzed for both gamma-producing nuclides and the beta-producing nuclide ^{90}Sr (Table 5).

During June 1986, in addition to obtaining water samples from leachate collectors and Moisture Cup 3, water samples were taken from Moisture Cup 5 (the one nearest the soil surface) of each soil lysimeter. Those samples were then combined for use as a composite sample. Because Moisture Cup 5 is located above the waste forms, the composite water sample serves as a control to detect nuclides which might originate from sources other than the waste forms. Radionuclides were not found above background quantitites in those samples and sampling of those cups was not performed after these data.

Gamma-producing nuclides were not found in the first two samplings. However, in April 1986, ^{60}Co was discovered in water samples from the moisture cup of ANL 5 (Table 3); ^{137}Cs was found in the leachate of ANL 5 (the sand-filled lysimeter); and ^{125}Sb was found in the moisture cup of ORNL 5 (Table 6) (also a sand-filled lysimeter). In addition, ^{90}Sr was found in significant quantities in the moisture cups of ANL 4 and 5 and in the leachate of all ANL and ORNL lysimeters during the April sampling. The concentration of ^{90}Sr in the ORNL lysimeter leachate was almost two orders of magnitude higher than the ANL samples.

Analysis of water samples obtained in June 1986 (Tables 5 and 6) showed that ^{60}Co still persisted in Moisture Cup ANL 3-3, with a substantial increase in ^{125}Sb in ORNL Moisture Cup 5-3. The origin of ^{125}Sb is not known, but it is assumed to be the waste forms. Original evaluation of radionuclide content of the prefilters from which this resin was taken identified ^{125}Sb in quantities of 0.1% of the total nuclide content, although it was absent in a subsequent resin analysis. Cobalt-60 was also found for the first time in ORNL 5-3 in June. Also, ^{90}Sr was detected in moisture cups at ORNL (ORNL 3-3 and 5-3) and in two additional cups at ANL (ANL 3-3 and 5-1), while there was none detected in this sampling of ANL 4-3. The concentration of ^{90}Sr in ANL 5-3 was more than double that found in the April sample.

TABLE 5—*Results of beta and gamma analysis of ANL-E soil moisture and leachate samples obtained in 1986.*

	Concentration,[a] pCi/L								
	^{60}Co			^{137}Cs			^{90}Sr		
Identification	April	June	Oct.	April	June	Oct.	April	June	Oct.
1-1[b]	NS[d]	...	...	NS	...	...	NS	...	...
1-3	...[e]	...	...	...	...	...	...	...	...
2-3	...	...	...	...	...	109 ± 19	...	...	...
3-1	NS	NS	...	NS	NS	...	NS	NS	...
3-3	11 ± 7	13	...	...	...	...	...	11.3 ± 1.4	...
4-3	...	...	...	...	...	...	2.7 ± 1.8	...	...
5-1	NS	...	...	NS	...	...	NS	349.6 ± 11.3	99 ± 5
5-3	...	...	...	...	...	...	55.6 ± 6.7	127.6 ± 6.7	647 ± 1[illegible]
5-4	NS	NS	...	NS	NS	...	NS	NS	6 ± 5
1[c]	...	...	...	...	...	...	0.5 ± 0.3	...	...
2	...	...	...	...	...	...	0.5 ± 0.2	...	...
3	...	...	...	...	...	...	0.4 ± 0.1	...	...
4	...	...	...	...	...	...	0.6 ± 0.3	...	...
5	...	...	...	5.4 ± 1.1	...	...	1.0 ± 0.4	5.8 ± 0.3	...

[a] Concentraction ± 2 sigma.
[b] Lysimeter and moisture cup identity number.
[c] Lysimeter leachate collector identity number.
[d] NS indicates that moisture cup was not sampled at this time.
[e] Ellipsis (...) indicates that radionuclite concentration was not above background.

TABLE 6—*Results of beta and gamma analysis of ORNL soil moisture and leachate samples obtained in 1986.*

Sample Identification	Concentration,[a] pCi/L											
	^{60}Co			^{137}Cs			^{125}Sb			^{90}Sr		
	April	June	Oct.	April	June	Oct.	April	June	Oct.	April	June	Oct.
1-3[b]	...[d]	...	...	...	...	...	...	...	...	...	...	64.9 ± 10.8
2-3	...	...	...	...	...	...	...	...	...	...	...	64.9 ± 27.0
3-3	...	...	...	...	...	...	...	...	...	...	32.4 ± 18.9	24.9 ± 7.8
4-3	...	...	...	...	...	...	...	...	...	...	...	6.2 ± 3.2
5-3	...	89.2 ± 32.4	...	...	...	...	351 ± 5	540 ± 81	568 ± 108	...	17.6 ± 12.2	16.5 ± 10.3
5-1	NS[e]	NS	...	NS	NS	...	NS	NS	...	NS	NS	70.3 ± 16.3
1[c]	...	...	...	...	...	...	...	...	...	62.2 ± 8.1	9.2 ± 3.5	29.7 ± 5.4
2	...	...	...	...	...	...	...	...	...	27.0 ± 5.4	...	20.8 ± 5.1
3	...	...	...	...	...	...	...	...	...	4.9 ± 2.7	...	108 ± 10.8
4	...	...	...	...	...	...	...	...	...	54.1 ± 8.1	11.6 ± 4.6	...
5	...	...	...	...	...	...	...	...	64.9 ± 24	45.9 ± 8.1	...	3.5 ± 2.7

[a] Concentration ± sigma.
[b] Lysimeter and moisture cup identity number.
[c] Lysimeter leachate collector identity number.
[d] Ellipsis (...) indicates that radionuclide concentration was not above background.
[e] NS indicates that moisture cap was not sampled at this time.

ANL 5-1 is the moisture cup located directly below 5-3, and the water from this moisture cup was analyzed for the first time in June in an attempt to detect movement of ^{90}Sr through the silica sand profile. The concentration of ^{90}Sr in ANL 5-1 was almost three times that of 5-3. In general, occurrence of ^{90}Sr in the leachate sample at both sites was down sharply from the April 1986 sampling, with measurable amounts being found only in ANL 5 and ORNL 1 and 4.

Results from the analyses of the October 1986 water samples (Tables 5 and 6) showed that ^{60}Co, which was found during the previous two sampling periods in ANL 3-3, was absent. Cesium-137 was discovered in the water samples from ANL 2-3. Cobalt-60 was not found in ORNL 5-3 this sampling period; however, ^{125}Sb not only persisted in ORNL 5-3, but was detected in the leachate collection tank of Lysimeter 5 (the sand-filled lysimeters). At ANL, ^{90}Sr was detected only in the moisture cups of Lysimeter 5 (ANL 5-1, 5-3, and 5-4) and was absent from that lysimeter's leachate collectors. On the other hand, ^{90}Sr was found in all the moisture cups and four of the five leachate collectors of the ORNL lysimeters. There was no ^{90}Sr in ORNL-4 at this sampling.

Occurrence of nuclides in the water samples from both the soil and inert sand lysimeters in such a short period of time (months rather than years) was unexpected. While ^{90}Sr is known to be soluble in soil solution and does move through the soil column almost unhindered by the soil matrix, it appears that leaching and movement of the nuclides occurs at a more accelerated rate in the soil than was thought possible. Appearance of ^{90}Sr in the April leachate of all lysimeters at both sites and at ORNL in October indicates that small quantities were readily leached from all the waste forms irrespective of their formulation (cement or VES) or initial ^{90}Sr concentration. The higher initial concentration at ORNL could reflect the greater pore volume of water that has passed through these lysimeters. This could also be the case for the occurrence in October 1986. When comparing ^{90}Sr data from April and June 1986 samplings, it appears that ^{90}Sr moved through the lysimeters as an initial slug which, in the case of ORNL, has been washed out or, as at ANL-E, is in the process of being flushed. Data from ANL 5 would support this hypothesis, since it appears that a plume of ^{90}Sr movement has been detected in this lysimeter, with the trailing edge

TABLE 7—*Radiation intensity with depth in EPICOR-II field lysimers.*

Depth from Soil Surface cm	Radiation Intensity mR/h									
	ANL-E Lysimeter No.					ORNL Lysimeter No.				
	1	2	3	4	5	1	2	3	4	5
15.2	...[a]	...	...	...	...	...	...	...	...	...
30.5	...	...	...	...	...	...	...	...	...	...
45.7	...	...	...	...	...	...	0.02	0.04	0.01	0.04
61.0	...	...	...	0.3	1.0	0.005	0.04	0.18	0.29	0.21
76.2	...	...	...	1.0	3.0	0.16	0.24	0.65	1.6	1.4
91.4	0.4	...	0.5	1.3	6.0	0.81	1.0	2.7	1.6	1.4
106.7	0.7	2.0	2.5	7.0	10.0	2.9	3.5	13.7	25.8	20.8
121.9[b]	3.5	18.0	5.0	35.0	12.0	6.1	11.2	29.0	52.4	40.3
137.2	6.0	28.0	20.0	43.0	12.0	11.2	16.1	39.5	70.2	48.3
152.4	7.5	39.0	18.0	67.0	12.0	11.2	17.7	40.3	70.9	36.3
167.6	7.0	32.0	17.0	65.0	8.0	8.1	12.9	30.6	54.8	20.9
182.9	3.6	21.0	10.0	38.0	6.0	2.4	4.6	14.5	25.8	5.9
198.1	1.8	10.0	2.5	18.0	5.0	0.73	1.5	3.7	9.7	1.1
213.4	...	...	...	...	...	0.16	0.31	0.89	1.4	0.16

[a] Ellipsis (...) indicates readings were not above background.
[b] Location of waste form is indicated by boxes.

showing up in the area of ANL 5-3, the bulk near ANL 5-1, and a leading edge moving into the leachate collector. The October ORNL data would indicate that there is a source of mobile ^{90}Sr and that continued leaching of this element is to be expected. These data are also supported by the October ANL data, in which it is seen that ^{90}Sr has moved from the location of the waste form (Cup 1) to near the bottom of the sand profile (Cup 3). Though ^{90}Sr has been detected, the total quantity leached is only a small fraction of that available in the waste forms (Table 1).

Finally, because it is apparent that the soil in the lysimeters has subsided (very evident at ANL), it was decided to determine if the movement had caused a shift in the position of the waste forms. This was accomplished by lowering a radiation-detecting probe down the access tube which leads into the leachate holding tank. Readings were taken every 15.2 cm in all lysimeters, and radiation intensity with depth was recorded. Readings of the soil lysimeters were then compared with readings from the sand-filled controls. At ORNL, the intensity of radiation readings for each lysimeter approximated the known depth of the waste forms (Table 7). However, at ANL, some settling has occurred; radiation readings from the waste forms in the soil-filled lysimeters (1 to 4) were still high at the 182.9-cm depth, whereas the activity in the inert control had moderated by that depth, indicating a downward movement of the waste forms of about 7.5 cm in the soil-filled lysimeters. There is no evidence that the movement has impacted the experiment except for minor damage to some moisture/temperature probes.

References

[*1*] "Licensing Requirements for Land Disposal of Radioactive Waste," 10 CFR 61, *Code of Federal Regulations,* Office of Federal Register, U.S. Government Printing Office, Washington, DC, 24 July 1981.

[*2*] Neilson, R. M. and McConnell, J. W., Jr., *Solidification of EPICOR-II Resin Waste Forms,* GEND-INF-055, EG&G Idaho, Inc., Idaho Falls, IO, Aug. 1984.

[*3*] Rogers, R. D., McConnell, J. W., Jr., Davis, E. C., and Findley, M. W., *Field Testing of Waste Forms Containing EPICOR-II Ion Exchange Resins Using Lysimeters,* NUREG/CR-4498, Nuclear Regulatory Commission, June 1986.

[*4*] Schmitt, R. C. and Reno, H. W., *Program Plan of the EPICOR and Waste Research and Disposition Program of the Technical Support Branch,* EGG-TMI-0521 revised, EG&G Idaho, Inc., Idaho Falls, IO, Dec. 1983.

[*5*] McConnell, J. W., Jr., *EPICOR-II Resin/Liner Research Plan,* EG&G-TMI-6198, EG&G Idaho, Inc., Idaho Falls, IO, March 1983.

[*6*] Rogers, R. D., McConnell, J. W., Jr., Findley, M. W., and Davis, E. C., *Lysimeter Data from EPICOR-II Waste Forms—Fiscal Year 1986,* EG&G-TMI-7417, EG&G Idaho, Inc., Idaho Falls, IO, Oct. 1986.

Roger D. Spence[1] *and Tsuneo Tamura*[2]

In Situ Grouting of Shallow Landfill Radioactive Waste Trenches

REFERENCE: Spence, R. D. and Tamura, T., "***In Situ*** **Grouting of Shallow Landfill Radioactive Waste Trenches,"** *Environmental Aspects of Stabilization and Solidification of Hazardous and Radioactive Wastes, ASTM STP 1033,* P. L. Côté and T. M. Gilliam, Eds., American Society for Testing and Materials, Philadelphia, 1989, pp. 418–429.

ABSTRACT: A backfilled trench containing low-level radioactive waste was grouted with a particulate grout. The accessible void volume of the waste zone was estimated to be 20% or 28 m^3, but 30.6 m^3 of grout was injected into the trench. Part of the grout forced a path outside the trench. The water permeation into the trench from monitoring wells was reduced by two orders of magnitude. Site characterization and monitoring costs were outside the scope of this paper, so the only costs presented are the grouting costs. The grout costs only $0.055/L; but most of the cost was in manpower, 1220 man-hours for this demonstration.

KEY WORDS: remedial action, shallow land burials, grouting, particulate grouts, cement grouts, radioactive waste, low-level waste

Waste buried in shallow land burials (SLB's) combined with water intrusion may cause problems for many years in the resulting contaminant migration in ground and surface water. In addition, waste and container disintegration causes trench subsidence, leading to even more water intrusion. Long-term monitoring programs are required to detect waste migration. Remedial action may be required if waste migration is detected at a disposal site. Subsidence has proven to be a chronic problem at some sites, requiring routine surveying and maintenance many years after a site has been closed. Subsidence is caused by collapse and settling of the overburden into large voids in the matrix of the waste and backfill. Some of these voids result from sheltered areas around the waste. Most come from voids inside the waste packaging and decay of the waste material (for example, biodegradation of paper). As the containers and waste materials decay, the loose backfill falls into and fills these voids, causing subsidence and the formation of depressions in the trench cover. Such depressions usually lead to water intrusion through a cover designed to direct surface water away from the trench. *In situ* grouting of commercial and municipal landfills has been used to control subsidence and to help direct ground and surface water away from the waste.

This remedial action has been proposed for SLB's containing waste contaminated with low-level radioactivity. *In situ* grouting may be done with solution (chemical) grouts or particulate (slurry) grouts. As the names imply, solution grouts consist of dissolved chemicals (sodium silicate or a monomer or a polymer, such as polyacrylamide) that will form a gel after injection, and particulate grouts (cement-based or lime-flyash) consist of a liquid suspension of particles that harden into a solid mass after injection. Solution grouts are

[1] Chemical Technology Division, Oak Ridge National Laboratory, Oak Ridge, TN 37831.
[2] Environmental Science Division, Oak Ridge National Laboratory, Oak Ridge, TN 37831.

available with viscosities approaching that of water. These grouts can penetrate almost anywhere water can and may subsequently fill a waste trench the way a bathtub is filled—grouting the voids, soil matrix, and waste. Solution grouts are too expensive for most commercial applications and are usually used only for temporary water seals in tight matrices. The long-term stability of these grouts is unknown but is being studied. Injection of a low-level waste trench with a sodium silicate grout was demonstrated at Maxey Flats, Kentucky [*1*].

Particulate grouts cannot penetrate voids or passages too small to accommodate the suspended particles and, thus, do not have the penetrating power of the solution grouts. In general, these grouts will be limited to the larger voids and will not penetrate the backfill soil material. The Maxey Flats demonstration illustrated that there are significant hydraulic connections within a burial site, and the large voids are likely to be present around the waste. Thus, grouting with a particulate grout should fill the large voids (preventing subsidence) and may redirect water away from the waste (helping control radionuclide migration and water contamination). In addition, these grouts offer the advantage of using natural materials with an established history of long-term stability. This paper describes a full-scale demonstration of the injection of a particulate grout into a low-level waste trench at Oak Ridge National Laboratory (ORNL) during August–September 1986. First, the trench selection and characterization are discussed, then the grout development, and finally the grout injection. Monitoring and assessment are continuing, so not all results are in, but the results to date are presented.

Trench Characterization and Site Preparation

The main objective of this *in situ* demonstration was to grout, with particulate slurry, a backfilled trench filled with low-level radioactive waste and to evaluate the cost, implementation, and percentage of voids filled. Establishing an immediate definitive correlation between the selected trench and any known radionuclide seeps was beyond the scope of this demonstration; nevertheless, it was hoped that a beneficial effect could eventually be established.

During the wet season, a small stream flows in a gully west of the 49 trench area of Solid Waste Storage Area 6 (SWSA-6) of ORNL. This gully can be located on the topographical contour map of Fig. 1 for SWSA-6. A seep in this gully just south of the 49 trench area discharges small quantities of strontium-90 into the stream during the wet season. Several new wells have been installed between the seep and the 49 trench area and a monitoring program has sampled these wells, as well as some of the older wells in the area (most of the older wells are located inside the backfilled trenches). A sampling survey in September 1982 established that high levels of strontium-90 were present in the trench waters of Trenches 150 and 152 (see Fig. 2). Based on this data, Trenches 150 and 152 were speculated to be the source of the strontium-90 in the seep water. Thus, Trench 150 was selected for the grouting demonstration and its companion, Trench 152, is being used as a comparative control.

The proposed strontium-90 distribution illustrated in Fig. 2 is consistent with the surface topology of the 49 area and the strontium-90 activities found in the 1982 survey. However, the geology and hydrology of an area generally lead to ground-water flowpaths (and subsequent radionuclide migration) more complicated than apparent from the surface topology, and the limited number of monitoring wells in the area precludes establishing definitely the suspected correlation between the two trenches (150 and 152) and the seep. Hydrology more complex than that suggested in Fig. 2 was confirmed from the differences in subsequent samples of the tritium activity between these two trenches and the seep. This sampling (part

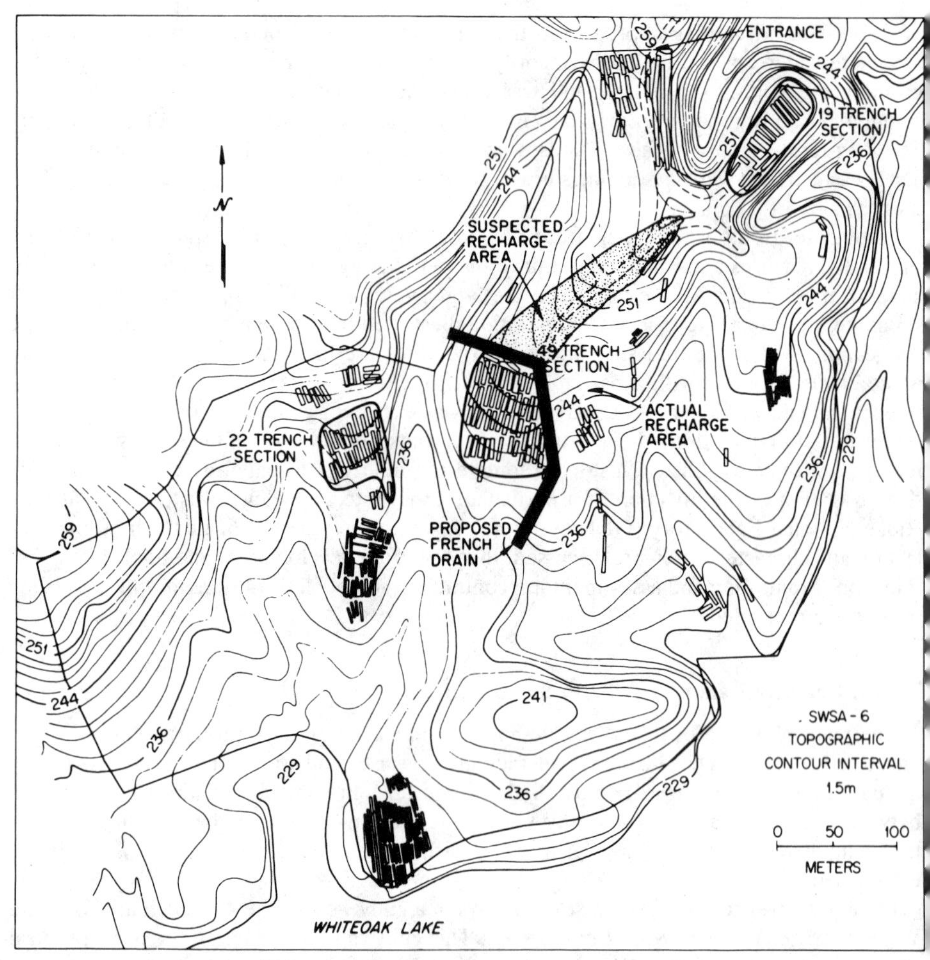

FIG. 1—*Trench locations within SWSA No. 6.*

of the demonstration effort) began in November 1984 and is to be continued into the near future as the effects of the grouting of Trench 150 are assessed and evaluated. Table 1 lists the average of the analytical results for the samples taken from November 7, 1984 through March 4, 1985. In general, tritium is considered to flow as fast as the ground water and does not interact with the soil and rocks any differently than water. Thus, tritium is an excellent tracer for evaluating radioisotope migration and the effects of remedial action if a constant tritium leakage is occurring (not the case for Trench 150). Other radioisotopes, including strontium-90, do interact and any effects caused by remedial action may be delayed at locations remote from the source; that is, it may be years before remedial action at Trench 150 causes a change at the seep (assuming that these two trenches are sources for the seep).

A water quality analysis of the trench water found calcium and magnesium (<100 mg/kg and <25 mg/kg, respectively) to be the main cations. Based on this analysis, no retardation or acceleration of the normal grout reactions was expected upon contact of the grout with trench water.

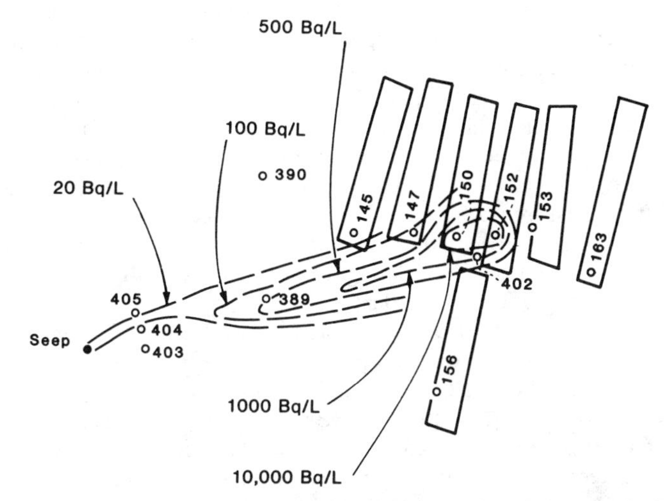

FIG. 2—*Speculation on the Sr-90 flowpath based on the September 1982 survey by E. C. Davis.*

TABLE 1—*Averaged analytical results for water samples taken in the 49 trench area during the period 11-7-84 to 3-4-85 (standard deviations in parentheses).*

Trench/Well	pH	Conductivity, μS/cm	Strontium-90, Bq/L	Tritium, Bq/L
T145	6.8	250	16 300 (16 000)	1 200
T147	6.9	200	486 (241)	32 000
T150	6.9 (0.3)	516 (86)	14 700 (6 570)	367 (52)
T152	6.7 (0.3)	514 (77)	20 700 (12 200)	3 240 (3 230)
T163	6.8	248	65 (43)	1 700
W389	7.8 (0.8)	442 (153)	5 050 (8 030)	13 300 (13 200)
W390	6.6	62	195	120
W403	7.3 (0.1)	246 (30)	32 (24)	12 900 (10 100)
W404	7.3 (0.6)	613 (30)	22 (20)	1 050 000 (574 000)
W405	7.2 (0.2)	673 (125)	6 (4)	1 560 000 (711 000)
Seep	7.2 (0.5)	311 (27)	156 (143)	183 000 (38 600)

In general, the trenches were originally excavated so the trench bottom was about 0.6 m above the water table. The trenches were subsequently filled with waste contaminated with low-level radioactivity, classified as combustibles and noncombustibles (150 and 152 had about 15 to 20% combustibles). Once the trenches were filled with waste to within 0.6 m of the surface they were backfilled with excavated soil (coarse weathered shale). Later, the original 0.6-m cover was removed, replaced with bentonite mixed with soil, and seeded with grass (150 and 152 were closed in 1975). Trench 150 measures 17 m long by 2.8 m wide by 3.6 m deep and Trench 152 measures 18.4 by 3.05 by 4.0 m. In preparation for the grouting demonstration, five new wells were installed in Trenches 150 and 152 and 0.3 m of overburden was removed from each trench (see Fig. 3). This excavated area acted as a collection and containment reservoir for Trench 150 during the grout injection. Trench 152 was excavated to enhance its use as a control. Thus, during the demonstration, the actual depth of Trench 150 was 3.3 m, including 0.3 m of overburden.

Prior to the grout injection, Trench 150 was filled with water through one of the new monitoring wells. As the trench was saturated, records were kept of the quantity of water added and the water level in the trench. This test indicated that 41.7 m^3 of void space was present. This void volume represented both accessible void (which is penetrable by particulate grout) and soil pore void (which is accessible to water but not to particulate grout). To estimate the accessible void volume, it was assumed that the larger voids accepted water as fast as it could be pumped and that the soil pores represented a much slower rate of filling. Using this assumption, it was estimated that approximately ⅔ of the total voids represented the larger voids, and ⅓ the smaller soil pores. Thus, about 28.0 m^3 represent larger voids accessible to the particulate grout and 13.8 m^3 represent the smaller soil pores which would not be penetrated by the grout.

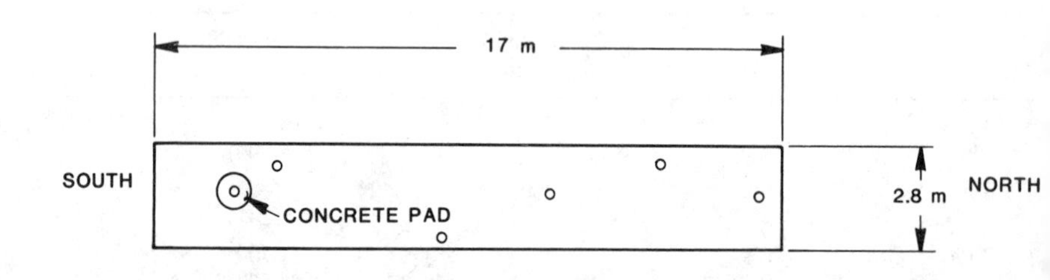

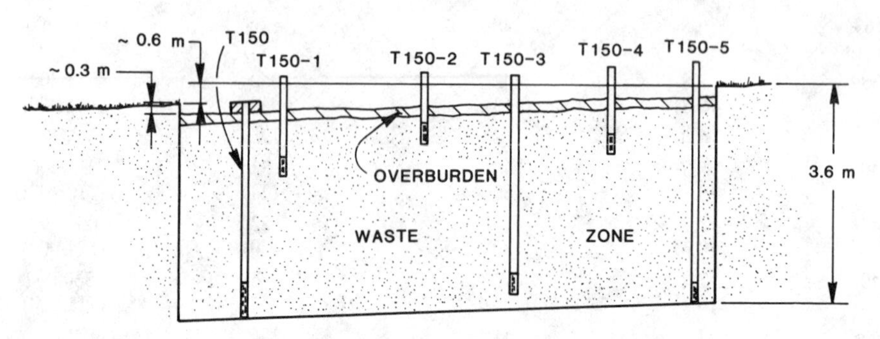

FIG. 3—*Monitoring well locations in Trench 150.*

Grout Development

The main emphasis was to develop a grout that could flow from one end of the trench to the other if a suitable flowpath existed, that would remain fluid long enough for a relatively large batch to be injected, that would set into a strong solid mass, and that would not spread the waste contamination beyond its present limits. The criteria selected to meet these goals were

1. Apparent viscosity at $\leq$ 50 mPa · s.
2. 10 min gel strength of $\leq$ 48 Pa.
3. 28*d* phase separation of 0 volume present.
4. 28*d* compressive strength of $\geq$ 414 kPa.

The ingredients of the grouts investigated were Type I portland cement, eastern Classes C and F (ASTM) flyash, bentonite, and water. Grouts passing the above criteria were prepared from the following two dry blends:

	Blend 1 (wt %)	Blend 2 (wt %)
Type I portland cement	36	41
Class C flyash	58	54
Bentonite	6	5

The acceptable grouts were prepared at mix ratios (ratio of the mass of dry blend to volume of water) of 1.4, 1.6, and 1.8 kg dry blend/litre water. This gave a broad operating range of grout composition for the field demonstration. The target dry-blend composition used for the field demonstration was 39 weight percent Type I portland cement, 55.5 weight percent Eastern Class C flyash, and 5.5 weight percent bentonite. In general, the lower mix ratios were more fluid and cheaper. Thus, the target mix ratio for the field demonstration was 1.5 kg/L. This target gave a tolerance of 0.1 kg/L on the low side, with 0.3 kg/L on the high side.

Grouting Operation

Dry-Solids Blending

The cement, flyash, and bentonite were blended to make a homogeneous powder. The blending tanks consisted of three pneumatic tanks, each capable of holding 18 160 kg, located at an existing facility at ORNL and remote from SWSA-6. One of these tanks rests on a set of scales. The batch composition was measured by adding each ingredient to the weighing tank and noting the tank weight after each addition. Next, the batch was blended by blowing the entire tank contents from one tank to another three times, ending in the weighing tank for dispensing as needed (this procedure resulted in uniform mixing in the past). One 18 000-kg batch was sufficient for three days, or more, of grout injection.

Grout Mixing

The dry blend was mixed with water by blowing a preweighed quantity of the blend into the rotating drum (which contained the water) of a concrete mixing truck. This technique, which was atypical of commercial grouting operations, worked well for this demonstration, but may not work if the resulting grout is not fluid (that is, sticks to drum walls). However, the grout used was designed to be fluid while retaining other desired properties.

The dry blend to be used for a given day's operations was blown from the weighing tank to a holding tank. The weight of blend transferred was determined from the weight before and after the transfer. A predetermined quantity of water was metered into the truck's drum and the set-retarder/dispersing agent (delta gluconolactone) added. Finally, all of the blend in the holding tank was blown into the drum while it was rotating. After adding the solids and rotating the drum for several minutes, the density of the mix in the drum was measured and compared with the expected density. An acceptable density verified the proper mix ratio and acceptable mixing. Only one batch of grout was made each day and consisted of 5.7 m^3 or 2.8 m^3 of grout. At the beginning, the large batch was injected during the workday. Later, as the trench voids filled, the smaller batch lasted through the workday.

Lance Placement

The lances consisted of nominal 5-cm-diameter (2-in.), Schedule 80 pipe with a disposable, conical-shaped point placed in one end. The lances were driven into the trench with a portable 54-kg pneumatic hammer. The points helped penetrate the backfill and prevented soil from plugging the grout exit. The lances were driven to the trench bottom, or as close as possible. (Most did not reach the bottom because of too much resistance to driving the pipe; a smaller pipe would have helped.) Once in place, the lance was pulled up 0.15 m, the point knocked out (and left), and the lance pulled up another 0.15 m. The next step was grout injection.

Grout Injection

The grout was pressure injected into the trench using a progressive cavity pump. The concrete mixing truck served as the grout reservoir for a given day's operation, with the drum rotating to keep the grout slurried. Two 0.26-m^3 tanks with slurry agitators supplied the pump with grout for injection. These tanks were filled as needed from the truck. The pump propelled the grout through 15.2 m or 30.4 m of 5-cm-diameter high-pressure hose, past a pressure gage, through the lance, and into the trench. The pump rate was set so the injection pressure was 34 to 69 kPa. In general, initial grout pump rates were high, but had to be decreased as the area became saturated with grout. Some initial rates were as high as 0.76 L/s. One motor drove both the grout pump and slurry mixers. To keep the grout slurry well mixed required a pump rate of at least 0.19 L/s. (Using separate motors for the pump and slurry mixers would have allowed a larger pump rate range.) Thus, once the rate reached 0.19 to 0.32 L/s, the pressure was allowed to increase. Injection was stopped if the pressure approached 138 kPa and the lance pulled up 0.3 m before restarting injection. The lance was moved to a different location if the injection depth was moved to less than 0.6 m or the grout broke through to the surface during injection at a given location.

Results

The *in situ* grouting of Trench 150 was done in two sets of injections. The primary injections were done in a diamond pattern with a 3.0 m center-to-center basis. The secondary injections were done in another diamond pattern offset 1.5 m from the primary injection locations, giving an overall center-to-center basis of 1.5 m and a total of 36 injection locations. Figure 4 illustrates these injection locations and the location of grout breakthroughs outside the trench. Table 2 summarizes the grouting operation. The depths reported in this table (and

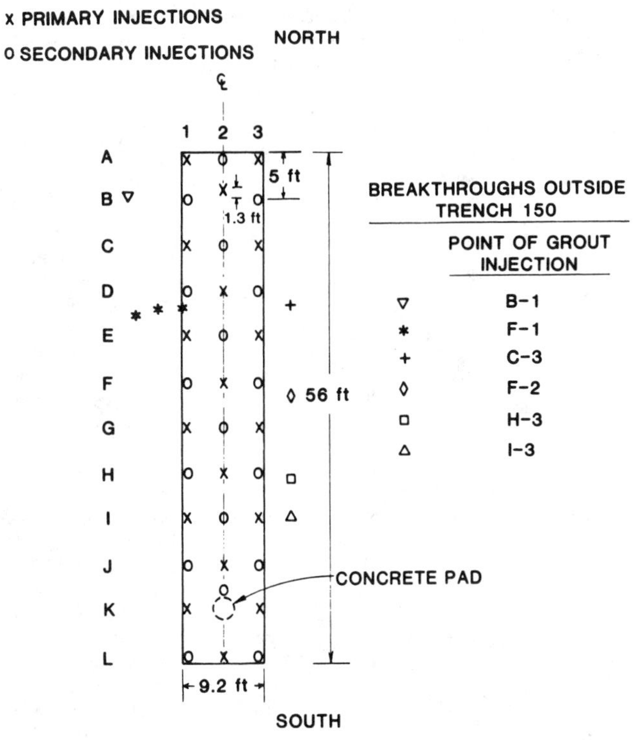

FIG. 4—*Grid locations for grouting Trench 150 in SWSA-6.*

throughout this paper) refer to the depth of the end of the lance below ground at the time of the grout injection, that is, the depth below the excavated surface.

A total of 30.3 m^3 of grout was injected into the trench in 1325 min. Over 85% of the total was injected in the primary pattern and most of the grout was injected into only a few holes. Referring to the primary pattern of Fig. 4, F-2, C-3, I-3, and B-2 took 38.4%, 21.4%, 7.7%, and 4.8% of the total, respectively, cumulatively 72.4%. Including the 5.9% injected in F-1 during the secondary pattern, 78.3% of the grout was injected in these five locations.

Based on the location of the grout takes, the trench can be roughly divided into three regions, North, Central, and South. Once a region became saturated with grout, further addition pushed grout out of a previous penetration or breakthrough. The Northern region was saturated by the injections into B-2, C-3, and C-1. The Central region was saturated by the injection into F-2, and the Southern region by the injection into I-3. For the secondary pattern, only the first injection (into F-1 in the Central region) took a large amount. Based on these results, the recommended pattern for future grouting operations is to inject first at center-to-center distances of about 6 m. Follow-up injections at closer center-to-center distances would still be required to ensure complete grouting of the trench.

The quantity of grout injected was 30.3 m^3, or 21% of the trench waste zone, compared to the 20% accessible void volume estimated from the hydraulic behavior of the trench. This implies that most or all of the accessible voids were grouted. Since several breakthroughs occurred just outside the trench boundaries (see Fig. 4), not all of the grout remained within the trench volume. Exactly where grout is and is not within the trench is uncertain at this

TABLE 2—*Summary of the grout injection of Trench 150.*

No.	Location	Depth, m	Volume, L	Time, min	Uptake Rate, L/min	Reasons for Stopping
1	B-2	2.9	1 363	38	36	breakthrough 0.5 m S of B-1
		2.3	95	2	46	changing depth no help
2	C-3	0.8	6 586	300	22	breakthrough between 150 and 152
3	C-1	2.1	783	30	26	breakthrough at B-2
4	D-2	1.6	0	0	—	high pressure
		1.2	0	0	—	breakthrough at C-1
5	E-3	1.8	227	5	45	breakthrough at C-1
6	E-1	1.7	144	5	29	breakthrough at D-2
7	F-2	2.6	697	64	10	high pressure, low rate
		2.3	2 237	80	28	grout thickening
		2.0	6 378	260	25	high pressure, low rate
		1.5	681	40	17	high pressure, low rate
		0.9	1 817	75	24	breakthrough between 150 and 152
8	G-3	1.8	0	0	—	breakthrough at F-2
9	G-1	3.0	322	10	32	breakthrough 0.9 m S of D-2
10	H-2	1.2	0	0	—	breakthrough at G-3
11	I-3	1.9	2 366	65	36	breakthrough between 150 and 152
12	I-1	1.0	568	17	33	breakthrough within 0.3 m I-1
13	J-2	1.3	303	11	28	breakthrough at I-3
14	K-3	1.2	265	10	27	breakthrough at K-3 wall
15	L-2	1.5	568	15	38	breakthrough at K-1 and pad
16	K-1	2.2	114	3	38	breakthrough within 0.3 m K-1
17	A-1	1.8	700	45	16	breakthrough 0.3 m N of F-1
18	A-3	1.8	26	1	26	breakthrough at A-1
19	F-1	2.6	1 817	60	30	breakthrough W 150 and S D-1
20	F-3	2.5	30	10	30	breakthrough S of D-1
21	G-2	1.1	0	0	—	breakthrough at G-1 wall
22	H-1	1.6	76	3	25	breakthrough 1.2 m N of H-1
23	H-3	1.4	416	15	28	breakthrough between 150 and 152
24	E-2	1.6	38	2	19	breakthrough at F-3
25	D-3	0.3	0	0	0	too shallow
26	D-1	0.8	38	2	19	breakthrough at D-3
27	A-2	1.4	76	12	6	breakthrough 0.6 m N C-2 and C-3
28	B-1	1.2	95	7	14	breakthrough W of 150
29	B-3	1.21	208	15	14	breakthrough at C-2
30	C-2	1.8	473	29	16	breakthrough at B-3
31	I-2	0.9	114	6	19	breakthrough 0.3 m S of H-1
32	J-1	2.0	151	8	19	breakthrough 0.3 m N of L-1
33	J-3	1.0	38	2	19	breakthrough J-3, 0.3 m S K-3
34	K-2	0.9	19	1	19	same as J-3
35	L-3	0.7	511	45	11	breakthrough 0.9 m W of L-3
36	L-1	0.6	26	2	13	too shallow
		Total	30 316	1325		

time. Future assessment should give a better idea of where the grout is inside the trench. It is reasonable to assume that most of the grout is within the trench boundaries with some extrusions of the grout along water flowpaths in the surrounding ground. Although the exact location of the grout is unknown, it is useful to illustrate how only a few injection points could fill the trench. This was done assuming a 20% accessible void volume distributed uniformly throughout the trench. To simplify the illustration, the grout injected at each location was assumed to occupy the entire trench depth and to be roughly symmetrical about the injection point. This estimation is illustrated in Fig. 5, where the quantity of grout

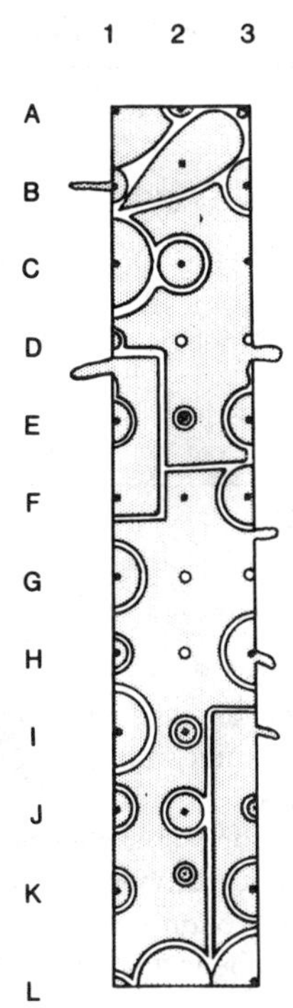

FIG. 5—*Quantities of grout injected represented as shaded areas on a sketch of Trench 150.*

injected at each point is represented by a shaded area bounded within the trench but roughly symmetrical about each point.

Preliminary Assessment

The amount of grout injected, 30.3 m^3, compares favorably with the estimated accessible void, 28.0 m^3. The difference of about 2.6-m^3 excess grout may be rationalized to be the amount of grout present outside the trench, as well as the amount on the surface within the trench border (Table 2), not to mention the uncertainty in estimating the accessible void volume.

The water injection test to obtain void volume calculations also provided information on the rate of water flow from the well into the trench before grouting. This flow rate was a measure of the local resistance to water flow and was used as a site-specific parameter for comparison before and after grouting. Before grouting, periodic measurement of the pump rate showed that greater than 62 L/min could be injected without overfilling of the well. After the grouting, the average of four wells was 0.79 L/min, with values ranging from 0.32 to 1.20 L/min. The fifth well, located at the lower south end (T150-1), took water at a rate of 0.010 L/min. From these results, it appears that the average resistance to water permeation increased by approximately two orders of magnitude.

In Table 3 lists some of the measured properties of water collected from Trenches 150 and 152 prior to and after (1/16/87) grouting. The most striking change is in the increase in pH of the water after grouting. Since cement reactions do lead to an alkaline pH, the value of 10.13 is not surprising; however, chemical analysis of the solution revealed that the

TABLE 3—*Selected properties of samples collected from Trenches 150 and 152.*

Trench No.	Sampling Date	pH	Ca, mg/L	Mg, mg/L	Na, mg/L	EC,[a] μS/cm	^{90}Sr, Bq/L	^{3}H, Bq/L
T-150	4-17-85	(insufficient water in well)						
	8-29-85	6.80	...	...	...	500	21 345	...
	11-05-85	6.90	...	...	...	510	21 125	5 200
	1-16-87[b]	10.13	2.8	0.3	95	530	8 000	12 000
T-152	4-17-85	6.65	80	23	2.0	580	25 675	33 000
	8-29-85	6.70	...	...	...	420	35	...
	11-05-85	6.90	...	...	...	230	230	2 200
	1-16-87[b]	6.58	79	12	2.7	400	21	11 000

[a] Electrical conductivity.
[b] After grouting Trench 150.

major cation was sodium (Table 3) rather than the calcium and magnesium commonly found in cement solutions. Since no cationic analysis of Trench 150 water prior to grouting was available, the source of the sodium was uncertain, although the bentonite (sodium-form) in the grout may have been a contributor.

To observe any further changes in the water chemistry and radionuclide concentration, follow-up samples have been taken but analyses have not been completed to date. Future studies are also planned whereby the volumes of contaminated water in the trenches will be determined and changes in trench behavior will be followed to evaluate this mode of remedial action.

Economic Analysis

The delivered cost of the cement, flyash, and bentonite was \$0.066/kg, \$0.044/kg, and \$0.11/kg, respectively. The target mix ratio of 1.50 kg/L had a grout density of 1.62 kg/L, meaning that each cubic metre of grout contained 390 kg cement, 540 kg flyash, and 50 kg bentonite. Thus, a cubic metre of this grout cost \$26 for cement, \$24 for flyash, and \$5 for bentonite, or a total of \$55/m^3 grout. Thus, the 30.3 m^3 of grout injected into Trench 150 cost \$1667.

The total cost of renting the grouting equipment for eleven days was \$3449 or about \$314/day (four days were used for training and experimentation). Expressed in terms of the amount of grout injected, the equipment cost was \$114/m^3 grout.

The estimated number of man-hours used for this grouting demonstration was 1220, including a foreman and engineer. Thus, 111 man-h/day or 40 man-h/m^3 grout was used. These manpower and cost estimates are for the grout injection only. No costs were included for the site characterization or follow-up evaluation.

In summary, it took 1220 man-h to inject 30.3 m^3 of grout into Trench 150 at a cost of \$1697 for the grout and \$3449 for the equipment, a total for grout and equipment of \$5146. Thus, the grout and equipment was \$468/day and labor was 111 man-h/day, or \$170/m^3 grout and 40 man-h/m^3 grout. Dividing the totals by the total trench volume of 158 m^3 gives 7.7 man-h/m^3 trench and \$33/m^3 trench. In general, most of the cost (about 90%) for the demonstration was manpower cost. The manpower used was higher for this demonstration than a typical commercial grouting operation, and no profit is reflected in the above costs. Inexperience contributed to much of the higher manpower use, but, also, a commercial operation would have used fewer people. Although, this would have led to a longer operation, the overall man-hours would have been fewer, maybe by a factor of two or more if the inexperience is included.

Summary

The *in situ* grouting of Trench 150 demonstrated that particulate grouts can be used to fill the accessible voids of radioactive low-level waste trenches. Trench 150 is suspected as one source of strontium-90 in the water from a nearby seep. Because of strontium-90 interaction with soil, only long-term monitoring will be able to detect any beneficial effects on strontium-90 migration from grouting the trench.

A cement-flyash-bentonite grout was developed that was both fluid and stable for injecting into the trench. Blending of the cement, flyash, and bentonite powders was done in pneumatic tanks remote from the trench. The blended powders were mixed with water to make the grout at the blending site in a concrete mixing truck, and then the grout was transported to the trench in this truck. The grout was injected under pressure through lances into the trench with a progressive cavity pump. The lances were driven into the trench with a pneumatic hammer at a total of 36 locations.

A total of 30.3 m^3 of grout was injected into Trench 150, representing 21% of the waste zone. The accessible void volume had been measured as 20%, implying that most or all of the accessible voids were grouted. Much of the trench was grouted from only a few holes, abot 80% from five holes. The grout cost $55/$m^3$, and it took 1220 man-hours and eleven days to completely grout Trench 150. The total cost of renting the grouting equipment for this job was $3449. The resistance to water permeation into the trench from the wells increased by two orders of magnitude.

Preliminary assessment indicates successful grouting of accessible voids and improved resistance to water intrusion. Only the ongoing evaluation over several years will determine if the grouting prevents trench subsidence and reduced radionuclide migration. The economic information presented allows comparison of *in situ* grouting with other remedial action alternatives (for example, solution grouting) as similar information is generated. The site characterization and follow-up evaluation costs should be similar for different alternatives.

Acknowledgments

The research described herein was sponsored by the Office of Defense Waste and Transportation Management, U.S. Department of Energy, under Contract DE-AC05-84OR21400 with Martin Marietta Energy Systems, Inc.

The authors gratefully acknowledge the laboratory and field support of T. T. Godsey and O. M. Sealand, the help in grout development from O. K. Tallent and E. W. McDaniel, and the well-monitoring data of E. C. Davis.

References

[*1*] "In Situ Waste Grouting Demonstration, Maxey Flats Nuclear Waste Disposal Site," Fleming County, Kentucky, Natural Resources and Environmental Protection, Maxey Flats Branch, Commonwealth of Kentucky, Frankfort, Kentucky, U.S. Department of Energy Grant No. DE-FG07-811D12223, Jan. 1985.

Author Index

Subject Index

Q

R

S